IET ENERGY ENGINEERING SERIES 174

Silicon Solar Cell Metallization and Module Technology

Other volumes in this series:

Volume 1 **Power Circuit Breaker Theory and Design** C.H. Flurscheim (Editor)
Volume 4 **Industrial Microwave Heating** A.C. Metaxas and R.J. Meredith
Volume 7 **Insulators for High Voltages** J.S.T. Looms
Volume 8 **Variable Frequency AC Motor Drive Systems** D. Finney
Volume 10 **SF_6 Switchgear** H.M. Ryan and G.R. Jones
Volume 11 **Conduction and Induction Heating** E.J. Davies
Volume 13 **Statistical Techniques for High Voltage Engineering** W. Hauschild and W. Mosch
Volume 14 **Uninterruptible Power Supplies** J. Platts and J.D. St Aubyn (Editors)
Volume 15 **Digital Protection for Power Systems** A.T. Johns and S.K. Salman
Volume 16 **Electricity Economics and Planning** T.W. Berrie
Volume 18 **Vacuum Switchgear** A. Greenwood
Volume 19 **Electrical Safety: a guide to causes and prevention of hazards** J. Maxwell Adams
Volume 21 **Electricity Distribution Network Design, 2nd Edition** E. Lakervi and E.J. Holmes
Volume 22 **Artificial Intelligence Techniques in Power Systems** K. Warwick, A.O. Ekwue and R. Aggarwal (Editors)
Volume 24 **Power System Commissioning and Maintenance Practice** K. Harker
Volume 25 **Engineers' Handbook of Industrial Microwave Heating** R.J. Meredith
Volume 26 **Small Electric Motors** H. Moczala *et al.*
Volume 27 **AC-DC Power System Analysis** J. Arrillaga and B.C. Smith
Volume 29 **High Voltage Direct Current Transmission, 2nd Edition** J. Arrillaga
Volume 30 **Flexible AC Transmission Systems (FACTS)** Y-H. Song (Editor)
Volume 31 **Embedded generation** N. Jenkins *et al.*
Volume 32 **High Voltage Engineering and Testing, 2nd Edition** H.M. Ryan (Editor)
Volume 33 **Overvoltage Protection of Low-Voltage Systems, Revised Edition** P. Hasse
Volume 36 **Voltage Quality in Electrical Power Systems** J. Schlabbach *et al.*
Volume 37 **Electrical Steels for Rotating Machines** P. Beckley
Volume 38 **The Electric Car: Development and future of battery, hybrid and fuel-cell cars** M. Westbrook
Volume 39 **Power Systems Electromagnetic Transients Simulation** J. Arrillaga and N. Watson
Volume 40 **Advances in High Voltage Engineering** M. Haddad and D. Warne
Volume 41 **Electrical Operation of Electrostatic Precipitators** K. Parker
Volume 43 **Thermal Power Plant Simulation and Control** D. Flynn
Volume 44 **Economic Evaluation of Projects in the Electricity Supply Industry** H. Khatib
Volume 45 **Propulsion Systems for Hybrid Vehicles** J. Miller
Volume 46 **Distribution Switchgear** S. Stewart
Volume 47 **Protection of Electricity Distribution Networks, 2nd Edition** J. Gers and E. Holmes
Volume 48 **Wood Pole Overhead Lines** B. Wareing
Volume 49 **Electric Fuses, 3rd Edition** A. Wright and G. Newbery
Volume 50 **Wind Power Integration: Connection and system operational aspects** B. Fox *et al.*
Volume 51 **Short Circuit Currents** J. Schlabbach
Volume 52 **Nuclear Power** J. Wood
Volume 53 **Condition Assessment of High Voltage Insulation in Power System Equipment** R.E. James and Q. Su
Volume 55 **Local Energy: Distributed generation of heat and power** J. Wood
Volume 56 **Condition Monitoring of Rotating Electrical Machines** P. Tavner, L. Ran, J. Penman and H. Sedding
Volume 57 **The Control Techniques Drives and Controls Handbook, 2nd Edition** B. Drury
Volume 58 **Lightning Protection** V. Cooray (Editor)
Volume 59 **Ultracapacitor Applications** J.M. Miller
Volume 62 **Lightning Electromagnetics** V. Cooray
Volume 63 **Energy Storage for Power Systems, 2nd Edition** A. Ter-Gazarian
Volume 65 **Protection of Electricity Distribution Networks, 3rd Edition** J. Gers
Volume 66 **High Voltage Engineering Testing, 3rd Edition** H. Ryan (Editor)
Volume 67 **Multicore Simulation of Power System Transients** F.M. Uriate
Volume 68 **Distribution System Analysis and Automation** J. Gers
Volume 69 **The Lightening Flash, 2nd Edition** V. Cooray (Editor)
Volume 70 **Economic Evaluation of Projects in the Electricity Supply Industry, 3rd Edition** H. Khatib
Volume 72 **Control Circuits in Power Electronics: Practical issues in design and implementation** M. Castilla (Editor)
Volume 73 **Wide Area Monitoring, Protection and Control Systems: The enabler for Smarter Grids** A. Vaccaro and A. Zobaa (Editors)
Volume 74 **Power Electronic Converters and Systems: Frontiers and applications** A. M. Trzynadlowski (Editor)
Volume 75 **Power Distribution Automation** B. Das (Editor)
Volume 76 **Power System Stability: Modelling, analysis and control** A.A. Sallam and B. Om P. Malik
Volume 78 **Numerical Analysis of Power System Transients and Dynamics** A. Ametani (Editor)
Volume 79 **Vehicle-to-Grid: Linking electric vehicles to the smart grid** J. Lu and J. Hossain (Editors)
Volume 81 **Cyber-Physical-Social Systems and Constructs in Electric Power Engineering** S. Suryanarayanan, R. Roche and T.M. Hansen (Editors)
Volume 82 **Periodic Control of Power Electronic Converters** F. Blaabjerg, K.Zhou, D. Wang and Y. Yang

Volume 86 **Advances in Power System Modelling, Control and Stability Analysis** F. Milano (Editor)
Volume 87 **Cogeneration: Technologies, Optimisation and Implentation** C. A. Frangopoulos (Editor)
Volume 88 **Smarter Energy: from Smart Metering to the Smart Grid** H. Sun, N. Hatziargyriou, H. V. Poor, L. Carpanini and M. A. Sánchez Fornié (Editors)
Volume 89 **Hydrogen Production, Separation and Purification for Energy** A. Basile, F. Dalena, J. Tong and T.N.Veziroğlu (Editors)
Volume 90 **Clean Energy Microgrids** S. Obara and J. Morel (Editors)
Volume 91 **Fuzzy Logic Control in Energy Systems with Design Applications in Matlab/Simulink®** İ. H. Altaş
Volume 92 **Power Quality in Future Electrical Power Systems** A. F. Zobaa and S. H. E. A. Aleem (Editors)
Volume 93 **Cogeneration and District Energy Systems: Modelling, Analysis and Optimization** M. A. Rosen and S. Koohi-Fayegh
Volume 94 **Introduction to the Smart Grid: Concepts, technologies and evolution** S.K. Salman
Volume 95 **Communication, Control and Security Challenges for the Smart Grid** S.M. Muyeen and S. Rahman (Editors)
Volume 96 **Industrial Power Systems with Distributed and Embedded Generation** R Belu
Volume 97 **Synchronized Phasor Measurements for Smart Grids** M.J.B. Reddy and D.K. Mohanta (Editors)
Volume 98 **Large Scale Grid Integration of Renewable Energy Sources** A. Moreno-Munoz (Editor)
Volume 100 **Modeling and Dynamic Behaviour of Hydropower Plants** N. Kishor and J. Fraile-Ardanuy (Editors)
Volume 101 **Methane and Hydrogen for Energy Storage** R. Carriveau and D. S-K. Ting
Volume 104 **Power Transformer Condition Monitoring and Diagnosis** A. Abu-Siada (Editor)
Volume 106 **Surface Passivation of Industrial Crystalline Silicon Solar Cells** J. John (Editor)
Volume 107 **Bifacial Photovoltaics: Technology, applications and economics** J. Libal and R. Kopecek (Editors)
Volume 108 **Fault Diagnosis of Induction Motors** J. Faiz, V. Ghorbanian and G. Joksimović
Volume 110 **High Voltage Power Network Construction** K. Harker
Volume 111 **Energy Storage at Different Voltage Levels: Technology, integration, and market aspects** A.F. Zobaa, P.F. Ribeiro, S.H.A. Aleem and S.N. Afifi (Editors)
Volume 112 **Wireless Power Transfer: Theory, Technology and Application** N.Shinohara
Volume 114 **Lightning-Induced Effects in Electrical and Telecommunication Systems** Y. Baba and V. A. Rakov
Volume 115 **DC Distribution Systems and Microgrids** T. Dragičević, F.Blaabjerg and P. Wheeler
Volume 116 **Modelling and Simulation of HVDC Transmission** M. Han (Editor)
Volume 117 **Structural Control and Fault Detection of Wind Turbine Systems** H.R. Karimi
Volume 119 **Thermal Power Plant Control and Instrumentation: The control of boilers and HRSGs, 2nd Edition** D. Lindsley, J. Grist and D. Parker
Volume 120 **Fault Diagnosis for Robust Inverter Power Drives** A. Ginart (Editor)
Volume 121 **Monitoring and Control using Synchrophasors in Power Systems with Renewables** I. Kamwa and C. Lu (Editors)
Volume 123 **Power Systems Electromagnetic Transients Simulation, 2nd Edition** N. Watson and J. Arrillaga
Volume 124 **Power Market Transformation** B. Murray
Volume 125 **Wind Energy Modeling and Simulation Volume 1: Atmosphere and plant** P.Veers (Editor)
Volume 126 **Diagnosis and Fault Tolerance of Electrical Machines, Power Electronics and Drives** A.J. M. Cardoso
Volume 128 **Characterization of Wide Bandgap Power Semiconductor Devices** F. Wang, Z. Zhang and E.A. Jones
Volume 129 **Renewable Energy from the Oceans: From wave, tidal and gradient systems to offshore wind and solar** D. Coiro and T. Sant (Editors)
Volume 130 **Wind and Solar Based Energy Systems for Communities** R. Carriveau and D. S-K. Ting (Editors)
Volume 131 **Metaheuristic Optimization in Power Engineering** J. Radosavljević
Volume 132 **Power Line Communication Systems for Smart Grids** I.R.S Casella and A. Anpalagan
Volume 139 **Variability, Scalability and Stability of Microgrids** S. M. Muyeen, S. M. Islam and F. Blaabjerg (Editors)
Volume 145 **Condition Monitoring of Rotating Electrical Machines** P. Tavner, L. Ran, C. Crabtree
Volume 146 **Energy Storage for Power Systems, 3rd Edition** A.G. Ter-Gazarian
Volume 147 **Distribution Systems Analysis and Automation 2nd Edition** J. Gers
Volume 151 **SiC Power Module Design: Performance, robustness and reliability** Alberto Castellazzi and Andrea Irace (Editors)
Volume 152 **Power Electronic Devices: Applications, failure mechanisms and reliability** F Iannuzzo (Editor)
Volume 153 **Signal Processing for Fault Detection and Diagnosis in Electric Machines and Systems** Mohamed Benbouzid (Editor)
Volume 155 **Energy Generation and Efficiency Technologies for Green Residential Buildings** D. Ting and R. Carriveau (Editors)
Volume 156 **Lithium-ion Batteries Enabled by Silicon Anodes** Chunmei Ban and Kang Xu (Editors)
Volume 157 **Electrical Steels, 2 Volumes** A. Moses, K. Jenkins, Philip Anderson and H. Stanbury
Volume 158 **Advanced Dielectric Materials for Electrostatic Capacitors** Q Li (Editor)
Volume 159 **Transforming the Grid Towards Fully Renewable Energy** O. Probst, S. Castellanos and R. Palacios (Editors)
Volume 160 **Microgrids for Rural Areas: Research and case studies** R.K. Chauhan, K. Chauhan and S.N. Singh (Editors)
Volume 166 **Advanced Characterization of Thin Film Solar Cells** N Haegel and M Al-Jassim (Editors)
Volume 167 **Power Grids with Renewable Energy** Storage, integration and digitalization A. A. Sallam and B. OM P. Malik
Volume 170 **Reliability of Power Electronics Converters for Solar Photovoltaic Applications** Frede Blaabjerg, Ahteshamul Haque, Huai Wang, Zainul Abdin Jaffery and Yongheng Yang (Editors)
Volume 171 **Utility-scale Wind Turbines and Wind Farms** A. Vasel-Be-Hagh and D. S.-K. Ting

Volume 172 **Lighting interaction with Power Systems, 2 volumes** A. Piantini (Editor)
Volume 193 **Overhead Electric Power Lines: Theory and practice** S. Chattopadhyay and A. Das
Volume 199 **Model Predictive Control for Microgrids: From power electronic converters to energy management** J. Hu, J. M. Guerrero and S. Islam
Volume 905 **Power system protection, 4 volumes**

Silicon Solar Cell Metallization and Module Technology

Edited by
Thorsten Dullweber and Loic Tous

The Institution of Engineering and Technology

Published by The Institution of Engineering and Technology, London, United Kingdom

The Institution of Engineering and Technology is registered as a Charity in England & Wales (no. 211014) and Scotland (no. SC038698).

First published 2021

The Institution of Engineering and Technology
Michael Faraday House
Six Hills Way, Stevenage
Herts, SG1 2AY, United Kingdom

www.theiet.org

British Library Cataloguing in Publication Data
A catalogue record for this product is available from the British Library

ISBN 978-1-83953-155-2 (hardback)
ISBN 978-1-83953-156-9 (PDF)

Typeset in India by Exeter Premedia Sevices Private Limited

Contents

About the editors

Thorsten Dullweber leads the R&D group Industrial Solar Cells at the Institute for Solar Energy Research Hamelin (ISFH), Germany. His research focuses on high efficiency industrial silicon solar cells with screen-printed metal contacts. Before joining ISFH, Dr. Dullweber worked for 9 years as a project leader at Siemens AG and later Infineon Technologies AG. He is a member of the Scientific Committee of the EU-PVSEC conference and of the Editorial Board of Photovoltaics International.

Loic Tous is the R&D team leader of the PV Cells and Module team at the Interuniversity Microelectronics Centre (IMEC), Belgium. His research areas include silicon solar cells, advanced interconnection technologies semiconductor processing and fabrication, and the development of PV modules for various applications. Since 2016, he has co-organized the Workshop on Metallization & Interconnection for Crystalline Silicon Solar Cells. He is coordinator of a large Horizon 2020 project on photovoltaics.

Biographies

Dr. Jonas Bartsch studied chemical engineering at the University of Karlsruhe and received his diploma in 2007. He joined Fraunhofer ISE to pursue a Ph.D. in the field of advanced front contacts for silicon solar cells with plating technology. After receiving his Ph.D. from the Albert-Ludwigs University of Freiburg in 2011, he continued to work with plating at ISE and is currently co-head of the group "Electrochemical Processes"

Dr. Armand Bettinelli received his PhD in 1987 for his work on cofiring of alumina and tungsten at the Strasburg University. He worked in the industry as technical manager in the field of High and Low Cofired Ceramics then Plasma Display Panels, fields all using high-level screen-printing. In 2005 he joined CEA-INES where he works as senior expert for c-Si solar cell Metallization and Interconnection. From 2005, beside the optimization of the screen-printing for PV cells using mesh screens and stencil, Armand introduced and optimized at CEA the smartwire™ and the Ribbon gluing interconnections for heterojunction cells, more recently the shingling, taking always care of the link between cell metallization and cell interconnection quality.

Dr. Andreas Büchler studied Physics and Mathematics at the Albert-Ludwigs-University of Freiburg. In 2011 he joined the Fraunhofer ISE. He worked in the group of Kluska and Bartsch with a focus on the imaging techniques for characterization of contact properties in micro and nano level. In 2019 he finished his dissertation on the interface of laser-structured plated contacts with honors at the technical faculty of the Albert-Ludwigs-Univesity of Freiburg.

Dr. James Bullock completed his PhD in Engineering, as an Australian Renewable Energy Agency Fellow, at the Australian National University. Following that he was a postdoctoral researcher in the Electrical Engineering and Computer Sciences department at Berkeley and an affiliate in the Materials Sciences Division at the Lawrence Berkeley National Laboratory. James now heads the Sustainable Energy and Environmental Devices group in the Electrical and Electronic Engineering Department at the University of Melbourne. This group focuses on the development of novel materials-based solutions for optoelectronics, focusing on passivated contacts for silicon solar cells and two-dimensional material photodetectors.

Eng. Perrine Carroy holds a Master of Engineering in material science from the National Institute of Applied Sciences (INSA) of Lyon (France). She has been working in the field of Transparent Conducting Oxides (TCOs) for photovoltaics (PV) since 2008. She worked 3 years as a process engineer at the former PV equipment manufacturer Oerlikon Solar on the development of ZnO-based transparent electrodes for thin film solar cells. She joined the Laboratory for Heterojunction solar cells at CEA-INES in 2012. Since then, she has been in charge of the development of TCO materials as a researcher and project leader, participating to several EU-funded projects (DISC, AMPERE, HERCULES), French research projects as well as industrial projects. She holds several publications in the field of TCO materials for PV, as well as conference communications. She has also joined this year the working group in charge of developing perovskite/silicon tandem devices.

Haifeng Chu performed his PhD work at International Solar Energy Research Center Konstanz (ISC), Germany, on the topic of development and characterization of IBC silicon solar cells. After finishing his thesis in 2019, he continues to work at ISC Konstanz as a research engineer in the N-type Solar Cells Group.

Dr. Thorsten Dullweber studied physics at the Leibniz University Hannover, Germany, and received his Ph. D. degree in 2002 from the University of Stuttgart for research on $Cu(In,Ga)Se_2$ solar cells. From 2001 till 2009 he worked in the microelectronics industry first as R&D engineer at Siemens AG and later as technology project leader at Infineon Technologies AG developing CMOS memory chips. Since 2009, Thorsten is leading the R&D group Industrial Solar Cells at the Institute of Solar Energy Research Hamelin (ISFH). His research work focuses on high efficiency industrial-type PERC silicon solar cells and bifacial PERC+ solar cells. Thorsten authored more than 100 Journal and Conference publications, which have been cited more than 2000 times. He was awarded with the enercity energy efficiency price in 2015 for developing a world-record-efficient PERC solar cell. In 2017, Thorsten received the price of the German Foundation for Industrial Research for developing the bifacial PERC+ solar cell. Thorsten is member of the Scientific Committees of the EU-PVSEC and SNEC Conferences and of the Editorial Advisory Board of Photovoltaics International.

Dr Shubham Duttagupta is the Deputy Director of the Silicon Solar Cells & Modules (SSCM) Cluster at SERIS – Singapore's national laboratory for applied

solar energy research. Dr Duttagupta also heads the Advanced Silicon Solar Cell Group inside the SSCM Cluster. The scientific-technical aim of his research group is to develop technologies that targets the limit of silicon based single-junction cell efficiency while keeping the overall process cost-competitive. The research group collaborates with several solar cell/module, equipment, automation and material companies to jointly develop next-generation processes. Dr Duttagupta's PhD research focused on the development of advanced multifunctional passivation materials required for high-efficiency crystalline silicon wafer solar cells. In the last 12 years of his career in photovoltaics, Dr Duttagupta has achieved several awards, notably the 'Institute of Microelectronics Prize' by the National University of Singapore (NUS) in 2015, runner-up for the prestigious 'SolarWorld Junior Einstein Award' in 2014 and most recently the 'Young Scientist Award' at the PVSEC 2016. His research team have won APVIA Technological Achievement Award at SNEC 2019 and Research Partner Award at World Solar Congress (WSC) 2019 – both for the monoPoly© high-efficiency silicon solar cell technology.

Dr. Abasifreke Ebong received his Ph.D. in Electrical and Computer Engineering from the University of New South Wales, Australia in 1995. He then joined Samsung Electronics in South Korea served as a Postdoctoral Fellow. In September 1997, he joined the University Center of Excellence for Photovoltaic Research and Education (UCEP), Georgia Tech., Atlanta, as a Research Faculty. At UCEP, he worked on the development, design, modelling, fabrication, and characterization of low-cost, high-efficiency belt line multi-crystalline, Cz, and Fz silicon solar cells. In 2001 he joined GE Global Research as Electrical Engineer, working on Solid State Lighting (LED-light emitting diodes) based on III-V semiconductors. While at GE, he developed current spreading model for light emitting diodes, which enhanced the evaluations of several conceptual designs without actually fabricating them. In 2004, he returned to the UCEP at Georgia Tech as the Assistant Director of the center, responsible for sponsored research in crystalline and amorphous silicon solar cells. Dr Ebong joined the Faculty of the University of North Carolina at Charlotte as a Professor in February 2011. He has published over 160 papers in the field of Photovoltaics.

Dr. Antonin Faes is responsible of the metallization and interconnection of c-Si solar cells at CSEM PV-Center. He graduated as materials science engineer at EPFL in Lausanne and got his PhD in 2010. From 2012, he developed the advanced Smart-Wire Connection Technology (SWCT®) with Meyer Burger. He is author and co-author of more than 30 peer reviewed papers cited about 2000 times, few patents, and a book.

Since leaving Southampton Technical College, **Tom Falcon** has accrued over 35 years engineering experience in the electronics industry. This includes wire-bond R&D with IBM, SMT process engineering with Nortel, materials development with Cookson, semiconductor packaging with DEK, and Solar cell metallisation with ASM. He is currently a Senior Systems Engineer developing novel equipment for mass deposition of electronic materials and has a number of patents granted in this field.

Since June 2019, **Dr. Susanne Herritsch** has been a Project Manager for the VDMA Sector Group Photovoltaic Equipment, which is one of the four groups of the VDMA Sector Association Electronics, Micro and New Energy Production Technologies (EMINT), beside Battery Production, Productronics and Micro Technologies. In particular, the analysis of statistics and the involvement in the International Technology Roadmap for Photovoltaic (ITRPV) are among the areas of work. In addition, the organization of various events, face-to-face or in the form of online formats, is on their agenda alongside the project "Carbon-neutral Production". Dr. Susanne Herritsch studied chemistry at the Philipps University in Marburg and completed research internships at Imperial College in London as well as in an industrial environment. The focus of her PhD was on the development and synthesis of precursors for application in photovoltaics and in the field of *Smart Windows*.

Dr. Matthias Hoerteis studied Physics at the University of Augsburg, Germany and ANU (Australia). He started his career at Fraunhofer ISE in Freiburg (Germany) in 2005 where he worked on aerosol printed and fired metal contacts for silicon solar cells. After he received his PhD in 2009 from the University of Konstanz, he became head of the "metallization Group" at Fraunhofer ISE. In 2011, he joined Heraeus to grow the local solar paste development team to serve European customers. Between 2015 to 2018 he lived with his family in the USA and was responsible for Heraeus global solar paste Innovation center at Conshohocken (Philadelphia). During this time, several paste generations got developed and Heraeus market share increased to 30%. He authored more than 60 conference and journal papers and holds more than 30 patents in the field of solar cell metallization.

Jaap Hoornstra has a long career in Energy Research with Netherlands Energy Research Foundation (ECN) as researcher, project manager, and business developer. His main effort focused on Photovoltaic R&D, where Hoornstra developed technologies for crystalline silicon solar cell technologies, with an emphasis on metallization. Hoornstra also worked in the field of combustion, and was assigned at Combustion Research Facilities of Sandia

National Laboratories. In the field of photovoltaics Hoornstra was founder of SunLab, a subsidiary of ECN, where as managing director, he co-developed and marketed worldwide instruments for characterization of solar cell metallization, such as Corescan. Hoornstra was initiator, co-organizer and chair of the Workshop on Metallization and Interconnection of Crystalline Silicon Solar Cells. Hoornstra published more than 50 papers, reviewed journal papers, and contributed to several patents.

Dr. Pei-Chieh Hsiao studied electrical engineering at the National Taiwan University, Taiwan, and received his Ph. D. degree in 2015 from the University of New South Wales (UNSW) on Eutectic Sn-Bi Alloy for Interconnection of Silicon Solar Cells. Since 2015, he worked as a postdoctoral research fellow at UNSW. His research work focuses on copper-plated metallisation of PERC/PERT and heterojunction solar cells, innovative cell interconnection techniques, and analysis of the mechanical integrity of silicon solar modules involved using finite element analysis in the modeling of thermo-mechanical stresses in silicon solar cells resulting from thermal processing or mechanical loads. His publication in the Journal of the Electrochemical Society, 'Electroplated and Light-Induced Plated Sn-Bi Alloys for Silicon Photovoltaic Applications', was selected by the Renewable Energy Global Innovations as a key scientific article contributing to the excellence in renewable and clean energy research. He is an active manuscript reviewer of Solar Energy, Solar Energy Materials and Solar Cells, IEEE Journal of Photovoltaics and Journal of the Electrochemical Society.

Ankit Khanna is the head of the Emerging Silicon Solar Cells Group at the Solar Energy Research Institute of Singapore (SERIS). His research focuses on next-generation silicon solar cell technologies and industrial manufacturing processes. He received a Master of Technology degree in Engineering Physics from the Indian Institute of Technology-Banaras Hindu University, India in 2010 and a PhD degree in Electrical and Computer Engineering from the National University of Singapore in 2015.

Dr. Sven Kluska studied Physics the Albert-Ludwigs University of Freiburg and received his diploma in 2010. He joined Fraunhofer ISE to pursue a Ph.D. in the field of laser chemical processing for silicon solar cells and received his Ph.D. with honors from the Albert-Ludwigs University of Freiburg in 2011. His research interests include electrochemical processing for solar cell applications with focus on plating metallization. He is currently co-head of the group "Electrochemical Processes".

Dr. Saskia Kühnhold-Pospischil studied physics at the university of Tübingen, Germany and the university Oviedo, Spain and received her Ph. D. degree in 2017

from the University of Freiburg (IMTEK) for research at the Fraunhofer Institute for Solar Energy Systems (ISE) in Germany on Si surface passivation with Al2O3. Since the completion of her PhD, she has been working as a project leader at the Fraunhofer ISE on various topics. In addition to the passivation of Si surfaces, she has worked on the epitaxial fabrication of Si wafers and currently on the laser-induced metallization of solar cells.

Prof. Alisson Lennon is an academic in the School of Photovoltaic and Renewable Energy Engineering at UNSW Sydney, Australia. She holds PhDs in Biophysical NMR (University of Sydney, 1995) and Photovoltaic Engineering (UNSW, 2010), was awarded a University Medal (University of Sydney, 1991), an Australian Postgraduate Fellowship (1995) and an ARC Future Fellowship (2017). She has published more than 130 scientific papers and is an inventor of 29 granted US patents. Prior to her employment at UNSW in 2010, she was employed as a research scientist at Canon Information Systems Research Australia, where she was involved in a range of research ranging from device simulations, development of materials/technology for printing, imaging, display and medical applications. She currently conducts research into silicon solar cell metallisation and interconnection, optical modelling for photovoltaics and high-rate electrochemical storage materials.

Dr. Yang Li earned his PhD of Photovoltaic Engineering from University of New South Wales (UNSW) in 2016 with Australian International Postgraduate Research Scholarship. During his PhD, he fabricated new light trapping structures and developed the Angular Matrix Framework, which is a unique modelling method for solar applications at different scales. Before that, he received the B.Eng in Opto-Electronics Engineering from Zhejiang University and the B.Eng in Photovoltaics and Renewable Energy from UNSW with First Class Honor in 2011.

He had presentations on multiple conferences and proposed the optimization methodology of solar cell and modules based on more realistic yield simulation. Since 2018, he is an Australian Renewable Energy Agency (ARENA) research fellow in UNSW working on development of new module interconnection processes, performance modelling and power forecasting of solar systems.

Dr. Andreas Lorenz studied printing and media technology at Stuttgart University of Applied Sciences (HdM Stuttgart). Andreas was awarded with the Heidelberger Druckmaschinen Award "Student of the Year 2007" and a price of the VDM Baden-Württemberg e.V. for outstanding academic achievements. After having received his diploma degree in 2006, he worked for 6 years as process engineer and junior product manager for a global printing machine manufacturer. In 2012, he joined Fraunhofer ISE and worked from 2014 to 2017 as a Ph.D student on the development of rotary printing processes for solar cells. He received his Ph.D degree in 2018. Since 2018, Andreas is responsible for the project lead for several funded research and industry projects in the field of solar cell metallization, focusing primarily on the development and optimization of rotary printing methods, flatbed screen printing and new applications. Since 2020, he is head of the group "Printing Technology" at Fraunhofer ISE.

José de Jesús Ibarra Michel is currently a PhD candidate in the University of Melbourne, working on the development of advanced contacts for silicon solar cells. Jesús completed his Masters in the Erasmus Mundus Nanoscience and Nanotechnology program. He spent the first year at the Katholieke Universiteit Leuven in Belgium and the second at the Technical University of Dresden in Germany, where he followed a specialization in Nanoelectronics. During his Master's thesis research he worked at Namlab GmbH, studying the formation of fixed charge density for silicon solar cells.

Dr. Delfina Muñoz Cervantes is the principal researcher in the heterojunction solar cells laboratory at CEA-INES. After finishing her studies in industrial engineering in Barcelona in 2003, she started the heterojunction solar cell adventure in her PhD at Universidad Politécnica de Cataluña developing laboratory scale solar cells, processes and improving characterization skills. In 2008, she joined CEA-INES as a postdoc and since then, she has been improving heterojunction technology from the lab to the fab. She has been involved in several European (FP7, H2020) and French research projects (ANR) and she was the coordinator of the successful FP7 HERCULES project with the first heterojunction industrial transfer in Europe. She has more than 40 WoS publications and more than 60 conference presentations related to heterojunction technology. She is in the steering committee of the nPV and tandemPV workshops; she is reviewer in several high impact journals and conferences where she is chairing and invited regularly. She is also in the strategic team on European projects of CEA and works in the roadmap definition of next PV technologies as tandem photovoltaics. She still combines her project activity with the laboratory, directing PhD students and developing heterojunction solar cells for the next tandem photovoltaic technology.

Linda Ney received her diploma in the field of renewable energy systems from the technical university of Dresden. Since 2017 she works at Fraunhofer ISE, focusing on the optimization of screens for solar cell manufacturing. In 2019 she started her Ph.D. about the development of industry-related production processes for PEM-fuel cells, as well stability studies of catalyst inks and pastes.

Mr. Pradeep Padhamnath is a research scholar at Department of Electrical and Computer Engineering, National University of Singapore. He is attached to Solar Energy Research Institute of Singapore, where his research focuses on the development of advanced metallization of high-efficiency passivated contact solar cells fabricated using industrial processes. He completed his Masters in Technology from IIT Bombay in Energy Systems engineering, during which he was attached to National Centre of Photovoltaic Research and Education. His master's thesis focused on advanced wafer slicing methods using WEDM and development of advanced functional layers for Ni-Ci metallization of c-Si solar cells.

Maximilian Pospischil studied mechanical engineering at Technical Universities of Munich (GER) and Delft (NL) and received his Diploma degree from the TU Munich in Dec. 2009. In 2010, he joined Fraunhofer ISE in Freiburg as a Ph.D. student and completed his thesis about front side metallization of Silicon Solar cells using parallel dispensing in the year 2016. Afterwards he was responsible for further enhancements in the technology as head of team dispensing technology until March 2020 and supervised two more PhD thesis in the field of printing technologies. In March 2019, he co-founded HighLine Technology GmbH as managing director and technology expert with the goal to commercialize the parallel dispensing technology. In his eyes, metallization in PV should not only remain a question between screen printing and plating …

Dr. Armin Richter is Senior Scientist at the Fraunhofer Institute for Solar Energy Systems ISE, Germany. His current research interests include atomic layer deposition (ALD) of functional thin films (e.g. silicon surface passivation, electron/hole transport layers for perovskite solar cells or TCOs), the in depth characterization of dielectric surface passivation layers as well as the development of high efficiency silicon solar cells along the whole process chain in order to demonstrate the potential of different cell architectures or process technologies. This topic includes also 3d device simulations of silicon solar cells. He received his Ph.D. in physics in 2014 with work on the development of n-type silicon solar cells and Al2O3 based silicon surface passivation. He published more than 70 scientific papers, thereof more than 30 in scientific journals.

Dr. Torsten Roessler, née Geipel studied information system technology at the Technical University of Dresden, Germany. Between 2007 and 2011 he worked with Solon SE, a photovoltaic module manufacturer based in Berlin, Germany. Since 2011 he is with Fraunhofer Institute for Solar Energy Systems ISE in the department "Photovoltaic Modules" where he leads a team "Electrically Conductive Adhesives". Roessler wrote a doctoral thesis entitled "Electrically Conductive Adhesives for Photovoltaic Modules".

Angela De Rose studied physics at RWTH Aachen and University of Valencia and received her master's degree in 2015 from RWTH Aachen. In 2016, she joined Fraunhofer ISE as a Ph.D. student in the field of solder interconnection technology for silicon PV modules. Her studies focus on the interconnection of SHJ solar cells and soldering on aluminum surfaces. Since 2020, she is head of the team Soldering in the department Module Technology at Fraunhofer ISE.

Dr. Charles ROUX, head of CEA INES heterojunction solar cells laboratory and pilot line, holds a Ph.D. on II-VI semiconductors from Grenoble University (France). He previously spent 10 years at AMAT as process engineer in PECVD, PVD, dry etch processes for the semiconductor industry. In 2007 he contributed to the start-up of PV thin film turnkey Sunfab at T-Solar in Spain. Then in 2009 he joined CEA INES.

Dr. Henning Schulte-Huxel studied physics in Leipzig, Germany, and Bucharest, Romania, as well as laser technology in Jena, Germany. He joined the Institute of Solar Energy Research Hamelin, Germany in 2010, and became a project manager in 2014, with a focus on the module integration of high-efficiency solar cells. In 2015 he received his Ph.D. at Leibniz University Hannover on laser welding of silicon solar cells. During 2017 he stayed as postdoc fellow at the National Renewable Energy Laboratory (NREL) and worked on the module integration of 3 terminal tandem solar cells. Since 2016 he has been the vice head of the module technologies workgroup.

Sebastian Tepner studied electrical engineering and information technology at the University of Bremen in Germany. In 2016 he joined the Fraunhofer Institute for Solar Energy Systems. His research focuses on the process development for printing technologies such screen-printing and dispensing. In 2021, he became the head of the team "Printing Process Modeling & Rheology". His team develops solutions which accelerate the transition of printing technologies into the digital age.

Dr. Loic Tous is leading the PV Cell and Module activities at imec since early 2020. He received his Ph. D. degree in 2014 from the KU Leuven (Belgium) for his research on plated metallization of industrial high-efficiency c-Si solar cells. Between 2014 and 2020, his research focused on integration, metallization, interconnection of advanced PERC/PERT silicon solar cells. He has authored and co-authored more than 80 Journal and Conference publications. Since 2016, he is the co-organizer of the Workshop on Metallization & Interconnection for Crystalline Silicon Solar Cells.

Dr. Jutta Trube works as Division Manager of VDMA Sector Group Photovoltaic Equipment since December 2015 and is engaged especially in production equipment for the solar cell and module production. With this position at VDMA she is responsible for the yearly updates and publication of the International Technology Roadmap for Photovoltaic (ITRPV). Furthermore, she is the Vice Managing Director of VDMA Sector Association Electronics, Micro and New Energy Production Technologies. Prior to her engagement at VDMA she worked as Research and Development Director and Director of New Technologies in vacuum equipment manufacturing. The markets concerned photovoltaic, architectural glass as an energy saving element, storage technology, display technology, and other areas. Dr. Trube studied physics at the Georg-August University in Göttingen and got her doctorate in electrical engineering at the Technical University in Berlin.

Eng. Anthony Valla studied physics at the Polytech Clermont-Ferrand engineering school, France. Since 2011 he has been working on Transparent Conducting Oxides as a researcher-engineer at CEA at the National Solar Energy Institute (INES) located near Chambery. His research interests include the development of characterization methods (standard and advanced) and optical simulations to optimise the role of TCO specially focused on heterojunction solar cell and tandem.

Chapter 1
Introduction

Thorsten Dullweber[1] and Loic Tous[2]

According to the International Energy Agency (IEA) World Energy Outlook 2020, solar photovoltaics (PV) has become the new king of electricity generation thanks to sharp cost reductions over the past decade [1]. Solar PV is now consistently cheaper than new coal- or gas-fired power plants in most countries and offers some of the lowest levelized cost of electricity (LCOE) ever seen with values below 2 US cents/kWh in sunny locations. Solar PV is largely dominated by crystalline silicon technologies. Several books have already been published covering fundamentals to applications [2–4] or important topics such light management in solar cells [5] or surface passivation in industrial crystalline silicon solar cells [6]. This book is dedicated to another very important topic that is the metallization and interconnection of crystalline silicon solar cells.

The metal contacts play a key role in obtaining high-efficiency and low-cost crystalline silicon solar cells. The metal contacts have to extract charge carriers with low-ohmic contact resistance for both polarities, electrons and holes, at the respective silicon surfaces. The metal contacts are typically designed as a so-called finger grid consisting of multiple line conductors to transport the photo-generated current to the busbars with low resistive losses. At the busbars, the metal contacts have to ensure a reliable soldered electrical contact to the ribbons or wires for interconnecting the solar cells in the module. The metal contacts should cover only a small fraction of the silicon wafer in order to minimize shadowing losses of the incident sunlight and in order to minimize recombination of minority charge carriers at the metal surface. Finally, the manufacturing costs of the metal contacts must be continuously reduced for the industry to continue along its historical learning rate of reducing costs per watt peak by 24% for every doubling of the production capacity [7]. This has always made metallization and interconnection of crystalline silicon solar cells a very challenging topic, and it will continue to be the case in the foreseeable future.

In microelectronics technology, metal contacts typically consist of a sputter-deposited multi-layer stack applying different materials to obtain minimum contact

[1]Institute for Solar Energy Research Hamelin (ISFH), Germany
[2]Interuniversity Microelectronics Centre (IMEC), Belgium

resistance, excellent barrier properties and very high line conductivity thanks to subsequent use of copper (Cu) as the main conductor. Multiple structuring steps by photolithography are required to form nanometer-wide contact lines. The use of photolithography patterning is much too expensive for solar cell production and hence it is only used by research institutes developing high-efficiency lab-type solar cells mostly based on evaporated metals such as aluminum (Al), silver (Ag) or a stack of titanium/palladium/silver (Ti/Pd/Ag). In the past decades, many different materials and deposition technologies have been evaluated for the metallization of crystalline silicon solar cells, e.g. scaling up the evaporation of contacts to large production tools or depositing Ni/Cu/Ag stacks by electroplating. Overall, metallization approaches based on screen-printing of Ag- or Al-containing pastes have been the most successful when bringing cell concepts from laboratories to mass production.

Since decades most industrial solar cells apply silicon wafers and screen-printed Ag front and Al rear metal contacts. The advantage of this technology is that Ag establishes a good alloyed electrical contact to n-type silicon and Al to p-type silicon while screen-printing enables a very cheap and high-throughput structured deposition of the μm-wide metal finger grid with excellent line conductivity. Thereby, the screen-printing technology tremendously advanced over the past 20 years making PV the most demanding screen-printing industry in the world. By optimizing the rheology of Ag pastes and by developing finer screen meshes and advanced emulsions, the Ag finger width has been reduced from around 150 μm to presently around 30 μm thereby drastically minimizing shadowing loss and the consumption of expensive Ag paste. Until 2012, screen-printed Al pastes contacted the whole rear side of industrial silicon solar cells. Since then, Al pastes have been further developed, e.g. by adding Si to enable high-quality locally alloyed Al contacts through laser-contact openings of the rear passivation layer of passivated emitter and rear cell (PERC) solar cells which since then have become the mainstream industrial solar cell technology. By adjusting the Al paste viscosity and by applying high-performance screens, printing of an Al finger grid has been introduced to industrial PERC solar cells thereby enabling light absorption of both wafer surfaces which can significantly increase the energy yield of PV power plants. With these measures, the advances in screen-printed metal contacts strongly contributed to increase the conversion efficiency of industrial silicon solar cells from around 15% in the year 2000 to around 23% in the year 2020 and to further increase the energy yield by around 10% through bifacial light absorption. Reduced Ag paste consumption and cheap high-throughput screen printers contributed to reduce manufacturing costs for a solar module from around 5 US$/Wp in 2000 to below 0.25 US$/Wp in 2020.

For the coming years, new challenges arise for the metallization of silicon solar cells. Silver is already now a key cost component in manufacturing of silicon solar cells and may become even more expensive as the fast-growing PV industry consumes relevant shares of the worldwide yearly Ag production. According to recent calculations [8], the PV industry needs to reduce the consumption of Ag per cell from around 20 mg/W in 2019 to less than 5 mg/W by 2028 in order to achieve material sustainability. This can be achieved by introducing alternative printing techniques or by replacing screen-printing of Ag by Cu-based metallization approaches. The throughput of one single screen-print head is limited by the motion of the squeegee to around 3 000 wafers per hour. In the future, solar cell production lines will apply processing tools with a throughput above 10 000 wafers per hour [7]. Hence, high-throughput metallization approaches and equipment have to be developed. In addition to PERC, new high-efficiency silicon solar cell designs applying amorphous silicon or polysilicon-based heterojunctions are entering the PV industry. These new cell designs require new metallization concepts such as modified pastes or transparent conducting oxides (TCOs). Finally, the metal contacts and metallization techniques need to be adapted to the new module interconnection designs such as multi-busbar or shingling that are being introduced to improve performance and aesthetics.

Overall, this book aims at providing a detailed overview of the state of the art of different metallization and interconnection technologies for silicon solar cells. Chapter 2 describes the main requirements of the metallization of silicon solar cells such as the electrical and optical properties of the finger grid as well as cost and sustainability considerations. The basic fundamentals of metal-semiconductor contacts and charge carrier transport related to the material properties such as the work function and bandgap are summarized in Chapter 3. Chapter 4 provides an overview of the various metallization approaches for different silicon solar cell designs and summarizes their advantages and challenges. The evaporation of metal contacts is covered in Chapter 5 in terms of lab-type as well as industrial approaches and with respect to different materials. The industrial mainstream screen-printing technology is described in Chapter 6 with respect to Ag and Al paste properties, screen-printers as well as screen and stencil design. Alternative printing technologies, namely dispensing, rotary printing and laser transfer printing, are introduced in Chapter 7. Plating is a well-known low-cost metal deposition technology and has been intensively studied for solar cell metallization as discussed in Chapter 8. Chapter 9 provides an overview of the main industrial silicon solar cell concepts such as PERC, Heterojunction Technology (HJT), passivating contacts and Interdigitated Back Contact (IBC) and describes their specific metallization technologies. Chapter 10 explains different module interconnection technologies based on soldering, conductive adhesives and welding. Chapter 11 provides an overview of important design criteria for high-efficiency and reliable PV modules.

Each chapter of this book was written by recognized experts in the field whom we thank for their excellent contributions. We wish you a pleasant and insightful read.

Best regards,
Thorsten Dullweber and Loic Tous

References

[1] Birol F. *World Energy Outlook 2020*. Paris, France: International Energy Agency; 2020.

[2] Mertens K. *Photovoltaics: Fundamentals, Technology, and Practice*. 2nd edn. Hoboken, USA: John Wiley & Sons; 2018.

[3] Reinders A., Verlinden P., Sark Wvan., Freundlich A. *Photovoltaic Solar Energy: From Fundamentals to Applications*. 1st edn. Hoboken, USA: John Wiley & Sons; 2017.

[4] McEvoy A., Markvart T., Castaner L. *Practical Handbook of Photovoltaics: Fundamentals and Applications*. 2nd edn. Amsterdam, Nederlands: Elsevier; 2012.

[5] John J. *Surface Passivation of Industrial Crystalline Silicon Solar Cells*. Michael Faraday House, Stevenage, UK: Institution of Engineering and Technology; 2018.

[6] Enrichi F., Righini G. *Solar Cells and Light Management Materials, Strategies and Sustainability*. 1st edn. Amsterdam, Nederlands: Elsevier; 2019.

[7] Verlinden P.J. 'Future challenges for photovoltaic manufacturing at the terawatt level'. *Journal of Renewable and Sustainable Energy*. 2020;**12**(5):053505.

[8] Chen Y., Altermatt P.P., Chen D., *et al*. 'From laboratory to production: learning models of efficiency and manufacturing cost of industrial crystalline silicon and thin-film photovoltaic technologies'. *IEEE Journal of Photovoltaics*. 2018;**8**(6):1531–8.

Chapter 2
Main requirements for solar cells

Jian Wei Ho[1], *Thorsten Dullweber*[2], *Markus Fischer*[3], *Susanne Herritsch*[4], *and Jutta Trube*[4]

This chapter describes the general design considerations and requirements of a photovoltaic (PV) solar cell for efficiently converting radiant energy from the sun into electrical energy. The ideal converter is first considered in light of the solar spectrum and the detailed balance limit. Causes for departure from ideal behaviour are then presented and form the background for solar cell analysis. For better understanding of the requirements of solar cells, basic device principles are introduced through the course of this chapter. In addition to the physical and technological requirements, the economic and environmental impacts are finally considered, especially in light of solar cell metallization.

2.1 Operation of a PV device

The solar cell is the basic unit of solar PVs, converting solar radiant energy directly into electrochemical potential energy. The effectiveness of the device depends on the content of the incident solar energy as well as the cell design. The solar irradiance and spectrum at the Earth's surface are influenced by factors such as location, time, elevation and weather. For performance quantification, standard test conditions (STCs) of 1 000 Wm^{-2} and air mass 1.5 (AM1.5) have been defined in the IEC 60904 standards. The reference spectrum is illustrated in Figure 2.1. While these are not necessarily representative of actual field conditions, STC is important for comparison and analysis of PV device performance. In this chapter, performance quantification pertains to operation under STC global radiation or AM1.5 G.

For the absorbing material of the solar cell, the PV energy conversion process is frequently described in association with the concept of an energy gap E_g which separates excited states from the ground state. This serves to maintain excited

[1]Solar Energy Research Institute of Singapore (SERIS), Singapore
[2]Institute for Solar Energy Research Hamelin (ISFH), Germany
[3]Hanwha Q CELLS GmbH, Germany
[4]VDMA Photovoltaic Equipment, Germany

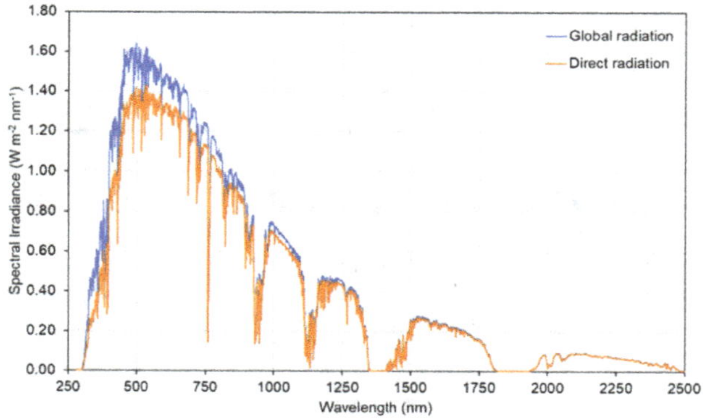

Figure 2.1 *Reference air mass 1.5 (AM1.5) solar irradiance spectrum (IEC 60904-3). Global radiation refers to the combined direct, diffuse and albedo components corresponding to an integrated 1 000 Wm^{-2} under specific atmospheric conditions and albedo.*

carriers at a higher energy level longer than the thermal relaxation time so they may be collected. In addition, the energy gap should be large compared to the characteristic thermal energy kT, where k and T are the Boltzmann's constant and temperature, respectively. Such a system can be fulfilled by a semiconductor, e.g. silicon.

The PV energy conversion process may be analysed in terms of the following steps:

1. Absorption of photons with energy $E \geq E_g$ in the photovoltaic material.
2. Excitation of charge carriers (e.g. electrons) into higher energy states/bands. Thermal relaxation of carriers within the bands, establishes a quasi-thermal equilibrium. For a two-band system, the increase in electrochemical energy is given by $N\Delta\mu$, where N is the number of carriers promoted and $\Delta\mu$ is the difference in chemical potentials between the excited and ground state populations.
3. Movement of excited carriers, e.g. by diffusion, within the material.
4. Separation of charges, e.g. by electric fields associated with p-n and high-low junctions, or selective contacts, such that charge carriers of opposite polarities are collected at different contacts.

The build-up of charges establishes a potential difference between the terminals of the cell. Connection to an external circuit will allow the charges to be transported out to do useful electrical work. Factors that influence the above steps will affect the efficiency of the solar cell and lead to certain requirements. Prior to this, the solar cell is first examined under ideal conditions.

2.2 The detailed balance limit of a p-n junction solar cell

The fundamental physical limitation of a solar cell can be analysed in terms of the detailed balance limit. The p-n junction (e.g. Si) solar cells are the most common form of solar energy converters and are invariably used for analysis. The current density-voltage characteristics of an ideal solar cell, neglecting series and shunt resistances, can be described with a simple one-diode model:

$$J(V) = J_0\left[\exp\frac{qV}{nkT} - 1\right] - J_L \tag{2.1}$$

where J_0 is the saturation current density, n is the diode ideality factor, q is the electronic charge and J_L is the light generated current density. The equation is based on Shockley's ideal diode equation of a p-n junction [1]. J_0 characterizes the diffusion and recombination in the device and determines the open-circuit voltage V_{oc}. When recombination in the space charge region and other non-ideal recombination effects are neglected, n is taken to be 1 [2], and V_{oc} can be related to J_0 as:

$$V_{oc} = \frac{kT}{q}\ln\left(\frac{J_L}{J_0} + 1\right) \tag{2.2}$$

where q is the elementary charge. The relationship outlines the basic principle of increasing the V_{oc} and hence efficiency of a solar cell. That is, maximizing J_L (or light absorption) and minimizing J_0 (or recombination losses).

In the ideal limiting case of a p-n solar cell, as presented in the seminal work of Shockley and Queisser in 1961, *radiative recombination* is the only recombination mechanism of charge carriers. They showed that the efficiency limit for a single junction solar cell at 300 K under black body radiation at 6000 K to be about 30% for an E_g of 1.1 eV [3]. Recent calculations place the optimal E_g to be about 1.32 eV [4]. For non-concentrated photovoltaic conversion which takes into account light absorption and scattering in the atmosphere, STC (AM1.5 G and $T = 298.15$ K or 25 °C), is more

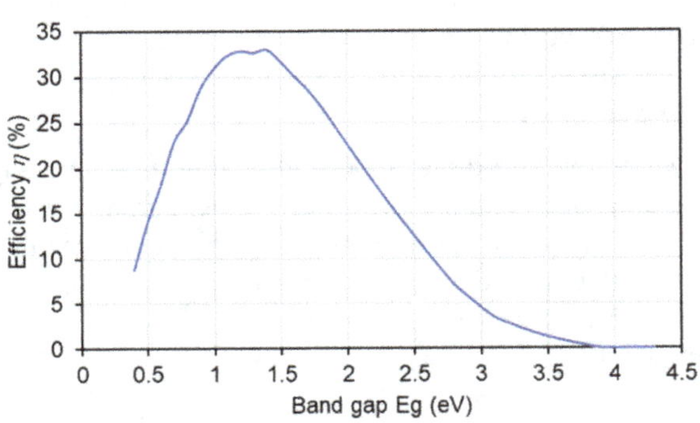

Figure 2.2 *Photovoltaic conversion efficiency η (detailed balance limit) for a solar cell under STC (AM1.5 G spectral irradiance and 25°C) as a function of the band gap energy E_g. The curve is not completely smooth due to atmospheric absorption.*

relevant, for which the maximum efficiency is found to be ~ 33.2% for an E_g of 1.34 eV (see Figure 2.2). The computation of the limiting efficiency shows the influence of the incident spectrum and energy gap, and is a fundamental consideration for PV material selection. A more complete list of the assumptions used in the calculations are listed as follows.

1. A well-defined energy gap E_g exists and that all incident light with $E \geq E_g$ is absorbed, while absorption is zero for $E < E_g$. Reflection losses are not considered.
2. Each absorbed photon photogenerates exactly one electron-hole pair.
3. Electrons and holes do not recombine except radiatively, as required by detailed balance.
4. The charge carriers are completely separated and collected. This and the prior assumption indicate that the external quantum efficiency is 100% for $E \geq E_g$.
5. Charges are transported to the external circuit without losses. That is, resistive losses (due to series and shunt resistance) are zero.

While the assumptions listed above may not necessarily be representative of actual conditions, understanding them allows PV scientists and engineers to be cognizant of the limiting requirements to achieve high PV conversion efficiency. Similar (more involved) considerations exist for multi-junction solar cells.

2.3 Practical solar cells

Practical solar cells detract from the ideal behaviour presented earlier, with the departure manifesting as power losses. The importance of analysing such losses in solar cells cannot be overemphasized and presents a critical step towards identifying performance-limiting factors and solar cell requirements. In this section, the frequently used two-diode model with its associated resistances is introduced along with an overview of loss channels.

2.3.1 Two-diode model

Situations arise where the one-diode model is unable to adequately describe the behaviour of solar cells. As such, the two-diode model parameterized by two diode ideality factors has been adopted as a more general representation of the solar cell [5]. The corresponding circuit diagram and the electrical current sign convention used in this chapter is illustrated in Figure 2.3.

The mathematical representation is as follows.

$$J(V) = J_{01}\left[\exp\frac{q(V-JR_s)}{n_1kT} - 1\right] + J_{02}\left[\exp\frac{q(V-JR_s)}{n_2kT} - 1\right] + \frac{V-JR_s}{R_{sh}} - J_L \tag{2.3}$$

where J_{01} and J_{02} are the saturation current densities, and n_1 and n_2 are the diode ideality factors for the first and second diodes, respectively. Recombination mechanisms for which the diode ideality factor is 1 are lumped into the *first* diode. This

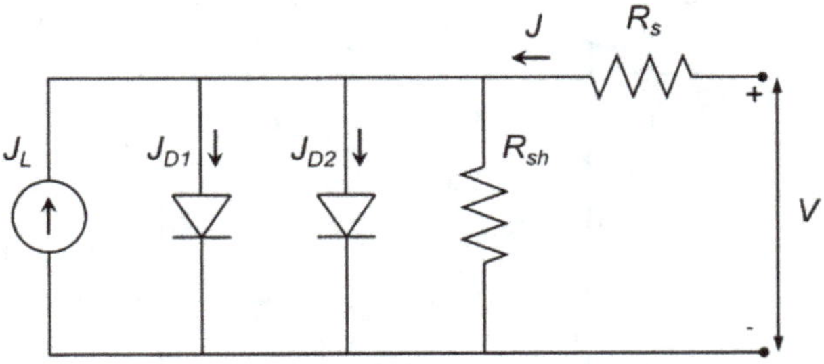

Figure 2.3 Circuit diagram corresponding to the two-diode model of a solar cell

typically comprises recombination in the quasi-neutral bulk and the two cell surfaces of the solar cell. Recombination mechanisms with n_2 are lumped into the *second* diode. $n_2 = 2$ recombination has been attributed to Shockley-Read-Hall (SRH) recombination occurring in the space charge region [6] and at the edges [7]. Cases for which $n_2 > 2$ have also been reported for recombination at localized regions with high defect density [8].

An example showing the influence of each diode ($n_1 = 1$ and $n_2 = 2$) at different operating conditions is given in the Suns-V_{oc} plot of Figure 2.4. As light intensity or V_{oc} decreases, the characteristics are in turn controlled by the n_1 diode, n_2 diode and then R_{shunt}. This suggests that, as least mathematically, while a single diode is unable to describe the solar cell over its full range of operation, there exists a limited regime where this can hold and over which the J_0 values can be extracted.

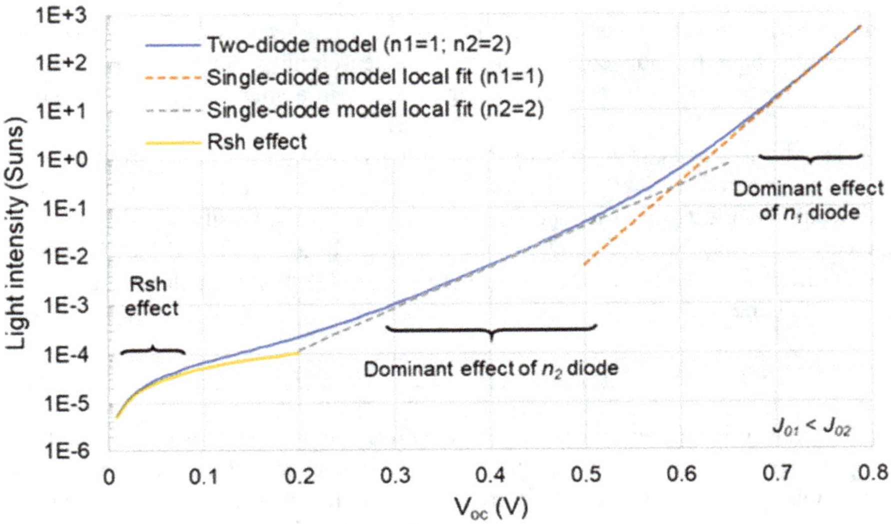

Figure 2.4 Suns-V_{oc} characteristics of a solar cell obeying the two-diode equation (where $J_{01} < J_{02}$)

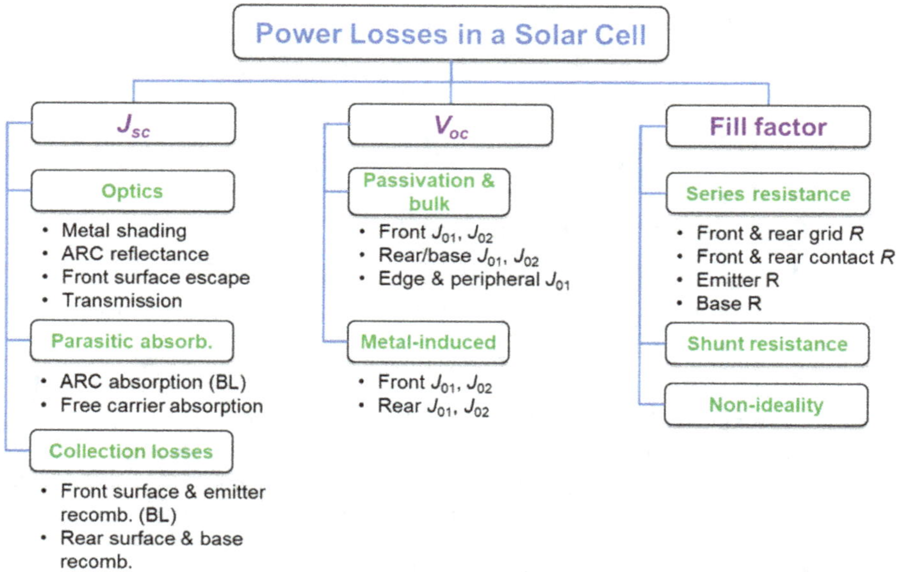

Figure 2.5 *Deviations from ideal solar cell behaviour categorized as power losses associated with open-circuit voltage (V_{oc}), fill factor and short-circuit current density (J_{sc}). (ARC, anti-reflection coating; BL, blue loss).*

2.3.2 Solar cell loss categories

The performance of a solar cell is typically parameterized by its current-voltage parameters, namely, short-circuit current (I_{sc}), open-circuit voltage (V_{oc}) and fill factor (FF). Departure from ideal behaviour or losses in a solar cell are described in relation to this and are categorized in Figure 2.5.

The short-circuit current density J_{sc} represents the highest output current density from a solar cell. Practically, when connected across a load, the output current density would be a fraction of it. The J_{sc} is controlled by the intensity of light reaching the cell, its absorption and resultant photogeneration rate $G(x)$, and the efficiency/probability of carrier collection $P(x)$, where x is the depth in the solar cell and W is the cell thickness.

$$J_{sc} = q \int_0^W G(x) \cdot P(x)\, dx \tag{2.4}$$

Factors that influence these processes will affect the J_{sc}. Shading from metallization, imperfect anti-reflection coating (ARC), light escape and transmission reduce the amount of light that can be absorbed. Competing parasitic absorption, e.g. in the ARC layer and due to free carriers, further diminish the number of photons available for photogeneration. Upon photogeneration, carriers would then need to travel to the respective electrodes for collection. A larger W provides more material for light absorption but would inevitably mean carriers need to travel a longer distance (higher resistance and also increased chance of recombination) to be collected. Recombination in various parts of the solar cell regions (which also affect V_{oc}) would need to be minimized to increase $P(x)$ and hence, J_{sc}.

The V_{oc} of a solar cell is fundamentally limited by the absorber material's energy gap and directly influenced by the extent of charge carrier recombination within it. As can be observed from the diode equations, V_{oc} is logarithmically related to the saturation current density. This can be broken down into the components J_{01} and J_{02} (based on the two-diode model) associated with the different parts of the solar cell, e.g. passivated front surfaces, base, passivated rear surface, metal surfaces, etc. A key requirement to improve the V_{oc} would thus be to reduce the component J_0 values. The practical considerations and relations are briefly elaborated later.

While the detailed balance limit assumes that separated carriers are transported to the external circuit without losses, practical solar cells have non-zero series resistance R_s, finite shunt resistance R_{sh} and non-ideality which affect carrier transport and contribute to FF losses. R_s can be traced to the metallization grid, metal-semiconductor contact, emitter and base resistances. R_{sh} describes the degree of 'leakiness' or rectifying behaviour of the diode. Apart from R_s and R_{sh}, non-ideality associated with second diode recombination, such as due to recombination in the metal-semiconductor interface, edge and peripheral regions presents additional pathways for FF losses. Requirements for solar cell design need to target low R_s, high R_{sh} and minimal non-ideality.

2.4 Practical considerations and requirements

In this section, the practical considerations in relation to basic solar cell device physics are elaborated with the consequent requirement on solar cell design.

2.4.1 Optics

The optical design of the solar cell plays a key role in coupling of light into it for absorption. Approaches to address this involves front surface texturing, metallization grid optimization and application of ARC. However, these require careful optimization in relation to other considerations.

2.4.1.1 Metallization grid shading

For front and back contacted solar cells, the front metallization grid forms the conduit for extraction of carriers from the solar cell but inevitably shades it. The extent of shading loss is proportional to the metallization fraction. Efforts to reduce the metallization area also reduce metal recombination but need to balance the resulting increase in series resistance, and hence FF loss.

2.4.1.2 Texturing and ARC

Minimization of reflectance is an important requirement in solar cell design and is accomplished through surface texturing and ARC. Texturing or roughening of the front surface of a solar cell is invariably employed to increase the chance of light reflected off an angled surface onto another to be coupled into the cell for absorption. Refraction at an angled surface further increases the effective path length of light within the cell for absorption and photogeneration. In Si solar cells, where both the front and rear surfaces are suitably (pyramidally) textured with rear surface reflective, light trapping (through internal reflection) close to the Lambertian limit (absorption enhancement factor of ~ $4a^2$, where a is the refractive index of the optical medium) can be achieved [9, 10].

The reflection loss is large if there is a large mismatch in the refractive indices of two adjacent optical media. An ARC reduces the optical reflectivity through interference effects. To minimize reflection, a basic design principle for the thickness d_1 of the ARC with refractive index a_1 is given by:

$$a_1 d_1 = \lambda_0/4 \tag{2.5}$$

where λ_0 is the peak wavelength of the incident light spectrum. Reflection can be further minimized by choosing an ARC with refractive index a_1 equal to the geometric mean of the refractive indices of the optical media which it bridges.

$$a_1 = \sqrt{a_0 a} \tag{2.6}$$

where a_0 and a are the refractive indices of air and the semiconductor (silicon), respectively.

Single layer ARC minimizes the reflection only over a narrow range of wavelengths. Increasing the range will require a graded ARC or several layers of ARC with refractive indices increasing from a_0 (ARC adjacent to air) to a (adjacent to the semiconductor).

A metric for evaluating the reflectance of solar cells is the reflectance weighted against the AM1.5 G solar spectrum. The weighted average reflectance (WAR) is given as follows:

$$\text{WAR} = \frac{\int_{\lambda_1}^{\lambda_2} \phi(\lambda) \cdot R(\lambda)\, d\lambda}{\int_{\lambda_1}^{\lambda_2} \phi(\lambda)\, d\lambda} \tag{2.7}$$

where λ_1 and λ_2 are the lower and upper limits of the wavelength range of interest, $\phi(\lambda)$ is the incident photon flux and $R(\lambda)$ is the total reflectance of the solar cell. All else constant, minimization of WAR is desired.

2.4.1.3 Front surface escape and transmission

IR light tends to be weakly absorbed in indirect bandgap silicon due to the low absorption coefficient of the latter at such wavelengths. For typical thicknesses of modern Si solar cells (<180 μm) which are too thin to fully absorb the IR light, the unabsorbed light undergoes multiple reflection within it before eventually exiting. Exit from the front side is known as front surface escape. Light trapping by appropriate texturing of the front and rear surfaces, with the rear surface made reflective, can increase the effective path length of the light within the cell and chances for absorption. For bifacial solar cells, the unmetallized regions of the rear surface presents another route for light to escape or be transmitted out.

2.4.2 Parasitic absorption

2.4.2.1 ARC absorption

A low reflecting ARC (low WAR) is desired to optimize light coupling into the solar cell. However, this should also be selected/designed to have low absorption as well. A classic example is the use of SiN_x as the ARC. By varying the deposition conditions such as temperature, gas flow rates and chamber pressure during plasma-enhanced chemical vapour deposition (PECVD), the refractive index of SiN_x can be tuned from less than 1.9 to over 2.2. While Si-rich films have higher refractive index, it has higher absorption, as compared to N-rich films. The optical and electrical properties are related to the structure and chemistry of the Si-H, N-H and Si-N bond densities [11]. Hence, although higher refractive index contributes to lower WAR, parasitic absorption in the ARC may negate the reduction in reflectance losses. A low reflecting and low absorption ARC is thus required.

2.4.2.2 NIR parasitic absorption

Near-infrared (NIR) parasitic absorption by free carriers can constitute notable current losses. Free carrier absorption (FCA) arises when incident light (usually NIR) excites a carrier, which is already in an excited state, to another unoccupied state in the same electronic band. This is facilitated by higher carrier concentrations such as in the doped semiconductor layers forming the back surface fields in Al-back surface field (BSF) and n^+ poly-silicon layers in tunnel oxide passivated contact (TOPCon) solar cells. FCA is also evident in transparent conductive oxide (TCO) layers used in Si heterojunction (HJT) solar cells. Eventually, NIR light which does not escape the solar cell and does not result in photogeneration contributes to its heating up and needs to be minimized.

2.4.3 Collection losses

Following photogeneration in the solar cell, it is not granted that the photogenerated carriers will reach and be collected by the electrodes as assumed in the detailed balance limit. The collection probability depends on the distance of the minority carrier from the p-n junction compared to its diffusion length. At the depletion region, the collection probability is high, since electron-hole pairs are quickly separated by the electric field. Away from the junction, the collection probability drops due to bulk recombination (in the emitter and base). For carriers photogenerated at distances greater than the diffusion length from the junction, the collection probability is low. In addition, at the surfaces (front and rear), surface recombination further acts to reduce the collection probability. Recombination is examined later but in general, low bulk and surface recombination are requirements for efficient solar cells. In

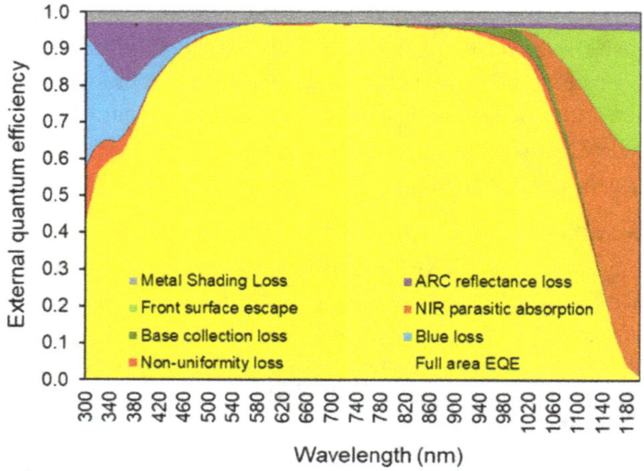

Figure 2.6 *Current loss analysis based on external quantum efficiency and reflectance measurements delineating the various loss categories. Non-uniformity loss arises due to spatial differences in external quantum efficiency. (ARC, anti-reflection coating).*

addition, while thin solar cells mean that photogenerated carriers are closer to the junction for collection, insufficient absorbing material or inadequate light trapping can adversely reduce the extent of photogeneration. Hence, this needs to be appropriately optimized.

The general categories relating to current losses can be broadly quantified by external quantum efficiency (EQE) and reflectance measurements [12]. An example of this is shown in Figure 2.6 for a Si passivated emitter and rear cell (PERC) showing the losses due to front metallization grid shading, reflectance, front surface escape, NIR parasitic absorption, base collection, short wavelength effects (blue loss due to ARC absorption, front surface and emitter recombination) and non-uniformity.

2.4.4 Energy gap

The detailed balance limit calculations show that the energy gap of a single junction solar convertor has a notable influence on its efficiency through J_{sc} and V_{oc} effects. The III–V compound semiconductors, GaAs and InP have direct band gaps (1.42 eV and 1.35 eV, respectively) near the optimum and have found application in high-efficiency cells. In particular, GaAs holds the record for single junction cell efficiency under STC at 29.1% [13]. Si having a band gap of 1.12 eV, and which is indirect, is less optimal. A caveat in the detailed balance limit analysis for indirect band gap materials is that they are limited by *Auger recombination* rather than radiative recombination, and have a lower detailed balance limit due to lower V_{OC} [14]. Despite this, optimized Si PV devices have achieved very respectable efficiencies up to 26.7% [15] and command the largest market share in PV due to the abundance and relative low cost of Si, as well as mature technology.

For every material, its band gap therefore introduces a fundamental limit to it conversion efficiency, determines the potential of the solar conversion device and needs to be judiciously considered among other factors. Notwithstanding the material system, a measure of experimental device quality is the bandgap-voltage offset at open circuit, W_{OC} [16].

$$W_{OC} = \frac{E_g}{q} - V_{OC} \tag{2.8}$$

W_{OC} has a weak dependence on E_g resulting in nearly constant values across different materials. Practical and good values of W_{OC} are typically below 0.4 V. For thin GaAs solar cells at one sun, values ~ 0.29 V have been attained [13, 17], while Si HJT solar cells have surpassed the 0.4 V barrier with W_{OC} ~ 0.37 V [18]. It can be shown that W_{OC}, when limited by intrinsic recombination, i.e. Auger and radiative recombinations, is given by

$$W_{OC,int} = \frac{kT}{q} ln\left(\frac{N_V N_C qw(C\Delta n+B)}{J_L}\right) \tag{2.9}$$

where N_V and N_C are the densities of states in the valence and conduction band, respectively, Δn is the excess minority carrier density, w is the thickness, C is the ambipolar Auger recombination coefficient and B is the radiative recombination coefficient [14]. While N_V, N_C, B and C are material intrinsic parameters, w and J_L can be optimized (minimized and maximized, respectively) from the perspective of solar cell design. However, before this intrinsic limit is reached, other practical qualities of the solar cell need to be refined. This relates to improving the V_{OC} of the cell. Most notably, recombination due to bulk and surface defects needs to be minimized through use of high-quality bulk material, and excellent surface and bulk passivation.

2.4.5 Recombination

Carrier recombination affects the build-up of charge carriers and hence V_{oc}. Radiative and Auger recombinations are intrinsic to the bulk material and depend on the doping concentration. Besides these, other categories of recombination have also been defined. These include SRH recombination through defects and impurities which form energy levels within the energy gap in the bulk, surfaces, emitter or high-low junction, and metal-semiconductor interface. The defects can be imperfections in the material crystal structure (e.g. vacancies, interstitials and dislocations), while impurities can be oxygen-related (e.g. resulting in thermal donors), oxide precipitate, complex (e.g. boron-oxygen), metallic (e.g. Cu, Au), or non-metallic (e.g. C) components. In any case, concentration of defects and impurities and hence associated recombination activity needs to be kept to a practical and economic minimum.

2.4.5.1 SRH recombination in the bulk

SRH recombination in the bulk exerts further impact on the diffusion of minority charge carriers to the respective contacts in the solar cell where they are separated and collected. Bulk recombination is described by the diffusion length of the minority carrier which is the average distance it can move between generation and recombination, and is in turn related to the minority carrier lifetime τ through

$$L = \sqrt{D\tau} \tag{2.10}$$

where D is the diffusion constant. Longer L is indicative of higher τ and lower recombination activity, and are basic requirements for efficiency improvement. These need to be considered when evaluating semiconductor materials for solar cells. The effect of Auger, radiative and SRH recombinations in the bulk can be combined and represented in the form of the bulk lifetime τ_{bulk}.

$$\frac{1}{\tau_{bulk}} = \frac{1}{\tau_{Auger}} + \frac{1}{\tau_{Rad}} + \frac{1}{\tau_{SRH}} \tag{2.11}$$

For the specific case of the one-diode model, where the diode ideality factor is one and *the diode is long such that the diffusion lengths of minority carriers are less than the diode dimensions*, the saturation current density J_0 is given by the following equation [1]:

$$J_0 = qn_i^2 \left(\frac{D_p}{L_p N_D} + \frac{D_n}{L_n N_A} \right) \tag{2.12}$$

where n_i is the intrinsic carrier concentration. The subscripts 'p' and 'n' pertain to holes and electrons, respectively. N_D and N_A are the donor and acceptor concentrations of the n-type and p-type regions of the solar cell p-n junction. The equation demonstrates the *general* inverse relationship between J_0 and the minority carrier diffusion length L (and lifetime τ). The raw semiconductor material and subsequent solar cell fabrication process have major impact on L. J_0 and L need to be respectively minimized and maximized through various cell design approaches which will be described latter in the chapter. The above equation further shows the dependence of J_0 on band gap and temperature through n_i.

2.4.5.2 Surface recombination

In general, the surface of a semiconductor is an abrupt discontinuation of the crystal lattice and is decorated by dangling bonds. These manifest as defect states within the bandgap of the material and provide additional pathways for defect-assisted recombination. This is often characterized by the surface recombination velocity S (units: cm/s) and sometimes by the surface lifetime τ_s [19]. The effective carrier lifetime τ_{eff} is typically measured in photoconductance-based techniques and is thus a more practical parameter to describe the material or device. For samples with non-diffused surfaces, τ_{eff} is given as

$$\frac{1}{\tau_{eff}} = \frac{1}{\tau_{bulk}} + \frac{S_{front}}{w} + \frac{S_{back}}{w} \tag{2.13}$$

where w is the sample thickness, and the subscripts 'front' and 'back' refer to the front and back surfaces of the material. For samples with diffused surfaces (or high-low junctions such as BSF), the relevant expression is

$$\frac{1}{\tau_{eff}} = \frac{1}{\tau_{bulk}} + \frac{J_{0,front}\left(N_{dop} + \Delta n\right)}{qn_i^2 w} + \frac{J_{0,back}\left(N_{dop} + \Delta n\right)}{qn_i^2 w} \tag{2.14}$$

where N_{dop} is the background dopant density, $J_{0,front}$ and $J_{0,back}$ are the saturation current densities associated with the combined effects of Auger, SRH and surface recombinations in the highly doped regions at the front and back surfaces of the solar cell. Minimizing/optimizing $J_{0,front}$ and $J_{0,back}$ in relation to other properties is thus a basic requirement/consideration in solar cell design.

2.4.5.3 Metal-induced recombination

Similarly, the contact between metal and semiconductor is a region of high interface defect density and introduces significant recombination. For industrial Si solar cells, the recombination at the contacted Si region is further increased by the damage incurred during the high temperature firing step for contact formation using metal paste. Reasons for the enhanced contact recombination include:

1. Introduction of electronic defects below the metal-silicon interface, e.g. diffusion of impurities from the metal paste.
2. Removal of silicon, and formation of local pits and metal spikes when aluminium containing paste is used [20].
3. Degradation of the metal-silicon interface by aggressive glass frit in the metal paste [21].
4. Non-optimum work function difference between the metal (4.26 eV for Ag, 4.05 eV for Al) and Si (~5.17 eV for p-type Si) resulting in increased minority carrier concentration at the interface.
5. Local inhomogeneities of doping, where lower doping concentrations in localized regions will result in locally increased minority carrier concentration [22].

As contact is made at the emitter and surface fields (where applicable), the metal-induced recombination is typically lumped with the recombination at the local doped regions and described in terms of saturation current densities associated with the metal contact, e.g. that at the p$^+$ emitter-metal region is lumped together as $J_{0,e,metal}$ or $J_{0,p+(metal)}$. Various methods have been employed to experimentally determine the recombination. This typically involves varying the metallization fraction, measuring the respective Suns-V_{oc} and fitting to a one-diode model [21, 23]. For instance, in the case of bifacial n-PERT solar cells with diode ideality factor of 1

$$J_0 = J_{0,bulk} + J_{0,p^+} + J_{0,n^+} \tag{2.15}$$

where J_0 is the total saturation current density which can be extracted based on Suns-V_{oc} measurements, $J_{0,p+}$, $J_{0,bulk}$ and $J_{0,n+}$ present the contribution from the p$^+$ surface (emitter), bulk and the n$^+$ surface (BSF), respectively. The contributions from the passivated and metallized surfaces can be further distinguished as follows:

$$\begin{aligned} J_{0,p^+} &= J_{0,p^+(pass.)} \times (1 - F_{metal,p^+}) \\ &+ J_{0,p^+(metal)} \times F_{metal,p^+} \end{aligned} \tag{2.16}$$

and

$$J_{0,n^+} = J_{0,n^+(pass.)} \times (1 - F_{metal,n^+}) + J_{0,n^+(metal)} \times F_{metal,n^+} \tag{2.17}$$

where $F_{metal,p+}$ and $F_{metal,n+}$ refer to the metallization fraction at the p$^+$ and n$^+$ diffused surfaces. $J_{0,p+(pass.)}$, $J_{0,p+(metal)}$, $J_{0,n+(pass.)}$ and $J_{0,n+(metal)}$ represent the J_0 contributions from the passivated diffused and metallization regions at the respective surfaces and are target parameters to reduce. At the same time, reducing the metallization fraction in contact with the semiconductor forms a key consideration in solar cell design optimization.

Apart from the above technique, metal recombination measurement by discerning the FF losses has also been employed [24]. A more rigorous approach using photoluminescence measurement and finite element modelling to further account for the effects of lateral balancing currents and localized recombination currents near the wafer edges has also been reported [25, 26]. The consideration of the doping profile in relation to recombination and resistance is discussed in the next section.

2.4.6 Series resistance

Series resistance R_s impedes the flow of current in a solar cell, resulting in the dissipation of electrical power as heat and loss in FF/power conversion efficiency. Therefore, minimizing R_s is a key requirement to achieving efficient solar cells. The various contributions to R_s along the path of the charge carriers are illustrated in Figure 2.7.

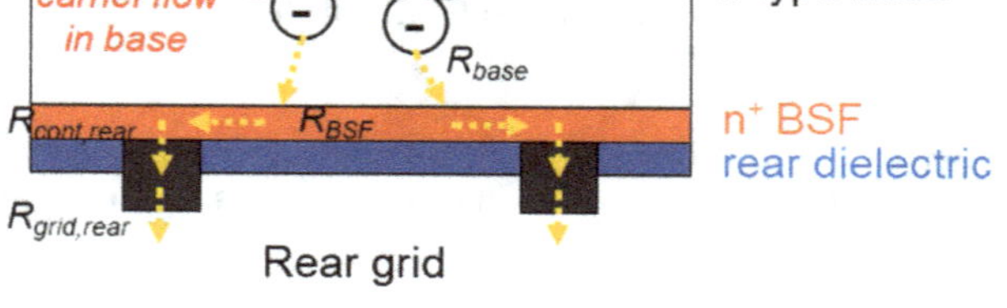

Figure 2.7 Schematic cross-section of a Si solar cell (n-PERT) showing the simplified paths of carrier movement and associated resistances. (BSF, back surface field).

2.4.6.1 Base resistance *R*base

For front and back contacted solar cells, photogenerated carriers in the base move in the general direction normal to the cell plane towards the surface doped layers for collection by the metallization grid. The base resistance is expressed as

$$R_{base} = \frac{\rho_b w}{A} \tag{2.18}$$

where ρ_b is the bulk resistivity of the cell material, w is the width of the bulk (approximated as the cell thickness) and A is the cell area. Both lower ρ_b and w contribute to lower R_s but have to be optimized with respect to other considerations. Lower ρ_b is achieved by increasing the bulk doping concentration, but leads to higher ionized impurity scattering (reduces current density), and Auger and defect-related recombinations (reduces minority carrier lifetime and hence V_{OC}). A semi-empirical metric for assessing Si wafers is the ratio of the bulk carrier lifetime τ_{bulk} to the bulk resistivity ρ_b. A higher τ_{bulk}/ρ_b ratio has been correlated to higher power conversion efficiency [27] and is sought. A suitably low w is desired to reduce resistance but not excessively as it reduces the absorption the cell.

2.4.6.2 Surface doped layer

In the surface doped layers, that is, the emitter surface field or BSF, current flows laterally and is collected by the metallization grid. These layers are described by their sheet resistance R_{sheet} and are determined by the doping profile $N(x)$ as follows.

$$R_{sheet} = \frac{1}{q\int_0^d \mu_M(x)\,N(x)\,dx} \tag{2.19}$$

where d is the junction depth, x is the depth from the solar cell surface, $\mu_M(x)$ is the mobility of the majority carriers and $N(x)$ is the net ionized doping concentration. The considerations for the doping profile can be elaborate because in addition to influencing lateral conductivity in the surface doped layer, it directly affects the *p-n* junction properties and metal-semiconductor contact resistance.

Low R_{sheet} improves FF and is realized by higher doping concentration and deeper junction depths. High doping concentration also makes for easier contacting by the metallization grid (low contact resistance). However, very high doping can result in a 'dead' layer at the surface that is highly defective and recombination active. On the other hand, reducing the depth to create shallow junctions increase susceptibility to shunting (during firing of screen-printed metallization) and increased metal recombination. Tuning the doping profile is thus important. For phosphorus doped layers which are contacted by Ag paste, high surface concentration ~2–4×10^{20} cm^{-3} have been used with moderate junction depth ~0.3–0.5 μm [25, 26]. The high surface concentration shields minority carriers from reaching the surface due to field-effect passivation [23] helping to reduce surface recombination. For boron doped layers and contacting by Ag-Al paste where the inclusion of Al helps to reduce contact resistance, Al spiking into the diffused layer can lead to highly pronounced metal

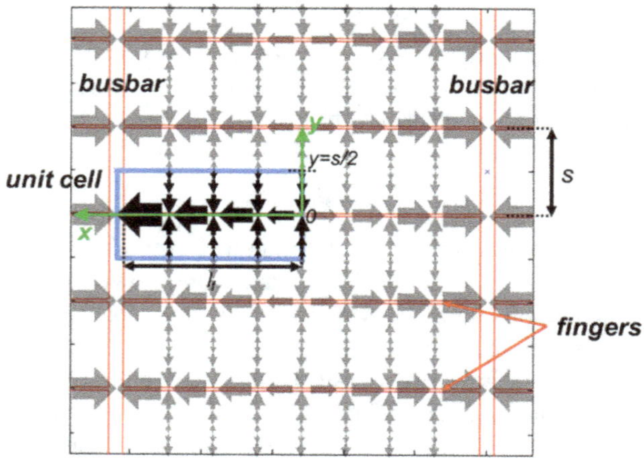

Figure 2.8 *Schematic of simplified lateral current flow in the surface doped layer and the metallization grid*

recombination. For deep spiking, deep and flat doping profiles with boron density above ~10^{18} cm^{-3} are recommended [28, 29].

The analysis of lateral current flow in the surface doped layer is performed along with the metallization grid pattern it is in contact with. A simplified current flow is shown in Figure 2.8 and can be analysed in terms of a unit cell (outlined in blue). Under uniform illumination, the current is taken to be zero at the mid-point between metal fingers ($y = s/2$) and increases linearly until the finger ($y = 0$). The finger collects a linearly increasing amount of current along its length from $x = 0$ to $x = l_f$ at the busbar.

At the maximum power point, the fractional ohmic power loss $f_{p,doped}$ associated with the surface doped layer is given by:

$$f_{p,\,doped} \approx \frac{R_{sheet} s^2 J_{mp}}{12 V_{mp}} \tag{2.20}$$

where s is the pitch of the fingers in the metallization grid, and J_{mp} and V_{mp} are the current density and the voltage at the maximum power point, respectively.

2.4.6.3 Contact resistance

A finite electrical contact resistance is inevitably present at the interface between two materials. For conventional Si solar cells, this usually refers to the metal-semiconductor contact between Si and the metallization grid. A Schottky barrier exists due to the difference in the work function of the metal and the electron affinity of the semiconductor. When the silicon is lowly doped (<10^{15} cm^{-3}) such that the depletion region is wide, carriers can only overcome the barrier by thermionic emission. With increasing doping level (~10^{17}–10^{18} cm^{-3}), the depletion region narrows such that carrier tunnelling becomes possible. With sufficiently high doping (> 10^{19} cm^{-3}), a low resistance ohmic contact can be formed [30]. Hence, the metal contacted regions of a solar cell (e.g. the emitter surface field and BSF) are usually highly doped. This also reduces metal recombination at these regions. In some cases, high doping selectively at such regions has been employed, as this helps to decouple the moderate doping requirement at the non-metal contacted region allowing lower recombination (Auger and SRH) and parasitic absorption here.

2.4.6.4 Metallization grid pattern

From the prior analysis, the finger pitch s should be small to minimize ohmic power loss due to R_{sheet}. However, this will come at the expense of increased shading by the metallization grid. The optimal design is thus a compromise. In addition, the resistance of the metallization grid has to be minimized. In particular, the power loss at maximum power point in a finger of length l_f, width w_f and height h_f with resistivity ρ_f is given by

$$P_{loss,\ finger} \approx \frac{l_f^3 J_{mp}^2 s^2 \rho_f}{3 w_f h_f} \tag{2.21}$$

Reducing the finger resistivity ρ_f and increasing the finger height h_f (or aspect ratio h_f/w_f to avoid excessive shading) are approaches to reduce resistance losses but are limited by practical metallization material and the process capability of laying down high aspect ratio fingers. A more comprehensive treatment of different metallization grid patterns is given in [31]. Reducing the finger length l_f through incorporation of more busbars (multi-busbar designs) can notably help to reduce series resistance losses and improve FF.

2.4.7 Shunt resistance

Shunt paths in a solar cell provide leakage routes for the photogenerated current. Usually, this is a consequence of practical process design or fabrication defects. For instance, interruption to the continuity of the emitter (due to scratches or metal spiking during the metallization process) can result in the direct contact of the top metal with the base, bypassing the junction. Inadequate removal of emitter wraparound at the edges of the solar cell can similarly provide a shunt path. The effect of shunting is more pronounced at low illumination levels since the photogenerated current is smaller. Further, at lower voltages, the effective series resistance is relatively higher, exacerbating the impact of shunts. As such, shunt resistance should be maintained as high as possible.

2.4.8 Non-ideality

After discounting for series and shunt resistances, the practical upper limit for FF is frequently considered to be set by recombination currents in the quasi-neutral bulk and the two surfaces. This is described by the J_{01} diode. Further FF losses have generally been lumped under non-ideality or J_{02} recombination. As mentioned earlier in the two-diode model, J_{02} recombination has been attributed to SRH recombination occurring in the space charge region [6] and at the edges [7], as well as, at

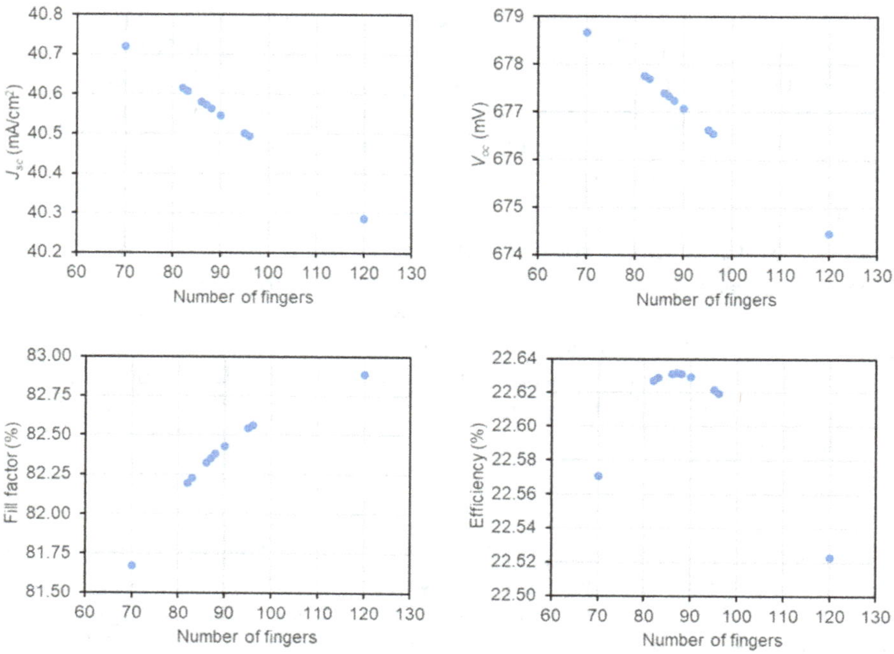

Figure 2.9   General influence of the number of fingers in the front metallization grid on I–V parameters of a Si solar cell as an example

localized regions with high defect density [8]. As such, considerations for improving V_{oc} through reducing J_{02} recombination, will similarly apply to reduce FF loss related to non-ideality.

In industrial Si solar cells, the metallization grid is formed by a screen-printing process for which the screen can be easily modified. As such, the grid is one of the most straightforward and heavily optimized aspects of solar cell design. Two-dimensional solar cell modelling software such as Griddler [32] can significantly reduce the time and effort required to optimize the metallization grid pattern, taking into account metallization grid resistance, contact resistance, surface doped layer resistance, metal-semiconductor recombination and optical shading. An example showing the general influence of the number of fingers in the front metallization grid on the I–V parameters of a Si solar cell is shown in Figure 2.9. As the number of fingers increases, both J_{sc} and V_{oc} reduce, but the FF rises. There is thus an optimal number of fingers where the power conversion efficiency is maximum.

2.5 Economic and environmental aspects of solar cell metallization

In addition to the physical and technological requirements to enable high solar cell conversion efficiencies as described in the previous subchapters, in order to be economically attractive for a solar cell manufacturer, the metallization technology has to be very cost-effective also in high-volume production. Hence, production tool investments (capital expenditure, CapEx) and daily production costs such as e.g. Ag paste consumption (operational expenditure, OpEx) have to be as low as possible. The throughput of a production tool is another very important economic parameter as today's solar cell fabs process thousands of silicon wafers per hour in order to supply the market and in order to minimize costs per piece. These cost aspects will be addressed in the following section on production cost requirements. Afterwards, the section on sustainability requirements will describe environmental metallization aspects such as the worldwide availability of raw materials for metallization (e.g. Ag) in sufficient amounts to support hundreds of GWp of yearly solar cell production. Another important aspect is the usage of ecologically friendly materials where hazardous materials such as lead (Pb) should be avoided.

2.5.1 Production cost requirements

PV modules from different manufacturers for solar power plants or rooftop applications typically do not differentiate much in design or technology and hence are considered as a commodity product today. Whereas in the 1970s PV modules were sold at between 50 US$/$W_p$ and 100 US$/$W_p$, average spot market prices at the end of 2019 were at 0.23 US$/$W_p$. Figure 2.10 shows the learning curve for module price

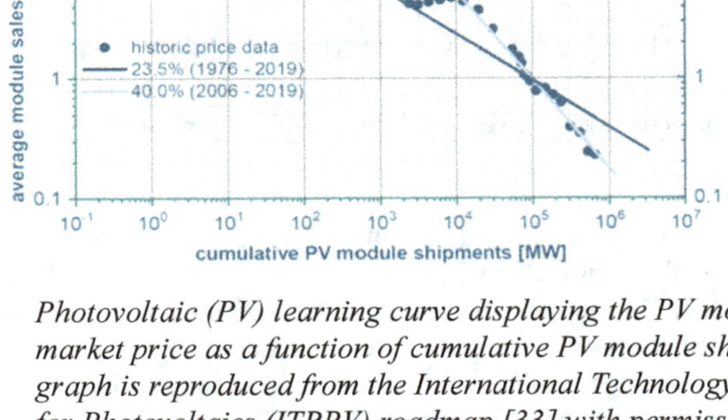

Figure 2.10 Photovoltaic (PV) learning curve displaying the PV module spot market price as a function of cumulative PV module shipment. This graph is reproduced from the International Technology Roadmap for Photovoltaics (ITRPV) roadmap [33] with permission.

as a function of cumulative module shipments between 1976 and 2019 plotted in a log-log scale. The average learning rate (LR) is 23.5% – this is an average price reduction of 23.5% with doubling of the cumulated module shipments. In the period between 2006 and 2019, the LR increased to 40%. This development was possible thanks to an increase in poly-Si production capacity, the increase of crystalline silicon (c-Si) cell and module production capacity utilizing further-developed highly productive processing tools as well as the introduction of new high-efficiency c-Si cell and module technologies. So today 95% of the worldwide PV module market is supplied by c-Si solar cell technologies. Thin film PV technologies account for only about 5% [33].

In order to stay cost-competitive in the market, the International Technology Roadmap for Photovoltaics (ITRPV) [33] proposes the following strategy to solar cells manufacturers:

- increase module power conversion efficiency without significantly increasing production costs
- continue cost reduction per piece by more *efficient equipment utilization* (e.g. higher production throughput per tool) *and material utilization* (e.g. reducing Si and Ag consumption per wafer).

On top, there is a permanent increase in PV production capacity. Annual shipments in 2019 were about 130 GWp while the production capacity at the end of 2019 was about 200 GWp. Production capacity will increase further. This will result in oversupply situations. In case of oversupply of modules in the PV market, the PV module price drops drastically and only solar cell manufactures with the lowest production costs and competitive products survive in the PV market. This economic mechanism enforces a very tough cost competition and an accelerated introduction of advanced production technologies in order to improve the performance and to further reduce the costs of PV cell and module production.

According to the ITRPV [33], at the end of 2019 the average module price of 0.24 US$/Wp did consist of the following three main contributions:

1. 30%: manufacturing of the Czochralski (Cz) silicon wafer (poly-Si, Cz crystal growth, sawing)
2. 17%: manufacturing of solar cells (texturing, diffusion, surface passivation, metallization)
3. 53%: manufacturing of solar modules (half-cut, interconnection, lamination, framing).

Solar cell metallization contributes to the costs of solar cell manufacturing. The PV market today mostly manufactures PERC solar cells with screen-printed and furnace-fired Ag front metallization and Al rear metallization including Ag pads on the rear as described in Subchapter 9.1 in detail. Whereas screen-printing tools are relatively cheap and provide a high production throughput, the consumption of Ag paste is one of the biggest cost contributors to the metallization costs of today's silicon solar cells

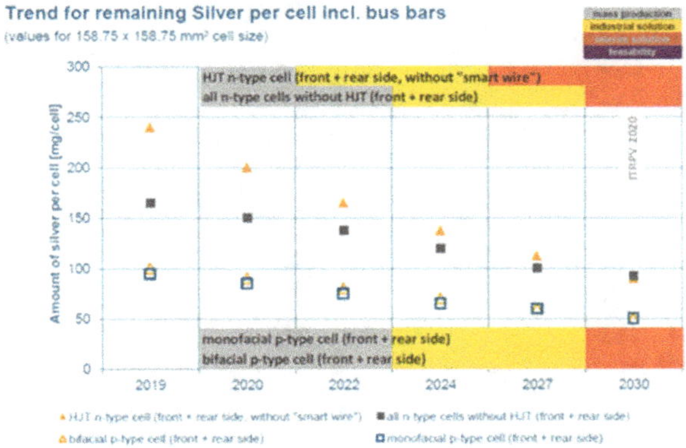

Figure 2.11 *Ag consumption per 158 mm Cz Wafer. Taken from [33] with permission. (HJT, Heterojunction Technology).*

due to the high price of raw Ag which is in the range of 400 US$/kg to 1400 US$/kg in the past 10 years. Figure 2.11 shows the trend of printed Ag per Si wafer, which is around 80 mg today for monofacial (PERC) and bifacial (PERC+) p-type cells. Solar cell manufacturers, Ag paste suppliers, screen suppliers and screen-printer suppliers are carrying out extensive research to further reduce the Ag consumption in the next years towards 50 mg per cell as shown in Figure 2.11, e.g. by printing narrower Ag fingers at maximum throughput. However, as can be seen by the colour code, suitable mass production technologies still have to be developed in order to reach this target. Figure 2.11 also shows that other c-Si cell technologies like HJT or n-type TOPCon concepts require even more Ag per wafer since they use Ag metallization on both wafer sides. A more disruptive approach to reduce the consumption of Ag is to completely replace Ag, e.g. by screen-printed Al contacts (see Subchapter 6.2. on screen-printed Al pastes) or by electroplated NiCuAg contacts as described in Chapter 8. Aluminium paste for the rear side of PERC cells is relatively cheap. Nevertheless, bifacial PERC+ cells reduced the Al paste consumption from 850 mg/cell for PERC to 200 mg/cell for PERC+ by applying Al fingers instead of a full-area-Al layer; see Subchapter 9.1 on PERC cells and [33].

The throughput of processed wafers per hour of a production tool is another important economic parameter determining the productions costs of the solar cell metallization. A higher throughput means that fixed production costs such as CapEx or labour costs are distributed over a larger number of wafers, thereby lowering the production cost per solar cell. Hence, over the past decade the throughput of solar cell production tools for all process steps such as wet chemistry, diffusion, deposition and metallization increased from a few hundred wafers per hour to around 8 000 wafers per hour in 2020 as shown in Figure 2.12 [33, 34]. As a consequence, the CapEx required to build a new 1 GWp solar cell production fab reduced by more than 70% [34]. By further reducing the process times, increasing the batch sizes and other measures, it is expected that the throughput will further increase beyond 10 000 wafers per hour as shown in Figure 2.12 [33] thereby leveraging further

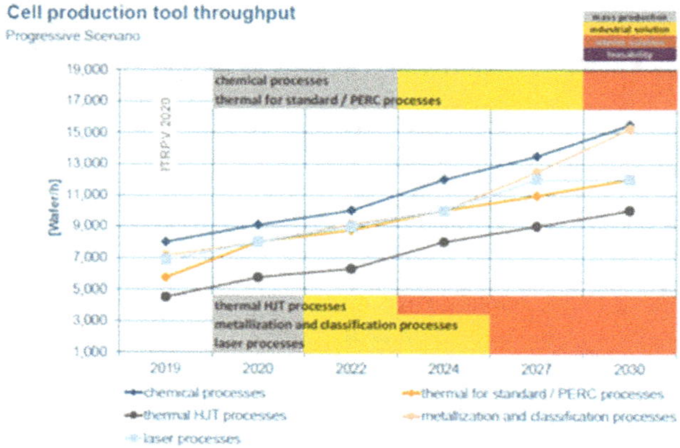

Figure 2.12 *Throughput of production tools for new solar cell production lines. Figure reused from [33] with permission. (HJT, Heterojunction Technology; ITRPV, International Technology Roadmap for Photovoltaics; PERC, Passivated Emitter and Rear Cell).*

production cost reductions. Hence, for a solar cell metallization technology, it is mandatory to fullfil today's and future production throughput requirements.

2.5.2 Sustainability requirements

Besides production costs, another important aspect of high-volume manufacturing is the sustainability of production. For example, all required raw materials have to be available in sufficient amounts. Since the photovoltaics industry grows very fast, it is important to project the future global production volume in order to assess if materials could become scarce. Figure 2.13 shows the 'broad electrification scenario' of ITRPV [33] for the annual PV production capacity which is expected to increase from presently around 100 GWp to beyond 2 500 GWp in the next 25 years. Altermatt *et al.* [35] calculated that if PV wants to contribute to limiting global warming to below 2°C until 2050 according to the Paris climate agreement by mitigating CO_2 emission, PV production capacity must be increased annually by 18% over the next 10 years until 2030. This scenario by Altermatt *et al.* [35]. corresponds to an annual PV market of around 800 GWp in 2030 which is below the 'broad electrification scenario' ITRPV scenario in Figure 2.13.

A future annual production capacity of around 1 000 GWp in 2028 according to Figure 2.13 corresponds to around 160 000 000 000 silicon solar cells produced per year, assuming that in the future one solar cell delivers 6 Wp (presently 5 Wp). Taking into account the projected Ag consumption of 50 mg/solar cell as shown in Figure 2.11, this corresponds to a future annual Ag consumption of 8 000 tons per year just for solar cell production. On the other hand, the worldwide production of Ag is in the range of 30 000 tons per year [36]. So potentially, PV will consume about one-third of the annual Ag market which may further increase the price of

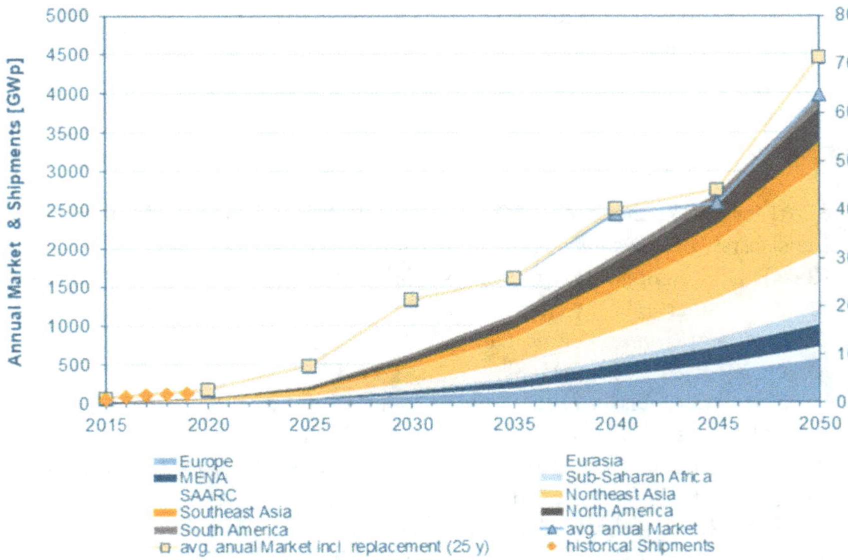

Figure 2.13 *Forecast of worldwide annual PV production and cumulative global PV installations in the 'broad electrification scenario' of the International Technology Roadmap for Photovoltaics (ITRPV) 11th edition. Figure reproduced from [33] with permission.*

Ag. These are reasons why PV institutes and companies perform research to further reduce the Ag consumption per solar cell or to replace Ag with other metals such as aluminium or copper, which are available in much larger quantities and at a much cheaper price. Also, recycling and reusing Ag and other materials from old PV modules which exit field operation will become very important in order to enable a sustainable production of PV modules. A similar situation was present in the 1990s when Ag-based photography also accounted for about 30% of the Ag world market. Recycling of photographic Ag via development labs was the method of choice in that time.

Last but not least, the materials used in solar cell production should be environmentally friendly. In particularly the final PV module should not contain hazardous or toxic materials which could be released to the environment in case of fire, module breakage, etc. Whereas typical metals such as Ag, Al, Cu used for solar cell metallization fullfil this requirement, in the past Ag pastes designed for screen-printing and alloying with the phosphorus-doped emitter did contain small amounts of Pb. The Pb lowers the eutectic temperature of the Al–Si phase diagram and hence enables Ag–Si alloying already at around 800°C firing temperature. In the past years, Ag paste manufacturers have developed alternative Pb-free Ag pastes which fullfil the 'Restriction of (the use of certain) Hazardous Substances in electrical and electronic Equipment' (RoHS) requirements. Meanwhile these Pb-free Ag pastes are widely used in the market and cells will become more and more Pb-free in the future according to the ITRPV [33].

References

[1] Shockley W. 'The theory of *p-n* junctions in semiconductors and *p-n* junction transistors'. *Bell System Technical Journal*. 1949;**28**(3):435–89.

[2] Rose R.D., Magnone P., Zanuccoli M., Sangiorgi E., Fiegna C. 'Loss analysis of silicon solar cells by means of numerical device simulation'. *2013 14th International Conference on Ultimate Integration on Silicon (ULIS)*, 19–21 March 2013; 2013. pp. 205–8.

[3] Shockley W., Queisser H.J. 'Detailed balance limit of efficiency of *p-n* junction solar cells'. *Journal of Applied Physics*. 1961;**32**(3):510–19.

[4] Rühle S. 'Tabulated values of the Shockley–Queisser limit for single junction solar cells'. *Solar Energy*. 2016;**30**:139–47.

[5] Wolf M., Rauschenbach H. 'Series resistance effects on solar cell measurements'. *Advanced Energy Conversion*. 1963;**3**(2):455–79.

[6] Sah C.-tang., Noyce R., Shockley W. 'Carrier generation and recombination in p-n junctions and P-N junction characteristics'. *Proceedings of the IRE*. 1957;**45**(9):1228–43.

[7] Mclntosh and Keith. Lumps, humps and bumps: Three detrimental effects in the current-voltage curve of silicon solar cells. *PhD, Centre for Photovoltaic Engineering*; Sydney: University of New South Wales; 2001.

[8] Steingrube S., Breitenstein O., Ramspeck K., Glunz S., Schenk A., Altermatt P.P. 'Explanation of commonly observed shunt currents in c-Si solar cells by means of recombination statistics beyond the Shockley-Read-Hall approximation'. *Journal of Applied Physics*. 2011;**110**(1):014515.

[9] Campbell P., Green M.A. 'Light trapping properties of pyramidally textured surfaces'. *Journal of Applied Physics*. 1987;**62**(1):243–9.

[10] Yablonovitch E., Cody G.D. 'Intensity enhancement in textured optical sheets for solar cells'. *IEEE Transactions on Electron Devices*. 1982;**29**(2):300–5.

[11] Sopori B. 'Silicon nitride processing for control of optical and electronic properties of silicon solar cells'. *Journal of Electronic Materials*. 2003;**32**(10):1034–42.

[12] Wong J., Duttagupta S., Stangl R., Hoex B., Aberle A.G. 'A systematic loss analysis method for rear-passivated silicon solar cells'. *IEEE Journal of Photovoltaics*. 2015;**5**(2):619–26.

[13] Green M.A., Dunlop E.D., Hohl-Ebinger J., Yoshita M., Kopidakis N., Hao X. 'Solar cell efficiency tables (version 56)'. *Progress in Photovoltaics: Research and Applications*. 2020;**28**(7):629–38.

[14] Augusto A., Herasimenka S.Y., King R.R., Bowden S.G., Honsberg C. 'Analysis of the recombination mechanisms of a silicon solar cell with low bandgap-voltage offset'. *Journal of Applied Physics*. 2017;**121**(20):205704.

[15] Yoshikawa K., Yoshida W., Irie T., *et al.* 'Exceeding conversion efficiency of 26% by heterojunction interdigitated back contact solar cell with thin film Si technology'. *Solar Energy Materials and Solar Cells.* 2017;**173**(17):37–42.

[16] King R.R., Bhusari D., Boca A., *et al.* 'Band gap-voltage offset and energy production in next-generation multijunction solar cells'. *Progress in Photovoltaics: Research and Applications.* 2011;**19**(7):797–812.

[17] Kayes B.M. 27.6% Conversion efficiency, a new record for single-junction solar cells under 1 sun illumination. *In 2011 37th IEEE Photovoltaic Specialists Conference*, 19–24 June 2011; 2011. pp. 000004–8.

[18] Masuko K., Shigematsu M., Hashiguchi T., *et al.* 'Achievement of more than 25% conversion efficiency with crystalline silicon heterojunction solar cell'. *IEEE Journal of Photovoltaics.* 2014;**4**(6):1433–5.

[19] Sproul A.B. 'Dimensionless solution of the equation describing the effect of surface recombination on carrier decay in semiconductors'. *Journal of Applied Physics.* 1994;**76**(5):2851–4.

[20] Lago R., Pérez L., Kerp H., *et al.* 'Screen printing metallization of boron emitters'. *Progress in Photovoltaics: Research and Applications.* 2010;**18**(1):20–7.

[21] Edler A., Mihailetchi V.D., Koduvelikulathu L.J., Comparotto C., Kopecek R., Harney R. 'Metallization-induced recombination losses of bifacial silicon solar cells'. *Progress in Photovoltaics: Research and Applications.* 2015;**23**(5):620–7.

[22] Kiefer F., Peibst R., Ohrdes T., Krügener J., Osten H.J., Brendel R. Emitter recombination current densities of boron emitters with silver/aluminum pastes. *In 2014 IEEE 40th Photovoltaic Specialist Conference (PVSC)*, 8–13 June; 2014. pp. 2808–12.

[23] Fellmeth T., Born A., Kimmerle A., Clement F., Biro D., Preu R. 'Recombination at metal-emitter interfaces of front contact technologies for highly efficient silicon solar cells'. *Energy Procedia.* 2011;**8**(6):115–21.

[24] Hoenig R., Glatthaar M., Clement F., Greulich J., Wilde J., Biro D. 'New measurement method for the investigation of space charge region recombination losses induced by the metallization of silicon solar cells'. *Energy Procedia.* 2011;**8**(7):694–9.

[25] Shanmugam V., Mueller T., Aberle A.G., Wong J. 'Determination of metal contact recombination parameters for silicon wafer solar cells by photoluminescence imaging'. *Solar Energy.* 2015;**118**:20–7.

[26] Ananthanarayanan D., Wong J., Wang E.-C., Chen N., Ho J.W., Aberle A.G. 'Determination of metal-induced recombination of N-type bifacial Si solar cells using special print patterns'. *IEEE Journal of Photovoltaics.* 2019;**9**(3):643–51.

[27] Zhao J. '>24% Silicon heterojunction solar cells on meyer burger's on mass production tools and how wafer material impacts cell parameters'. *2018 IEEE 7th World Conference on Photovoltaic Energy Conversion (WCPEC) 2018 (A Joint Conference of 45th IEEE PVSC, 28th PVSEC & 34th EU PVSEC)*, 10–15 June; 2018. pp. 1514–19.

[28] Li M., Ma F.-J., Stangl R.A., Aberle A.G. 'Investigation on the Ag-Al metal spiking into boron-diffused P+ layer of industrial bifacial N-type silicon wafer solar cells by numerical simulation'. *33rd European Photovoltaic Solar Energy Conference and Exhibition, Amsterdam*; 2017. pp. 251–5.

[29] Rodriguez J., Wang E.-C., Chen N., *et al.* 'Towards 22% efficient screen-printed bifacial n-type silicon solar cells'. *Solar Energy Materials and Solar Cells*. 2018;**187**:91–6.

[30] Ebong A., Chen N. 'Metallization of crystalline silicon solar cells: a review'. *High Capacity Optical Networks and Emerging/Enabling Technologies*, 12–14 Dec; 2012. pp. 102–9.

[31] Serreze H.B. 'Optimizing solar cell performance by simultaneous consideration of grid pattern design and interconnect configurations'. *13th IEEE Photovoltaic Specialists Conference*; Washington, D.C., USA; 1978.

[32] Wong J., Raj S., Ho J.W., Wang J., Lin J. 'Voltage loss analysis for bifacial silicon solar cells: case for two-dimensional large-area modeling'. *IEEE Journal of Photovoltaics*. 2016;**6**(6):1421–6.

[33] ITRPV. *International technology roadmap for photovoltaic (ITRPV), results 2019 including maturity report, 11th edition* [online]. 2020. Available from https://itrpv.vdma.org/en/ [Accessed Oct 2020].

[34] Chen Y., Altermatt P.P., Chen D., *et al.* 'From laboratory to production: learning models of efficiency and manufacturing cost of industrial crystalline silicon and thin-film photovoltaic technologies'. *IEEE Journal of Photovoltaics*. 2018;**8**(6):1531–8.

[35] Altermatt P.P., Yang Y., Chen Y. 'Requirements of the Paris climate agreement for the coming 10 years on investments, technical roadmap, and expansion of pv manufacturing'. *37th European Photovoltaic Solar Energy Conference and Exhibition*; 2004. pp. 1999–2004.

[36] The Silver Institute. *Metals focus world silver survey 2020* [online]. 2020. Available from www.silverinstitute.org/all-world-silver-surveys/ [Accessed Apr 2020].

Chapter 3
Fundamentals of metallization

Abasifreke Ebong[1]

3.1 Introduction

Every electronic device requires the metallic contacts to the two surfaces of the semiconductor bulk to transfer carriers from semiconductor to the metal contacts. The interfaces between the semiconductor and the metal contacts can pose a barrier to the flow of carriers into and out of the metal contacts. Since the minority and majority carriers are present at both interfaces, the minority carrier should be blocked at each interface for effective transport of the majority carrier to the respective contacts. In this chapter, therefore, the fundamentals of metallization of a silicon solar cell are explored starting with: (i) the barrier heights, (ii) current transports, (iii) contact types – selective and passivated and (iv) characterization of the contacts thereof.

3.2 Barrier height

When a metal is in contact with the semiconductor, a metal–semiconductor interface is created as shown in Figure 3.1b, where a band bending and the alignment of the Fermi energy level in both metal and semiconductor occur. At thermal equilibrium, electrons would flow from the semiconductor into the lower energy states in the metal through the interface. Positively charged donor atoms remain in the semiconductor, creating a space charge region. Figure 3.1a shows the metal and the semiconductors before they are brought in contact with each other. Before the contact to each other, the Fermi level in the semiconductor is above that in the metal.

The parameter ϕ_{BO} is the ideal barrier height of the semiconductor contact; i.e., the potential barrier seen by electrons in the metal trying to move into the semiconductor. On the metal side, this barrier is given as $\phi_{BO} = \phi_m - \chi$, is the Schottky barrier. On the semiconductor side, V_{bi} is the built-in potential barrier, which is similar to that of a p–n junction. This is the barrier seen by electrons in the conduction band trying to move into the metal. The built-in potential is thus, $V_{bi} = \phi_{BO} - \phi_n$. This

[1]Energy Production and Infrastructure Center (EPIC), Department of Electrical and Computer Engineering, The University of North Carolina at Charlotte, USA

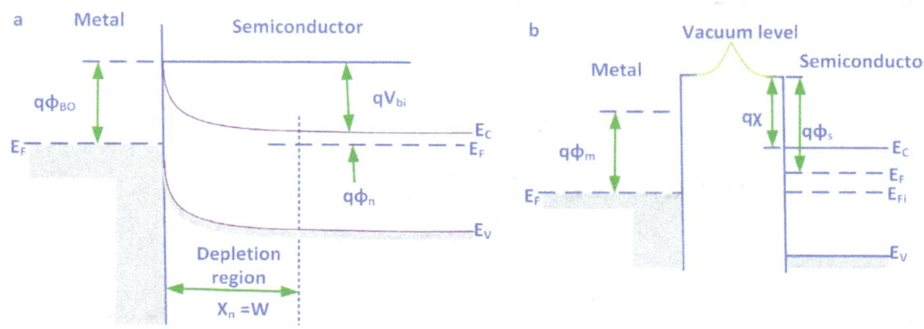

Figure 3.1 *(a) Metal and semiconductor energy band diagram before contact and (b) in contact for metal–n-type semiconductor junction for $\phi_m > \phi_s$ [1]*

implies that the built-in potential is a slight function of the doping as in the case of a p–n junction because $\phi_n = V_t \ln\left(\frac{N_C}{N_D}\right)$, where V_t is the thermal voltage, N_c is the effective density of state function in the conduction band and N_D is the donor doping concentration. The barrier height is, therefore, defined as the potential difference between the Fermi energy of the metal and the band edge where the majority carrier resides.

The impact of the barrier height is often manifested in the electrical output parameters of a solar cell through the contact resistance, which, in turn, impact the fill factor (FF) as shown in Figure 3.2. Figure 3.2a is a two-dimensional modelling of a solar cell (Ag–Si partial contact at 0.78 eV and Ni – 0.6–0.85 eV full contact) showing the impact of Schottky barrier height on the FF for a full aluminium back surface field (Al-BSF) silicon solar cell. While Figure 3.2b shows the effect of the Schottky barrier height on the open circuit voltage (V_{OC}) of the same Al-BSF silicon solar cell. It can be seen from Figure 3.2a,b that the FF and the V_{OC} are opposing

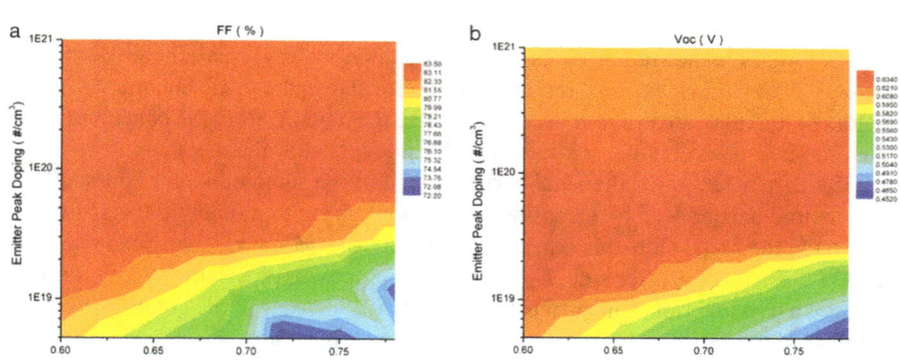

Figure 3.2 *Effect of Schottky barrier on (a) fill factor (FF) and (b) open circuit voltage (V_{OC})*

each other as the barrier height increases with respect to the emitter peak doping concentration. This is because the V_{OC} is peaked at 634 mV at the peak emitter doping concentration as the opacity of the emitter to the photons reached its peak, maximizing recombination.

3.3 Carrier transport

There are three carrier transport mechanisms at the metal–semiconductor interface depending on the doping concentrations in the semiconductor including:

1. Thermionic emission, where the electrons in the semiconductor trying to flow into the metal must do it thermally by possessing high enough kinetic energy to overcome the built-in potential barrier in the semiconductor region.
2. Thermionic field emission, where the depletion width in the semiconductor is of a size that permits electrons to thermally flow as well as tunnelling through the thin part of the barrier.
3. Field emission, where the doping in the semiconductor is large so that the depletion width is very small and electrons can tunnel through
   (i) When a lowly doped ($<10^{15}$ cm^{-3}) [2] semiconductor is in contact with a metal, a Schottky barrier height between the metal and the semiconductor is formed. This Schottky barrier height is the difference between the work function of the metal and the electron affinity of the semiconductor, as discussed in Section 3.1. The depletion region in the semiconductor is wide so that the electrons trying to flow into the metal must overcome the potential barrier by thermionic emission, as illustrated in Figure 3.3a.
   (ii) As the doping level in the semiconductor increases to ~10^{17}–10^{18} cm^{-3} (Figure 3.3b), the depletion width decreases such that the electrons can cross the barrier by tunnelling through the thin barrier (2–3 nm) in the upper end of the conduction band as well as thermionically.
   (iii) As the doping in the semiconductor increases above 10^{19} cm^{-3}, the depletion region is very thin so that the electrons tunnel through this region into the metal depicting an ohmic contact as shown in Figure 3.3c with no voltage drop.

The third option is desirable for every electronic device, including solar cells. However, the technique to make an ohmic contact includes:

1. Choosing a metal with work function close to that of semiconductor, so that there would be very small Schottky barrier height between metal and semiconductor.
2. Inserting a thin layer of narrow bandgap material between metal and semiconductor.
3. Increasing the semiconductor surface peak doping concentration to enable electron tunnelling. Of course, this is detrimental to V_{OC}, and short circuit current (I_{SC}).

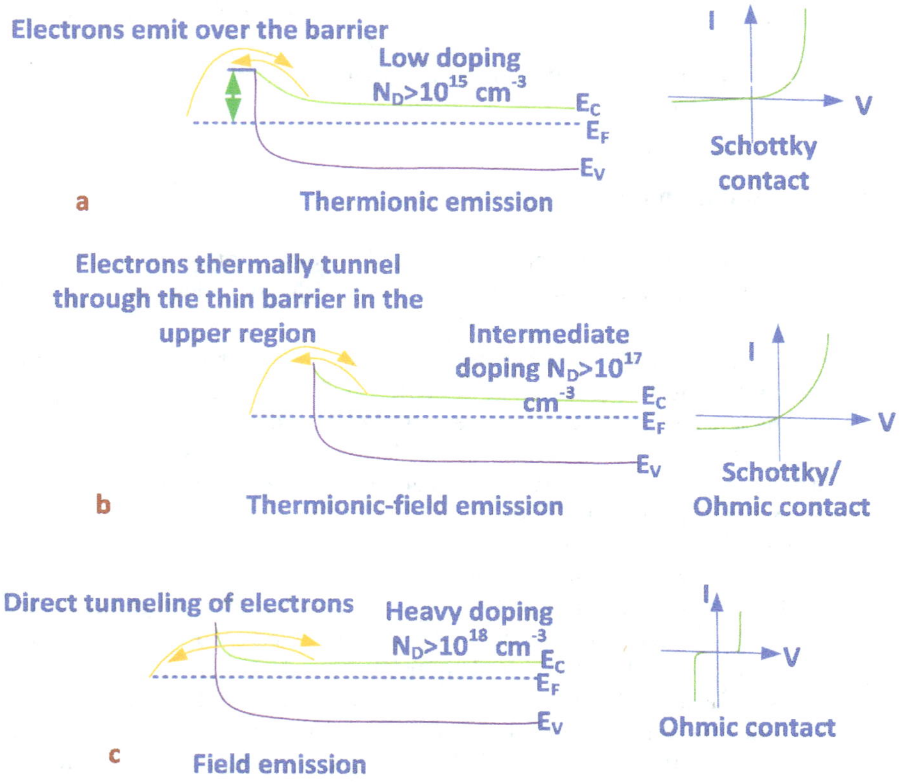

Figure 3.3 *(a) Carrier transport through metal/n-semiconductor contacts as a function of barrier height for the case of thermionic emission. (b) Carrier transport through metal/n-semiconductor contacts as a function of barrier height for the case of thermionic field emission. (c) Carrier transport through metal/n-semiconductor contacts as a function of barrier height for the case of field emission.*

For a metal–semiconductor contact to exhibit minimal resistance across it, the work function of the metal must be very close to or smaller than the electron affinity of the n-type semiconductor. However, for a p-type semiconductor, the work function of the metal should be close to or smaller than the sum of the electron affinity and the bandgap energy [3, 4]. Since most metals have a work function of <5 eV, it is very challenging to find a metal that can provide an ohmic contact to a p-type semiconductor with a large bandgap. For example, Ag has a work function of 4.26 eV and that of Al is 4.28 eV, and most semiconductor materials have electron affinity, typically, of ~4 eV (Si – 4.01 eV) [5]. The experimentally measured work function for different metals have been reviewed and plotted for easy reference [6].

Now focusing on the carrier transport for commercial solar cell technology, which is based on screen-printed contact on the homogeneous emitter. Figure 3.4 is

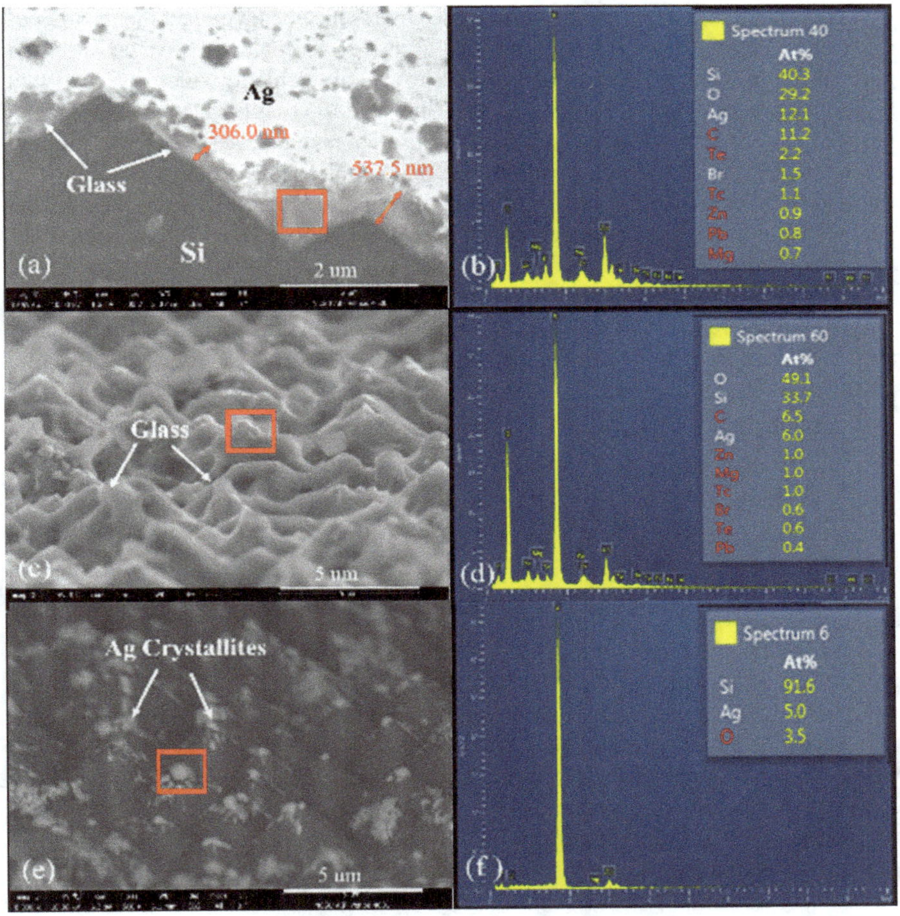

Figure 3.4 *SEM images (a, c, e) and EDS analysis results (b, d, f) of the selected areas (in red boxes) of standard cell (a, b) with Ag gridlines, (c, d) after Ag metal removal and (e, f) after glass removal*

the cross-sectional view of the metal/Si interface, depicting (i) glass layer, (ii) Ag crystallites, (iii) Ag bulk and (iv) Si. It is generally acceptable that the carrier at the Ag/Si contact transports in several different paths from the Si under the emitter to the Ag gridline [7–12] including:

1. Through Ag crystallites direct contact with Si emitter
2. Tunnelling through an ultrathin glass layer
3. Through Ag nanoparticle assisted tunnelling
4. Through the thick glass layer embedded with semimetal colloids resulting in resonant tunnelling.

The EDS images of Figure 3.4b,d show the presence of several elements, including Te and Pb, in the glass at the interface of the Ag/Si contact. However, after the glass is removed, the EDS of Figure 3.4f showed no element, which suggests that all these elements were in the glass only. This suggests that the additives in the Ag paste can enhance the interface glass conductivity and decrease the barrier height to present low resistant to majority carriers at that contact. In this case, the addition of TeO_2 glass in the front Ag paste (i) accelerates the flow of the molten glass during contact firing [13], (ii) causes uniform etching of the SiNx [14], (iii) forms semimetals of Ag_2Te and PbTe, which bandgaps are very small [15] and (iv) increases the interface glass conductivity.

The redox reactions that resulted in the formation of the semimetals (Ag_2Te and PbTe) are given below in reactions (3.1)–(3.6). The calculated thermodynamic potential for these reactions is reported in [16] as reproduced here:

$$6PbO_{(\text{in glass})} + Si_3N_4 \rightarrow 6Pb + 3SiO_{2(\text{in glass})} + 2N_2 \tag{3.1}$$

$$3TeO_{2(\text{in glass})} + Si_3N_4 \rightarrow 3Te + 3SiO_{2(\text{in glass})} + 2N_2 \tag{3.2}$$

$$2PbO_{(\text{in glass})} + Si_{3(\text{wafer})} \rightarrow 2Pb + SiO_{2(\text{in glass})} \tag{3.3}$$

$$TeO_{2(\text{in glass})} + Si_{(\text{wafer})} \rightarrow Te + SiO_{2(\text{in glass})} + 2N_2 \tag{3.4}$$

$$Pb_{(s)} + Te_{(s)} \rightarrow PbTe_{(s)} \tag{3.5}$$

$$2Ag_{(s)} + Te_{(s)} \rightarrow Ag_2Te_{(s)} \tag{3.6}$$

These reactions establish the possibility of the formation of the compounds Ag_2Te and PbTe as supported by the EDS, which showed the presence of Te and Pb in the glass. This leads to the improved contact between the Ag and Si and the lower contact resistance exhibited by PERC cells in today's manufacturing floor. This modifies the transport mechanism of the carriers from the Si underneath the emitter to the Ag/Si contact as shown in Figure 3.5.

Because of the formation of the semimetals – Ag_2Te and PbTe, in the interface glass, which bandgaps are very small, 0.014 eV and 0.30 eV [17], respectively, the barrier height of 0.1 eV was calculated [12]. Figure 3.5 is the resulting energy band diagram; the conduction band aligns between the semimetals and Si with a low barrier height between the Ag and the semimetals. Thus, the carriers are conducted unimpeded through the semimetals at the interface to the Ag contact.

3.4 Selective contacts

A contact for any electronic device can be defined as the region between the highly recombination active metal interface and the bulk semiconductor. As in every electronic device, a solar cell has two metallic regions at which voltage and current are extracted. It is therefore necessary that these two regions have only one Fermi level for ease of carrier transfer. Also, these metallic interfaces are highly defective and recombinative, which prohibit any type of passivation. However, the bulk semiconductor should be defects free with a bandgap suited for high optical absorption and highest quasi-Fermi-level splitting. Based on these key regions common to all solar cells, a contact can be defined as the region starting from the recombination-active

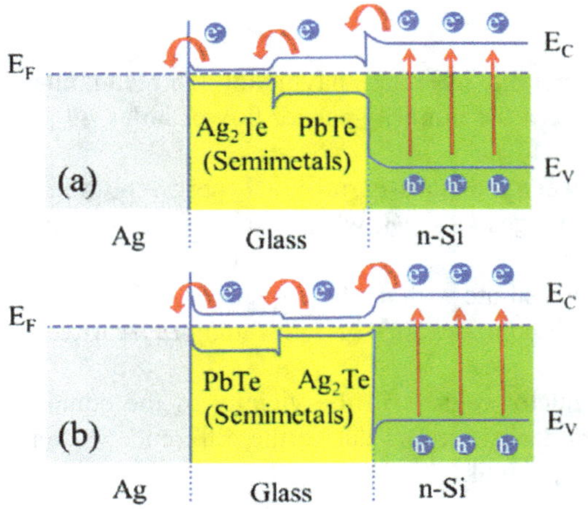

Figure 3.5 *Energy band diagram of Ag/interface glass layer with nanoparticles)/ Si contact: (a) Ag/Ag2Te/PbTe/Si and (b) Ag/PbTe/Ag2Te/Si*

metallic interface and ending at the surface of the light-absorbing bulk semiconductor as shown in Figure 3.6 [18]. From Figure 3.6, the high resistivity to electrons inherent to the hole contact is responsible for the contact's performance and enhanced by the induced effects in the bulk. In general, for two adjacent layers having differing work functions, an electric field with accompanied built-in potential will be present in the vicinity of the junction. It is noted that the separation of carriers in the junctions is not caused by the built-in potential and the electric field, rather the asymmetry in carrier concentration on either side of the junction that is responsible for the separation [18–20]. Thus, when analysing a contact, two components are necessary: (i) inherent minority-carrier resistivity in the hole contact itself and (ii) induced minority-carrier resistivity in the adjacent bulk material [18].

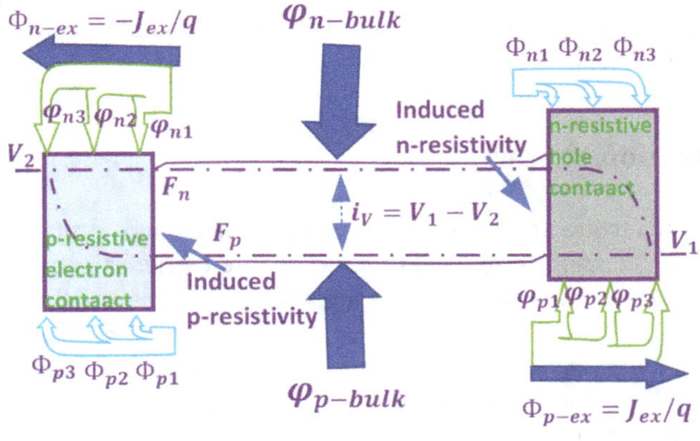

Figure 3.6 *Contact under illumination for a generalized solar cell with electron and hole contacts [18]*

Figure 3.6 shows:

1. Left and right metallic interfaces having infinite recombination velocities.
2. The magnitudes of the minority-carrier flux of holes (Φ_{p1}, Φ_{p2}, Φ_{p3}) and electrons (Φ_{n1}, Φ_{n2}, Φ_{n3}).
3. Difference between the quasi-Fermi-level energy splitting (*iv*).
4. The external voltages $V_1 - V_2$, which depends on the contact resistance of the majority carriers.
5. $\Phi_{n/p-\text{bulk}}$ commensurate with the bulk quality.
6. Recombination components of the majority-carrier fluxes, $\varphi_{n1}, \varphi_{n2}, \varphi_{n3}$ and $\varphi_{p1}, \varphi_{p2}, \varphi_{p3}$.
7. Bulk semiconductor regions in the vicinity of the contacts showing induced resistivity to minority carriers due to the difference in work function between the contact and the bulk.

The fundamental essential requirement for a solar cell is that the photon be absorbed by exciting electrons from a low-energy state to a high-energy state [19]. After the excitation, (i) the electron is extracted from the high-energy state and transported to high-energy contact (negative potential) and (ii) replenishing of the vacated low-energy state from the low-energy contact (positive) [19]. The critical requirement is that the high-energy contact should not be able to supply readily to the low-energy state, or the low-energy state should not be able to extract carriers readily from the high-energy contact. Otherwise, it will result in the wastage of the excited energy of the electrons. Thus, the p–n junction works well as a solar cell because the n-type region allows selective contact to the conduction band (excited states), and the p-type region allows selective contact to the valence band – the ground state [19]. As depicted in Figure 3.6, the (i) inherent minority-carrier resistivity in the hole contact and (ii) induced minority-carrier resistivity in the adjacent bulk material are critical in the selectivity process. Solar cells, including the Perovskite, Dye Sensitise and other organic solar cells, rely on the selective contacts to enhance the cell FF. The recombination rates in these types of cells are governed by the selective contacts [21, 22].

3.5 Passivation contacts

Passivating contacts incorporate thin films within the contact structure that simultaneously supress recombination and promote charge-carrier selectivity. The placement of a self-passivating layer between metal and silicon is an effective approach to overcome contact recombination. The main objective of passivation is to minimize the recombination fluxes at the electron and hole contacts as depicted in Figure 3.6. The passivation enhances the quasi-Fermi-level splitting while maintaining high implied voltage (i_V) as shown in Figure 3.6 [18]. The minority-carrier fluxes of holes $\left(\Phi_{p1-3} = \Phi_{p1} + \Phi_{p2} + \Phi_{p3}\right)$ at the electron contact and electrons $\left(\Phi_{n1-3} = \Phi_{n1} + \Phi_{n2} + \Phi_{n3}\right)$ at hole contact are used in the illustration of the passivation contact described by Koswatta *et al.* [18]. The hole contact passivation can be achieved by minimizing the minority-carrier recombination flux components (Φ_{n1}, Φ_{n2}, Φ_{n3}) or the majority-carrier recombination flux components (φ_{p1}, φ_{p2}, φ_{p3}). Since the recombination fluxes of electrons and holes of each region are coupled together, the smaller of the two determines the other.

It is noted in Figure 3.6 that Φ_{n1-3} depends on the electron resistivity (n-resistivity) of the preceding regions as well as the recombination-active defect densities of each region [18]:

1. Φ_{n1} is determined by induced n-resistivity in the bulk and the defect density at the interface between the bulk and the hole selector. This is also very close to the surface recombination velocity.
2. Φ_{n2} is determined by a combination of induced resistivity of the bulk and the inherent n-resistivity of the hole contact, and the density of bulk defects in the hole contact.
3. Φ_{n3} is dependent only on the total n-resistivity of the contact, since the defect density at the outermost interface is high.

The majority-carrier recombination fluxes (φ_{p1}, φ_{p2}, φ_{p3}) match the minority-carrier fluxes (Φ_{n1}, Φ_{n2}, Φ_{n3}); thus, the total current output of the solar cell ($J_{ex} = \Phi_{p-ex}$) and the amount of recombination are interdependent. Equal passivation is also achievable by limiting the majority carrier instead. Overall, low majority-carrier resistance in the contact is key to achieving high FF, and hence increased efficiency as attributed in the cells described by Yan *et al.* [23].

It should be noted that the interface passivation reported by Feldmann [24] is characterized by the implied voltage at 1 sun (iV_{OC}) calculated according to $iV_{OC} = \frac{kT}{q}\ln\left(\frac{1}{n_i^2}\left(\Delta n + \frac{n_i^2}{N_D}\right)(\Delta n + N_D)\right)$ as indicated in Figure 3.6.

3.6 Characterization of solar cell contacts

By achieving the above qualities for the two adjacent contacts, the series resistance, which is a critical parameter that limits the short circuit current density (J_{SC}) and hence the FF, can be minimized. Other factors in conjunction with the series resistance (R_S) are the shunt resistance (R_{SH}), ideality factor (n) and reverse saturation current density (Jo_2). However, R_s has a direct impact on J_{SC}, while the other factors are induced by high-temperature sintering, which can result in excessive metal particle penetration of the p–n junction and manifest in low R_{SH}, high n factor and high J_{O2}, which are detrimental to the open circuit voltage (V_{OC}). The figure of merit of a solar cell, efficiency, is the product of these three parameters, J_{SC}, V_{OC} and FF. That is why all these three parameters must be simultaneously maintained at high values in order to achieve high efficiency.

There are six components of R_S, including (i) gridline, (ii) contact, (iii) emitter, (iv) semiconductor, (v) busbar and (vi) back contact as shown in Figure 3.7:

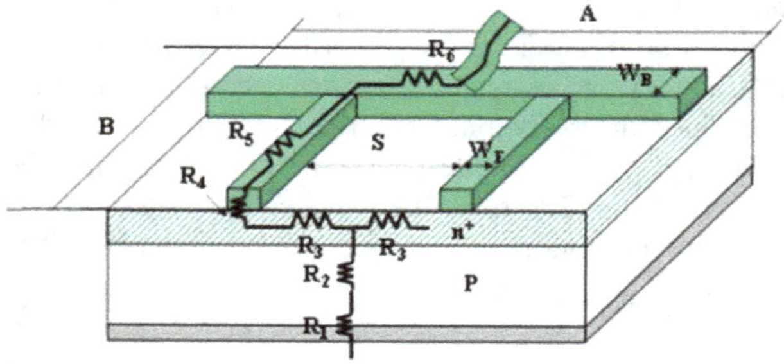

Figure 3.7 *p–n junction solar cell structure showing the six resistance components from the front to the rear contact*

1. The gridline resistance depends on the bulk metal conductivity, which for evaporated/plated as in lithographically or buried or selectively defined contacts are very low. While for the screen-printed contacts, which uses Ag paste with ≤90%, Ag metal exhibits somewhat higher gridline resistance.
2. Contact resistance for the lithographically or buried or selectively defined contact is lower than the paste-based Ag because the gridline is in direct contact with the semiconductor at a surface concentration of ≥1E19 atoms/cm³ (for an n-type) or ~1E17 atoms/cm³ for p-type [1]. The contact in the case of the selectively contacted gridline is almost ohmic, which means a low majority-carrier resistance in the contact is achieved by the heavily doped region.
3. Emitter resistance for lithographically defined contact is also smaller than the paste-based Ag, because the gridline spacing is narrower and therefore the travel distance for carriers is shorter.
4. The resistivity of the base material is common to all contact technologies.
5. The busbar resistance follows the gridline resistance according to the contact technology.
6. The back contact resistance is also common to the contact technology depending on the cell architecture.

Figure 3.8a shows typical series resistance (R_S) components for a solar cell having efficiency of 17.4%, FF of 78.2% and R_s of 0.89 Ω cm². The components of the series resistance in Figure 3.8a,b follow Meier *et al.* [16] analysis. The finger width for a typical ink-jetted Ag ink is about 46 μm wide and 15 μm tall, that is why the R_{finger} component is as high as the R_{sheet}. When the number of fingers was increased from 86 to 103 the R_{finger} dropped to only 0.07 Ω cm². Thus, the FF increased to 80% and the total series resistance dropped to 0.58 Ω cm².

3.7 Summary

The contact resistance is critical to the overall performance of a solar cell. Thus, blocking the minority carrier at the appropriate metal contacts and achieving a low

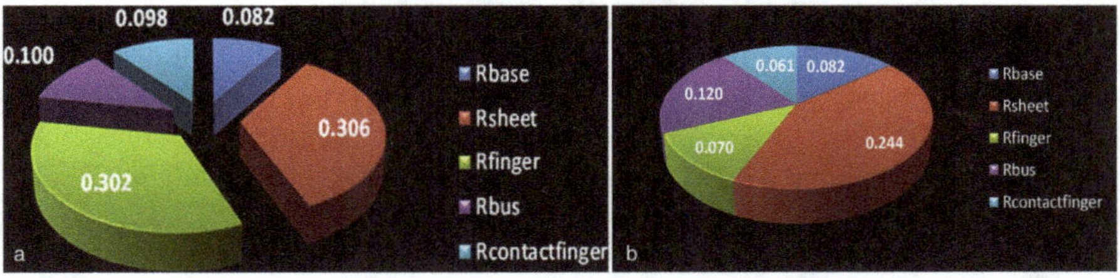

Figure 3.8 *(a) Series resistance (R_S) component for a fire through dielectric full Al-BSF cell with ink-jetted Ag contacts having series resistance of 0.89 Ω cm². (b) Series resistance (R_S) component for a fire through dielectric full Al-BSF cell with ink-jetted Ag contacts having series resistance of 0.58 Ωcm².*

majority-carrier resistivity for low contact resistance are keys to achieving the low contact resistance. This is the basis of the functionality of a p–n junction solar cell. The insertion of a thin passivated layer in the metal contact minimizes the contact recombination and enhance the carrier selectivity.

References

[1] Neamen D. *An Introduction to Semiconductor Devices*. Boston, USA: McGraw Hill; 2006. p. 195.

[2] Singh J. *Semiconductor Devices: Basic Principles*. New Jersey, USA: John-Wiley and Sons, Inc; 2001.

[3] Sze S.M., Ng K.K. *Physics of the Semiconductor Devices*. 3rd edn. New York, USA: John Wiley & Sons, Inc.; 2007.

[4] Rhoderick E.R., Williams R.H. *Metal-semiconductor Contacts (Monographs in Electrical and Electronic Engineering)*. 2nd edn. USA: Oxford University Press; 1988.

[5] Sze S.M. *Physics of Semiconductor Devices*. 2nd edn. New York: John Wiley & Sons; 1981.

[6] Ebong A., Chen N. 'Metallization of crystalline silicon solar cells: a review'. *Conference Proceeding of IEEE HONET-ICT*. Istanbul: IEEE; 2012. p. 102.

[7] Ballif C., Huljić D.M., Willeke G., Hessler-Wyser A. 'Silver thick-film contacts o*n* highly doped *n*-type silico*n* emitters: structural and electronic properties of the interface'. *Applied Physics Letters*. 2003;**82**(12):1878–80.

[8] Grupp G., Huljic D., Preu R., Willeke G., Luther J. 'Peak firing temperature dependence of the microstructure of Ag thick-film contacts on silicon solar cells–a detailed AFM study of the interface'. *EU PVSEC Proceedings: WIP GmbH & Co Planungs-KG*; 2005. pp. 1379–82.

[9] Kontermann S., Willeke G., Bauer J. 'Electronic properties of nanoscale silver crystals at the interface of silver thick film contacts on N-type silicon'. *Applied Physics Letters*. 2010;**97**(19):191910.

[10] Kontermann S., Preu R., Willeke G. 'Calculating the specific contact resistance from the nanostructure at the interface of silver thick film contacts on N-type silicon'. *Applied Physics Letters*. 2011;**99**(11):111905.

[11] Lin C.-H., Tsai S.-Y., Hsu S.-P., Hsieh M.-H. 'Investigation of Ag-bulk/glassy-phase/Si heterostructures of printed Ag contacts on crystalline Si solar cells'. *Solar Energy Materials and Solar Cells*. 2008;**92**(9):1011–15.

[12] Ren K., Han D., Ye T., Zhang Y., Ebong A. 'The impact of semimetal nanoparticles on the conduction of thick glass layer at Ag/Si contact interface'. *Journal of Applied Physics*. 2020;**127**(22):225302.

[13] Prunchak R. 'Glass Frit, US patent'. 2010:773654B2.

[14] Ebong A., Bezawada N., Unsur V., Ren K., Chowdhury A. 'Investigation of Telluride based fire-through dielectric (TTD) Ag paste contacts for Si Solar cell with high sheet resistance emitter'. *Proceedings of the 7th Workshop on Metallization and Interconnection for Crystalline Silicon Solar Cells*; 2017.

[15] Ren K., Ye T., Zhang Y., Ebong A. 'The role of nano-crystallites on conduction mechanisms of current through Ag gridlines of Si solar cells'. *MRS Advances*. 2019;**4**(5–6):311–8.

[16] Ahn K., Li C., Uher C., Kanatzidis M.G. 'Improvement in the thermoelectric figure of merit by La/Ag cosubstitution in PbTe'. *Chemistry of Materials*. 2009;**21**(7):1361–7.

[17] Green M.A. 'Photovoltaic principles'. *Physica E: Low-dimensional Systems and Nanostructures*. 2002;**14**(1–2):11–17.

[18] Koswatta P., Boccard M., Holman Z. 'Carrier selective contacts in silicon solar cells'. *Conference Proceedings of IEEE PVSC*; 2015.

[19] Meier D., Good E., Garcia R. 'Determining the components of series resistance from measurements on a finished cell'. *Conference Proceedings of the 31st IEEE PVSC*; 2006. pp. 1315–8.

[20] Wurfel U., Cuevas A., Wurfel P. 'Charge carrier separation in solar cells'. *IEEE Journal of Photovoltaics*. 2015;**5**(1):461–9.

[21] Juarez-Perez E.J., Wuβler M., Fabregat-Santiago F., *et al*. 'Role of the selective contacts in the performance of lead halide perovskite solar cells'. *The Journal of Physical Chemistry Letters*. 2014;**5**(4):680–5.

[22] Bisquert J., Cahen D., Hodes G., Rühle S., Zaban A. 'Physical chemical principles of photovoltaic conversion with nanoparticulate, mesoporous dye-sensitized solar cells'. *The Journal of Physical Chemistry B*. 2004;**108**(24):8106–18.

[23] Yan D., Cuevas A., Wan Y., Bullock J. 'Passivating contacts for silicon solar cells based on boron-diffused recrystallized amorphous silicon and thin dielectric interlayers'. *Solar Energy Materials and Solar Cells*. 2016;**152**(Part A):73–9.

[24] Feldmann F., Simon M., Bivour M., Reichel C., Hermle M., Glunz S.W. 'Carrier-selective contacts for Si solar cells'. *Applied Physics Letters*. 2014;**104**(18):181105.

Chapter 4
Metallization approaches

Loic Tous[1]

Together with wet processing, junction formation and passivation, metallization has always been and remain a key area of development for crystalline silicon (c-Si) solar cells. As the development of metallization approaches is intimately linked to the race towards the 25% cell efficiency milestone (see Figure 4.1) that took place between the early 1940s and 2000s [3], we first start with a brief history of metallization approaches for laboratory-type c-Si cells. We then give a short history of the main industrial c-Si cell concepts with a focus on screen-printed passivated emitter and rear cell (PERC), which is the dominant technology today in mass production [2].

In the subsequent sections, we introduce the main metallization approaches for industrial-type c-Si cells, which are based on the use of either of evaporated contacts, screen-printed contacts, alternative printing techniques or plated contacts. Importantly, the continuous progress of metallization approaches based on screen-printed contacts has led to certain metallization techniques or approaches falling out of favour and others being 're-discovered' as we will explain below. For each of those main approaches, we refer the readers to the other chapters of this book where significant further insights are given. In Section 4.8, we briefly compare the different metallization approaches based on general techno-economic aspects (typical finger widths, throughputs, costs, etc.) and finish with an outlook on metallization trends.

4.1 A brief history of metallization approaches for laboratory-type c-Si solar cells

The first silicon solar cell was reported in a US patent filed on 27 May 1941 and consisted of a melt grown junction with electroplated rhodium contacts on both sides [4]. Although no cell efficiency was reported, analysis of the data available suggests results well below 1%. The refinement of techniques, mainly at Bell Laboratories, to grow single crystal silicon wafers and to form the p–n junction via high-temperature diffusion of dopants (first using lithium [5] and then boron [6]) led to champion cell efficiencies quickly rising above 10% in the 1950s [7]. These cells had a wraparound

[1]IMEC, Belgium

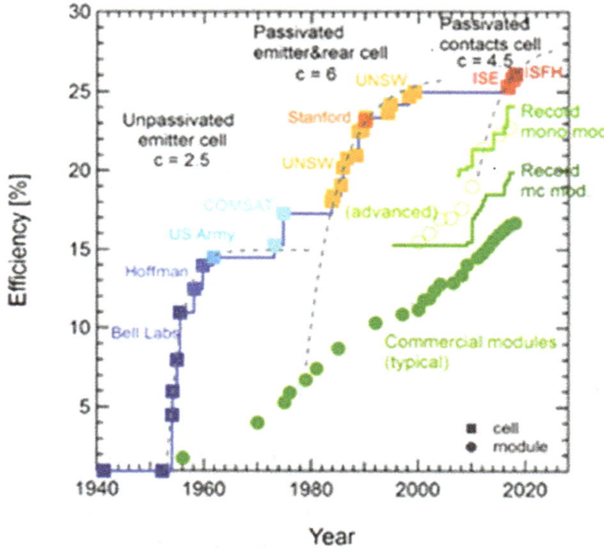

Figure 4.1 *Evolution of the efficiency of champion laboratory c-Si solar cells (square symbols), record modules (green solid lines) and typical commercial modules (circle symbols) over time. Heterostructures, such as silicon heterojunction back contact solar cells which have led to the current 26.7% world record efficiency [1], are not shown here for simplicity. Reprinted with permission from [2].*

structure (see Figure 4.2a) with plated contacts at the rear side, which helped the interconnection of cells into modules but also led to high resistive losses. The next improvement to efficiencies around 14%, achieved in the 1960s at the Hoffmann Laboratories, came from the development of photolithography patterning to form a finger grid contact pattern at the top side to drastically reduce resistive losses, while keeping shading losses as low as possible [8].

At about the same time in the 1960s, demand for space applications led to a shift to p-type substrates because of its improved radiation resistance. At the front side, the plated nickel (Ni) fingers were replaced by a stack of titanium (Ti), palladium (Pd) and silver (Ag) deposited by evaporation. Ti was chosen for its excellent contact properties, which allowed to reduce the doping in the phosphorous diffused n^+ region. Pd was chosen as interlayer to improve moisture resistance in pre-flight environment, while Ag was chosen for its excellent conductivity. By the early 1970s, the experimental benefits of using evaporated aluminium (Al) to form a heavily doped p^+ region at the rear side, thereby reducing rear side recombination and resistive losses, became well understood and led to COMSAT Laboratories producing 'violet' cells (see Figure 4.2c) with efficiencies over 15% [9]. The term 'violet' was coined due to the front antireflective coating (ARC) based on titanium oxide (TiO_x), which gave them their characteristic colour. A subsequent improvement

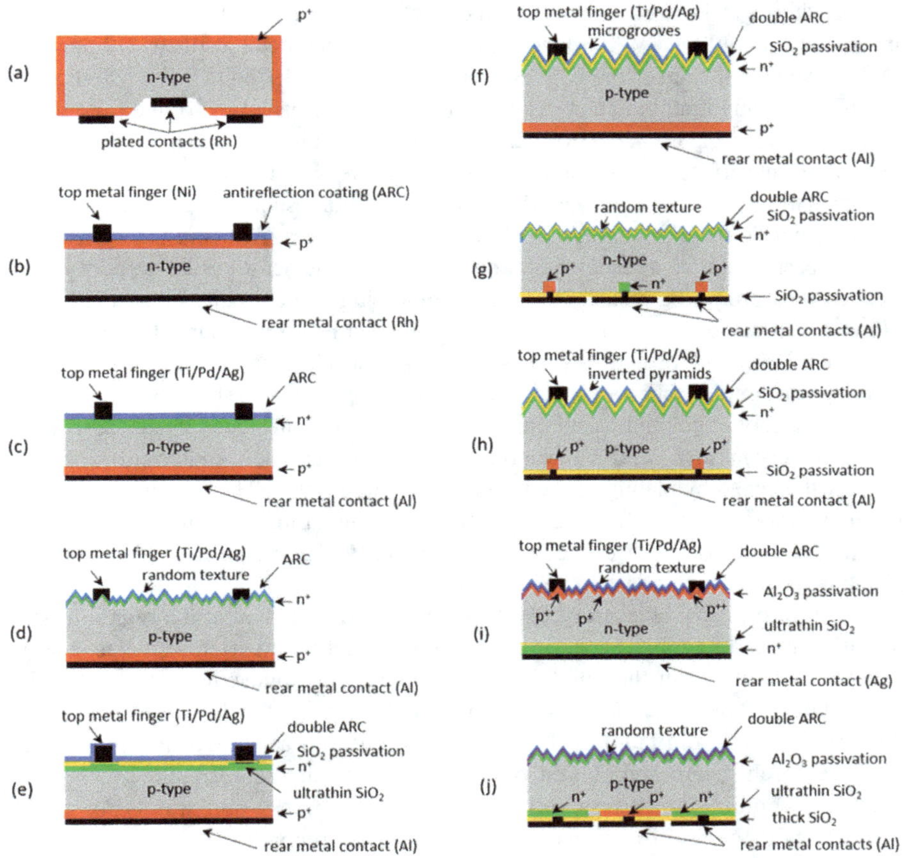

Figure 4.2 *Schematics showing the evolution in the design of champion crystalline silicon solar cells are listed in Figure 4.1. (a) First modern cell made at Bell Laboratories. (b) Two-side contacted solar cell fabricated at Hoffman Laboratories. (c) 'Violet' cell made at COMSAT Laboratories. (d) 'Black' cell made at COMSAT Laboratories. (e) Metal insulator NP junction (MINP) cell made at UNSW. (f) Passivated emitter solar cell (PESC) made at UNSW. (g) Interdigitated back contact (IBC) cell made at Stanford University. (h) Passivated emitter and rear locally diffused (PERL) cell made at UNSW. (i) Tunnel oxide passivated contact (TOPCON) cell made at ISE. (j) Polycrystalline on oxide (POLO) IBC cell made at ISFH.*

by COMSAT Laboratories was the 'black' cell (see Figure 4.2d), which relied on random pyramid texturing of the front side to drastically reduce reflection losses [10].

As shown in Figure 4.1, the efficiencies of the COMSAT 'black' cell remained unchallenged until UNSW introduced the concept of the metal insulator NP junction (MINP) solar cell to cross the 18% efficiency threshold in 1983 [11]. The MINP solar cell relied on an ultrathin tunnel oxide layer, grown after patterning of the front oxide passivation (see Figure 4.2d), to reduce surface recombination under the front contacts. As front surface passivation was easier to achieve on polished surfaces, the MINP concept employed a double layer ARC consisting of zinc sulphide (ZnS) and magnesium difluoride (MgF_2) to minimize front reflection losses. At about the same time, contact passivation approaches based on polysilicon and semi-insulating polysilicon also showed promising results in terms of improving open circuit voltages [12, 13]. However, these approaches fell out of favour at most research groups due to the process complexity and the narrow process window. Instead, research groups like UNSW developed a simpler concept named passivated emitter solar cell (PESC), which relied on a single photolithography patterning step to open the front silicon oxide passivation and form very narrow Ti/Pd/Ag-evaporated front contacts. These contacts were subsequently thickened by plating of Ag to reduce resistive losses. The optimization of the different processes and the addition of microgrooves texturing at the front led to PESC efficiencies crossing the 20% efficiency threshold in 1985 using lab-size (4 inch) expensive boron-doped float-zone (FZ) Si wafers [14].

As microgroove laboratory PESC were limited by losses at the rear side, the attention turned to applying rear surface and contact passivation schemes. Major progress was made in the late 1980s at UNSW, who focused on standard two-side contacted p-type cells, and at Stanford University, who focused on n-type cells with an interdigitated back contact (IBC) design as shown in Figure 4.2g. Both research groups employed oxide passivation at the rear side (to reduce surface recombination losses and improve light trapping) and small area point contacts with heavy diffusions (to reduce contact recombination and resistive losses). Optimization of the different cell concepts led to the Stanford University IBC design crossing the 22% cell efficiency threshold in 1988 [15]. The UNSW research group then re-took the world-record cell efficiency crown in the following years. It did it first with its PERC design and then with its passivated emitter and rear locally diffused cell design, shown in Figure 4.2h, which crossed the 24% cell efficiency threshold in 1994 and then the 25% milestone in 1999 [3]. Interestingly, it took the industry over 20 years of continuous developments to approach 24% cell efficiencies using an industrial version of the PERC design on 6-inch Cz–Si substrates [16].

Approaches based on polysilicon-based passivated contacts regained broad interest in the 2010s when they were 're-discovered' by multiple research groups who made rapid progress in terms of cell efficiencies [17]. This culminated in Fraunhofer ISE reporting efficiencies up to 25.8% on n-type FZ substrates, and more recently up to 26.0% on p-type FZ substrates [18], using its tunnel oxide passivating contact (TOPCon) approach in two-side contacted cells as shown in Figure 4.2i. As for the current world record on p-type FZ substrates, it is held by ISFH using its polycrystalline on oxide IBC cell design [19] schematically shown in Figure 4.2j.

4.2 A short history of the main industrial c-Si cell concepts

Bringing cell concepts from expensive small-area (typically ~4 cm^2) laboratory cells to low-cost large-area cells and modules for terrestrial applications required the development of simple, robust and low-cost metallization techniques and approaches. The US Government was a significant driving force in this with its 'Block' purchases of cells and modules between 1975 and 1983 with increasingly stringent acceptance criteria, which led to the modern era of terrestrial module development [20]. The first 'Block I' purchase in 1975 saw one company supplying 7.5-cm-diameter cells essentially based on space technology (similar to COMSAT 'black' cell), a second company supplying cells with metal contacts formed by Ni plating followed by tin solder dipping and a third company using screen printing and high-temperature firing of metal pastes (Ag to contact the front n^+ and Al to form the rear p^+ by alloying with the Si substrate), which was a new approach for the metallization of solar cells at the time [21]. The approach based on screen-printed contacts quickly became the preferred metallization approach for the subsequent US Government Block purchases and for all industrial silicon solar cells in general. This is because screen-printing technology is relatively simple, requires minimum waste treatment, benefits from robust and low-cost equipment, and can be continuously improved with minimum capital investments as mentioned in Section 4.3 and further explained in Chapter 6. In the next paragraphs, we give a short history of the main industrial cell concepts (see schematics in Figure 4.3) that have been put into mass production between the 1970s and today. For more detailed explanations, we refer the readers to several excellent review papers that cover the different cell designs [22–26].

Screen-printed full-area 'Al-BSF' cells on p-type multicrystalline silicon (see Figure 4.3a) were the mainstream technology between the 1970s and the mid-2010s. It is interesting to point out that Al-BSF cells are essentially a simplified PESC with screen-printed contacts on both sides. While PESC lab-type cells crossed 20% efficiency in 1985, it took another 30 years of incremental improvements in materials and processes for the industry to make 20% efficient Al-BSF cells on industrial-size wafers [27]. The main developments for multi Al-BSF cells included: improving material quality, introducing isotropic texturing at the front using acidic solutions (so-called 'iso-texturing'), replacing the TiO_x front ARC by hydrogenated silicon nitride (SiN_x:H), developing front Ag metallization pastes that could fire through the front ARC to contact n^+ emitters with increasingly low doping (as Ag pastes improved) and improving the quality of rear p^+ formation. Altogether the continuous improvements in Al-BSF cell efficiencies contributed to commercial module efficiencies progressing from under 5% in the 1970s to over 18% in the mid-2010s as shown in Figure 4.1. As explained in Section 4.1, PESC or Al-BSF cells are limited by losses at the rear side and so today most manufacturers have transitioned to producing screen-printed PERC with mean cell efficiencies around 23%.

With the development and widespread adoption of screen-printed contacts in the 1970s–1980s, the use of plated contacts declined until BP Solar revived this

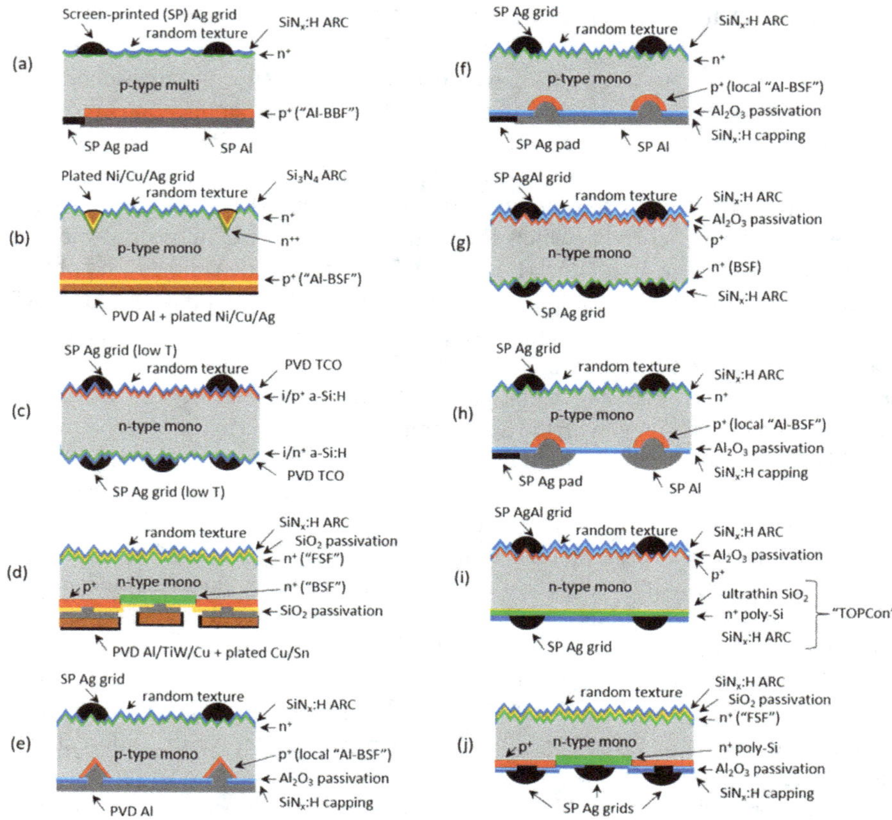

Figure 4.3 *Schematics showing the different designs of industrial c-Si solar cells: (a) screen-printed (SP) full area 'Al-BSF' cells on p-type multicrystalline silicon, which was the mainstream technology between the 1970s and the mid-2010s, (b) 'Saturn' cell produced at BP Solar between 1990 and 2008, (c) n-type heterojunction technology (HJT) pioneered by Sanyo in the 1990s, (d) Gen1 IBC cell pioneered by SunPower in the mid-2000s, (e) p-type PERC with PVD Al rear contact introduced in the mid-2000s, (f) p-type PERC with SP contacts on both sides mass produced since the early 2010s, (g) bifacial n-type PERT cell mass produced since the early 2010s, (h) bifacial p-type PERC mass produced since the mid-2010s, (i) bifacial n-type TOPCon cell mass produced since the late-2010s and (j) example of industrial IBC-TOPCon cell concept with rear n*$^+$ *poly BSF and SP Ag grids*

approach in the 1990s with their 'Saturn' cells (see schematic in Figure 4.3b) based on a laser-grooved buried contact (LGBC) design licensed from UNSW [28]. Compared to screen-printed Al-BSF solar cells produced at that time (with front Ag finger widths over 100 μm wide), the early 'Saturn' cells combined several benefits. First, the selective doping design, with a heavy n^{++} doping under the contacts and a shallow n^{+} doping in between, allowed to minimize resistive and recombination losses under the front contacts while achieving excellent blue response in between. Second, the narrow (20–30 μm wide) and buried (30–40 μm deep) Ni/Cu/Ag-plated front contact allowed the use of more fingers (to reduce resistive losses in the emitter and in the contacts) without drastically increasing shading losses. Third, the process sequence resulted in solderable Ni/Cu/Ag contacts on both sides contrary to Al-BSF cells, which require the use of additional SP Ag pads at the rear (since Al paste cannot be soldered using conventional approaches). However, the electroless plating sequence implemented by BP Solar was not without its problems (see details in [25]), which together with other considerations (incremental efficiency improvement of screen-printed cells, competition from Asia, etc.) led to BP Solar exiting the PV manufacturing industry altogether in 2008.

Silicon heterojunction technology (HJT) consists of thin hydrogenated and doped amorphous silicon layers (a-Si:H) that are deposited on both sides of c-Si wafers (mostly high-quality n-type monocrystalline). HJT was pioneered by Sanyo in the 1990s under the trademark HIT (for heterojunction with intrinsic thin-layer structure), which was later acquired by Panasonic who continues to produce HIT cells and modules to this day. The key feature of HJT is that the metal contacts, which are highly recombination-active in conventional cells, are electronically separated from the crystalline surface by evaporated transparent conductive oxides (TCOs) layers on top of a stack of intrinsic and doped a-Si:H layers (see schematic in Figure 4.3c). This leads to extremely high open-circuit voltages (V_{oc}) and fill factors (FF), while achievable short-circuit currents (I_{sc}) are somewhat more limited due to parasitic absorption losses in the a-Si:H and TCO layers. Major advantages of HJT include its short processing sequence, which can be highly automated, its compatibility with thin wafers (allowing further cost reductions) and its symmetric structure with random pyramid texture on both sides (allowing high bifaciality). Major challenges with HJT include process control (with surface preparation, wafer handling and deposition of a-Si:H and TCO layers being particularly sensitive steps), the availability of low-cost equipment, the cost of consumables (low-temperature Ag pastes, TCO materials) and temperature limitations (typically <250°C for cell and module processing). As a result, a lot of R&D efforts have been spent in the last three decades towards improving metallization approaches for HJT as further explained in Section 9.3. Today, a few companies are reporting mean cell efficiencies >24% in mass production making HJT one of the major candidates to replace PERC as the mainstream technology in the near future.

The IBC design developed at Stanford University in the late 1990s was further developed by SunPower who started pilot production of its Gen1 IBC cells in 2003 using 125-mm semi-square wafers [29]. Key features of the SunPower Gen1 IBC (see Figure 4.3d) included the absence of front metallization (as the back contact design implies), local point contact openings and a back side metallization based on physical vapour deposition (PVD) metal layers (see details in Section 4.3). This led to excellent aesthetics ('all-black' front side) and high performance with the champion cell at 21.5% efficiency and the best module achieving aperture area efficiency of 18.2%, which was outstanding in the mid-2000s. Major challenges with IBC cells include the lengthy and complicated process sequence and the need for custom-made equipment. Impressively, SunPower reached mean cell efficiencies of 22.4% in 2007 with its Gen2 IBC thanks to improved patterning [30] and a total area cell efficiency above 25% in 2016 for its Gen3 IBC featuring passivating contacts [31]. Today, SunPower is manufacturing IBC modules with total area efficiency up to 22.7% using plated contacts, while a few other companies (LG, SPIC, Jolywood, Sharp, etc.) are producing IBC modules with efficiencies in the range of 22% using screen-printed contacts. In general, IBC modules are mainly targeted at residential applications (which offer higher average selling prices). For IBC modules to gain significant market share in the future, further developments are needed to reduce equipment and processing costs, in particular the ones associated with IBC cell metallization and interconnection.

With industrial Al-BSF cells limited by losses at the rear side, several institutes and companies started to develop industrial PERC cells in the mid-2000s. The first approaches were based on using thin (typically ~2 μm) PVD Al as rear contact (see schematic in Figure 4.3e) with the local Al p^+ regions either formed using the laser-fired rear contact approach pioneered by Fraunhofer ISE [32] or using laser contact opening (LCO) prior to PVD Al as pioneered by imec [33]. However, it quickly became apparent that using PVD Al was a non-economic proposition compared to using screen-printing Al paste (as further explained in Section 4.3). As a result, R&D efforts were mostly focused towards improving PERC cells with screen-printed contacts on both sides (see schematic in Figure 4.3f). The company Schott Solar was the first one to report screen-printed industrial PERC cell efficiencies above 21% in 2014 on 6-inch p-type monocrystalline wafers [34]. Since then, further optimization of materials and processing have led to champion 6-inch PERC cells crossing the 24% efficiency threshold [35] and to several companies reporting mean cell efficiencies around 23% in mass production using industrial PERC technology. Today, industrial PERC is seen as the industry workhorse as it is the dominant technology in mass production [36].

Although the concept of bifacial cells and modules (which can capture light from both sides) is not known, several companies (Yingli, PVGS, LG and bSolar, among others) started to pick it up again when they put various industrial bifacial n-type cell concepts in mass production in the early 2010s [37]. Most of them opted for a front junction (FJ) n-type passivated emitter and rear totally diffused (PERT) cell design with narrow screen-printed Ag contacts on both sides (as shown schematically in Figure 4.3f) for several reasons. First, it allowed to avoid the light-induced degradation effect commonly present in p-type Al-BSF and PERC cells. Second, it helped achieving high bifaciality values (ratio between rear and front illuminated *I–V* parameters). Third, it only required the use of specific AgAl pastes (containing a few % of Al) to contact the p^+ region and a few additional tools compared to PERC. While several institutes and companies were able to quickly achieve bifacial FJ n-type PERT cell efficiencies beyond 21.5%, this cell design quickly became limited by the high recombination losses under the contacts and by its typically higher cost of ownership vs PERC. This became even more problematic in the mid-2010s when PERC manufacturers started to tap into the growing bifacial PV market simply by adapting a few processing steps to produce bifacial PERC (see schematic in Figure 4.3h) with cell efficiencies not far behind what can be achieved with monofacial PERC [38].

Different approaches have been developed over the last two decades to address the high recombination losses in n-type PERT cells and their high cost of ownership delta vs industrial p-type PERC cells. One promising approach is the back junction n-type PERT design because mean cell efficiencies above 23% have been demonstrated using a process sequence having a lot in common with the one used for bifacial p-type PERC cells [39]. Another more promising approach is to replace the rear n^+ BSF in FJ n-type PERT cells by a TOPCon design. The advantage of such bifacial n-type TOPCon cells (see schematic in Figure 4.3i) is that recombination losses at the rear side are practically eliminated allowing median cell efficiencies to approach 24% and champion cell efficiency to exceed 24.5% [40]. This has led bifacial n-type TOPCon cells to be the other major candidate (with HJT) to replace PERC as the mainstream technology in the near future. Compared to HJT, n-type TOPCon present the advantage that only a few additional processing tools are required to upgrade existing PERC lines. On the other hand, the total number of processing steps required is typically higher than with HJT and the bifaciality values are typically lower than with HJT.

Finally, the different cell designs can be combined into so-called industrial IBC-TOPCon (see one possible example in Figure 4.3j) or IBC-HJT cell concepts with SP Ag grids at the rear side. The idea is to try to combine the advantages of the IBC design (high aesthetics, high I_{sc} due to absence of front grid shading) with the ones of industrial TOPCon (high V_{oc} and FF, process sequence similar to PERC) or HJT (high V_{oc} and FF, few highly automated steps). However, doing so without drastically increasing process complexity and cost of ownership is extremely challenging. So far, the most promising results on large area n-type substrates include: (i) the 25% total area efficiency reported by Trina Solar using a IBC-TOPCon design [41], (ii) the 25% on large area (90.25 cm^2) reported by Meyer Burger using a 'tunnel' IBC-HJT design [42] and (iii) the current world-record efficiency of 26.7% reported by Kaneka using a IBC-HJT design [43]. In the next sections, we provide a brief introduction of the different approaches that have been developed for the metallization of industrial silicon solar cells.

4.3 Metallization approaches based on PVD contacts

PVD is routinely used in many industries (integrated circuits, displays, architectural glass, etc.) to deposit thin layers of metals, metal alloys, dielectrics, etc. PVD takes place in vacuum systems because the mean free path of particles, meaning the distance they can travel before collision with another particle, is inversely proportional to the system pressure. This allows faster deposition rates and a better control of the material properties (composition, etc.). PVD allows excellent thickness control from the nanometre (nm) scale (e.g. barrier in integrated circuits) up to hundreds of microns (thermal barrier coatings). For industrial silicon solar cells, PVD layer thicknesses are more in the range of 1–1 000 nm due to constrains in terms of throughput (limited by pumping/venting time, size of the PVD system, material deposition rate, etc.) and cost. Many different PVD techniques and approaches have been developed for the metallization of silicon solar cells as further discussed in Chapter 5. In the next paragraphs, we briefly introduce metallization approaches using either evaporation or direct current (DC) sputtering techniques.

Evaporation techniques rely on the fact that metals will evaporate in the same way as water when heated to sufficiently high temperatures in vacuum. The metal vapour will then condense on any nearby cold surface (substrate, vacuum chamber walls, etc.) to form a thin high purity metallic layer. Common methods of heating the material to be deposited include resistive heating (thermal evaporation), focusing an electron beam (e-beam evaporation) or focusing a laser beam (pulsed laser deposition). Evaporation was and continues to be routinely used in high-efficiency laboratory cells to form narrow contacts at the front (using expensive photolithography patterning) and to deposit the full area Al at the rear (or other metals depending on the chosen cell design). With large area industrial cells requiring thick metal layers and low processing costs, industrial evaporation equipment was mainly developed to deposit the full area Al contact at the rear. One example of such equipment is the ATON 500 pilot tool from Applied Materials, which made use of an Al wire continuously fed onto a boron nitride boat (to avoid reactions with molten Al) as shown in Figure 4.4a. This allowed to achieve throughputs in the range of 500 wafers per hour using a moving carrier tray containing nine 6-inch wafers [44]. Apart from the limited throughput, other limitations with this approach included significant wafer heating during processing and the need for alternative interconnection schemes since the Al surface readily oxidizes in air which prevents the use of standard soldering techniques. Altogether, this made the use of evaporated Al a non-economic proposition for industrial PERC particularly as the industry made significant progress with screen-printed PERC (see [45] for a detailed review and Chapter 9 for further insights).

Sputtering is a thin film PVD coating technique, which makes use of a plasma. Different variations of the sputtering process exist including DC sputtering, radio frequency (RF) sputtering, pulsed DC sputtering and DC magnetron sputtering (shown schematically in Figure 4.4b) among many others [46]. In the basic configuration, the material to be coated is a solid plate (called the target) that is placed in a vacuum chamber parallel to the substrate to be coated. The vacuum chamber is evacuated to low pressure and then backfilled with a high-purity inert gas, usually Argon (Ar). Typically, a large negative bias is applied to the target, which becomes the cathode of the electrical circuit, while a positive bias is applied to the substrate, which becomes the anode. This leads to a small amount of the Ar gas to be ionized with the Ar^+ ions accelerated towards the target material (due to the high potential

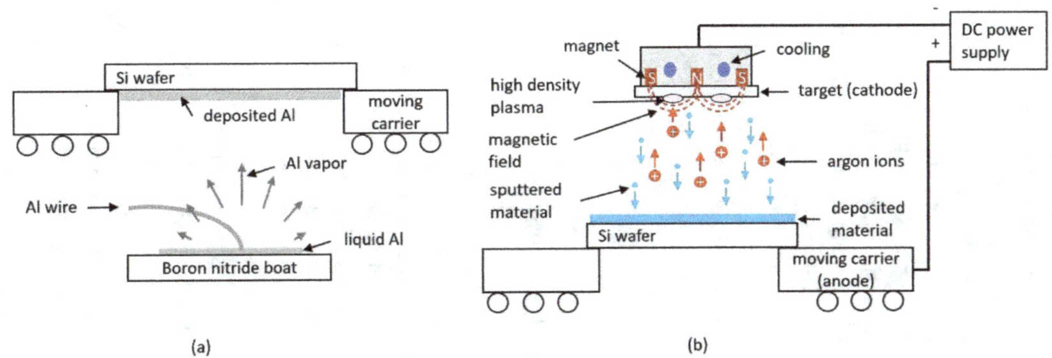

Figure 4.4 *(a) Schematic drawing for inline high-rate thermal evaporation of aluminium (Al). Al vapour is generated in vacuum (typical pressure <5.10^{-4} mbar) by feeding an Al wire onto the heated boron nitride boat. (b) Schematic drawing for inline direct current (DC) magnetron sputtering, which is routinely used to deposit ceramic materials like indium tin oxide (ITO) or metals like molybdenum (Mo). The chamber is evacuated to low pressure (typically <5.10^{-3} mbar) and backfilled with high purity argon (Ar) gas. DC magnetron sputtering uses magnets behind the negatively charged target material (cathode) to trap electrons so that they are not free to bombard the positively charged substrate to be coated (anode). This allows faster deposition rates and improved target utilization.*

difference) and new free electrons sustaining the ionization reaction. The impact of the high-energetic Ar^+ ions onto the target provokes cascade reactions with atoms or molecules being released from the target, which will condense on the substrate and its surrounding to form a thin layer, as well as radiation (giving the plasma its typical purple colour) and heat being generated. More complex films can be deposited using reactive sputtering, which consists in adding a small amount of additional gas (nitrogen, oxygen, hydrogen, etc.) during the sputtering process. The primary limitation of DC sputtering is that non-conducting dielectric materials will build up charge over time, which can lead to issues like arcing or even a complete stop of the deposition process. Such limitations can be overcome using more complex sputtering techniques such as pulsed DC sputtering or RF sputtering. As with evaporated layers, the use of sputtered layers in industrial silicon solar cells is limited to thin layers, typically well below 1 μm, due to throughput and cost constraints. A successful approach is the use of thin (typically ~100 nm thick) sputtered TCOs in silicon heterojunction cells as further explained in Chapter 4. Another successful approach is to sputter thin multi-layer metal stacks prior to contact thickening using plating. One such example is the SunPower Maxeon IBC cell, which reportedly uses a thin sputtered stack consisting of Al (for contacting and internal reflection properties), titanium tungsten (as barrier against Cu diffusion) and Cu (as seed layer for the subsequent Cu plating step) according to a patent filed in 2008 [47].

4.4 Metallization approaches based on screen-printed contacts

Approaches based on screen-printed contacts are the most routinely used in industrial silicon solar cells because screen-printing equipment is cheap, robust, easy to automate and because the industry as a whole has continued to make incremental improvements to the cost and performance of screen-printed cells since the 1970s. Today, most commercial screen printing lines have a net throughput over 4 500 wafers per hour using a dual lane configuration, a breakage rate below 0.2%, are compatible with a wide variety of solar cell sizes and are capable of printing narrow (30–40 μm wide) front Ag contacts with excellent aspect ratios (AR; ratio between finger height and finger width). Commercial screen printing lines typically consist of multiple screen printers and drying ovens in sequence to print the front and rear contacts. Depending on the cell design and the type of pastes used, the front and rear contacts are then sintered simultaneously in a curing oven (e.g. n-type silicon heterojunction cells) or fired simultaneously in a high-temperature firing furnace (e.g. p-type PERC requiring the front Ag paste to fire through the front passivation layers and the rear Al paste to form local p^+ regions by Al alloying). Finally, testing/sorting equipment is used to characterize the finished cells (I–V curve, electroluminescence to detect finger breaks, etc.) and to sort them in different bins.

The same type of industrial screen printer is typically used to print the front and rear contacts (see example in Figure 4.5a). The 'blue' wafers (due to ARC colour) are first moved onto a printing table using a conveyor belt or walking beam. A camera is used to recognize the wafer edges or a specific pattern inside the wafer (e.g.

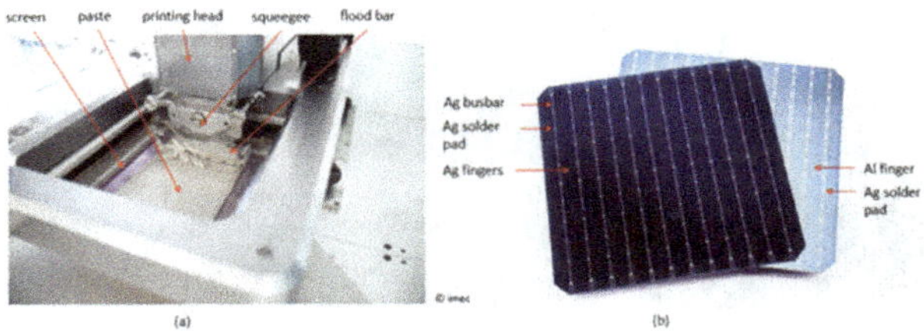

Figure 4.5 *(a) Picture of an industrial screen printer from Applied Materials. Source: [48]. (b) Picture of an industrial 6-inch bifacial n-type cell with 23.2% efficiency featuring a 12 busbar grid design on both sides produced by screen printing. Source: [49].*

alignment marks). Then the wafer is precisely positioned under the printing section and at a specific distance from the screen (snap-off distance). In the next step, a squeegee is moved with a defined pressure and speed (controlled by the printing head) locally pressing the screen against the wafer. The screen typically consists of a stainless steel mesh mounted onto an aluminium frame. It has areas blocked off by an emulsion and areas that are open for the paste to go through. Due to the high mesh tension, the screen snaps off in all areas where the squeegee is not pressing against the wafer, which improves the print definition and prevents the wafer from sticking to the screen. After printing, the squeegee is raised and moved to its initial position, while a floor bar fills the screen openings with paste in preparation for printing the next wafer. The printed wafer is transported to the unloading section (where a camera can be mounted to check print quality) and then into a drying furnace so that the printer wafer can go through the next sections. A discussion of the impact of the different printing parameters on the quality of the print is given by Holmes and Loabsy [50]. The current understanding of contact formation for different types of pastes is further discussed in Chapter 6. It also contains insights regarding the design of standard stainless steel mesh screens, screen printer design and stencil printing. Finally, Chapter 9 provides more information regarding the different screen-printing approaches that have been optimized for specific cell designs.

4.5 Metallization approaches based on alternative printing techniques

Due to the perceived limitations of early screen-printed contacts (typical Ag finger widths >150 μm in the 1980s, limited AR, limited throughput, etc.) many alternative printing approaches have been developed over the last four decades for the metallization of silicon solar cells. In many cases, those alternative printing approaches adapted know-how and equipment initially developed for other applications such as displays, flexible electronics, jewelleries, etc. One way to divide those different approaches is whether they require direct contact with the substrate (like in screen printing) or whether they can be contactless (also called direct-writing technology

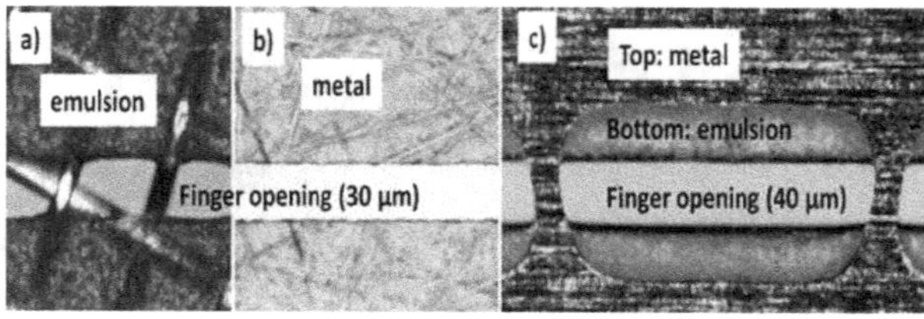

Figure 4.6 *(a) Emulsion screen with 30 µm opening and 14 µm stainless steel wires. (b) Single layer electroformed nickel stencil with 30 µm opening. (c) Hybrid stencil with 40 µm opening, the stainless steel at the top is chemically etched while the bottom emulsion layer (wafer side) is patterned by photolithography. Reprinted with permission from [55].*

[51]). Another way to divide is between analogue techniques (requiring a permanent printing master like the patterned emulsion screen in screen printing) and digital techniques (meaning that the pattern can be changed digitally via a software interface). An excellent introduction to over 20 different analogue and digital printing techniques for organic and printed electronics is given in [52]. In the next paragraphs, we focus on briefly introducing the main alternative printing approaches for industrial silicon solar cells. Chapter 7 provides further insights for some of those approaches.

Stencil printing was first envisaged for fine line printing in production in 1998 [53]. The printing image is typically formed by nickel electroforming, pulse laser processing or chemical etching of a metal foil. The advantage of stencil printing over screen printing lies in the improved paste release, non-wear character of the stencil and the ability to print fingers with high AR. Stencils can be categorized into single layer stencils or double layer (also called hybrid) stencils [54]. In both cases, the opened fraction is much higher than with conventional emulsion screens. In single layer stencils, as shown in Figure 4.6b, the open fraction is 100% and hence a second printing step is required to print the busbars (often referred to as dual printing). In a double-layer stencil, as shown in Figure 4.6c, the fingers and busbars are fully opened in one layer, while in the other connections are made between fingers and busbars to prevent the stencil from falling apart. Excellent results have been achieved with stencil printing [56, 57]. However, industrial adoption has been limited, mainly because of the recent progress made using emulsions screens [58] and also because the manufacturing of stencils remains expensive.

Ink-jet printing and aerosol jet printing are two contactless techniques allowing very high line resolution (line width <20 µm), which are well suited for the metallization of thin or uneven wafers. There are different ink-jet printing techniques but the most common one is drop-on-demand where a piezoelectric transducer generates on demand tiny droplets, which are ejected through a nozzle tip towards the substrate. Throughput and resolution can be improved by using multiple nozzle tips in parallel, which can all be addressed separately. The achievable printed line widths depend on many parameters like the droplet volume, the ink properties (viscosity, etc.), the printer resolution and the surface conditions. For printing the front Ag fingers, dedicated inks consisting of nano-sized Ag particles with a low metal content (and a high amount of solvents) are required since the ink needs to pass through the narrow nozzle tips. This leads to challenges both in terms of contact formation and line conductivity. As a result, ink-jet printing of Ag fingers has been mostly used in combination with a subsequent Ag-plating step to increase line conductivity [59] or with multi-wire interconnection schemes to relax line conductivity requirements [60]. Another approach is to use ink-jet printing of resist as a low-cost patterning method in combination with plated contacts [61]. Aerosol printing is different from ink-jet printing because the paste is atomized into an aerosol and focused through a nozzle using a sheath gas. Thanks to this, conventional Ag particles (~1–10 μm in size) can be passed through the nozzle without risks of clogging. Major developments in the formulation of inks for aerosol printing have been done to achieve low contact resistance on lowly doped emitters [62, 63]. However, despite their respective advantages, both aerosol and ink-jet printing have had difficulties to find their way to mass production due to issues related to contact formation, line conductivity and throughput but also because of progress made with metallization approaches based on screen-printed contacts.

In dispensing, the printing material (e.g. Ag or Al pastes) is extruded through a tip, which is moved across the substrate at a fixed height and speed. Dispensing is a contactless, high-throughput, single-step metallization technique, which allows to achieve very narrow line widths (<30 μm wide) with superior AR and line homogeneity compared to screen printing [64, 65]. As discussed in [66], parallel dispensing units were in use for many years at the former Schott Solar AG manufacturing facilities, while Solar World Innovations GmbH also developed a production tool capable of reaching a rated throughput of 2 700 wafers/h. More recently, excellent results have been reported for different cell designs using high-throughput parallel dispensing. This included the parallel dispensing of 150 Ag contacts with a finger width of only 17 μm for p-type PERC, the dispensing of AgAl and Al pastes for n-type PERT cells and the dispensing of Ag and AgCu pastes for n-type SHJ cells showing the high versatility of the approach [67]. Further insights on dispensing technology are given in Subsection 7.1.

Transfer pad printing is one of the favoured printing techniques for 3D objects, and it has been extensively investigated for the metallization of silicon solar cells in the last two decades [68]. The printing process can be summarized as follows. First, a photopolymer gravure plate (called the cliché) with the print image etched into the surface is filled with ink by means of a doctor blade. Due to the formulation of the ink, its top layer becomes tacky when exposed to air. In the next step, a silicone pad is moved downward and pressed onto the cliché. When the pad moves upward some of the ink is transferred to the pad and the ink layer exposed to the air continues to dry (making it even tackier). At this point, the pad is moved downwards onto the substrate where the ink adheres more strongly than onto the pad. Finally, the pad is lifted off the substrate for the next printing cycle. For mass production of silicon solar cells, rotary pad printing (also called rotary printing) has advantages over pad printing in terms of applying more even pressure onto the wafer surface and in terms of achievable throughputs. Rotary printing can be applied to metallize the full rear side (e.g. Al paste for PERC) or a specially structured front as further described in detail in Subsection 7.2.

4.6 Metallization approaches based on plated contacts

Metallization approaches based on plated contacts regained interest in the PV manufacturing industry at regular intervals as companies sought ways to reduce metallization costs associated with screen-printing expensive Ag paste. Historically, most institutes/companies focused on direct plating of Ni to Si (to achieve low contact resistances) followed by plating of copper (Cu) as it is ~100× cheaper than Ag while being only slightly more resistive (1.68 vs 1.59 μΩ cm) [69]. As further explained in detail in Chapter 8, the main plating techniques include electroless plating, immersion plating and electrolytic plating. In electroless plating, deposition is achieved using a reducing agent present in the electrolyte, which provides the electrons to reduce metal ions at the silicon surface. In immersion plating, no reducing agent is present, the substrate is the source of electrons for the deposition and the process is self-terminating (only thin layers can be formed). In electrolytic plating, the source of electrons can be provided by an external power supply (e.g. electroplating and forward-bias induced plating) or by the light-induced current of an illuminated solar cell immersed in the electrolyte (light-induced plating).

The first laboratory silicon solar cells made at Bell Laboratories in 1950s and at Hoffman Laboratories in the 1960s relied on plated contacts for both polarities. Plated contacts were also used by one company for the first 'Block I' US Government purchase in 1975. However, with the development and widespread adoption of screen-printed contacts in the 1970s–1980s, the use of plated contacts declined until BP Solar revived this approach in their 'Saturn' LGBC cells (see Section 4.2). These cells were manufactured between the 1990s and 2008 when BP Solar exited the solar business altogether. In 2009, Suntech introduced the 'Pluto' cell, which featured simpler processes to form the front Ni/Cu/Ag-plated contacts using a combination of higher throughput laser doping and electrolytic plating equipment [70]. However, this production was also discontinued in 2014 [71].

Even though BP Solar and Suntech both discontinued the production of Cu-plated cells, significant progress has been made in in the last two decades to address challenges linked to Cu-plated metallization as reviewed by Alison *et al.* [72]. It is worth mentioning that SunPower has been commercializing high-efficiency Cu-plated IBC cells and modules since 2008 with excellent performance and reliability. For this SunPower is reportedly using a seed-and-plate approach based on a thin sputtered stack that is patterned (using low-cost techniques) and subsequently plated with a very thick layer of Cu that is capped by a thin layer of tin (Sn) to avoid surface oxidation. Similarly, many research groups and companies have recently been showing excellent efficiency and reliability results together with reduced cost-of-ownership using Cu-plated contacts in various types of high-efficiency

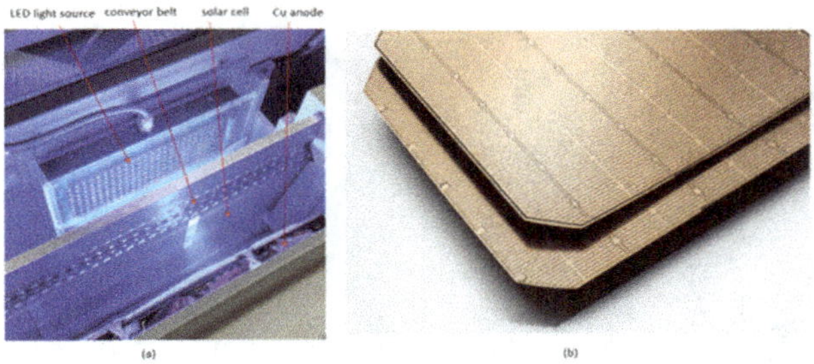

Figure 4.7 *(a) Light-induced plating of Cu on two-side contacted cell in Meco DPL tool. Source: [75]. (b) Large area IBC cells obtained using a thin sputtered stack + electrolytic plating approach. Source: [76].*

n-type cell concepts (IBC, SHJ, Tunnel Oxide Passivated Contact) [73, 61, 74]. Therefore, approaches based on plated contacts are still expected to play an important role in the future as further discussed in Section 4.7 and in Chapter 8, which also covers the different industrial plating techniques such as light-induced plating or electrolytic plating (see Figure 4.7).

4.7 Comparison of the different metallization approaches

First, it is important to mention that the desired metallization properties (finger dimensions, specific contact resistance, line resistance, etc.) are strongly dependent on: (i) the chosen cell design (Al-BSF, PERC, TOPCon, HJT, IBC, etc.), (ii) specific features in the cell design (surface morphology, doping level under the contact regions, etc.) and (iii) the properties of the pastes (composition, rheology, etc.) or other materials (TCO and metals) used for the metallization process. As a result, it is extremely difficult to provide a detailed comparison of the different metallization approaches for today's industrial c-Si cells. Therefore, in the following paragraphs we focus on briefly comparing the different metallization approaches based on more general techno-economic aspects (typical finger widths, throughputs, costs, etc.) that have been compiled in Table 4.1.

The first category is PVD, which is mainly employed today for blanket depositions of metals or TCO in high-efficiency cell designs (mainly HJT and IBC) which have an estimated market share below 3% according to the latest edition of International Technology Roadmap for Photovoltaics [77]. For PVD of metals, the typical specific contact resistivity values strongly depend on the chosen materials (with Al allowing values down to 0.01 mΩ cm^2) and the underlying doping level in Si [78]. As for TCO, the typical specific contact resistivity values strongly depend on the activation energy in the underlying doped layers and on the doping concentration in the TCO [79]. In both cases, the latest generations of PVD tools are capable of throughputs up to 10 000 wph, which helps reducing the cost per piece [80]. Still, PVD tools remain more expensive compared to more basic

Table 4.1 Comparison of the different metallization approaches in terms of: typical finger dimensions (the range includes front and rear metallization), typical specific contact resistivity (the range includes results on homogenously or selectively doped regions), availability of mass production tools ('+' for <5 tool vendors, '+++' for >10 tool vendors), typical throughput, typical tool and consumables costs ('+' for high cost, '+++' for low cost) and estimated market share

Category	Sub-category	Typical finger width [μm]	Typical finger height [μm]	Typical aspect ratio (AR)	Typical specific contact resistivity [mΩ cm²]	Availability of mass production tools	Typical throughput [wph]	Typical tool costs	Typical consumables costs	Estimated market share [%]
PVD	Metals	–	–	–	0.01–2	+++	3 000–10 000	+	++	<3
	TCO	–	–	–	15–200	+++	3 000–10 000	+	++	<3
Screen printing of metal pastes	Mesh screens	30–80	20–40	0.25–0.67	0.1–10	+++	3 000–4 500	+++	++	>90
	Stencil	30–80	20–40	0.25–0.67	0.1–10	+++	3 000–4 500	+++	+	<2
Alternative printing of metals pastes or inks	Ink-jet of metals	30–80	1–5	0.03–0.1	0.1–10	+	3 000–4 500	++	+	<1
	Aerosol jet	30–80	5–10	0.1–0.2	0.1–10	+	3 000–4 500	++	+	<1
	Laser transfer	20–80	20–40	0.25–1.0	0.1–10	+	3 000–4 500	++	+	<1
	Dispensing	20–80	20–40	0.25–1.0	0.1–10	+	3 000–6 000	++	++	<1
	Rotary printing	30–80	10–20	0.13–0.25	0.1–10	+	5 000–10 000	++	++	<1
Plating	Directly on Si	15–40	5–20	0.13–0.50	0.1–2	++	3 000–4 500	+	+++	<1
	Seed-and-plate	15–40	15–40	0.25–1.0	0.1–2	++	3 000–4 500	+	++	<2

screen-printing equipment and the consumable costs (targets, spare parts, etc.) can also be significant.

The second category is screen printing of metal pastes, which can be performed with either standard mesh screens or stencils. While stencils initially had an edge in achievable finger widths and AR, drastic progress has been made in the last decades using cheaper mesh screens [81], which explains why they are largely preferred over stencils. Today, the typical finger widths in mass production are in the range of 30–80 μm, mainly depending if it is for the front or rear metallization and also on the type of pastes used (Ag, AgAl, Al or low-T Ag). As for typical AR, it again depends on the type of pastes used as well as the chosen screen and printing parameters (with knotless screens and print-on-print approaches allowing higher AR values). As with PVD, typical specific contact resistivity values strongly depend on the materials used (in this case the type of pastes) and the doping level in the underlying material (Si or TCO material). Importantly, significant progress has been made with firing-through Ag pastes, and so it is now feasible to contact lowly doped phosphorous emitters with surface concentration below 5E19 cm^{-3}, which was seen a blocking point in the past [82]. A very wide range of low-cost screen-printing tools is available on the market today with typical throughputs in the range of 3 000–4 500 wph. Furthermore, there is a potential to increase throughputs to 7 500 wph by reducing cycle time [83]. Overall, the main limitation of screen-printing technology is the amount of paste (and associated consumable costs) needed per cell.

The third category is alternative printing of metal pastes or inks, which includes different sub-categories like ink-jet printing, aerosol jet printing, laser transfer printing, dispensing and rotary printing, among others (see Chapter 7 for further insights). Ink-jet and aerosol jet typically suffer from low AR due to the low viscosity of the metallic inks (which are also typically more expensive than conventional pastes). Laser transfer and dispensing are promising techniques to achieve narrow fingers with high AR but they require the use of custom pastes. Laser transfer printing also suffers from the additional consumable costs linked to the dedicated carrier foil that is needed. Rotary printing shows promises for reaching very high throughput. However, like ink-jet and aerosol printing, it suffers from lower AR due to the low viscosity pastes needed. Overall, alternative printing techniques currently suffer from the low availability of mass production tools which are also more expensive than the ones used for screen printing (due to the lack of scale and competition pushing prices down). Nevertheless, some of these techniques might gain market share in the coming years with further developments.

The last category is plating, which includes direct plating on Si (typically used in PERC, PERT or TOPCon cells) and seed-and-plate (mostly used in IBC and HJT cells). Direct plating on Si presents the advantage that extremely narrow contacts can be achieved (typically determined by the LCO width) and that no additional masking consumables are required (since the plating is self-aligned using the existing ARC as masking layers). Despite the relatively low consumable costs, limitations for direct plating on Si include the relatively poor AR (as layers plate in both vertical and horizontal directions) and the relatively high cost of the equipment (since at least two tools are needed for laser patterning and plating). As for the seed-and-plate approach, it typically consists of depositing a thin PVD seed, patterning it (typically using ink-jet or screen printing of a masking layer) and plating it (typically with Cu followed by a thin Sn or Ag capping layer). Major advantages include the excellent AR (as plating can be contained to the vertical direction only) and the possibility to achieve excellent long-term reliability (as demonstrated by SunPower). However, this approach also includes major challenges mainly due to the number of specialty equipment needed (PVD, patterning and plating) and due to the higher consumable costs (sputtering targets, masking layer, etc.). Nevertheless, both plating approaches have the potential to gain market share in the coming years as companies seek ways to reduce metallization costs.

4.8 Outlook on metallization trends

As explained in detail in [2], the c-Si PV industry has been able to sustain a high learning rate (LR) of 24% (which represents the percentage change of manufacturing costs in \$/Wp vs cumulative production) for many reasons. Among others, these include: (i) a very large number of companies competing to increase market share, (ii) standardized cell and module designs with efficiencies improving at a rate of ~0.5%/year and (iii) continuous reductions in the capital required to build new cell lines (mainly due to the low price of tools made in China and because of the fast development of big tools now reaching 10 000 wph for some steps). One consequence of this is that the PV industry is now in a better financial situation than ever to grow and build new production lines with the cash it generates from the sales of products. Recent publications indicate that the industry could grow to annual production rates around 1 000 GW/year by the end of the decade [84, 85]. Another consequence of the high LR is that the barrier for entry for new technologies (including metallization) is increasing because they need to: (i) fit within relatively

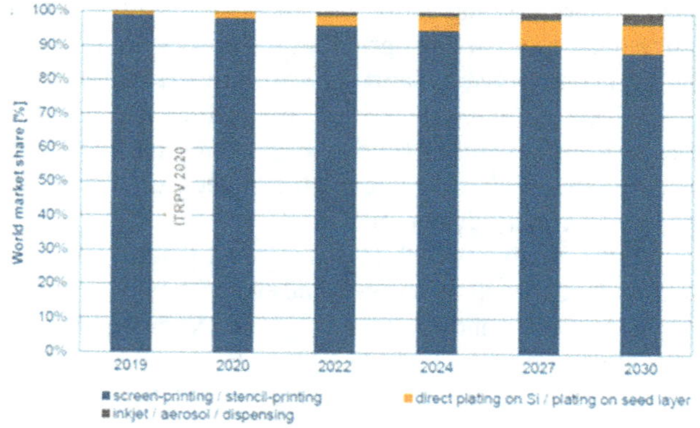

Figure 4.8 Project world market share for different front side metallization technologies according to the latest edition of the ITRPV. Reprinted with permission from [77].

standardized cell and module designs (e.g. the metallization needs to be compatible with the chosen cell format and interconnection technology) and (ii) quickly show potential for high cell efficiency (preferably >24%), high-throughput (preferably >4 500 wph) and low processing costs (which includes capital and consumable costs). Altogether, this explains why so many companies are focusing on making incremental improvements to existing technology and why people are becoming increasingly conservative when asked to predict the future market of different metallization techniques [86]. Similar observations are also made in the latest edition of the ITRPV [77], which projects that screen printing (of metal pastes) will remain by far the dominant technology until at least 2030 for both the front side metallization (see Figure 4.8) as well as for the rear side metallization. Although some publications mentioned that demand from PV industry could represent 100% of the global Ag production by the end of the decade if nothing changes [84], it is important to mention that technological solutions exist to reduce demand from the current level of about 70–90 mg/cell to below 50 mg/cell by 2030 [77, 85].

References

[1] Green M.A., Dunlop E.D., Hohl-Ebinger J., Yoshita M., Kopidakis N., Hao X. 'Solar cell efficiency tables (version 56)'. *Progress in Photovoltaics: Research and Applications*. 2020;**28**(7):629–38.

[2] Chen Y., Altermatt P.P., Chen D., *et al.* 'From laboratory to production: learning models of efficiency and manufacturing cost of industrial crystalline silicon and thin-film photovoltaic technologies'. *IEEE Journal of Photovoltaics*. 2018;**8**(6):1531–8.

[3] Green M.A. 'The path to 25% silicon solar cell efficiency: history of silicon cell evolution'. *Progress in Photovoltaics: Research and Applications*. 2009;**17**(3):183–9.

[4] Ohl R.S. 'Light-sensitive electric device. US Patent No 2,402,662'. 25 June 1946.

[5] Chapin D.M., Fuller C.S., Pearson G.L. 'A new silicon *p-n* junction photocell for converting solar radiation into electrical power'. *Journal of Applied Physics*. 1954;**25**(5):676–7.

[6] Prince M.B. 'Silicon solar energy converters'. *Journal of Applied Physics*. 1955;**26**(5):534–40.

[7] 'Bell laboratories, record'. May 1955:166.

[8] Green M.A. 'High-efficiency silicon solar cell concepts'. *McEvoy's Handbook of Photovoltaics*. Academic Press; 2018. pp. 95–128.

[9] Lindmayer J., Allison J. 'The violet cell: an improved sílicon solar cell'. *COMSAT Technical Review*. 1973;**3**:1–22.

[10] Haynos J., Allison J., Arndt R., Meulenberg A. 'The COMSAT nonreflective silicon solar cell: a second-generation improved cell'. *International Conference on Photovoltaic Power Generation*; 1974.

[11] Green M.A., Blakers A.W. 'Advantages of metal-insulator-semiconductor structures for silicon solar cells'. *Solar Cells*. 1983;**8**(1):3–16.
[12] Lindholm F.A., Neugroschel A., Arienzo M., Iles P.A. 'Heavily doped polysilicon-contact solar cells'. *IEEE Electron Device Letters*. 1985;**6**(7):363–5.
[13] Yablonovitch E., Gmitter T., Swanson R.M., Kwark Y.H. 'A 720 mV open circuit voltage SiOx: c-Si: SiOx double heterostructure solar cell'. *Applied Physics Letters*. 1985;**47**(11):1211–13.
[14] Blakers A.W., Green M.A. '20% efficiency silicon solar cells'. *Applied Physics Letters*. 1986;**48**(3):215–17.
[15] King R.R., Sinton R.A., Swanson R.M. 'Front and back surface fields for point-contact solar cells'. *Conference Record of the 20th IEEE Photovoltaic Specialists Conference*; 1988. pp. 538–44.
[16] Chen R., Tong H., Zhu H., *et al*. '23.83% efficient mono-PERC incorporating advanced hydrogenation'. *Progress in Photovoltaics: Research and Applications*. 2020;**28**(12):1239–47.
[17] Feldmann F., Bivour M., Reichel C., Steinkemper H., Hermle M., Glunz S.W. 'Tunnel oxide passivated contacts as an alternative to partial rear contacts'. *Solar Energy Materials and Solar Cells*. 2014;**131**:46–50.
[18] Richter A., Benick J., Feldmann F., *et al*. 'Both sides contacted silicon solar cells: options for approaching 26% efficiency'. *In 36th European Photovoltaic Solar Energy Conference and Exhibition*; 2019. pp. 9–13.
[19] Hollemann C., Haase F., Schäfer S., Krügener J., Brendel R., Peibst R. '26.1%-efficient POLO-IBC cells: quantification of electrical and optical loss mechanisms'. *Progress in Photovoltaics: Research and Applications*. 2019;**27**(11):950–8.
[20] Green M.A. 'Silicon photovoltaic modules: a brief history of the first 50 years'. *Progress in Photovoltaics: Research and Applications*. 2005;**13**(5):447–55.
[21] Ralph E.L. 'Recent advancements in low cost solar cell processing'. *Conference Proceedings of the 11th IEEE Photovoltaic Specialists Conference*; Scottsdale; 1975. pp. 315–16.
[22] Hermle M., Feldmann F., Bivour M., Goldschmidt J.C., Glunz S.W. 'Passivating contacts and tandem concepts: approaches for the highest silicon-based solar cell efficiencies'. *Applied Physics Reviews*. 2020;**7**(2):021305.
[23] Kerschaver E.V., Beaucarne G. 'Back-contact solar cells: a review'. *Progress in Photovoltaics: Research and Applications*. 2006;**14**(2):107–23.
[24] De Wolf S., Descoeudres A., Holman Z.C., Ballif C. 'High-efficiency silicon heterojunction solar cells: a review'. *Green*. 2012;**2**(1):7–24.
[25] Lennon A., Yao Y., Wenham S. 'Evolution of metal plating for silicon solar cell metallisation'. *Progress in Photovoltaics: Research and Applications*. 2013;**21**(7):1454–68.
[26] Neuhaus D.-H., Münzer A. 'Industrial silicon Wafer solar cells'. *Advances in OptoElectronics*. 2007;**2007**:1–15.

[27] Kim K.H., Park C.S., Lee J.D., *et al.* 'Record high efficiency of screen-printed silicon aluminum back surface field solar cell: 20.29%'. *Japanese Journal of Applied Physics*. 2017;**56**(8S2):08MB25.

[28] Bruton T., Mason N., Roberts S., *et al.* 'Towards 20% efficient silicon solar cells manufactured at 60 MWp per annum'. *Proceedings of the 3rd World Conference on Photovoltaic Energy Conversion*; 2003. pp. 899–902.

[29] Mulligan W.P., Rose D.H., Cudzinovic M.J., *et al.* 'Manufacture of solar cells with 21% efficiency'. *Proceedings of the 19th European Photovoltaic Specialists Conference (EUPVSEC)*; 2004. p. 387.

[30] De Ceuster D., Cousins P., Rose D., Vicente D., Tipones P., Mulligan W. 'Low cost, high volume production of >22% efficiency silicon solar cells'. *Proceedings of the 22nd European Photovoltaic Solar Energy Conference (EUPVSEC)*; 2007. pp. 816–19.

[31] Smith D.D., Reich G., Baldrias M., Reich M., Boitnott N., Bunea G. 'Silicon solar cells with total area efficiency above 25%'. *2016 IEEE 43rd Photovoltaic Specialists Conference (PVSC)*; 2016. pp. 3351–5.

[32] Schneiderlöchner E., Preu R., Lüdemann R., Glunz S.W. 'Laser-fired rear contacts for crystalline silicon solar cells'. *Progress in Photovoltaics: Research and Applications*. 2002;**10**(1):29–34.

[33] Agostinelli G., Szlufcick J., Choulat P., Beaucarne G. 'Local contact structures for industrial PERC-type solar cells'. *Proceedings of the 20th European Photovoltaic Solar Energy Conference*; 2005. pp. 942–5.

[34] Metz A., Adler D., Bagus S., *et al.* 'Industrial high performance crystalline silicon solar cells and modules based on rear surface passivation technology'. *Solar Energy Materials and Solar Cells*. 2014;**120**(13):417–25.

[35] Bhambhani A. Taiyangnews. 2019. Available from http://taiyangnews.info/technology/longi-24-06-efficiency-perc-cell-world-record/ [Accessed 1 Nov 2020].

[36] Altermatt P.P., Chen Y., Feng Z. 'Riding the workhorse of the industry: PERC'. *Photovoltaics International*. 2018;**41**:46–54.

[37] Kopecek R., Veschetti Y., Gerritsen E., *et al.* 'Bifaciality: one small step for technology, one giant leap for kWh cost reduction'. *Photovoltaics International*. 2014;**26**:32–45.

[38] Dullweber T., Schulte-Huxel H., Blankemeyer S., *et al.* 'Present status and future perspectives of bifacial PERC+ solar cells and modules'. *Japanese Journal of Applied Physics*. 2018;**57**(8S3):08RA01.

[39] Choulat P., Singh S., Tous L., *et al.* 'Exceeding 23% screen-printed rear-emitter bifacial n-PERT cells'. *Proceedings of the 36th European Photovoltaic Solar Energy Conference*; 2019. pp. 157–60.

[40] Chen D., Chen Y., Wang Z., *et al.* '24.58% total area efficiency of screen-printed, large area industrial silicon solar cells with the tunnel oxide passivated contacts (i-TOPCon) design'. *Solar Energy Materials and Solar Cells*. 2020;**206**(1):110258.

[41] Xu G., Deng M., Chen S., *et al.* '25% cell efficiency with integration of passivating contact technology and interdigitated back contact structure on 6″

wafers'. *2019 IEEE 46th Photovoltaic Specialists Conference (PVSC)*; 2019. pp. 1452–5.

[42] Lachenal D., Papet P., Legradic B., *et al.* 'Optimization of tunnel-junction IBC solar cells based on a series resistance model'. *Solar Energy Materials and Solar Cells*. 2019;**200**(4):110036.

[43] Yamamoto K., Yoshikawa K., Uzu H., Adachi D. 'High-efficiency heterojunction crystalline Si solar cells'. *Japanese Journal of Applied Physics*. 2018;**57**(8S3):08RB20.

[44] Nekarda J., Reinwand D., Grohe A., *et al.* 'Industrial PVD metallization for high efficiency crystalline silicon solar cells'. *In 2009 34th IEEE Photovoltaic Specialists Conference (PVSC)*; 2009. pp. 892–6.

[45] Dullweber T., Schmidt J. 'Industrial silicon solar cells applying the passivated emitter and rear cell (PERC) concept—a review'. *IEEE Journal of Photovoltaics*. 2016;**6**(5):1366–81.

[46] Baptista A., Silva F., Porteiro J.,*et al.* 'Sputtering physical vapour deposition (PVD) coatings: a critical review on process improvement and market trend demands'. *Coatings*. 2018;**8**(11):402.

[47] Mulligan W.P., Cudzinovic M.J., Pass T., Smith D. 'Solar cell and method of manufacture". U.S. patent no 7,339,110, 4 Mar 2008'.

[48] Applied Materials. Available from https://www.appliedmaterials.com/solar/products/screen-printing [Accessed 20 Jun 2021].

[49] Degans H. imec. 2019. Available from https://www.imec-int.com/en/articles/imec-and-jolywood-achieve-a-record-of-23-2-percent-with-bifacial-n-pert-solar-cells [Accessed 20 Jun 2021].

[50] Holmes P.J., Loasby R.G. *Handbook of Thick Film Technology*. Glasgow, Scotland, UK: Electrochemical Publications; 1976.

[51] Hon K.K.B., Li L., Hutchings I.M. 'Direct writing technology—advances and developments'. *CIRP Annals*. 2008;**57**(2):601–20.

[52] Organic Electronics Association. *OE-A Roadmap for Organic and Printed Electronics, White Paper*. 8th edn; 2020.

[53] Hoornstra J., Roberts S., de Moor H.H.C., *et al.* 'First experiences with double layer stencil printing for low cost production solar cells'. *Proceedings of the 2nd World PVSEC*; 1998. pp. 1527–30.

[54] Hoornstra J., Heurtault B. 'Stentil print applications and progress for crystalline silicon solar cells'. *Proceedings of the 24th European Photovoltaic Solar Energy Conference and Exhibition*; 2009. pp. 989–92.

[55] Tous L. *Nickel/Copper Plated Contacts as an Alternative to Silver Screen Printing for the Front Side Metallization of Industrial High Efficiency Silicon Solar Cells*. PhD thesis; Leuven; 2014.

[56] Falcon T. 'Ultra-fine line printing for silicon solar cells: mesh screens or metal stencils?'. *Proceedings of the 3rd Metallization Workshop*; Konstanz, Germany; 2011.

[57] Hannebauer H., Falcon T., Cunnusamy J., Dullweber T. 'Single print metal stencils for high-efficiency PERC solar cells'. *Energy Procedia*. 2016;**98**(98):40–5.

[58] Zhang W. 'Screen printing Ag metallization technology – the key driver and enabler for high efficiency industrial solar cells'. *Presentation at the Shanghai International Photovoltaic Power Generation and Smart Energy Exhibition & Conference (SNEC)*; 2018.

[59] Ebong A., Rounsaville B., Cooper I.B., *et al.* 'High efficiency silicon solar cells with ink jetted seed and plated grid on high sheet resistance emitter'. *Proceedings of the 35th IEEE Photovoltaic Specialists Conference*; 2010.

[60] Schube J., Fellmeth T., Jahn M., *et al.* 'Advanced metallization with low silver consumption for silicon heterojunction solar cells'. *AIP Conference Proceedings*. AIP Publishing LLC; 2019. p. 020007.

[61] Faes A., Lachowicz A., Bettinelli A., *et al.* 'Metallization and interconnection for high-efficiency bifacial silicon heterojunction solar cells and modules'. *Photovoltaics Bulletin*. 2018;**41**.

[62] Horteis M. *Fine-Line Printed Contacts on Crystalline Silicon Solar Cells*, PhD thesis; Freiburg; 2008.

[63] Kalio A., Richter A., Glatthaar M., Wilde J. 'Study of contact formation of printed contacts through different passivation layers on lowly doped emitters'. *Energy Procedia*. 2013;**38**:745–52.

[64] Specht J., Zengerle K., Pospischil M., *et al.* 'High aspect ratio front contacts by single step dispensing of metal pastes'. *In Proceedings of the 25th European Photovoltaic Solar Energy Conference and Exhibition*; 2010. pp. 1867–70.

[65] Pospischil M., Fellmeth T., Brand A., *et al.* 'Optimizing fine line dispensed contact grids'. *Energy Procedia*. 2014;**55**:693–701.

[66] Pospischil M., Klawitter M., Kuchler M., *et al.* 'High speed dispensing with novel 6″ print head'. *Energy Procedia*. 2016;**98**:61–5.

[67] Pospischil M., Riebe T., Jimenez A., *et al.* 'Applications of parallel dispensing in pv metallization'. *AIP Conference Proceedings*. 2019;**2156**(1).

[68] Huljic D.M., Thormann S., Preu R., Lüdemann R., Willeke G. 'Pad printed front contacts for c-Si solar cells - a technological and economical evaluation'. *In Proceedings of the 29th IEEE Photovoltaic Specialists Conference (PVSC)*; 2002. pp. 126–9.

[69] Letize A., Lee B., Cullen D. 'Wet chemical metallization of silicon solar cells: status and perspective of industrial application'. *In Proceedings of the 32nd European Photovoltaic Solar Energy Conference and Exhibition*; 2016. pp. 683–9.

[70] Shi Z., Wenham S., Ji J. 'Mass production of the innovative PLUTO solar cell technology'. *In Proceedings of the 34th IEEE Photovoltaic Specialists Conference (PVSC)*. IEEE; 2009.

[71] Cielsa A., Chen R., Ran, Wang S., *et al.* 'High-voltage p-type PERC solar cells with anchored plating and hydrogenation'. *Progress in Photovoltaics: Research and Applications*. 2018;**26**(6):397–401.

[72] Lennon A., Colwell J., Rodbell K.P. 'Challenges facing copper-plated metallisation for silicon photovoltaics: insights from integrated circuit technology development'. *Progress in Photovoltaics: Research and Applications*. 2019;**27**(1):67–97.

[73] O'Sullivan B.J., Jambaldinni S., Singh S., Kyuzo M., Debucquoy M., Szlufcik J. 'Loss analysis and design optimization of large-area high-efficiency back-contacted silicon solar cells'. *IEEE journal of photovoltaics*. 2016;**6**(4):810–16.

[74] Steinhauser B., Grübel B., Nold S., Arya V. 'Plating on TOPCON as a way to reduce the costs of i-TOPCON solar cells'. *Presented at the 37th European PV Solar Energy Conference and Exhibition*; 2020.

[75] Meco DPLBESI. Available from https://www.besi.com/products-technology/product-details/product/meco-dpl/ [Accessed 20 Jun 2021].

[76] Imec press release. Imec and besi demonstrate long-term reliability of Ni-Cu-Ag plated solar modules. Available from https://www.imec-int.com/en/articles/imec-and-besi-demonstrate-long-term-reliability-of-ni-cu-ag-plated-solar-modules [Accessed 3 Jan 2017].

[77] *International Technology Roadmap for Photovoltaic (ITRPV) 2019 results, 11th edition*; April 2020.

[78] Franklin E., Fong K., McIntosh K., *et al*. 'Design, fabrication and characterisation of a 24.4% efficient interdigitated back contact solar cell'. *Progress in Photovoltaics: Research and Applications*. 2016;**24**(4):411–27.

[79] Procel P., Xu H., Saez A., *et al*. 'The role of heterointerfaces and subgap energy states on transport mechanisms in silicon heterojunction solar cells'. *Progress in Photovoltaics: Research and Applications*. 2020;**28**(9):935–45.

[80] Schneiderlöchner E., Brükner J., Biedermann R., *et al*. 'High throughput magnetron sputtering technology at 10,000 wafers per hour for cost competitive manufacturing of transparent conductive oxides (TCO)'. *Presentation at the 3rd International Workshop on SHJ Solar Cells*, 19 October; 2020.

[81] Lorenz A., Linse M., Frintrup H., *et al*. 'Screen printed thick film metallization of silicon solar cells–recent developments and future perspectives'. *Proceedings of the 35th European Photovoltaic Solar Energy Conference and Exhibition*; 2018. pp. 819–24.

[82] Werner S., Lohmüller E., Maier S., Mourad S., Wolf A. 'Challenges for lowly-doped phosphorus emitters in silicon solar cells with screen-printed silver contacts'. *Energy Procedia*. 2017;**124**:936–46.

[83] Galiazzo M. 'Enabling 7,500 wph productivity by a novel screen printing process'. *Presentation at the 8th edition of the Metallization & Interconnection Workshop*; Konstanz, May 13; 2019.

[84] Verlinden P.J. 'Future challenges for photovoltaic manufacturing at the terawatt level'. *Journal of Renewable and Sustainable Energy*. 2020;**12**(5):053505.

[85] Altermatt P.P., Yang Y., Chen Y., *et al*. 'Requirements of the Paris climate agreement for the coming 10 years on investments, technical roadmap, and expansion of PV manufacturing'. *Proceedings of the 37th European PV Solar Energy Conference and Exhibition*; 2020.

[86] Schubert G., Beaucarne G., Tous L., Hoornstra J. 'Trends in metallization and interconnection–Results of the survey conducted during the 8th Metallization and interconnection workshop'. *AIP Conference Proceedings*. 2019;**2156**(1):020002.

Chapter 5
Evaporated contacts

Delfina Muñoz[1], *Armin Richter*[2], *Perrine Carroy*[1], *Anthony Valla*[1], *James Bullock*[3], *Jesús Ibarra Michel*[3], *and Charles Roux*[1]

5.1 Metallization by means of physical vapour deposition

Metal contacts to semiconductor (SC) surfaces prepared by means of physical vapour deposition (PVD) are one of the most classical metallization techniques for SC devices, which were intensively investigated especially for microelectronic devices. This led to a huge variety of different metallization schemes, which can be adapted to almost any application to form a high-quality contact, especially also in the field of Si photovoltaic (PV) devices. That is why PVD metal contacts are frequently used especially for R&D silicon solar cells to realize, for instance, high efficiency and record devices [1–6]. The realization of fine contact geometries, for instance, for a front side metal grid, requires structuring, e.g. by photolithography. Due to the relatively high process complexity and the high material waste, standard PVD tools are costly and not suited for the mass production of terrestrial silicon solar cells. However, for space application or concentrating systems, PVD metal contacts are still used also in production. There are also special PVD systems, which are suited for silicon solar cell mass production [7], which are presented in Section 5.4.

In this sub-section, the properties of different PVD metal contacts like Al, Ti or Ag are discussed, focusing not only on the electrical properties but also on properties like adhesion or optics, i.e., light trapping. Furthermore, the basics of the deposition methods are described together with issues that can arise from PVD deposition like degradation of surface passivation layers. Finally, also some basics on metal structuring processes are discussed, such as photolithography and the lift-off technique often used for R&D cells as well as more industrial feasible methods, e.g. based on electrochemical machining.

[1]CEA-INES, Le Bourget-du-Lac, France
[2]Fraunhofer ISE, Freiburg im Breisgau, Germany
[3]University of Melbourne, Melbourne, Australia

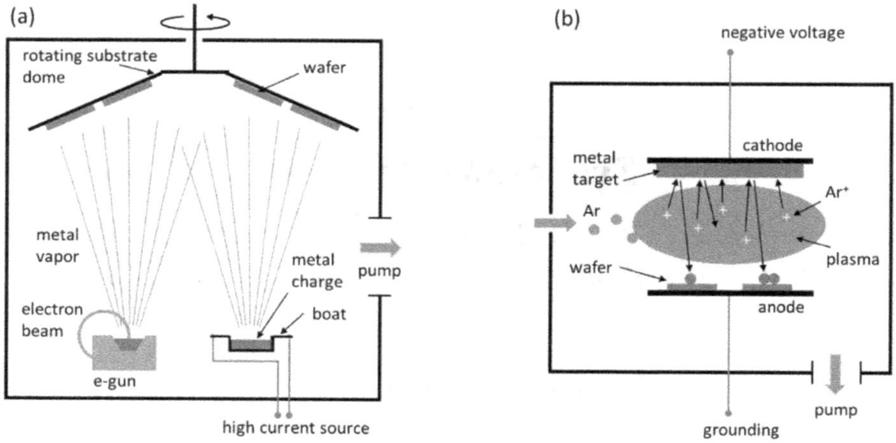

Figure 5.1 *(a) Schematic illustration of an evaporation system with an e-beam (left) and thermal (right) heating source. (b) Schematic illustration of a simple DC sputtering system.*

5.1.1 Deposition

In contrast to chemical vapour deposition (CVD), where gases are activated typically by temperature (heated substrate) or plasma and react at the sample surface, PVD consists of following steps:

- conversion of the source metal from a condensed phase (typically solid) into gas phase,
- transportation of the evaporated metal atoms from the source to the substrate and
- condensation of the metal atoms on the substrate surface, followed by the nucleation and the film growth.

In the following, the basics of the main two PVD deposition methods, namely, evaporation and sputtering are briefly described. More details on the deposition processes can be found in [8–10]. Some practical aspects regarding evaporation or sputtering of various materials can be found in [11].

5.1.1.1 Evaporation

In the case of evaporation, the solid metal, e.g. in form of pellets, is heated to a sufficiently high temperature where it melts and evaporates. Figure 5.1a shows a schematic illustration of typical evaporation system. The system consists of four basic components: (i) a chamber that can be evacuated to a sufficient vacuum for deposition, (ii) vacuum pumps, (iii) instrumentation, e.g. for monitoring the vacuum level and the growth rate and (iv) a method for heating the metal source. The wafers to be metallized are placed above the evaporation source at a distance of about 50 cm from the metal source. Ideally, the wafer position is not static but held on a rotating carrousel-type support during the deposition to ensure uniform film growth. A shutter is usually located on top of the metal source, which is opened after the heating has been started and the metal evaporation is stabilized. The growth rate is typically measured in situ with an oscillating quartz sensor, which is used to dynamically regulate the heating power.

Evaporation processes are performed at high-vacuum conditions at a pressure of about 10^{-6} mbar. At this vacuum level, the mean free path length is usually much longer than the distance between the metal source and the substrate to be metallized. This minimizes collisions and chemical reactions of the metal atoms with other particles and hence prevents a contamination of the metal layer, e.g. with oxygen. It further results in a line-of-sight path of the evaporated metal atoms from the metal source to the substrate. This results usually in a poor step coverage, which is actually a desired property for structuring the metal films using photolithography and the lift-off technique (cf. Section 5.1.3). In order to reach the low pressures, a mechanical pump (usually called roughing pump), which brings the system from atmospheric pressure to about 1 mbar, is combined with a turbomolecular pump, cryopump or an oil-diffusion pump, which requires further a liquid nitrogen trap for minimizing the oil contamination streaming back into the main chamber.

The source metal can be heated by different methods. The most usual methods used for silicon solar cell metallization are resistive or electron-bombardment heating, which are referred to as thermal or e-beam evaporation, respectively. Alternatively, inductive or laser heating is also used. Resistive heating is the simplest method, where the metal is load into evaporator (known as boat), which is heated with a high current until the metal is melt and evaporation starts (see Figure 5.1a, right). These boats are made of a conductive material with a high melting point, typically molybdenum, tungsten or tantalum. In the case of e-beam evaporation, electrons emitted from an incandescent filament are accelerated in an electric field and directed vertically onto the metal source in a water-cooled crucible using an electrostatic lens system for focusing and magnetic deflection (see Figure 5.1a, left). The beam energy is usually set to about 5–15 keV. In addition to the intense optical radiation emitted from the incandescent filament, the interaction of the electron beam with the melt generates X-rays, optical radiation (from UV to IR) as well as secondary and backscattered electrons. While most of the emitted electrons will not reach the substrate due to the magnetic deflection field and most of the optical radiation is reflected from the substrate as the metal film develops, the substrate is continuously exposed to X-rays with an energy up to the energy of the incident electron beam [12]. These X-rays can degrade, for instance, the surface passivation quality of layers being present on the wafers to be metallized (see Section 5.1.1.3).

5.1.1.2 Sputtering

In the case of sputtering, metal atoms are ejected from the target metal via bombardment with energetic ions, which condense on the substrate surface and form the film. The deposition via these ejected atoms has the advantage that it is independent of the melting point of the material to be deposited, thus even materials with a high melting point can easily be deposited. Figure 5.1b shows a schematic illustration of a direct current (DC) sputtering system, the simplest form of sputtering. It consists of a vacuum chamber with a cathode, which is the metal target, and an anode (often only the chamber walls), on which the substrate to be metallized is placed. During the sputtering process, the sputter gas, typically an inert gas like argon for pure metal deposition, is continuously entered into the chamber. A high voltage of usually several 100 V between the cathode and the anode ignites a low-pressure argon plasma at around 10^{-3}–10^{-2} mbar, which generates the argon ions required for sputtering. The electrical field accelerates the Ar^{+} ions towards the cathode, which can then eject metal atoms. The efficiency of sputtering processes, also known as sputtering yield, depends on the kinetic energy of the ions, the binding energy of the target metal atoms and on the mass ratio of the target metal and sputter gas atoms. Despite DC plasma excitation, also radio frequency (RF) can be used usually to reduce charging effects, which is, however, typically not an issue for the deposition of metals. In addition, permanent magnets can be located behind the metal target, which provide strong magnetic fields to confine the charged plasma particles close to the surface of the metal target. In the magnetic field, the electrons are forced on cycloidal trajectories, which, together with the confinement of the plasma, increases the argon ionization rate close to the target and hence, also increases the sputter rate. That is why these processes with permanent magnets, referred to as magnetron sputtering (MS), are often utilized for metal deposition especially with DC excitation.

Due to the required argon plasma, the pressure during sputtering is much higher than during evaporation. Therefore, the mean free path length of the sputtered atoms is much lower. This causes collisions of the sputtered atoms with atoms from the plasma or with other sputtered atoms. Therefore, their path to the substrate follows more a random walk rather than the line-of-sight path of evaporated atoms. This has the advantage that sputtered films exhibit a better step coverage. In addition, the sputtered atoms have a wide range of kinetic energies [13, 14]. For instance, the neutral particles have typically a maximum of their energy distribution at around half of the binding energy of the surface atoms of the target metal, which is usually up to tens of eV. The fact that these sputtered metal atoms hit the surface with a kinetic energy adds chemical activity to the growing film, which can result in changes of its crystallinity, morphology and could even improve the adhesion to the substrate. However, this kinetic energy can also cause damage of layers which are already present at the wafer surface [9].

5.1.1.3 Degradation effects during deposition

E-beam evaporation or sputtering of metals causes severe damage to the c-Si/SiO_2 interface, a problem which is also well known for microelectronic devices [15–18]. For Si solar cells, a strong degradation of the surface passivation after e-beam evaporation was observed for various layers such as SiO_2, Al_2O_3 or a-Si [19–22]. This degradation correlates with a strong increase of the interface state density, i.e., a degradation of the chemical interface passivation quality. Such degradation is not observed after thermal evaporation. The damage to the SiO_2 surface passivation can be, however, typically completely removed by a proper post-metallization anneal around 450 °C [19, 20]. A rather similar recovery after annealing was also observed for Al_2O_3-passivated silicon surfaces [22]. In the case of thermally grown SiO_2 coated with Al, hydrogen, which is released from the SiO_2/Al interface during such a thermal treatment [23], can contribute to the recovery of the chemical passivation of the c-Si/SiO_2 interface. For other layers like Al_2O_3, which contain a significant amount of hydrogen [24], hydrogen might play a similar role, as it becomes mobile during such annealing processes [25] and could, hence, also contribute to the recovery of the chemical interface passivation.

In the case of e-beam evaporation, the degradation is caused by photon radiation with energies from the UV to soft X-ray range, which is the bremsstrahlung caused by the energetic electrons when decelerated in the target metal, rather then, for instance, damage induced by energetic electrons from the beam that are scattered off the evaporating target metal [12, 16, 18, 19]. Kopp *et al.* found that a 1.8-μm-thick photoresist, as well as a thermally evaporated Al layer of either 0.1- or 1.6-μm thickness, does not protect from e-beam damage [20], although Blakers *et al.* reported an effective protection for 0.1 μm of thermally evaporated Al, Ti or Mg [19]. This disagreement might be explained by differences in the deposition parameters, e.g. in the e-beam acceleration voltage, because Mette found that the e-beam evaporation of Cr at low voltage caused a much lower damage than the e-beam evaporation of other metals at higher voltage [26].

In the case of sputtering, the substrate to be metallized is not only exposed to the radiation from the UV to soft X-ray range caused by the interactions of energetic electrons and ions with the target metal, but also to atoms or ions from the target metal or the sputtering gas as well as to energetic electrons from the plasma [8]. The energy distribution of the impinging species or radiation can strongly vary with the type of sputtering processes, its processes conditions as well as the geometry of deposition system. For DC sputtering, Reinwand *et al.* found a severe degradation of SiO_2 surface passivation after sputtering of different metals [27]. They found that a post-metallization anneal removes the damage of Ti or Cr sputtering, while for other metals like NiV or NiCr the damage did almost not recover. Katkhouda *et al.* reported sputtering processes, which reduce the damage, e.g. by sputtering at elevated sample temperatures or by introducing a proper capping layer (e.g. SiN_x) [28].

Most metals such as Ti, Pd, Ag or Al can also be thermally evaporated, for which no damage has been observed. Hence, this might be the deposition method of choice for more sensitive devices, for instance, for a-Si heterojunction devices due to their limited temperature tolerance for post-metallization anneal to recover the initial surface passivation quality.

5.1.2 Properties of PVD metal contacts

The primary functionality of silicon solar cell metallization is to provide a low resistivity current transport at the silicon/metal contact and the lateral transport to the external contacts. Besides these electrical properties, the metal contact also requires stability properties like a good adhesion to silicon and other layers, mechanical stability as well as stability in ambient air (e.g. low corrosion due to oxygen or humidity). Another important point is that the metal contact also needs to withstand post-metallization processes, such as thermal treatments around 400 °C, which are often required for contact formation, improvement of the metal adhesion and/or for the activation of the silicon surface passivation (e.g. for silicon oxide (SiO_2) [23], aluminium oxide (Al_2O_3) [29, 30] or tunnel oxide passivating contacts [31]). In this context, it is often also important that the metal contact exhibits a very low penetration depth into the silicon especially for shallow p–n junctions or passivating contacts based on thin films (TF) of amorphous silicon (a-Si) or polycrystalline silicon (poly-Si). With respect to the silicon bulk, it is quite important that the metal contacts do not cause a contamination via diffusion of metal atoms into the silicon, for instance, during a post-metallization thermal treatment or even at room temperature. Such an impurity contamination can result in a severe reduction of the charge carrier diffusion length due to the defect recombination in the silicon bulk, which can strongly reduce the device performance. Finally, for some solar cell configurations, also the light trapping (i.e. the reflectivity and absorption effects) at the metallized surface plays a role, especially in the case of a full-area metallization at the rear surface.

A huge variety of different PVD metal contact schemes based on elementary metals and metal alloys such as silicides or nitrides as well as multi-layer stacks of them have been intensively investigated, in particular for silicon microelectronic devices [9, 32, 33]. For silicon solar cells, especially for space application [34–37], also various different PVD metallization schemes have been developed. Table 5.1 summarizes some basic properties of typical elementary metals used for silicon devices, in particular also for silicon solar cells.

PVD metallization for silicon solar cells can be divided in two applications: (i) metal contacts, which are completely deposited via PVD and (ii) metal contacts, where only a seed layer is deposited via PVD and the bulk metal, which provides the high lateral conductivity, is deposited via plating. The later relaxes the conductivity required for the PVD metal. If, however, a metal grid is fully deposited via PVD the metal requires a low resistivity. Ag and Cu are the most conducting metals followed by Au and Al. However, Cu and Au have a quite high diffusivity in silicon, especially Cu [48], which is actually a fast diffusor having a diffusivity at room temperature, which is about seven orders of magnitude higher than that of typical dopants like B or P at high temperatures of diffusion processes [46, 47]. This together with the fact that both feature deep energy levels in silicon [49], i.e. form quite recombination-active defect states, can cause severe contamination effects when Au [50] or in particular Cu [34, 51] is directly applied to the silicon surface. If, however, a suitable barrier layer is located at the Si/metal interface, even Cu can be applied to silicon solar cells [52], which is interesting from a cost point of view.

Single layers of PVD Ag show a very good stability in air, but typically a poor adhesion to silicon [26] that is why Ag is often combined with another metal. With respect to adhesion, metals like Al, Ti or Ta show a good adhesion to silicon because they are very reactive (especially Ti) and form oxides with an enthalpy of formation

Table 5.1 *Properties of various metals: bulk electrical resistivity* ρb, *melting point* Tm, *coefficient of linear thermal expansion* α, *typical refractive index* n *(at a wavelength of 1 100 nm) for PVD metals [38–43], work function [44, 45], silicide formation temperature* T_{Sil} *[32] and diffusivity in Si at 600 °C [46, 47]*

Metal	ρ_b (μ Ω cm)	T_m (°C)	α (10^{-6}/K)	*n*	W (eV)	T_{Sil} (°C)	D (cm²/s)
Ag	1.6	960	19.7	0.16	4.3–4.4		1×10^{-12}
Al	2.7–3.0	660	23.6	1.01	4.17–4.25		2×10^{-18}
Au	2.2	1 063	14.2	0.18	4.8–5.1		4×10^{-10}
Cr	20	1 900	6.3	4.53	4.7	450 ($CrSi_2$)	
Cu	1.7	1 083	17.0	0.13	4.55		7×10^{-8}
Mg	4.4	923	24.8	2.70	3.7		
Mo	5.2	2 625	5.0	5.25	4.4	525 ($MoSi_2$)	
Ni	6.8	1 455	13.3	7.48	4.5–5.1	200–350 (Ni_2Si) 350–750 (NiSi) ≥750 ($NiSi_2$)	1×10^{-12}
Pd	10.8	1 552	11.0	3.09	5.0	100–300 (Pd_2Si) 850 (PdSi)	
Pt	10.6	1 774	9.0	0.95	5.3	200–500 (Pt_2Si) 300 (PtSi)	6×10^{-12}
Ti	40	1 668	8.6	2.95	4.0	500 (TiSi) 550 ($TiSi_2$)	2.3×10^{-13}
W	5.5	3 415	4.5	3.13	4.6	650 (WSi_2)	

higher than that of SiO_2, i.e. they can reduce thin SiO_2 layers [9]. For various applications, single layers of Al are an alternative to Ag, which fulfil almost a lot of the requirements specified above: low contact resistivity, high reflectivity, good adhesion and good optical properties [53]. However, Si is soluble in Al, and during a thermal treatment (starting at ~300 °C) Si dissolves in Al by diffusion, which forms pits in Si [9, 54]. This highly non-uniform process, also referred to as spiking, can result in pits with a size of several μm (depending, e.g. on the annealing temperature and time), which can be quite crucial especially for shallow p–n junctions or thin Si-based passivating contacts like a-Si or poly-Si. Ti, in contrast, shows no such strong reactions with silicon. Due to the low conductivity, Ti is often combined with Ag. Ti/Ag layer stacks are, however, not stable especially under humidity because of an oxidation of Ti and a peeling of Ag [35]. This is improved by introducing a thin layer of Pd between Ti and Ag, which prevents the oxidation and thus, makes this Ti/Pd/Ag stack a very reliable metallization [35]. This stack suffers, however, from poor optical properties due to the very high refractive index of Ti, which is crucial when applied on a large area fraction directly onto silicon [31, 55].

All this discussion shows that no metallization satisfies all the desired requirements specified above, thus the different metallization schemes need to be selected according to the specific requirements of a certain solar cell design. In the following, the properties of PVD metallization schemes typically used for silicon solar cells are reviewed in more detail with a focus on Al single layers and Ti/Pd/Ag layer stacks deposited via evaporation, which are used most often. Note that the wide topic of metal alloys, such as nitrides or metal silicides, which are extensively investigated for silicon contact applications, is only little touched here; more details can be found in [32, 44, 56, 57].

(a) Aluminium

PVD Al is a very classical metallization for silicon devices, in particular for silicon solar cells. Al exhibits a low barrier height to p-type Si [45] that is why it exhibits a very low specific contact resistivity ρ_c on p-type Si even for low surface doping concentrations N_{surf}. On highly doped n^+ surfaces, also quite low ρ_c is observed, which, however, strongly increases with decreasing N_{surf}. Exemplary experimental ρ_c values for evaporated Al contacts on n-type and p-type Si surfaces are shown in Figure 5.2. Fong *et al.* found that ρ_c can strongly depend on deposition conditions, such as the power for thermal evaporation [53]. In our labs, we found a lower ρ_c for e-beam evaporated Al compared to thermally evaporated Al. (Contact resistivity of evaporated Al (300 nm) and Ti/Pd/Ag (50 nm / 50 nm / 200 nm) contacts was measured with the TLM technique on different $POCl_3$ diffused n+ and BBr_3 diffused p+ surfaces on planar Si wafers. Direct before the evaporation, the samples were dipped in diluted HF. The TLM contacts were structured by means of photolithography and the lift-off technique. After deposition, some of the samples were annealed on a hotplate for 1 min at 350°C or annealed under forming gas (FGA) at 425°C for 25 min. The surface doping concentration was measured via ECV.) Annealing at temperature between around 300 and 425 °C is found to increase ρ_c on n^+ surfaces [53, 59], while such annealing decreases ρ_c on p^+ surfaces even after only 1 min at 350 °C [59].

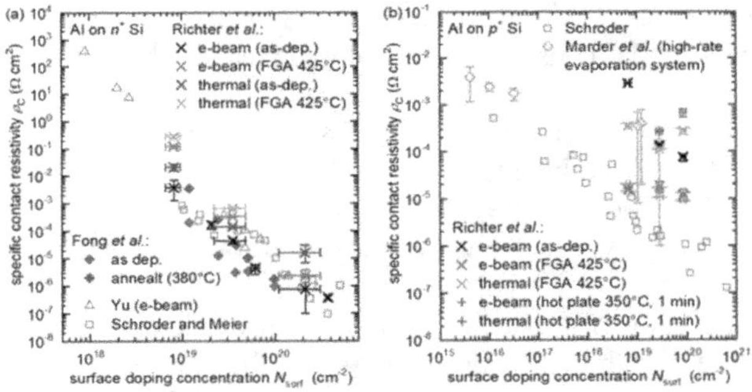

Figure 5.2 *(a) specific contact resistivity* ρ_c *of Al on* n^+ *Si surfaces as a function of P surface doping concentration from Schroder and Maier [45], Yu (e-beam evaporated, annealed at 400–500 °C in* N_2*) [58], Fong* et al. *(thermally evaporated, as-deposited and after annealing in forming gas (FGA) at 380 °C) [53] and Richter* et al. *(e-beam and thermally evaporated, as-deposited and after annealing) [59]. (b)* ρ_c *of Al on* p^+ *Si surfaces as a function of B surface doping concentration from Schroder [60], Marder* et al. *(high-rate inline evaporation system, as-deposited) [61] and Richter* et al. *(e-beam and thermally evaporated, as-deposited and after annealing) [59]. Note the different* ρ_c *scale of both graphs.*

Due to these low ρ_c values, PVD Al is used as a contact metallization for various silicon solar cell designs to contact p^+ [62–65] and n^+ [62, 66] surfaces, especially also for interdigitated back contact (IBC) solar cells where both contact polarities are present at the rear surface [67–69]. In the passivated emitter and rear cell (PERC) cell design, PVD Al can be even used to contact the p-type Si base, if the Si base doping is high enough [70]. If applied as full-area metal on the rear surface, PVD Al is a good optical reflector, especially when combined with a SiO_2 layer [65, 71] due to its low refractive index. If PVD Al is combined with SiO_2, it even helps to activate the SiO_2 surface passivation quality during a forming gas anneal (FGA), also referred to as 'alneal', because hydrogen is released from the SiO_2/Al interface, which contributes to the Si/SiO_2 interface passivation [72].

In p-type Si PERC and passivated emitter and rear locally diffused (PERL) cells, PVD Al can be used as a contact to the p-type Si base at the rear surface with or without a p^+ layer at the surface (i.e. a high–low junction) [19, 65, 70]. In this case, the above-mentioned Al spiking during annealing is less crucial as if Al is applied to a p–n junction, where the spiking can cause shunt effects. Al spiking can, however, be strongly reduced, e.g. by inserting a barrier layer at the Si/Al interface like PVD Ti or Al–Si [73]. PVD Al is also used to form a Al–Si alloy (eutectic point at 577 °C) during high-temperature treatments, which is utilized as a full-area or locally Al-doped p^+ back surface field (BSF) in p-type Si solar cells [64, 74–76]. The locally alloyed Al–Si BSF [64] was a predecessor of the Al–Si-alloyed p^+ region formed at the local contact openings in the dielectric passivation layer at the rear surface of industrial PERC solar cells, where the alloying usually from a screen-printed Al but also from PVD Al takes even place during very short contact firing processes (usually around 800 °C for around 1 s) [77, 78]. Actually, such an Al–Si-alloyed p^+ region can also be formed by an ultra-short laser heating as realized, for instance, by the laser fired contacts, where PVD Al is fired through a dielectric passivation layer (e.g. Al_2O_3/SiN_x layer stack) at the rear surface of PERC solar cells to form the contact [7, 79].

Evaporated Al was also used to form a metal insulator semiconductor (MIS) contact through a ultrathin SiO_2 layer to contact the front side inversion layer of p-type Si MIS solar cells [80]. Besides contacts to crystalline silicon (c-Si) surfaces, PVD Al is also successfully applied, for instance, to passivating contacts such as a-Si of silicon heterojunction (SHJ) solar cells [81], poly-Si [1, 82–84], a-Si/LiF_x [85], Cs_2CO_3 [86], MgO_x [87, 88], MgF_2 [89], Nb_2O_5 [90], TaN_x [91] or TiO_2 [92]. Annealing of a-Si/Al at temperatures above 200 °C can completely dissolve a-Si in Al, which can be utilized to form a contact to the underlying c-Si [93]. Reichel *et al.* applied PVD Al contacts also for various organic interface dipole layers [94].

With respect to module interconnection, PVD Al cannot easily be combined with the preferred soldering process in industry due to the rapid formation of a native Al_2O_3 layer [95]. This can be solved, however, either by ultrasonic tinning [96], wet-chemical treatments [97] or other PVD metals on top Al, such as NiV/Ag [98], TiN/Ti/Ag [99], TiN/NiV/Ag [100], NiV [101] or only pure Ni (about 100 nm) [95], which can be deposited in the same PVD system. For instance, the ultrasonic tinning and the pure PVD Ni layer were found to result in a PVD Al rear metallization for industrial PERC cells, which is compatible with high-throughput tabber stringer module interconnection and showed a very good stability under climate tests as required according to the standard IEC 61215 [95, 96].

(b) Titanium as contact metal of Ti/Pd/Ag layer stacks

The barrier height of Ti is lower to n-type Si than to p-Type Si [45] that is why it exhibits a rather low contact resistivity on n^+ surfaces, as can be seen from the exemplary ρ_c data of evaporated Si/Ti contacts in Figure 5.3a. c-Si solar cells with Ti/Pd/Ag contacts to n^+ emitters were found to be stable after annealing at up to 500 °C [26], where Ti starts to form silicides (see Table 5.1). If N_{surf} is high enough, Ti exhibits also a low contact resistivity to p^+ surfaces [55, 59] (see Figure 5.3b).

Due to these low contact resistivities, Ti/Pd/Ag stacks are not only applied to n^+ surfaces [55, 103–105] but also to p^+ surface surfaces of high-efficiency Si solar cells [2, 3, 31, 62, 106]. Although a poor adhesion was observed for Ti/Ag stacks without Pd after humidity tests [35], Ti/Pd stacks were also found showing a good performance [26, 107] even after annealing up to 500 °C and after storage of 24 months in air [26]. PVD Ti/Pd can also act as a diffusion barrier against plated Cu [34]. Beside contact to c-Si, PVD Ti/Pd/Ag stacks were also used to form, for instance, contacts to B-doped and P-doped poly-Si-based passivating contacts [31, 108–110], B-doped a-Si [103, 111], P-doped a-Si [111, 112] or indium tin oxide (ITO) [103, 109, 111, 112].

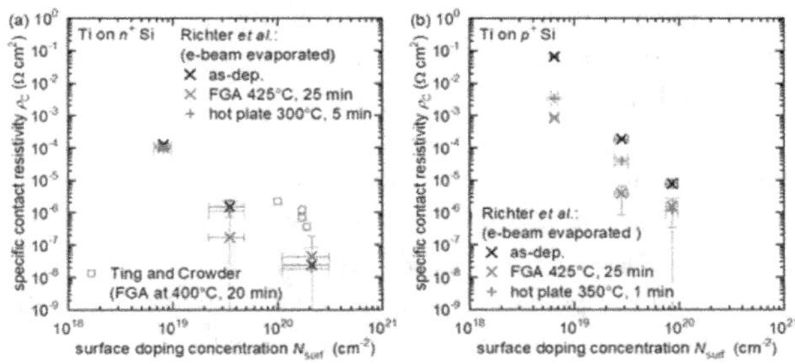

Figure 5.3 Contact resistivity of Ti on n⁺ Si and p⁺ surfaces as a function, the surface doping concentration for e-beam evaporated Ti (50 nm)/Pd (50 nm)/Ag (200 nm) stacks directly after deposition (as dep.) and after a FGA at 425 °C for 25 min or a short hot plate anneal in air [59] and for thermally evaporated Ti (100 nm)/Al (–500 nm) after a FGA at 400 °C for 20 min [102]

The low contact resistivity together with the good adhesion to Si and SiO_2 makes Ti-based contacts (especially Ti/Pd/Ag stacks) very suited to form very narrow contact grids for the solar cell front side, which are structured with photolithography and the lift-off technique, e.g. for fully evaporated contacts with a thick Ag film of typically 5–10-μm thickness [3, 106, 109, 112] (see Figure 5.4a) or as a seed layer which is afterwards thickened via plating [31, 37, 62, 104, 105, 113]. Full-area Ti-based contacts for the rear side are typically not used due to the poor light trapping of Ti when in direct contact to Si [31, 55].

(c) Silver

Although evaporated Ag exhibits typically no good adhesion to silicon, which, for instance, prevents a fine grid structuring via photolithography and lift off [26], it is often applied as a full-area contact at the rear surface or deposited through a shadow mask to various silicon solar cell devices. Song *et al*. reported contact resistivities of ~0.2 m Ω cm² for n⁺ surfaces and ~70 m Ω cm² for p⁺ surfaces, which reduced to ~20 m Ω cm² after annealing at 160 °C [107]. Although these ρ_c values

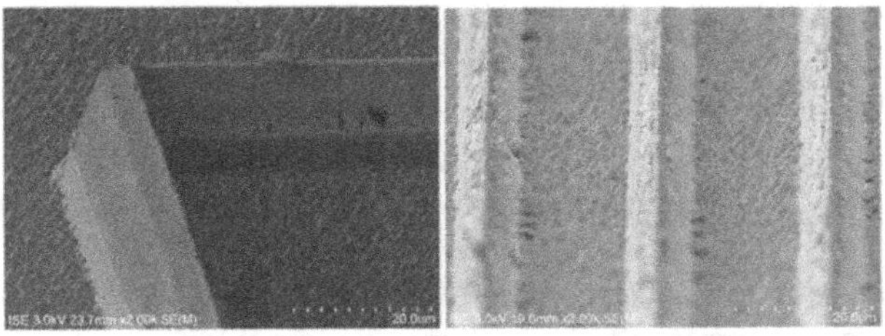

Figure 5.4 Fully evaporated metal grids structured via photolithography and the lift-off technique: (a) high-aspect ratio fingers used for high-efficiency silicon solar cells [2, 3] and (b) triangular-shaped fingers realized with maskless laser lithography [2]

are quite high when compared to the ρ_c values of Al and Ti, they are good enough for a full-area contact. One advantage of Ag is the excellent light trapping even without a dielectric layer, which resulted in the highest short circuit current density J_{SC} of >43 mA/cm^2 reported for silicon solar cells [3]. The light trapping of PVD Ag can even be improved by using an additional dielectric TF (e.g. SiO_x or SiN_x) at the rear interface [2, 114] or by introducing a photonic light trapping structure [115, 116]. The most typical application of full-area Ag contacts is for B-doped [84, 109, 117] and P-doped [2, 3, 31, 113] poly-Si-based passivating contacts, nc-SiC_x [118] and B-doped or P-doped a-Si [119]. Furthermore, full-area PVD Ag contacts are also successfully applied to ITO [31, 81, 114, 120, 121], AZO [81], PEDOT:PSS [86, 122] or a-Si/VO_x stacks [123].

d) Chromium

The low barrier height to p-type Si makes Cr particularly suited for p$^+$ surfaces [45]. This is shown by the application of PVD Cr/Pd/Ag stacks as contact to B-doped p$^+$ emitters of n-type Si solar cells [87, 91, 124]. Evaporated Cr/Pd/Ag stacks were also found to show a good contact to P-doped n$^+$ emitters of p-type Si solar cells, which were even stable after annealing up to 450 °C [26]. For annealing at higher temperatures, as significant reduction of the fill factor (FF) was observed indicating a contact degradation. This might be related with a silicide formation, which starts at such temperatures (see Table 5.1) and which is known to penetrate deep into c-Si [125, 126]. In addition to p$^+$ and n$^+$ surfaces, evaporated Cr/Pd/Ag stacks were, for instance, also successfully applied to contact ITO [119].

(e) Palladium

The Pd_2Si silicides have a rather low enthalpy of formation starting already at temperatures around 200 °C (see Table 5.1). An annealing at 275 °C for 20 min forms, for instance, a 30-nm-thick Pd_2Si layer. The very low barrier height to p-type Si makes Pd_2Si particularly suited for p$^+$ surfaces [45]. Mette applied Pd/Ag contacts to P-doped n$^+$ surfaces and found that it does not make a contact without annealing [26]. Annealing between 300 and 500 °C resulted in a good contact, however, not as good as Ti/Pd/Ag. In addition to c-Si, PVD Pd contacts were also successfully realized for passivating contacts, e.g. to CuO_x [127] or MoO_x [128].

(f) Nickel

PVD Ni is widely used as contact metal for silicon microelectronic devices [129] due to beneficial properties of Ni silicides such as a high electrical and mechanical reliability. Ni forms three different silicide phases below 800 °C, starting at around 200 °C with the formation of Ni_2Si followed by NiSi formation at around 350 °C (see Table 5.1). The barrier height of these two silicide phases to n-type Si was reported to be ~0.62 and 0.67 eV, respectively [130]. These silicides form low resistivity contacts to silicon. For instance, Stavitski *et al.* reported a systematic analysis of the contact resistivity of NiSi contacts to n-type and p-type Si as a function of the surface doping concentration N_{surf}, which revealed very low ρ_c values below 10^{-5} m Ω cm^2 for N_{surf} >10^{19} cm^{-3} for both polarities. Mette applied evaporated Ni contacts to P-doped n$^+$ emitters of p-type Si solar cells, which show a high and stable device performance for annealing at temperatures between 350 and 500 °C very similar to solar cells with the high-quality Ti/Pd/Ag stacks [26]. Xu *et al.* aimed to improve the contact resistivity of standard screen-printed Ag pastes for P-doped n^+ emitters of p-type Si solar cells using a sputtered Ni interlayer, which resulted in a significant reduction of the contact resistivity by more than one order of magnitude [131]. In general, however, PVD Ni has hardly been studied as contact metal to c-Si solar cells, while plated Ni is intensively studied due to its low-cost approach especially when combined with Cu plating (see Chapter 8). PVD Ni layers are recently more applied on the top of, e.g. PVD Al to improve module interconnection because it is well solderable (see the sub-section "Aluminium").

(g) Magnesium

Due to the low work function, Mg exhibits a low barrier height to n-type Si [45], which makes it well suited as an electron contact. This was utilized by Wan *et al.* to realize an a-Si/Mg/Al contact to the c-Si base of a n-type Si solar cell [132]. PVD Mg was further used for passivating contacts, e.g. Zn(O,S) [133] or stacks of ultra-thin Al_2O_3/TiO_2 [134]. Evaporated Mg was also used to induce a n^+ inversion layer through an ultrathin SiO_2 layer for a p-type Si MIS solar cell [135].

(h) Calcium

Ca is a low work function metal, which is used as an electron contact in organic PV devices [136]. Allen *et al.* showed that evaporated Ca can also exhibit a low contact resistivity to n-type Si (ρ_c = ~2 m Ω cm²), which was confirmed by n-type Si solar cells with local Ca contacts to the c-Si base [137]. It was also found that evaporated Ca has a low contact resistance to TiO_2 on similar n-type Si solar cells with local TiO_2/Ca contacts to the c-Si base [124].

5.1.3 *Structuring of PVD metallization*

There are different methods for the structuring of PVD metallization, e.g. to form a metal grid for the solar cell front side or the rear side of an IBC solar cell. The simplest method is the use of a shadow mask with a perforated strip pattern placed directly in front of the wafer to be metallized. This method has, however, the disadvantage that the smallest structure width is limited to around 50–100 µm. In order to realize finer structures, as required for high efficiency solar cells with minimal shading losses, photolithography and the lift-off technique can be used. This widely used method for R&D solar cells consists of a sequence of single process steps, which are conducted in a clean room with yellow lights due to the photosensitivity of photoresist. First, a TF of photoresist is spin-coated on the wafer. After a drying process to drive out excess solvents, a mask aligner is used to illuminate the photoresist through a glass mask with the desired pattern using short wavelength light (~400 nm). This light exposure chemically changes the photoresist, so that it can be removed by a special solution called developer. There are two different types of resist called positive and negative resist. In the case of a positive photoresist, the exposed area becomes soluble, while in the case of a negative photoresist, the unexposed area becomes soluble (see Figure 5.5a b). Next, the metal evaporation process takes place. Directly before the evaporation, usually a dip in diluted HF (usually 1% HF) is performed to remove a native oxide from the c-Si surface and, thus, to ensure a good contact resistivity. Finally, the remaining photoresist covered by the

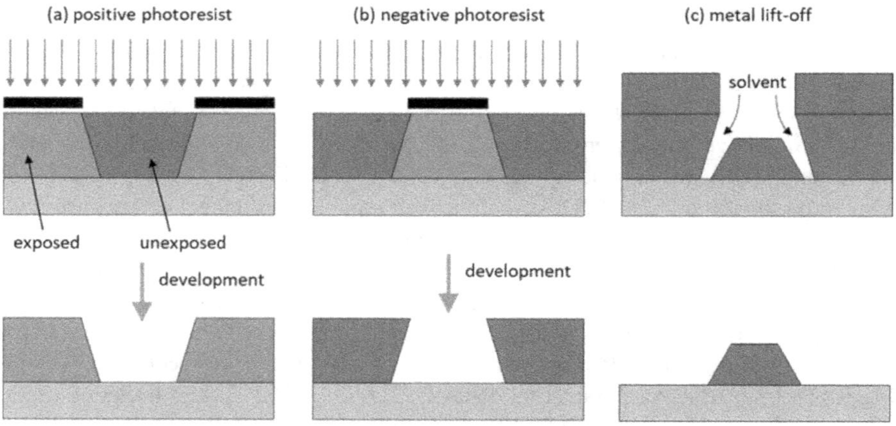

Figure 5.5 *Schematic illustration of a photolithography structuring of (a) positive and (b) negative photoresist and (c) a lift-off process using a negative photoresist*

excessive metal is removed in the lift-off process. In this step, a special solvent penetrates the photoresist through its edges at the openings. This causes a reaction with the photoresist and the metal together with photoresist can be lifted off, as illustrated in Figure 5.5c. Such a photolithographically structured PVD metallization has an accuracy in the order of 1 μm and a very good reproducibility.

This lift-off process can be used not only for depositing a thin metal layer of a few 100 nm as a seed layer for plating but also for structuring thick metal films up to around 10 μm of height. For the latter, it is especially important to use negative photoresists, which form an undercut at the openings where the grid metal is placed (cf. Figure 5.5b). This undercut strongly prevents metal deposition at the photoresist edges so that in the following lift-off process the solvents can penetrate into the photoresist. Furthermore, the thickness of the photoresist should be usually higher than the metal thickness, typically by a factor of 1.2–1.5. In this context, it could be difficult to combine this lift-off technique with sputtered metal deposition, due to the good step coverage of the sputtering process, which is exactly the opposite of what is being desired for the lift-off process. The good step coverage of the sputtering is result of the low vacuum pressure during sputtering causing a random-walk-like path of sputtered atoms to substrate where they arrive with more randomized angles of incidence. In contrast, the high vacuum during evaporation causes a mean free path length that is longer than the distance between the metal source and the substrate. Therefore, the evaporated atoms follow a line-of-sight path and, hence, have a much lower angle of incidence distributed, which is only determined be the geometry of the evaporation chamber, i.e., the position of the evaporating metal source and the position of substrates (cf. Section 5.1.1). Another point to consider for a successful lift-off is the temperature during metal deposition. If the temperature is too high, the photoresist can become soft, which can cause rounded edges at the photoresist opening and hence, a metal deposition at the opening edges that prevents the lift-off if it is too thick.

Figure 5.4a shows an example of fully evaporated high aspect ratio grid fingers with a width of about 10 μm and a height of about 9 μm (50 nm Ti/50 nm Pd/~9 μm Ag), which have been realized using photolithography, e-beam evaporation and lift-off. Such grid fingers have been used as a front side metal grid for high efficiency silicon solar cells with an energy conversion efficiency of 25.8% [2, 3]. A further optimization of the lift-off process using maskless lithography, where direct laser writing is used for the exposure of the photoresist with much smaller feature sizes, leads to triangular-shaped fingers [2] as shown in Figure 5.4b, which can help to reduce grid shading losses further. Very similar, triangular-shaped fingers have also been realized using a combination of nanoimprint lithography and metal lift-off [138].

Instead of using the lift-off technique, photolithography can be also used to structure a full area metallization after PVD deposition, where the structured photoresist acts as a masking for a metal etching, e.g. using a wet-chemical etching solution or dry etching, such as reactive-ion etching (RIE). An overview of chemical etches or gases and gas mixtures that can be used for wet-chemical etching or RIE of different metals can be found in [9]. This approach is often used to structure a PVD Al metallization at the rear surface of IBC solar cells [67, 69], where special Al wet-chemical etchant solutions, such as mixture of phosphorus acid (H_3PO_4), nitric acid (HNO_3) and acetic acid (CH_3COOH), were used.

These photolithography-based structuring methods are too complex for silicon solar cell mass production. However, there have been more other industrial feasible methods reported for the structuring of PVD metallization. For instance, a structured resist can be directly printed, e.g. using ink-jet printing [139, 140] or screen printing [141, 142], and used either for the lift-off technique or for the metal etching. In particular, for the separation of PVD Al contacts on the back surface of IBC solar cells, methods have also been reported, which require no resist at all. For instance, a PECVD SiO_x layer can be deposited on top of the PVD Al, which is structured using laser ablation and acts as a masking for an Al etch, e.g. in diluted boiling hydrochloric acid (HCl) [1, 61]. Kamp *et al.* proposed to locally anodize PVD Al, for instance, with a structured cathode or by printing an electrolyte, which is electrically contacted during printing to convert the Al into AlO_x during printing, and thus, electrically separates the PVD Al [143]. Kamp *et al.* also proposed electrochemical screen printing for structuring PVD Al in one process step (see Figure 5.6), where an electrochemical etching paste is screen printed through a contacted mesh acting as anode, while the PVD Al acts as a cathode and hence, an electrochemical Al etching takes place directly during the screen printing process [144, 145].

5.1.4 Conclusions

This sub-section gives an overview of the PVD metallization technology for PV applications, including deposition methods, properties and applications of various

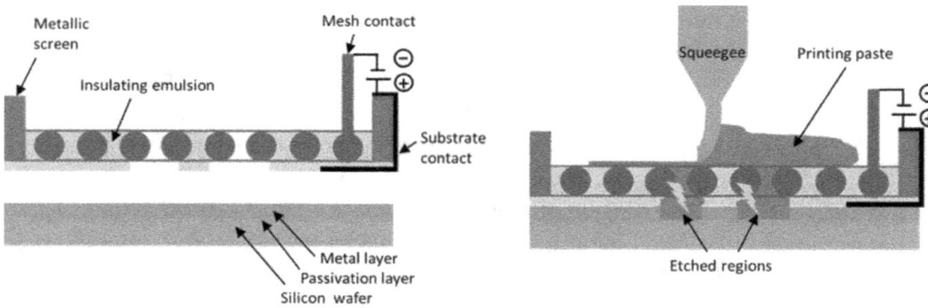

Figure 5.6 *Schematic illustration of the electrochemical screen printing process (ESP) after [144]. Left: silicon substrate with a metal layer and a screen with a contacted mesh as cathode and anodic substrate contact. Right: ESP during printing and etching process resulting in etched metal regions.*

mentalizations schemes, as well as structuring technologies for PVD metallization. PVD metal is usually deposited either via evaporation or via sputtering and both methods have many variants. Some of these variants have peculiarities that are important to note. For instance, electron beam evaporation or sputtering in general can cause damage to the surface of the substrate to be coated, which is important to consider in particular if surface passivation layers are already present. With respect to metallization schemes, aluminium and a layer stack of titanium/palladium/silver are often used especially due to their good contact properties and good mechanical adhesion. While Al has a wide variety of application especially as full-area rear side metallization, it has the drawback that spiking effects can appear after thermal treatments, which could be very crucial for p–n junction metallization. Ti/Pd/Ag stacks are more stable but also more costly, that is why this stack is typically used only for proto-type devices or for space solar cells. There is a huge variety of other metals like Cr, Ni or Mg, which cover a wide range of applications ranging from standard contacts to highly doped silicon surfaces over contacts for novel passivating contact schemes to PVD Al surface optimization for module interconnection. Finally, the metallization requires often structuring, for instance, for the local metal fingers at the solar cell front side. The photolithography-based lift-off technique is widely used to fabricate high-quality and advanced contacts to small-area prototype devices, but is much too complex for a mass production. Other structuring methods, e.g. based on printing, laser ablation or electrochemical processes, are much less complex and can be suitable for high throughput. More methods and technologies on high-throughput PVD metallization are presented in Sub-section 5.4.

5.2 Transparent conducting oxides in crystalline solar cells

5.2.1 Transparent conducting oxides

Transparent conducting oxides (TCOs) are generally defined by their dual characteristics, namely, their transparency in the visible (Vis) (and sometimes near-infrared, nIR) range as well as their electrical conductivity [146–151]. How do we explain this feat? To answer this question, we need to go back to the basics of the SC theory that allows to distinguish three material types: (1) insulators, (2) metals and (3) SCs.

TCOs are obtained by degenerately doping a wide band gap metallic oxide, thus inducing metal-like conductivity while preserving the optical transparency of the host material [146, 152]. There exists a wide variety of metallic oxides, displaying a broad range of properties. However, there are no rules to predict which oxide could exhibit transparent conductive properties [153]. Still, one can agree on some basic characteristics that a metallic oxide should have to become a potential TCO: a sufficiently wide band gap (>3.1 eV, about 400 nm) to allow a high optical transmission (from 80% to 90%) in the Vis and nIR spectrum, but not too wide so that doping is still possible to achieve good electrical properties (resistivity ρ between 10^{-4} and 10^{-2} Ω^{-1} cm^{-1}). Looking at these main properties, we can then conclude that TCOs belong to the third category of the above-mentioned materials.

So, once a proper metallic oxide is selected, doping is used in order to obtain better electrical properties. This allows to directly increase the charge carrier density N, thus increasing the conductivity σ (or inversely decreasing the resistivity ρ) according to the following equation:

$$\sigma = \frac{1}{\rho} = q.N.\mu \tag{5.1}$$

with q, the elementary charge of an electron, and μ, the electronic mobility. The doping consists in introducing defects in the material that induce discrete levels in the gap, located close to the host material conduction (n-type doping) or valence (p-type doping) band. These shallow levels can release electrons in the conduction band (CB) (n-type doping) [146] or holes in the valence band (p-type doping) [154]. Doping can be extrinsic (i.e. introducing foreign elements that will substitute to the atoms in the crystal lattice) or intrinsic (i.e. due to a deviation from stoichiometry produced by structure defects like interstitial atoms or vacancies). In the case of intentional doping with extrinsic impurities, it is important to know that the doping atom must have an atomic radius close to the host metal atom to ease its insertion in the crystal structure and to cause the least possible structural defects [155, 156].

At this point, we have answered our initial question: the double characteristic of TCOs is possible thanks to the heavy doping of a transparent material. Doping is thus a powerful way to control the properties of TCO materials and its adjustment through deposition parameters is essential. Indeed, as we will see in detail further in this text, the charge carrier density not only, understandably, influences the electrical properties of TCOs, but it also has a strong impact on their optical properties and requires finding a compromise between electrical and optical properties, depending on the desired application.

The first mention of TCO materials in the literature dates back to 1907. Bädeker investigated lead oxide (PbO), copper oxide (Cu_2O) and cadmium oxide (CdO) [157]. The latter had interesting electrical properties for that time, with ρ values as low as $1.2.10^{-3}$ Ω cm. Then publications on TCO materials are fairly scattered until the 1980s, when the number of reported works starts to intensify, showing the growing interest for these materials, which remains intact. This translates into a constant growth in the number of new materials and new deposition techniques, as well as in the diversity of applications.

There are several TCO material families. First, they are divided according to the majority charge carrier type:

- n-type materials that are doped with electron-donor impurities, where electrons provide the conductivity.
- p-type materials that are doped with electron-acceptor impurities, where holes provide the conductivity.

Then, within these two categories, there are several materials grouped into families depending on the host material to which the dopant impurities are added.

In the n-type TCO category, we find:

- the indium oxide family (In_2O_3): ITO (In_2O_3 :Sn), IOH (In_2O_3 :H), IWO (In_2O_3 :W), IZO (In_2O_3 :*Zn*), IZrO (In_2O_3 :Zr), IMO (In_2O_3 :Mo), ITiO (In_2O_3 :Ti), ICO (In_2O_3 :Ce), IFO (In_2O_3 :F), etc.
- the zinc oxide family (ZnO): BZO (ZnO:B), AZO (ZnO:Al), GZO (ZnO:Ga), etc.
- the tin oxide family (SnO_2): FTO (ZnO:F), SnO_2: Ta, ATO (SnO_2: Sb), etc.
- the cadmium oxide family (CdO): CdO: In, CdO: Sn, etc.
- families based on binary or ternary compounds combining the above-mentioned 'basic' oxides: ZnO–SnO_2 family, ZnO – In_2O_3 family, CdO–SnO_2 family, etc.

In the p-type TCO category, we essentially find ternary copper-based oxides: $CuAlO_2$, $CuGaO_2$, $CuInO_2$, $SrCu_2O_2$, etc.

It is important to know that the first category (n-type) commonly possess significantly superior performances than the second category (p-type), especially regarding electrical properties. Indeed, an n-type TCO can display $ZnO -_2 O_3$ values down to 10^{-4} Ω cm while a p-type TCO hardly reaches 10^{-2} Ω cm [147]. Currently many studies are ongoing to develop p-type TCOs with good-enough performances in order to expand the field of transparent electronics.

There are many deposition methods to fabricate TCO TF. They are generally classified according to the environment in which they take place (gaseous or liquid phase) and according to the processes involved (physical phenomena, chemical reactions). Among the techniques involving physical phenomena, we can distinguish the purely thermal methods and the energetic processes that involve additional kinetic energy by momentum transfer of bombarding species. As they take place in the gas phase, they are referred to as PVD techniques. Chemical methods can take place in the gas phase and are in this case referred to as CVD techniques, or in the liquid phase from chemical solutions.

One of the first common method used for fabrication TCO films was the post-oxidation of evaporated metallic films [158, 159]. Later, reactive evaporation (with oxygen addition during deposition) or direct evaporation of oxide sources were developed to obtain oxide films and are still commonly used [160].

Spray pyrolysis was another historically common method to deposit TCO films, as of its invention in the 1940s [161, 162]. It was one of the first widely spread technique for large-area deposition of TCOs and is still commonly used for the large-area fabrication of SnO^2-based films [163]. It consists in spraying a solution that is thermally pyrolyzed on the surface of a substrate heated at high temperatures (>200 °C). As such, it is an intermediary technique between CVD and liquid-phase chemical methods.

CVD of TCO films was somewhat later developed [164] and has become a very common method to obtain large-area TCO films [148]. In particular, TCOs are generally fabricated by metalorganic chemical vapour deposition (MOCVD), which is the thermal decomposition of a metalorganic compound and its reaction in an oxygen-rich atmosphere.

Nowadays, one of the most extensively used technique for TCO deposition is MS. Sputtering is an energetic PVD process where bombarding ions from a glow discharge eject by momentum transfer the cathode surface atoms that subsequently travel with high kinetic energy towards the anode (substrate) on which they impinge to form a solid film. While sputtering was discovered in the middle of the nineteenth century [165] and despite the fact that the first identified TCO had been fabricated by post-oxidation of sputtered metallic films [157], it was not much used for TCO deposition until the 1970s and the invention of planar magnetron [166]. By confining the electrons into cycloidal paths close to the target, MS solved most of the simple diode sputtering disadvantages, namely, low deposition rate, low ionization of the plasma and strong substrate heating.

Many PVD systems involving energetic species were later developed to fabricate films of even higher quality by using particle bombardment in different manners. Among them, TCO films were successfully deposited by ion beam sputtering [167], high-power impulse MS [168], ion-beam-assisted deposition (IBAD) [169] and different configurations of ion plating [170]. High-quality TCO films were also deposited by pulsed laser deposition (PLD), thanks to the high-energy species produced in the laser plasma plume [171, 172].

TCO films can also be deposited directly from liquid chemical solutions by spray, spin, dip coating of the solution on the substrate and subsequent chemical reaction of the residue by thermal oxidation or hydrolysis to produce the inorganic solid film [173]. They are simple and low-cost methods, however, leading to lower quality film than vacuum-based processes [148].

The peculiar properties of TCOs make them interesting for a large diversity of applications. Apart from their transparent/conducting duality, TCOs are widely used for their spectral selectivity [149]: TCOs are transparent in a well-defined range, that includes visible light, and become highly reflective above a certain wavelength that depends on the material and its deposition conditions (see Section 5.2.2.3). One of the largest applications of TCOs is indeed as energy-efficient architectural windows. Depending on the optical properties of the window TCO coating, it is possible to manage the reflection of the thermal radiation of the sun and/or of the inside in different ways and thus to either help keeping a warm inside space, which is desirable in cold climates, or rather help keeping a 'fresh' inside space, which is desirable in hot climates [149]. Other applications based on spectral selectivity include the coating of sky-facing windows (e.g. car windscreens) to avoid radiative cooling and thus mist or frost formation [174], the coating of oven windows [175] or incandescent lamp glasses [176] to serve as a heat mirror, thus increasing the efficiency of these objects and lowering their external temperature.

Naturally, many applications are based on the transparent/conducting duality of TCO materials. For these applications, high ρ and high transparency (T) are required. One of the first important applications of TCO films was as transparent heater for defrosting the windshields of high-altitude bomber aircrafts during the Second World War [161]. The use of TCO films as transparent heaters remains very common for window defrosting in modern aircrafts as well as in supermarket freezers, e.g. [175]. Today, TCO films are mostly used as transparent electrodes (TEs) in many devices. Among them, flat panel displays (FPDs) represent the major use of TCO films as TEs [177]. TCO films are also used as TEs in organic light-emitting diodes (OLEDs), touch screens, electrochromic windows and, of course, PV devices, including TF, perovskite (PK), SHJ, PK/SHJ tandem and carrier-selective passivated contact (CSPC) solar cells [178]. Other applications that rely on the transparent/conducting duality of TCOs include their use as electromagnetic interference shielding [179], antistatic films [153] and transparent TF transistors [180].

Apart from the basic properties on which the above-mentioned applications are based, TCO films shall meet other requirements, depending on the considered application: surface morphology, mechanical properties, fabrication constraints (low-temperature deposition, low bombardment by energetic species during deposition or 'low-damage' deposition, ability to be etched for patterning purposes), adapted interfaces with adjacent layers, environmental stability (stability in air, in humid environment, in damp heat (DH)), thermal stability, chemical stability and cost considerations. Table 5.2 summarizes these requirements, as well as basic optical and electrical properties, for the most common applications of TCO films.

5.2.2 Properties and characterization of TCOs

TCO properties can be classified in three categories: (1) structural and morphological properties, (2) electrical properties, (3) optical properties.

It is necessary to study and characterize the three categories of properties to have a good global understanding of TCO TF properties. Indeed these three groups

Table 5.2 Main requirements for the most common applications of TCO films

	Applications based on spectral selectivity			Applications based on transparent/conducting duality				
					Transparent electrodes			
	Winter window	Summer window	Heat mirrors (oven, lamps)	Transparent heaters	FPDs	OLEDs	Touch screens	Solar cells
Optical properties	High T_{vis} High R_{IR} ($3 < \lambda < 50$ μm)	High T_{vis} High R_{IR} ($0.7 < \lambda < 50$ μm)	High T_{vis} High R_{IR} ($0.7 < \lambda < 50$ μm)	High T_{vis}	High T_{vis}	High T_{vis}	High T_{vis} Low R_{vis}	High $T_{UV-vis-nIR}$ ARC (adapted refractive index and thickness)
Electrical properties				Low R_{sh} (<100 Ω/sq)	Medium R_{sh} (30–300 Ω/sq)	Low R_{sh} (<100 Ω/sq) low ρ_c with active layer (adapted Φ_w) low ρ_c with metal (high enough N)	Med-high R_{sh} (100–1 000 Ω/sq)	Low R_{sh} (<100 Ω/sq) low ρ_c with active layer (adapted Φ_w) low ρ_c with metal (high enough N)
Morphology	Smooth	Smooth	Smooth	Smooth	Smooth	Very smooth	Very smooth	Smooth for c-Si-and PK-based solar cells rough for TF solar cells
Mechanical properties	Mechanical resistance and durability	Mechanical resistance and durability	Mechanical resistance and durability	Mechanical resistance and durability	Bendability (on flexible substrates)	Bendability (on flexible substrates)	Mechanical resistance and durability bendability (on flexible substrates)	Bendability (on flexible substrates)
Fabrication constraints					Etchability	Low temperature (<200 °C) low damage	Etchability	Low temperature (<200 °C for SHJandTF,<150 °C for PK) low damage (SHJ, PK, CSPC)

(Continues)

Table 5.2 Continued

	Applications based on spectral selectivity			Applications based on transparent/conducting duality				
					Transparent electrodes			
	Winter window	**Summer window**	**Heat mirrors (oven, lamps)**	**Transparent heaters**	**FPDs**	**OLEDs**	**Touch screens**	**Solar cells**
Stability	Environment UV	Environment UV	High temperatures chemicals	Environment UV	Chemicals (common household chemicals)	Chemicals	Chemicals (common household chemicals)	Environment UV chemicals (PK) high temperatures (metallization process)
Costs	Low	Low	Low	Low	Low	Low	Low	Low

λ, wavelength; ARC, antireflective coating; IR, infrared; N, charge carrier concentration; R, reflectance; Rsh, sheet resistance; T, transmittance; UV, ultraviolet; nIR, near-infrared; vis, visible spectral range; Φw, work function.

of properties are strongly interrelated. The structural and morphological properties of TCOs are the basic material properties from which their electrical properties are derived, as well as their fundamental optical properties. In addition, the electrical properties are the source of major optical features of TCO materials. This is why we are going to present the different properties of TCOs in this logical order.

5.2.2.1 Structural properties of TCOs

The structural properties of TCOs mainly include the morphological aspect of the grains and their average size, the crystallinity of the film, the crystalline orientations if applicable, and the lattice parameter.

Crystallinity and crystallographic orientation

The crystallinity of a TCO TF is a primordial information: it describes the degree of structural order in the film. We can essentially distinguish two crystalline phases: amorphous phase and polycrystalline phase [181]. While the amorphous phase does not display any long-range order, the polycrystalline phase is constituted of many small crystals, named crystallites or grains separated by the so-called grain boundaries (GBs), which are very thin regions of high disorder located between two grains. Single-crystal TF is much more complicated to fabricate. A TF can be entirely constituted of one of the above-mentioned phases. But also, a coexistence between amorphous and polycrystalline phases is possible, e.g. in the case of incomplete crystallization. It is important to mention that an amorphous TCO film may have electrical and optical properties equivalent to those of a (poly) crystalline film.

Crystallographic orientation is the orientation of a crystallite with respect to the substrate (or to the growth direction). A limited distribution of crystallographic orientations, tending towards a single preferred orientation, indicates a better structural quality due to a good control of crystalline growth. On the contrary, a fully random distribution of crystallographic orientations may indicate the presence of many structural defects, in particular numerous GBs due to the presence of many small randomly oriented grains. This can be detrimental to the electrical properties (see Section 5.2.2.2c). X-ray diffraction technique (XRD) and scanning electron microscopy (SEM) (see Figure 5.7) allow to evaluate the crystallinity of a TCO film and determine the orientations of crystalline phases.

Lattice parameters

Lattice parameters are quantities used to describe the physical dimensions of a unit cell in a crystal lattice. They can also be determined by XRD by fitting a simulated diffractogram to the experimental measurement. For a cubic structure, the lattice parameter a is directly proportional to the lattice spacing d and can be simply calculated by a combination of the Bragg law with the plane spacing equation [183]. The calculated values of d and a can also help to estimate if the material is under stress or not, by comparing them to the ones of the theoretical powder pattern. If the calculated value of d (or a) is equal to that of the theoretical pattern, this indicates that the material has crystallized without stress, suggesting a good crystallographic quality potentially inducing better electrical properties. If T (or a) is different from the theoretical value, this means that the crystallized materials exhibit residual

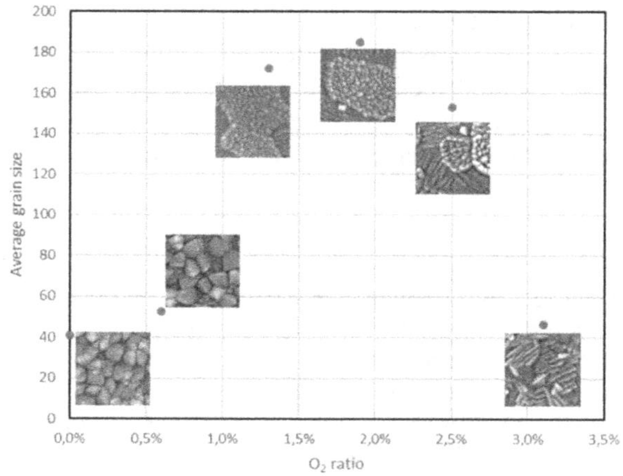

Figure 5.7 SEM images with ×300 000 magnification and grain size measured from SEM images of ITO films on c-Si substrates for different O_2 concentrations in gas mixture during deposition [182]

stresses (too high density, stress induced by the deposition technique) [184]. In this case, electrical properties may be negatively impacted.

Average grain size

The average grain size can be determined by SEM observations, evaluating the average number of grains per unit area. It is also possible, though less accurate, to evaluate the grain size from the shape of the peaks of an XRD measurement with the Debye and Scherrer method [183]. A large average grain size, in the range of several hundred nanometres, may contribute to a significant increase of the μ of a TCO film (see Section 5.2.2.2c) compared to a film constituted of smaller grains [179]. Figure 5.7 displays an example of SEM images and calculated grain size as a function the O_2 flow during ITO deposition.

5.2.2.2 Electrical properties of TCOs

(a) Resistivity ρ/conductivity σ

The conductivity σ and its inverse resistivity ρ are the fundamental quantities describing the electrical properties of a material. ρ (resp. σ) quantifies the ability of a material to resist (resp. conduct) electric current. For TF, another convenient quantity to describe this ability is the sheet resistance R_{sh} that is related to the resistivity and conductivity by:

$$R_{sh} = \rho/t = 1/\sigma t$$

with t the film thickness. In solar cells, TCO TF must have excellent electrical properties to collect and transport the photogenerated charges. To date, In-based TCOs display the lowest ρ values: R_{sh} values below 10^{-4} Ω cm have been demonstrated for ITO films deposited by radio-frequency magnetron sputtering (RF-MS) [185] and PLD [171, 172, 186, 187].

However, as we will see in more detail below, the electrical and optical properties of a TCO are strongly interrelated, so that defining the electrical behaviour of a TCO solely by its resistivity does not allow this complexity to be taken into account. As explained in the introduction, there are two important quantities that control ρ (or σ): the charge carrier mobility μ and concentration N (see (5.1)). Indeed, increasing μ and/or N directly decreases ρ (or increases σ).

μ, N and ρ can be determined by Hall effect measurement coupled with the Van der Pauw method knowing the thickness of the TCO. R_{sh} can also be measured by the four-point probe technique.

(b) Carrier concentration *N*

As mentioned in the introduction, the charge carrier concentration N is an essential parameter to control in order to obtain a TCO TF with the desired characteristics for a particular use.

For a doped oxide to have a metallic behaviour, its carrier concentration must exceed a critical density N_C, calculated from Mott's criterion with the effective Bohr radius of the material [188]. For ITO, N_C is about 2.3.10^{19} cm^{-3} [189]. Usually, this limit is reached regardless of the deposition method.

There are several doping sources to increase N that can be intrinsic or extrinsic. In n-type TCOs, an important intrinsic doping source comes from a deviation from stoichiometry, namely, oxygen deficiency [190]. Oxygen vacancies ($V_O^{\odot\odot}$ following the Kröger and Vink notation) in the lattice release each two electrons to the CB, thus increasing N. The more the TCO TF is reduced (with a precise control of the oxygen content in the atmosphere during deposition or post-deposition annealing), the more the number of N increases, inducing an increase in N. The other side of the coin with oxygen is that it may be integrated as interstitial oxygen $O_i^{''}$ into the structure, especially when introduced in excess, thus trapping two electrons per site, resulting in a decrease in N [190].

It is also possible to increase N by introducing extrinsic dopants, usually metallic atoms of higher valence than the host metallic cation. In the case of ITO, the host metallic cation is indium (In^{3+}) and the metallic dopant is tin (Sn^{4+}). The substitution of an indium atom by a tin atom releases an electron per site. It is possible to improve the doping efficiency by introducing impurities with an even higher valence than tin (like tungsten $W^{\odot\odot\odot}$ or molybdenum $Mo^{\odot\odot\odot}$) [181, 191, 192].

Many investigations on hydrogen doping have also been realized. They show that hydrogen influences N in multiple ways: interstitial hydrogen $H_i^{\odot}$ as well as substitutional hydrogen at an oxygen site $H_i^{\odot}$ may contribute to N. Hydrogen can also passivate structural defects possible electron traps, thus neutralizing them and allowing more electrons to be released to the CB [193, 194].

Depending on the material, one or the other doping source may prevail. In the case of ITO, it seems that electron generation is mostly caused by oxygen vacancies [190, 195, 196]. For other In_2O_3-based TCOs, it has been demonstrated that the single-charge donors (e.g. $Me^{\odot}$ with Me: Ti, Zr, Ce, $H_i^{\odot}$, $H_O^{\odot}$, etc.) are the main contributors to N, as they are energetically more favourable than μ_{gb} in these materials. It is interesting to mention that, since their conductivity is not that strongly related to oxygen, they are more stable and resistant to oxygen-rich treatments during or after deposition [181, 194].

However, N cannot be increased at any cost: as we will see further, increasing N generates more light absorption in the nIR range, and at some point may also impact the visible transparency of the TCO film (cf. optical properties §). If the considered application requires a high T over a broad range (which is the case for the TE of solar cells), N should not be too high and this may affect the μ. Hence, a compromise between electrical and optical properties is necessary and needs to be adapted depending on the considered application [192, 197, 198].

(c) Mobility, μ

Fortunately, there is an interesting way to get around or at least improve that compromise [199]. Indeed, it is possible to maintain a good σ level by compensating a reasonable N reduction by an improved μ_{ii}, thus releasing the constraint on optical properties.

Charge carrier mobility is a quantity that measures how easily charge carriers can move in a conductive material.

There are different scattering mechanisms limiting μ. Here, we will list only the most recurring scattering mechanisms that play a role in TCO films. The components of the mobility resulting from the different scattering mechanisms are linked together by Matthiessen's rule. This equation indicates that the lowest μ predominates:

$$\frac{1}{\mu} = \sum_i \frac{1}{\mu_i} \tag{5.2}$$

(i) Scattering by ionized impurities, μ_{ii}

Ionized impurities are essentially constituted of ionized dopant atoms or charged defects that disturbs the electronic transport by Coulomb interaction with the free carriers. They are an important source of scattering in degenerately doped SCs, impacting μ and thus ρ. The mobility component due to ionized impurity scattering ρ could be calculated by Brooks [200] and Dingle [201]. Then this model was improved by Pisarkiewicz *et al.* [202], Kim *et al.* [203], Young *et al.* [204] and Look *et al.* [205]. Thanks to this model, the independence of μ_x with the temperature was demonstrated as well as its dependence with$N$.

(ii) Scattering by neutral impurities, μ

Neutral impurities result from dopant atoms that remained inactive because they were not correctly incorporated into the lattice or because their ionization energy is too high to be activated at room temperature. Also, if the dopant concentration exceeds its solid solubility limit in the considered TCO material, part of the doping impurities will form neutral complexes in the structure [181, 191, 206]. Under specific conditions (high doping concentration, high oxygen incorporation), the mobility component due to neutral impurities μ is considered to significantly influence the total mobility [206]. This mechanism is not dependent on the temperature and on N so that the value of μ is not easily accessible by experimental measurements. It is, therefore, necessary to calculate it from theoretical equations [207, 208].

(iii) Scattering by grains boundaries, μ_{gb}

GBs are a complex structure composed of a few atomic layers of crystallographic disorder. This disturbance in the crystallographic lattice results in a large number of electronic defects in the SC band gap acting as trap states for free charge carriers. These traps become electrically charged after the capture of free carriers thus creating a potential barrier that disturbs the transport of charges and, as a consequence, limits μ. Seto was one of the first to propose a model explaining this mechanism [209]. This model, based on the thermionic emission principle, was later completed by tunnelling effects in order to better explain the charge transport in degenerate SCs [210–213].

For TF with a thickness of <200 nm, the GB density is high and the mobility due to GBs μ has a strong impact on the global μ. The influence of μ_{gb} depends on a lot on the deposition conditions that will have an impact on the GB density (through the crystallographic quality of the layer), as well as on the number of trap states. The latter is the key to obtain a high μ and a high σ in n-type TCOs [147, 214]. A grain size significantly higher than the width of GBs is also of course necessary to minimize the effect of μ_{gb} on the total mobility.

GB scattering is temperature dependent. It is possible to estimate its contribution to the mobility from a temperature-dependent Hall effect measurement. With the help of the above-mentioned models and material-specific inputs, the trap state density can be calculated, and thus μ [182, 215]. It is also possible to easily and quickly evaluate which scattering mechanism is predominant, just by analyzing the shape of the temperature-dependent mobility curve [215].

(iv) Other scattering mechanisms

There are other scattering mechanisms such as phonons, structural defects (dislocations, stacking faults), etc. They will not be detailed here but details can be found elsewhere [215–218]. Phonon scattering is temperature dependent and becomes increasingly significant with increasing temperature. At room temperature, it is the predominant scattering mechanism for highly crystalline films (with very large grains or single-crystal films) with limited carrier concentration [181, 219, 220]. Scattering by structural defects is rarely taken into account in the description of transport properties in TCO films, on one hand because its effect is not that well known (stacking faults, point defects) and on the other hand because dislocation density seems to be low in TCO films [163].

(v) Improving, μ

Many researchers consider μ_{gb} to be a major parameter to improve the contribution of TCOs in many applications that require high σ and high T over a large range, in particular solar cells. μ is strongly linked to the structural properties of the material. For example, an initial amorphous TCO can become a highly crystalline TCO TF with an optimized μ [220] after controlled post-annealing steps. In the case of a TCO TF that has already crystallized during deposition, annealing allows to reduce defects and thus improve μ [221, 222].

It is also possible to improve μ by choosing appropriate dopant impurities. For In_2O_3, much higher μ values (>80 cm^2/Vs) are obtained replacing Sn by W [223], Ti [224], Mo [191] and Zr [225]. It seems that these types of impurities allow to decrease the scattering effect of interstitial oxygen anions by polarizing their electronic charges more strongly towards the doping impurity, thus screening the charge of the oxygen anion [198, 226]. As previously mentioned, hydrogen doping is also beneficial as it passivates structural defects [193].

Finally, control of oxygen incorporation is also important to achieve good values: a low oxygen concentration induces a high number of $V_O^{\odot\odot}$ leading to a strong μ_{ii}. On the contrary, if the oxygen concentration is increased during deposition, with the hypothesis that the TCO film keeps the same structural properties, the number of μ decreases, leading to an improvement of μ. However, a high oxygen concentration in the layer may also favour the formation of complexes acting as neutral impurities.

5.2.2.3 Optical properties of a TCO

Optically, a TCO TF has two specific fundamental behaviours: (i) TCOs have a forbidden wide energy band (>3.1 eV) gap. As such, they will absorb high energy photons, with an absorption edge lying in the UV, while being highly T over the Vis and nIR range. (ii) TCOs exhibit metal-like conductivity. All metals feature an optical behaviour due to the free electrons in their CB, which is explained through the Drude model [227, 228]. The electrons in the CB absorb low energy photons by realizing intraband transitions. This generates absorption in the nIR. The absorption curve shows a peak at a particular wavelength called the plasma wavelength λ_p. For wavelengths higher than λ_p, the material becomes more and more reflective. As N increases, the nIR absorption becomes more important and λ_p shifts towards shorter wavelengths. For very high N, free-carrier absorption can thus also affect visible transparency. This is why trying to obtain high μ values is so important to keep moderate N values and thus limit free-carrier absorption while keeping a good σ.

Optical indices

The optical constants are intrinsic properties of TCOs that best represent their optical properties. There are two optical constants:

- The refractive index $n(\lambda)$ allows to estimate the (anti)reflective behaviour of the layer.
- The extinction coefficient $k(\lambda)$ allows to estimate the absorption of the layer.

The optical constants are directly linked to the complex dielectric function of the material, thus explaining the intrinsic relation between electrical and optical properties. They can be estimated from a spectroscopic ellipsometry measurement by fitting the experimental results with an adequate model and an optimization algorithm. An example of n and k extraction for ITO layers fabricated with different O_2 contents is shown in Figure 5.8a.

Transmission, reflection and absorption

Another way to estimate the optical properties of a TCO layer in a quick, direct but more approximate way is to measure the T and the R and to deduce from

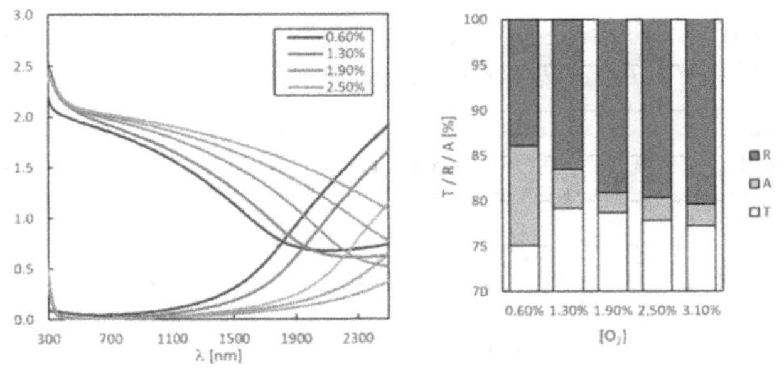

Figure 5.8 *(a) Optical indices (refractive index and extinction coefficient) extracted from ellipsometer measurements with a double Lorentz–Drude model on c-Si substrates for different O_2 concentration and (b) transmission, reflectivity and absorption measured by spectrophotometry on glass substrates for different O_2 concentration*

the previous measurements the absorption (A) of the TCO layer (see example in Figure 5.8b). The spectrophotometer is the common equipment used to measure these quantities. It is also possible to extract optical indices from spectrophotometer measurements [216, 229].

5.2.2.4 Thickness

We will finally shortly address the thickness, which is a fundamental parameter for thin layers that have an influence on all the properties that we have already described.

The layer thickness has a strong influence on structural properties. Indeed, a minimum thickness is generally needed for the layer to reach sufficient order in its structure. As a consequence, very thin layers (<20 nm) are generally more defect rich and present a higher GB density than thicker layers. As explained in Section 5.2.2.2, this will strongly impact the electrical transport. Not to mention the direct influence of the thickness value on the sheet resistance.

The influence of the thickness on optical properties is of course important. It has obviously a direct impact on the absorption of the layer. Also, if antireflective properties are required for the considered application, there is an optimal thickness that depends on the optical properties of the TCO layer itself, as well as on those of the substrate.

5.2.3 TCO materials for c-Si-based solar cells

While crystalline silicon (c-Si)-based technologies widely dominate the PV market [230], the main use of TCOs in PV applications relies on TF technologies: amorphous silicon (a-Si:H), amorphous/microcrystalline silicon (a-Si:H/μc-Si:H) tandems, cadmium telluride (CdTe), copper indium gallium selenide (CIGS), dye-sensitized and organic solar cells [231]. The reason for this is that the lateral transport of photogenerated charges in the two mainstream c-Si-based technologies, namely PERC and aluminium back surface field, is ensured by a highly conductive

Table 5.3 Most commonly used TCO materials and their deposition methods for different solar cell technologies, based on [163, 231]

Solar cell technology	TCO material	Deposition method
c-Si-based (SHJ, CSPC)	ITO	MS
Si-based TF	AZO	MS + post-etching
(a-Si:H, a-Si:H/μc-Si:H)	BZO, FTO, SnO_2	MOCVD
CIGS	AZO/ZnO	MS, MOCVD
CdTe	FTO	MOCVD
	ITO, Cd_2SnO_4	MS
Perovskite	ITO, FTO	MS

MOCVD, Metal-organic Chemical Vapor Deposition; MS, Magnetron Sputtering.

emitter. On the contrary, the electron and hole collection layers of TF devices generally display poor lateral conductivity and, therefore, require a transparent current-collecting electrode. Still, among c-Si-based technologies, TCOs are expected to play an increasing role in high-efficiency solar cells [232]. They have been already commonly used for years in a-Si:H/c-Si SHJ solar cells [233]. Recently, solar cells using CSPC based on polycrystalline silicon (poly-Si) layers is attracting increasing attention in the PV community and has rapidly shown world record efficiencies [106, 234] when implemented at the rear side with full-area metallization. The use of such contacts on both sides is thus a seducing idea [235]. But, like in SHJ solar cells, the high R_{sh} of the thin poly-Si films requires the use of a full-area transparent front electrode, role that can be fulfilled by a TCO film [236]. Other c-Si-based approaches might also need the use of TCOs as listed in [232]. Table 5.3 lists the main solar cell technologies relying on TCO materials for current collection, with the corresponding typically used materials and deposition methods.

5.2.3.1 Requirements for TCO films in c-Si-based solar cells

The main requirements of TCO films used as TE in solar cells are already listed in Table 5.2. Table 5.4 gives more details on the specific requirements for the TEs of c-Si-based solar cells. They are nearly the same for the two considered technologies, except for the specifications involving temperature limits: while the a-Si:H layers of SHJ cells set a rather low temperature limit (<200 °C) for all subsequent processing steps, poly-Si layers can perfectly withstand high temperatures. While this loosens the constraint on TCO deposition temperature, it also makes it necessary to address the thermal stability of TCO films, as high-temperature metallization process can be considered for this technology.

While TF solar cells make wide use of ZnO- and SnO_2-based TCOs [231], c-Si-based solar cells incorporating TCO layers, i.e., mainly SHJ cells, massively rely on In-based TCOs. Still, concerns regarding In cost and availability makes the In-free alternative interesting.

Table 5.4 Specific TCO requirements for TCOs as transparent electrodes in c-Si-based solar cells

	SHJ	Poly-Si/SiOx
Optical properties	High T, low k ($300 < \lambda < 1200$ nm) $n \approx 2$ and $t \approx 70$ nm for antireflective properties	
Electrical properties	Low R_{sh} (<100 Ω/sq)	
	Low ρ_c with active layer: → on n-side: low Φ_w → on p-side: high Φ_w	
	low ρ_c with metal → high enough	
Fabrication constraints	Low damage temperature <200 °C	Low damage
Durability	UV stability, stability under damp heat, humidity, air thermal stability for interfaces and metallization process:	
	≈200 °C	<350 °C
Costs	Low	Low

5.2.3.2 In-based TCO films

MS ITO is currently the most used TCO material for the front and back TE of SHJ cells. Apart from being one of the most transparent and conductive TCO materials, its reliability as well as its compatibility with large-scale production have been largely demonstrated with its extensive use in FPDs. ITO is generally deposit from mixed In_2O_3–SnO_2 ceramic targets, with different SnO_2 contents, depending on the properties required for the considered application. For PV applications, SnO_2 target contents between 3% and 10% are usually used. For such contents, T of the ITO films understandably increases with increasing SnO_2 content due to the higher Sn doping. μ remains in the 20–40 cm²/Vs range, being slightly higher at lower SnO_2 content, so that resulting ρ tends to be of the same order of magnitude [163, 237, 238]. On the other hand, nIR absorption increases with increasing SnO_2 content due to increased free-carrier absorption thus encouraging the use of low SnO_2 target contents for solar cell application.

Still, due to its limited μ values, sputtered ITO needs to display sufficiently high N values in order to be sufficiently conductive, thus leading to high parasitic absorption in the nIR range. Other high-quality deposition techniques like PLD [171] or ion plating [170] lead to μ values in the same range. In order to obtain a better compromise between optical and electrical properties, via high μ values, research has shifted towards alternative doping elements other than Sn (Table 5.5). High μ values were obtained with IWO (>80 cm²/Vs) [223] and IZrO (100 cm²/Vs) [240], e.g. besides, hydrogen, alone or in association with other dopants, leads to very high μ values: 130 cm²/Vs for IOH [194], up to 140 cm²/Vs for ICO:H [239]. These exceptional μ values allow to keep free-carrier densities below 3×10^{20} cm^{-3} while maintaining a very low R_{sh} in the range of 30–40 Ω/sq. As a consequence, the A of these films remains below 2% over the whole visible and nIR range [178]. Most of these materials have been successfully integrated in SHJ cell, showing an efficiency

Table 5.5 List of record μ values obtained on In-based TCO layers at low temperature (<200 °C)

Material	**Dopant**	**Deposition method**	**Temperature (°C)**	**Thickness (nm)**	**ρ (Ω.cm)**	**μ (cm^2/Vs)**	**N (cm^{-3})**	**References**
ICO:H	Ce, H	RPD	150	100	$2.2.10^{-4}$	141	$2.0.10^{20}$	[239]
IOH	H	RF sputtering	RT, post-annealing 200	240	$2.7.10^{-4}$	130	$1.8.10^{20}$	[194]
IZrO	Zr	RF sputtering	RT, post-annealing 200	100	$2.5.10^{-4}$	100	$2.5.10^{20}$	[240]
IWO	W	RPD	200	80	$2.3.10^{-4}$	89	$3.0.10^{20}$	[223]

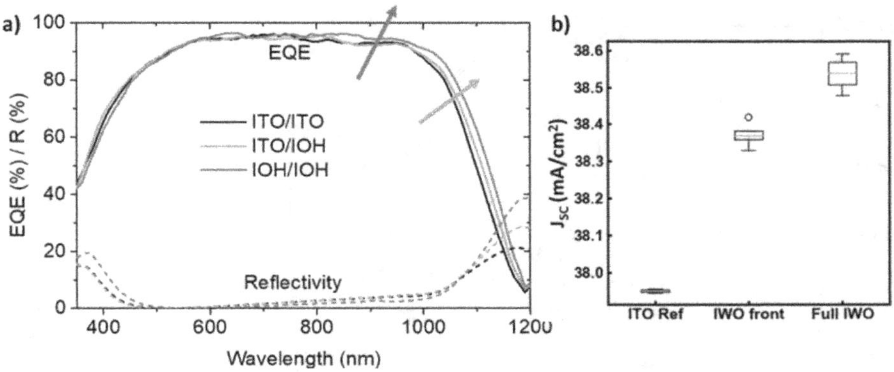

Figure 5.9 *(a) External quantum efficiency (solid lines), as well of front side reflectivity (dashed lines) of SHJ solar cells integrating ITO and/or IOH as transparent electrodes and (b) short circuit current density of SHJ solar cells integrating ITO and/or IWO as transparent electrodes [242]*

improvement compared to ITO TEs, essentially thanks to an increase in short-circuit current (J_{sc}). External quantum efficiency measurements could highlight the power conversion gain in the nIR range thanks to the lower absorption of these TCOs in these regions: this was observed for IOH [241], IZrO [240] and ICO [239]. IOH and IWO were also successfully implemented on CEA-INES cells, with similar conclusions (Figure 5.9).

Despite the very promising results achieved with these high materials, MS ITO has remained until now the most extensively used solution for the TEs of SHJ cells. The main reason for this being the lack of industrially viable solutions to deposit these alternative materials: sputtering target manufacturers have been facing difficulties to propose larger-area products based on these materials, either for technical or for economic reasons. Alternative deposition methods allowing more flexibility regarding source material like reactive plasma deposition (RPD) are still economically not competitive with the widely spread MS (see Section 2.4). But very recent results based on a commercially available In-based rotatable target have shown a promising increase of 0.4 point of the efficiency of SHJ cells when replacing standard MS ITO by this new material, whose exact composition remains undisclosed [243]. This gain was possible thanks to an improvement of the short-circuit current, as well as a reduction of the series resistance of the solar cells.

5.2.3.3 In-free TCO films

Despite the interest in removing costly In-based TCOs from SHJ solar cells, there are relatively few research works reporting the utilization of In-free TCOs as TEs of SHJ cells. Most of them rely on ZnO-based materials [244–247]. ZnO-based materials are attractive for large-scale applications like solar cells, as Zn is more than 1 000

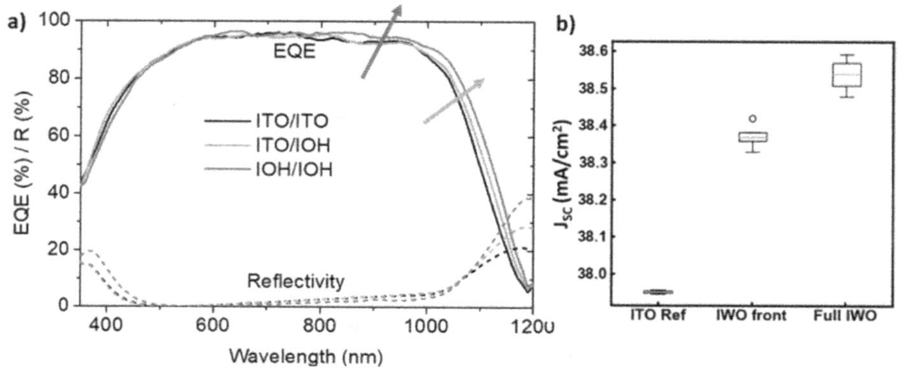

Figure 5.10 *Relative evolution of SHJ module performance depending on front/back TCO type and encapsulation material (polyolefin, PO: blue or ethylene-vinyl acetate, EVA: orange) after 1 000 h (light colours) and 2 000 h (dark colours) of DH treatment. The horizontal dashed line represents the IEC-61215 norm of <5% module performance degradation after 1 000 h of DH treatment [242].*

times more abundant than In in the earth crust [163]. However, Zn-based films have poorer electrical transport properties than In-based films, in particular due to the low μ they display when deposited as TF (<100 nm), as required in c-Si-based solar cells (Table 5.4) [248, 249]. Still, when integrated as front electrode of SHJ cells in a rear-junction configuration, ZnO:Al (AZO) performs similarly as ITO [249]. Good results with sputtered AZO were also obtained on poly-Si-based solar cells [250].

However, ZnO-based TCO films also suffer from stability issues. Its resistivity strongly increases upon annealing, even at low temperatures (<200 °C). The impact of annealing becomes stronger at higher annealing temperatures [232, 251]. This phenomenon is attributed to oxygen adsorption at the GB leading to an increased potential barrier and thus stronger GB scattering, resulting in strongly decreased μ [251]. This is critical for the use as TE in c-Si-based solar cells, as the TCO layers must withstand the curing step of the metallic ink used for the fabrication of the metal contacts (Table 5.4). Similarly, ZnO is sensible to humidity. As such, strong degradation of module performance under DH conditions is observed when replacing ITO by ZnO:Al [244], despite the encapsulation of the solar cell. Similar observations could be done at INES with different encapsulation materials (Figure 5.10).

5.2.4 Conclusions

Even if TCOs have been of enormous importance for the technological development of several applications in the last decades, in the PV field they were only used in TF technologies since traditionally c-Si solar cells used the bulk for lateral transport. The deployment of SHJ solar cell technology with the new lateral transport challenges puts in the scope of TCOs with different requirements than those of TF technologies and other applications.

SHJ is today a strong candidate for mass production due to its high-efficiency potential and the lean manufacturing process consisting of only four main steps [252]. Record single-junction efficiencies for crystalline silicon-based lab solar cells of 25.1% and 26.7% for two-side and all-back contacted devices, respectively, make this cell concept very attractive [253, 254]. This solar cell concept achieves high open-circuit voltages (V_{oc}) above 740 mV through passivating selective contacts using mainly a very thin stack formed by an intrinsic hydrogenated amorphous silicon (i)a-Si:H coupled with a n- or p-type-doped layer deposited on both sides of the silicon wafer acting as carrier selective contacts for holes and electrons, respectively [255]. Even if it has been always said that the key of the technology was the c-Si/a-Si:H interface and the respective stacks, the TCO layers play a major role to achieve very high efficiency devices and make them competitive in mass production. The requirements of the TCO layers as well as their interfaces with adjacent materials are extremely important to assure high efficiency and play a major role in all solar cell parameters, V_{oc}, FF and J_{sc}:

- minimizing series resistance: bulk, contact with metallization and with underlying layers
- minimizing parasitic absorption
- avoiding passivation degradation (or even improving passivation)
- improving carrier collection at each side of the device
- avoiding chemical reaction with encapsulant
- stability under light, UV, DH, humidity, air.

Moreover, the TCO growth and properties depend on the underlying layers [249] and their own constraints (for instance, temperature, bombardment, hydrogen effusion, etc.). Last but not least, cost and use of raw materials are drivers on SHJ deployment. All these requirements make the development of TCO specific for SHJ a challenge today. In Table 5.6, there is a summary of reported SHJ cells efficiencies >23% with detailed information on front and back TCO as well as metallization. First, it is important to comment that many publications do not specify the material nor the deposition technique, which reveals the importance of this development on the final efficiency. Second, the mostly used TCO for mass production of c-Si-based solar cells today is MS ITO, which makes sense because of the massive use of MS and ITO in other applications. Third, an interesting approach for high efficiency devices has been addressed using bilayers combining classical TCOs with antireflective coating, which allows improved efficiency with potential cost reduction for raw materials. Meyer Burger also presented this approach using thin ITO and SiN_x layers [264]. Finally, promising materials arise in research laboratories but implementation is still a challenge in mass production because of the evidence cited before. In any case, there is room for improvement in the whole TCO development combining material quality, deposition technique, mass production, cost and device and module performances.

Table 5.6 Best reported SHJ efficiencies (>23%) comparing front and back TCOs and metallization schemes

	Front TCO	Deposition technique	Front metallization	Back TCO	Deposition technique	Back metallization	η (%)	References
Hanergy	ITO 90/10	MS	Ag SP	ITO 90/10	MS	Ag SP	25.11	[253]
Kaneka	Not cited	Not cited	Cu plating	Not cited	Not cited	Ag	25.1	[254]
CEA-INES	ITO 97/3	MS	Ag SP	Not cited	MS	Ag SP	25	[256]
Panasonic	Not cited	Not cited	Ag SP	not cited	Not cited	Ag SP	24.7	[257]
CEA-INES	ITO 97/3	MS	Ag SP	ITO 97/3	MS	Ag SP	24.6	[258]
CSEM	IWO	MS	Cu plating	ITO	MS	Ag SP	24.5	[259]
EPFL	IZrO	MS	Ag evap.	ITO	MS	Ag MS	24.1	[260]
CIC	ICO	HPE	Ag SP	ICO	HPE	Ag MS	24.1	[239]
Meyer Burger	Not cited	Not cited	Ag SP	Not cited	Not cited	Ag SP	24	[261]
Mitsubishi	IOH	MS	Ag SP	IOH	MS	Ag SP	23.5	[262]
HZB	AZO/SiO_2	MS	Ag SP	AZO	MS	Ag MS	23	[244]
Fh-ISE	ITO 97/3	MS	Ag SP	ITO 97/3	MS	Ag SP	23	[263]

HPE, high-density plasma-enhanced evaporation; MS, magnetron sputtering; SP, screen-printing.

Concerning other c-Si-based technologies, the rush for TCO integration started in the last decade with the appearance of the passivated contacts development [265, 266]. TCOs appeared to be a good alternative to help in lateral conduction without degrading interface passivation due to classical firing. In this technology, TCOs play an important role to preserve and improve passivation thanks to high-temperature process not possible in SHJ (to avoid any epitaxy of a-Si:H layers). Again, MS seems to be the most common deposition technique and ITO the most used material. Even if efficiencies are still lower than for SHJ cells, the promising short-term results shown in Table 5.7 show that the implementation of TCOs on passivated contacts can also be a suitable option to achieve very high passivation and remarkable performances on final devices. Again, the interaction of the TCO within the subjacent layers seems to be a key parameter to define the best deposition technique, the best process and material properties for each technology [236], which makes the TCOs research open. Thus, many challenges and opportunities for TCO remain to be explored and implemented in the field of c-Si-based solar cells for the next generation of devices.

5.3 Thin metal compound contact interlayers

5.3.1 Introduction

The historical advancement of crystalline silicon (c-Si) solar industrial technologies can be linked to a series of key architectural upgrades, each of which addressed a loss mechanism dominating the device's performance at the time. It is widely accepted that one of the next major advancements will be to address the lossy silicon–metal contacts. As discussed in other chapters of this book, the directly metallized c-Si surface is known to introduce a range of negative effects, most prominently high surface recombination and strong Fermi-level pinning (FLP), which limit the ability of the contact regions to selectively collect just one of the two carrier types. Both of these issues are partially negated by introducing heavy doping underneath the metal contacts. This strategy reduces the surface minority carrier concentration (one of the limiting factors for contact recombination J_{0c}) and the surface potential barrier width (allowing carrier tunnelling across the silicon–metal interface barrier – reducing contact resistivity ρ_c). Despite the improvement heavy absorber doping affords, its efficacy is limited, particularly in relation to the J_{0c}, which is limited to a minimum value of ~400 fA/cm^2 due to the negative heavy doping effects such as Auger recombination and band gap narrowing [273, 274].

An alternative strategy to heavy doping is to address the metal–silicon interface issues directly, rather than minimizing their effects. This can be achieved by the insertion of thin interlayers between the outer metal contact and the c-Si to reduce/eliminate the effects of FLP and surface recombination. This strategy is known by many names including 'passivated contacts/junctions', 'heterocontacts/heterojunctions' and 'carrier-selective contacts'. It includes the well-known doped SHJ and doped polysilicon (poly-Si) junction technologies, discussed elsewhere in this book, both of which have been demonstrated in cells with efficiencies above 26% [275, 276]. Another option, which is the topic of this chapter section, involves the

Table 5.7 Best passivated contacts efficiencies (>20.5%) integrating TCOs

	Front collector	Front TCO	Rear collector	Rear TCO	Metal	η (%)	References
EPFL	SiO_x/SiC_x(n+)	IZrO (MS)	SiO_x/SiC_x(p+)	IZrO (MS)	Ag SP front/MS rear	22.6	[267]
ISFH	SiO_x/Poly-Si(n+)	ITO (MS)	SiO_x/Poly-Si(p+)	ITO (MS)	Ag SP	22.3	[268]
ISFH	SiO_x/Poly-Si(n+)	AZO (MS)	SiO_x/Poly-Si(p+)	AZO (MS)	Ag SP	21.6	[269]
CEA-INES	SiO_x/Poly-Si(n+)	ITO (MS)	SiO_x/Poly-Si(p+)	ITO (MS)	Ag SP	21.5	[270]
Fh-ISE	SiO_x/Poly-SiC_x(p+)	ITO (MS)	SiO_x/Poly-SiC_x(n+)	No	Ag SP	20.8	[271]
TU Delft	SiO_x/Poly-Si(n+)	IFO:H (MS)	SiO_x/Poly-Si(p+)	IFO:H (MS)	Ag SP	20.6	[272]

MS, magnetron sputtering; SP, screen printing.

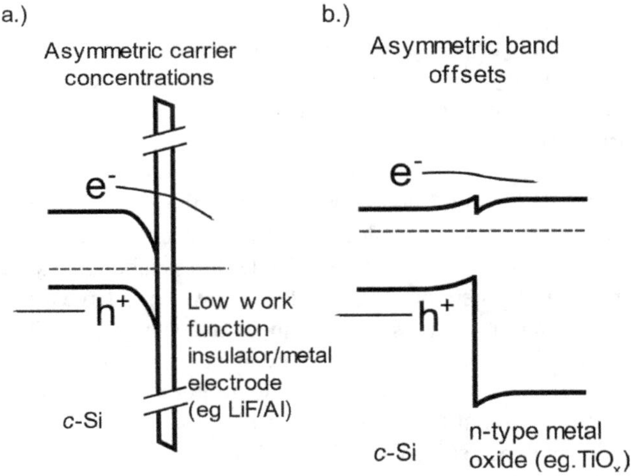

Figure 5.11 Metal compound interlayer concept. Band diagram depiction of contact carrier selectivity via (a) formation of asymmetric carrier concentrations and (b) formation of asymmetric band offsets.

insertion of thin metal compound interlayers, such as TiO_2, to mitigate metal–silicon interface losses [277].

5.3.1.1 Contact carrier selectivity

Whilst the exploration of possible metal compound interlayers has greatly expanded over the last 6 years, many of these trials are based on old ideas [278–285] and are assisted by previous advancements made in surface passivation understanding and techniques [286–289]. Others are borrowed from standard structures in other absorber type devices, particularly from organic electronics, where absorber doping is not so easily achieved and hence contact interlayers have been heavily developed [290]. A large diversity in the structure of these interlayers has arisen due to a wide range of mechanisms available to improve metal–silicon interfaces. Here we broadly divide these into two areas, which are commonly implemented simultaneously: manipulating surface carrier concentration profiles within the silicon absorber, an example of which is shown in Figure 5.11a, and creating asymmetries in conductivity presented to the two carriers at the silicon surface and through the interlayers, as depicted in Figure 5.11b.

5.3.1.2 Achieving asymmetric carrier profiles

The formation of a surface potential ψ_s, which alters the surface carrier concentrations, n_s and p_s, can arise due to several different mechanisms. Among these, the most accessible when fabricating contacts are (i) chemical potential differences (commonly known as work function differences) between the silicon absorber and the interlayer(s)/metal and (ii) a fixed charge density within the interlayer film(s) or at the surface of the c-Si wafer. These two elements can alter the magnitude and polarity of ψ_s, which if sufficiently large can result in the exclusive accumulation of one carrier type to the surface, similar to the case of heavy surface doping discussed above. A ψ_s that favours the build-up of one carrier (e.g. electrons) increases the conductivity for that carrier en route to the contact surface and can, to a small degree, assist in lateral transport [291, 292]. Conversely, the same ψ_s would result in a low surface concentration for the blocked carrier (e.g. holes) relative to the concentration within the bulk of the absorber, due to the law of mass action. This lower blocked carrier concentration reduces the rate of Shockley–Read–Hall surface recombination. Under illumination, the situation is more complicated and ψ_s will generally decrease [293]. However, a large asymmetry in the surface carrier concentrations can still exist under one sun illumination if the initial magnitude of the equilibrium ψ_s is sufficiently large. For example, under idealized conditions a Q_f of $>-2 \times 10^{12}$ cm^{-3} or a work function of >5 eV, both values that are easily attainable, would be required to maintain a surface p_s/n_s ratio > 10^8 (under one sun conditions on a lowly doped silicon absorber) [277].

In practice, however, larger values are required due to unavoidable non-idealities. These stem from the nature of the contact interface with the c-Si absorber, which dictates much of the recombination and resistive behaviour of the contact. Reducing recombination is of primary importance and creating asymmetries in carrier concentration is only one component. Surface recombination is also strongly dependent on the surface trap concentration N_{it} and their effectiveness at capturing the blocked carrier, quantified through the electron and hole capture cross sections c_e and c_h. The trap concentration can vary by orders of magnitude depending on how the c-Si surface is treated. For example, the directly metallized c-Si surface is known to host an extremely high concentration of surface traps, likely above 10^{13} cm^{-2} [294]. Fortunately, focused surface passivation research in the last 15 years has shown that a reduction in the interface trap density to 10^9–10^{11} cm^{-2}, i.e. one trap for every ~60 000 surface atoms, is achievable using several interlayer films with the assistance of hydrogenation [286, 289, 295, 296].

In a related manner, the nature of the interface also dictates how effectively we can manipulate the ψ_s using the contact's work function. For example, when a metal is used directly, the interface is known to suffer from FLP to c-Si's charge neutrality level (CNL, typically ~0.35 eV above the valence band) and hence the ψ_s cannot be well controlled [297]. Whilst the precise molecular origin of FLP remains an open discussion within the broader scientific community [298, 299], from a practical perspective, by inserting thin interlayers, as thin as 3 Å [298], between the outer metal contact and the c-Si, the effects of FLP (and surface recombination) can be reduced [300–303]. In this instance, part of the metal/c-Si work function difference will now fall across the interlayer, bringing into relevance the thickness of this interlayer and its dielectric constant. It is noted that due to the angstrom scale of interlayer thickness, the characteristics of this interlayer, including the dielectric constant, may not be reflected by the bulk properties of such films [304]. In addition to the above case, a semiconductor with a desirable work function, e.g. molybdenum trioxide MoO_3, can also be used in the place of the metal. In such a case, a substantial part of the balancing electric potential will also fall across this semiconductor unless it is degenerately doped or has a very large density of chargeable states. The use of an outer semiconductor rather than a metal has also been shown to reduce the effects of FLP [305] and can be considered particularly beneficial if it also acts as an additional source of hydrogen to assist in reducing the N_{it} [306, 307]. In many instances, however, the Fermi level remains partially pinned to the CNL of the passivating interlayer or outer semiconductor [308]. The degree of pinning is empirically quantified by the pinning factor S, taking a value between 0 and 1, where 1 signifies no pinning. Nevertheless, a number of interlayer contact systems have now been demonstrated on c-Si with pinning factors close to 1, including SiO_2 [305]. In addition, strong surface potentials can still be achieved in the case of non-ideal pinning factors by using metal compound materials/stacks with extreme work function values.

The use of a fixed charge density to manipulate the ψ_s at the c-Si surface also requires consideration of several additional effects to those discussed above. The fixed charge density is commonly described as a thin sheet of immobile charges within a dielectric film close to its interface with c-Si. These are generally attributed to molecular configurations, which either have a permanent fixed charge or can be charged by carriers from the underlying c-Si substrate [309]. The location of this charge is usually taken to be contained within the first few nanometres [291, 309], and hence in cases where an interlayer must be kept extremely thin to allow direct tunnelling, they may not develop the 'full charge' measured on thicker films. Regardless, this topic is being actively explored, including several preliminary studies focused on the use of polymer-based Lewis acids/bases to create a sheet of charge at the c-Si surface [310], which has been proposed for use in contact systems.

5.3.1.3 Introducing asymmetries at the interface

In addition to controlling the population of carriers at the c-Si contact surface, carrier selectivity can also be introduced, or further assisted, through the band structure and trap features of the contact itself. It has long been recognized that an ideal c-Si solar cell would comprise two wide band gap semiconductors asymmetrically straddled on either side of a c-Si wafer. In such a system, ideally the collected carrier of each heterocontact experiences no barrier en route to the outer metal contact, whilst the blocked carrier experiences a prohibitively large barrier for carrier transport, leading to obvious carrier selectivity. A perfect implementation of this approach is yet to be demonstrated, and typically, selectivity is achieved by film stacks [306, 307, 311], each layer of which performs a specific function. Whilst excellent surface passivation is achievable (indicating that the flux of the blocked carrier has been limited), typically small barriers to the collected carrier are unavoidable. Fortunately, c-Si solar cells with full/large-area contacts can tolerate relatively high ρ_c values reducing the impact of such barriers. Several mechanisms have been used to describe transport through/over these barriers, including thermionic emission, direct tunnelling and trap-assisted tunnelling whereby typically one mechanism dominates ρ_c. For example, in the simple case of thermionic emission, barriers below 0.15 eV result in a ρ_c with almost no impact on cell performance, and those above ~0.5 eV will limit ρ_c to at least 0.5 Ω cm^2 at room temperature, which will significantly detract from a cell's efficiency potential. In many cases, the use of wide band gap passivating interlayers is desired for their excellent temperature stability and passivation characteristics (e.g. SiO_2, Al_2O_3, etc.). In such a case, barrier heights > 0.5 eV are commonplace and hence transport cannot occur via thermionic emission. Instead, conduction can occur through the interlayer if it is sufficiently thin for direct tunnelling (i.e. <2 nm) or, e.g. if a large density of sub-gap states is involved [312, 313].

Several other strategies have been suggested to achieve selectivity at the contact including the asymmetric mobility contact [314], or the introduction of carrier selective defects into a passivating film [315, 316]. However, these ideas remain at the concept stage.

5.3.2 Materials

The earliest successful demonstration of interlayers for c-Si solar cells appears to be that of the MIS contacts utilized in c-Si MIS-inversion layer cells in the 1970s [281–284]. A diverse range of such devices was demonstrated with different metal and interlayer combinations [278, 285], but typically Al and SiO_x were the most successful. The SiO_x passivating interlayer was grown both chemically and thermally, with higher temperature oxides exhibiting greater device stability [317]. This standard c-Si(p)/SiO_x/Al design relied on both the low metal work function and fixed charges on the cell's front-side films to assist in carrier selectivity. The principal benefits of this technology are the simplicity of design, low-temperature fabrication and resultant low cost. However, despite reaching a conversion efficiency of 19.6% [318], focus was instead shifted to a parallel stream of research that combined MIS contacts and dopant diffused surfaces, commonly known as the metal-insulator-np or MINP cells [319]. This architecture held the world record efficiency for a short period in the 1980s [320] and shares some conceptual overlap with today's poly-Si/SiO_x-based contacts.

Since 2012, there has been a resurgence in interest towards the exploration of metal compound contact interlayers, motivated by the potential for reduced recombination, improved optics and simple low-temperature fabrication procedures. The improved optics originate from the wider range of available band gaps and refractive indices when using metal compound films, which can help to minimise parasitic absorption and possibly improve light trapping. This renewed interest has resulted in the emergence of several families of materials which have shown promise. Below is a brief discussion of some of the more successful families. Tables 5.8 and 5.9 provide a compilation of some of the best and electron and hole selective metal compound films demonstrated on c-Si to date, and Table 5.10 presents the best c-Si cell demonstrations using such films. Many of the higher performance devices were realized using dedicated c-Si surface passivating interlayers, which are also described in this section. It is interesting to note that despite the relatively short development time, all proof-of-concept cells in Table 5.10 were demonstrated at an efficiency above 20% and several have exceeded 22.5%. We note that compilations shown in Tables 5.8–5.10 do not include a description of similar approaches using organic or nanomaterial interlayers, a summary of such approaches can be found elsewhere [383, 384].

Table 5.8 Metal compounds successfully utilized in electron contacts for c-Si solar cells

Material	Dep. method	E_g	Stoich. (x=)	Underlayer(s)	Best passivation (thickness, wafer type, wafer doping)	Best ρ_c (thickness, wafer type, wafer doping)
TiO_x	ALD [321–325] Ebeam+Ox. [326]	3.35 eV[321]	x~2 [322]	c-Si	7 fA/cm^2 (10 nm, n-Si, 10 Ω cm) [324]	5 m Ω cm^2 (3.5 nm, n-Si, 1 Ω cm) [322]
				SiO_x (1.2 nm)	14 cm/s (3.5 nm, n-Si, 1 Ω cm) [323]	24 m Ω cm^2 (2.5 nm, n-Si, 1 Ω cm) [323]
LiF_x	Therm [327]	13.1 eV [328]	x~0.91 [329]	c-Si	7.7 × 10^4 fA/cm^2 (1.5 nm, n-Si, 1–3 Ω cm) [329]	1–3 m Ω cm^2 (1.5 nm, 1 Ω cm) [327, 329]
				TiO_x(1.5 - 6 nm) [321, 329]	15 fA/cm^2 (1.5 nm, n-Si, 1 Ω cm) [329]	5 m Ω cm^2 (1 nm, n-Si, 1 Ω cm) [321]
				a-Si (6 nm)	3.4 fA/cm^2 (1.5 nm, n-Si, 4.6 Ω cm) [329]	7m Ω cm^2 (1 nm, n-Si, 4.6 Ω cm) [329]
				a-Si:H (9 nm) /ZnO (140 nm) [330]	iV_{oc} = 735 mV (1.5 nm, n-Si, 2.1 Ω cm) [330]	390 m Ω cm^2 (1.5 nm, n-Si, 2.1 Ω cm) [330]
				a-Si:H (5 nm) /TiO_x (1.5 nm) [331]	iV_{oc} = 701 mV (0.6 nm, n-Si, 4 Ohm-cm) [330]	70 m Ω cm^2 (0.6 nm, n-Si, 4 Ω cm) [331]

(Continues)

Table 5.8 Continued

Material	Dep. method	E_g	Stoich. (x=)	Underlayer(s)	Best passivation (thickness, wafer type, wafer doping)	Best ρ_c (thickness, wafer type, wafer doping)
MgO_x	Therm [332] Ebeam [333] ALD [334]	<8 eV [308]	x–0.75 [332]	c-Si	580 μs @$\Delta n = 10^{15}$ cm^{-3} (10 nm, n-Si 3 Ω cm) [333]	17 m Ω cm^2 (1 nm, n-Si, 1 Ω cm) [332]
				a-Si:H (5 nm)	3.1 ms @$\Delta n = 10^{15}$ cm^{-3} (5 nm, n-Si 2 Ω cm) [334]	950 m Ω cm^2 (3 nm, n-Si, 1–3 Ω cm) [334]
NbO_x	ALD [90]	3.4 eV [90]	–	c-Si	25 fA/cm^2 (5 nm, n-Si, 3 Ω cm) [90]	70 m Ω cm^2 (1 nm, n-Si, 3 Ω cm) [90]
				SiO_x(1.3 nm)	30 fA/cm^2 (1 nm, n-Si, 3 Ω cm) [90]	–
TaO_x	ALD [335]	<4.5 eV [308]	x–2.44–2.53 [335]	c-Si	22.5 fA/cm^2 (6 nm, n-Si,1 Ω cm) [335]	300–350 m Ω cm^2(2–6 nm, n-Si, 1 Ω cm) [335]
TaN	ALD [336]	2.9 eV [336]	–	c-Si	128 μs @$\Delta n = 10^{15}$ cm^{-3} (4 nm, n-Si, 2.5 Ω cm) [336]	42 m Ω cm^2 (1 nm, n-Si, 0.1 Ω cm) [336]
TiN	Sput [337, 338]	3.3 eV [337]	–	c-Si	1500 fA/cm^2 (300 nm, n-Si, 1 Ω cm) [338]	11 m Ω cm^2 (300 nm, n-Si, 0.1–0.3 Ω cm) [338]
				SiO_x (1.7 nm)	500 fA/cm^2 (300 nm, n-Si, 1 Ω cm) [338]	10 m Ω cm^2 (300 nm, n-Si, 0.1–0.3 Ω cm) [338]

(Continues)

Table 5.8 Continued

Material	Dep. method	E_g	Stoich. (x=)	Underlayer(s)	Best passivation (thickness, wafer type, wafer doping)	Best ρ_c (thickness, wafer type, wafer doping)
TiO_xN_y	ALD [339]	3.4 eV [339]	–	c-Si	175 cm/s (3,8 nm, n-Si, 2.5 Ω cm) [339]	12 m Ω cm² (2.5 nm, n-Si, 0.3 Ω cm) [339]
				a-Si:H (4 nm)	2.7 cm/s (2.5 nm, n-Si, 2.5 Ω cm) [339]	78 m Ω cm² (2.5 nm, n-Si, 0.3 Ω cm) [339]
$Rb_2C_xO_y$	Therm [340]	–	$_X$–1.23 $_Y$–4.48 [340]	c-Si	–	138 m Ω cm² (1 nm, n-Si, 1 Ω cm) [340]
$Cs_2C_xO_y/CsO_x$	Therm [340, 341] Soln [342]	–	$_X$–0.87 $_Y$–3.09 [340]	c-Si	–	2 m Ω cm² (1 nm, n-Si, 1 Ω cm) [341]
SrC_xO_y	Therm [340]	–	$_X$–0.55 $_Y$–2.92 [340]	c-Si	–	190 m Ω cm² (1 nm, n-Si, 1 Ω cm) [340]
BaC_xO_y	Therm [340]	–	$_X$–0.59 $_Y$–2.51 [340]	c-Si	–	295 m Ω cm² (1 nm, n-Si, 1 Ω cm) [340]
$K_2C_xO_y$	Therm [340]	–	–	c-Si	–	83 m Ω cm² (1 nm, n-Si, 1 Ω cm) [340]
CaC_xO_y	Therm [340]	–	–	c-Si	–	169 m Ω cm² (1 nm, n-Si, Ω cm) [340]

(Continues)

Table 5.8 Continued

Material	Dep. method	E_g	Stoich. (x=)	Underlayer(s)	Best passivation (thickness, wafer type, wafer doping)	Best ρ_c (thickness, wafer type, wafer doping)
MgF_x	Therm [343]	~10 eV [344]	x~2 [343]	c-Si	1500 fA/cm^2 (1 nm, n-Si 10 Ω cm) [343]	35 m Ω cm^2 (1 nm, n-Si, 0.5 Ω cm) [343]
				a-Si:H (6.5 nm)	10 fA/cm^2 (1 nm, n-Si10 Ω cm) [343]	76 m Ω cm^2 (1 nm, n-Si, 0.5 Ω cm) [343]
BaO_x	Therm [345]	<4.5 eV [346]	–	c-Si	iV_{oc} = 607 mV @ 1 sun (2 nm, n-Si, 0.4 Ω cm) [345]	4.7 m Ω cm^2 (2 nm, n-Si, 0.4 Ω cm) [345]
CsF_x	Therm [329] Spin [347]	10 eV [328]	x~1 [347]	c-Si	–	2 m Ω cm^2 (1.5 nm, n-Si, 1–3 Ω cm) [347]
				a-Si:H (6 nm)	5.7 cm/s (1.5 nm, n-Si 1–5 Ω cm) [347]	13 m Ω cm^2 (1.5 nm, n-Si, 1–3 Ω cm) [347]
CsCl	Spin [347]	>8 eV [328]	–	c-Si	–	18 m Ω cm^2 (1.5 nm, n-Si, 1–3 Ω cm) [347]
				a-Si:H (6 nm)	10.5 cm/s (1.5 nm, n-Si, 1–5 Ω cm) [347]	70 m Ω cm^2 (1.5 nm, n-Si, 1–3 Ω cm) [347]

(Continues)

Table 5.8 Continued

Material	Dep. method	E_g	Stoich. (x=)	Underlayer(s)	Best passivation (thickness, wafer type, wafer doping)	Best ρ_c (thickness, wafer type, wafer doping)
CsBr	Spin [347]	7–8 eV [328]	–	c-Si	–	6 m Ω cm² (1.5 nm, n-Si, 1–3 Ω cm) [347]
				a-Si:H (6 nm)	6.4 cm/s (1.5 nm, n-Si, 1–5 Ω cm) [347]	39 m Ω cm² (1.5 nm, n-Si, 1–3 Ω cm) [347]
CsI	Spin [347]	>6.3 eV [328]	–	c-Si	–	13 m Ω cm² (1.5 nm, n-Si, 1–3 Ω cm) [347]
				a-Si:H (6 nm)	18 cm/s (1.5 nm, n-Si, 1–5 Ω cm) [347]	70 m Ω cm² (1.5 nm, n-Si, 1–3 Ω cm) [347]
ZnO_x	Li-doped Soln [348], B-doped ALD [349] Al-doped ALD [349]	3.25 eV [350]	–	c-Si	iV_{oc}= 724 mV (75 nm, n-Si, 3 Ω cm, after Al_2O_3 capping) [349]	18 m Ω cm² (22 nm, n-Si, 1–3 Ω cm) [348]
				a-Si:H (6 nm)	iV_{oc}= 701 mV @1 sun (22 nm, n-Si, 1–3 Ω cm) [348]	35 m Ω cm² (22 nm, n-Si, 1–3 Ω cm) [348]

(Continues)

Table 5.8 Continued

Material	Dep. method	E_g	Stoich. (x=)	Underlayer(s)	Best passivation (thickness, wafer type, wafer doping)	Best ρ_c (thickness, wafer type, wafer doping)
CaAcac	Soln [351]	–	–	c-Si	–	1.3 m Ω cm²(1, n-Si, 1–3 Ω cm) [351]
				a-Si:H (6 nm)	11 fA/cm²–, n-Si, 1–10 Ω cm) [351]	18.2 m Ω cm² (1, n-Si, 1–3 Ω cm) [351]
MgAcac	Soln [351]	–	–	c-Si	–	6 m Ω cm² (1, n-Si, 1–3 Ω cm) [351]
				a-Si:H (6 nm)	26 fA/cm² (–, n-Si, 1–10 Ω cm) [351]	55.1 m Ω cm² (1, n-Si, 1–3 Ω cm) [351]
$MgAcac_2$/Mg	Therm [352]	3.9 eV [352]	–	c-Si	–	7.5 m Ω cm² (2 nm, n-Si, 1 Ω cm) [352]
ZrAcac	Soln [351, 353]	–	–	c-Si	15 µs (4 nm, n-Si, 1–3 Ω cm) [353]	10.8 m Ω cm² (1, n-Si, 1–3 Ω cm) [351]
				a-Si:H (6 nm)	13 fA/cm² (–, n-Si, 1–10 Ω cm) [351]	64.9 m Ω cm² (1, n-Si, 1–3 Ω cm) [351]

(Continues)

Table 5.8 Continued

Material	Dep. method	E_g	Stoich. (x=)	Underlayer(s)	Best passivation (thickness, wafer type, wafer doping)	Best ρ_c (thickness, wafer type, wafer doping)
ZnAcac	Soln [351]	–	–	c-Si	–	18.1 m Ω cm^2 (1, n-Si, 1–3 Ω cm) [351]
				a-Si:H (6 nm)	32 fA/cm^2 (–, n-Si, 1–10 Ω cm) [351]	232.3 m Ω cm^2 (1, n-Si, 1–3 Ω cm) [351]
CeF_x	Therm. [354]	4.9 eV [354]	x-3 [354]	c-Si	-	11 mΩ cm2 (1.5 nm, n-Si, 1-3 Ω cm) [354]

ALD, atomic layer deposition; Acac, acetylacetonate; Ebeam, electron beam evaporation; Eg, bandgap; LPCVD, low pressure chemical vapor deposition; Ox, oxidation; Soln, solution/liquid deposition; Spin, Spin coating; Sput, Sputtering; Stoich, stoichiometry; Therm, thermal evaporation.

Table 5.9 Metal compounds successfully utilized in hole contacts for c-Si solar cells

Material	Dep. method	E_g	Stoich. (x=)	Underlayer	Best passivation (thickness, wafer type, wafer doping)	Best ρ_c (thickness, wafer type, wafer doping)
MoO_x	Therm [355–357] ALD [358, 359] Soln [360]	3.2 eV [356]	$_x$–2.5 [356] $_x$–2.87 [329]	c-Si	200 fA/cm^2 (10 nm, p-Si, 2.1 Ω cm) [355]	1 m Ω cm^2 (10 nm, p-Si, 2.1 Ω cm) [355]
				a-Si:H (5 nm)	10 fA/cm^2 (11 nm, p-Si, 1–5 Ω cm)	450 m Ω cm^2 (11 nm, p-Si, 1–5 Ω cm)
				AlO_x (1 nm)	146 fA/cm2 (4 nm, n-Si, 1 Ω cm) [359]	25 m Ω cm^2 (4 nm, p-Si, 1 Ω cm) [359]
V_2O_x	Therm [345, 356] ALD [361]	2.8–2.9 eV[356, 361]	$_x$–4.6 [356]	c-Si	41.8 fA/cm^2 (12 nm, p-Si, 4.5 Ω cm) [361]	95 m Ω cm^2 (4.5 nm p-Si, 0.1 Ω cm) [361]
				a-Si:H (5 nm)	15.5 ms @$\Delta n = 10^{15}$ cm^{-3} (4.5 nm, n-Si, 1.5 Ω cm) [361]	103 m Ω cm^2 (4.5 nm, p-Si, 0.1 Ω cm) [361]
WO_x	Therm [356, 362, 363]	3.1 eV [356]	$_x$–3.2 [356] $_x$–2.99 [364]	c-Si	V_{oc}~670 mV (-, p-Si, 1 Ω cm) [365]	54 m Ω cm^2(7–10 nm, p-Si, 5–10 Ω cm) [363]

(Continues)

Table 5.9 Continued

Material	Dep. method	E_g	Stoich. (x=)	Underlayer	Best passivation (thickness, wafer type, wafer doping)	Best ρ_c (thickness, wafer type, wafer doping)
NiO_x	Sput [366] Therm [367] PLD [368]	2.5–3.8 eV [369]	–	p^+-Si (epitaxial, 100 nm)	9000 fAm/cm^2 (5 nm, p-Si, 0.1 Ω cm [366]	<1 m Ω cm^2 (5 nm, p-Si, 0.01 Ω cm) [366]
				SiO_x (RCA, 1.74 nm)	54.45 µs @Δn = 10^{15} cm^{-3} (10 nm, n-Si, 3 Ω cm [367]	-
$Zn_xNi_{1-x}O$	ALD [370]	3.06 eV [370]	$_x$–0.62	c-Si	–	21 m Ω cm^2 (3.4 nm, p-Si, 2 Ω cm) [370]
Cu_2O_x 115	Sput [371–373] B doped [372] N doped [373]	2.55 eV [372]	–	c-Si	60 µs @ $\Delta n = 10^{16}$ cm^{-3} (15 nm, n-Si, 1 Ω cm) [371]	11 m Ω cm^2 (8 nm, p-Si, 1 Ω cm) [373]
CuI	Therm [374]	3.0 eV[374]	–	c-Si	–	30 m Ω cm^2 (20 nm, p-Si, 2 Ω cm) [374]
CrO_x	Therm [375, 376]	2.8–3.9 eV [376, 377]	68% $Cr(OH)_3$[356]	c-Si	–	45 m Ω cm^2 (4 nm, p-Si, 2 Ω cm) [356]

ALD, atomic layer deposition; Eg, bandgap; PLD, pulsed laser deposition; Soln, solution/liquid deposition; Sput, Sputtering; Stoich, stoichiometry; Therm, thermal evaporation.

Table 5.10 *Best reported c-Si cell efficiencies employing metal compound contact interlayers*

Structure	Electron selective	Hole selective	Area (cm^2)	Eff. (%)	Stability comments	Ref.
PRC	TiO_x/LiF_x / Al	c-Si (p^+) / Cr/Pd/ Ag	4	23.10	Cells stable up to 400 °C	[321]
PRC	TiO_x/Ca / Al	c-Si (p^+) / Cr/Pd/ Ag	4	21.80	–	[322]
PRC	LiF_x/Al	c-Si (p^+) / Cr/Pd/ Ag	4	20.60	Cell stable after 3 months	[327]
PRC	c-Si (n+) / Cr/Pd/ Ag	MoO_x / Ag	4	20.40	–	[378]
PRC	c-Si (n^+) / Ag	CrO_x/Ag/CrO_x /Ag	4	20.30	–	[375]
PRC	c-Si (n^+) / Ag	CuI / Ag	4	20.70	Cells stable up to 350 °C	[374]
PRC	CeF_3/Al	c-Si (p^+) / Ag	4	21.27	–	[354]
PRC	c-Si (n^+)	WO_x	4	20.50	Degradation @ 180 ˚C, dependent on contact fraction	[379]
Full	SiO_x/TiO_x / Al/Ag	c-Si (p^+) / Cr/Pd/ Ag	4	21.60	Passivation degrades @ 400 °C	[323]
Full	TiO_x/LiF_x/Al	c-Si (p^+) / Cr/Pd/ Ag	4	21.30	–	[380]
Full	a-Si:H (i)/CsF_x / Al	c-Si (p^+) / Cr/Pd/ Ag	4	21.80	Passed 1000-h 85 ˚C, 85% RH test	[347]
Full	a-Si:H (i)/a-Si:H (n) / μc-Si:H (n) / ITO / Ag	a-Si:H (i)/MoO_x / ITO/ Ag	4	23.50	Cell stable after 11 months (V_{OC} loss <1 mV)	[357]
Full	a-Si:H (i)/a-Si:H (n) / ITO / Ag	a-Si:H (i)/V_2O_x / Ag	4	21.60	Cell stable after 3 months	[361]

(Continues)

Table 5.10 Continued

Structure	Electron selective	Hole selective	Area (cm^2)	Eff. (%)	Stability comments	Ref.
Full	a-Si:H (i)/MgF_x / Al	c-Si (p^+) / Ag	4	20.10	–	[343]
Full	a-Si:H (i)/CaAcac / Al	c-Si (p^+) / Cr/Pd/ Ag	4	21.60	1 000-h 85 °C, 85% RH test; negligible loss observed. Cells stable up to 350 °C.	[351]
DASH	a-Si:H (i)/TiO_x/ LiF_x/Al	a-Si:H (i) / MoO_x / ITO / Ag	4	20.70	1 000-hour85 °C, 85% RH test with >96% of performance maintained.	[331]
DASH IBC	a-Si:H (i) / MgF_x / Mg / Al / Ag	a-Si:H (i) / MoO_x / Ag	4	22.10	–	[381]
DASH bifacial	a-Si:H (i)/ZnO / LiF / Al	a-Si:H (i)/MoO_x/ ITO / Ag	4	20.3 (21.0*)	Strong degradation at the e-contact side	[382]

*Bifacial efficiency value based on authors' calculations.

DASH, dopant free asymmetric heterocontact (both contacts use metal compound interlayers); DASH IBC, DASH interdigitated back contact; Eff, power conversion efficiency; Full, cell with one full area metal compound interlayer contact; PRC, cell with one partial area metal compound interlayer contact.

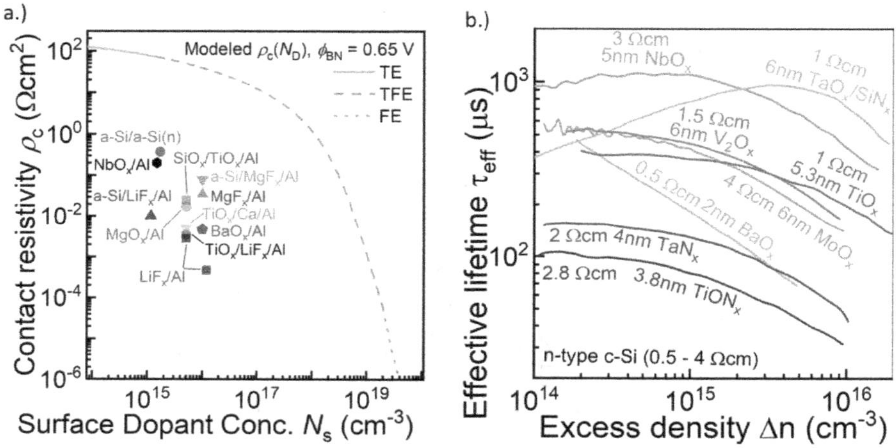

Figure 5.12 *(a) Contact resistivity ρ_c as a function of surface dopant concentration for a range of electron contact systems employing metal compound interlayers applied to n-type surfaces. The orange line provides a reference for the expected ρ_c behaviour of a directly metalized surface with a 0.65 eV barrier due to thermionic emission (TE), thermionic field emission (TFE) and field emission (FE). (b) Provides a compilation of effective lifetime results for metal compound interlayers directly deposited onto n-type c-Si (some points underwent additional thermal treatments post-deposition).*

5.3.2.1 Alkali and alkaline-earth metal compounds

These materials, which have been heavily explored for electron contacts, consist of a group I or II metal cation bonded to different anions, such as halogens, oxygen or carbonate [327, 332, 340, 343]. Typically, these are applied as nm-scale interlayers between the c-Si absorber and an outer metal contact, commonly Al. Within the wider electronic community there still exists some contention as to the mechanism of the improved electron injection/extraction afforded when inserting these materials. Three common explanations are (i) protection/isolation of the semiconductor layer from the metal layer (preventing surface damage and reducing FLP); (ii) chemical doping of the underlying semiconductor (such as Li doping from LiF_x) and (iii) formation of an exceptionally low work function electrode (such as that measured on LiF_x/Al contacts). Most studies attribute the latter to be the most important factor, but regardless of the mechanism, the insertion of these interlayers results in a dramatic decrease in ρ_c compared to the directly metalized surface. This is shown in Figure 5.12a, which provides a compilation of ρ_c values measured using different interlayer stacks on moderately doped n-type c-Si. The reference line shows the expected behaviour of a directly metalized surface with a surface barrier of 0.65 eV, which is a common value quoted for n-type surfaces. The large drop in ρ_c has permitted the development of otherwise unattainable cell structures, such as n-type cells with direct (i.e. non-heavily doped) partial rear electron contacts [327, 385].

The alkali and alkaline-earth metal salts are typically transparent across the visible region with wide band gaps (commonly > 5 eV), introducing potential optical benefits. However, thus far these materials have only proven effective when combined with opaque metal overlayers in c-Si cells, which negates this advantage. The wide band gaps are also the primary reason why nm-scale films are typically required due to the strict thickness constraints of direct carrier tunnelling, although conduction through >2 nm interlayers has been demonstrated, suggesting other conduction mechanisms possibly associated with sub-band states [329].

The most successful alternatives thus far include MgO_x, LiF_x and MgF_x, all of which have been integrated into cells with an efficiency of at least 20% [327, 332, 381]. A limited number of cases have shown a moderate level of surface passivation when directly applied to c-Si, such as MgO_x and BaO_x [333, 345]. Similarly, the stability of these materials against temperature and humidity conditions has been demonstrated in limited cases, including anneal stability up to >300 °C and unencapsulated survival of the 85 °C/85% relative humidity (RH) test [321, 331].

5.3.2.2 Low/moderate work function n-type transition metal oxides and nitrides

The oxides and nitrides of several of the transition metals, most prominently Ti, Ta, Zn and Nb, have also been utilized in electron contacts [321–323]. Most of these materials theoretically present a preference for electron collection when in contact with c-Si due to small CB offsets ΔE_c and large valence band offsets ΔE_v [308, 346], similar to the situation depicted in Figure 5.11b. However, the degree to which these offsets practically effects contact performance remains an open discussion, particularly in the context of amorphous oxides with large sub-band trap concentrations. Some studies have even shown effective collection of holes using H_2 plasma-treated TiO_x layers [386], suggesting that such oxides could be used as a passivating interlayer for both electron and hole contacts. The band gaps of these materials are commonly above 3 eV, suggesting their suitability as window layers on the front side of a solar cell, although a convincing transparent electron contact demonstration is yet to be made. This is again associated with the fact that these metal compounds are typically combined with opaque low work function electrodes, such as Ca, Mg, Al, Yb and LiF_x/Al [322, 323, 335, 387, 388]. These metal compounds are typically deposited in a manner that introduces intrinsic doping, through, e.g. oxygen vacancies, which also likely lowers their work function [389].

Interestingly, this family of materials is able to provide some level of surface passivation when applied directly to c-Si. Many examples of this are provided in Figure 5.12b, which compiles surface passivation results using a range of metal compound films directly applied to moderately doped n-type c-Si [90, 322, 335, 336, 339, 345, 361]. Whilst many of these surface passivating results were obtained using atomic layer deposition (ALD), a surprising number is deposited by thermal and electron beam evaporation [326, 333, 345, 355], all techniques which will be discussed in Sub-section 5.3.2.3. The most successful alternative to date is TiO_x, which has been demonstrated both with and without an additional passivation layer, in multiple cell structures above 20% [321–323, 380], including multiple demonstrations of thermal stability above 300 °C [321, 323]. It is worth noting that several other related oxides (including some from outside of the transition metal block), which also exhibit excellent surface passivation on c-Si and have been trialed in electron contacts, are yet to yield convincing results, such as AlO_x, HfO_x, ZrO_x and GaO_x [287, 390–392].

5.3.2.3 High work function n-type transition metal oxides

The most successful metal compound hole contacts demonstrated so far incorporate the high oxidation state oxides of Mo, W, V and less commonly Re and Cr. These oxides all have the unusual characteristic of very high work functions, much higher than that of the elemental metals [355, 356, 361, 362, 375, 393]. Whilst work functions > 6 eV have been experimentally measured for these materials in clean vacuum environments, such results have not been replicated on c-Si PV devices and instead values in the 5–6 eV range are typical (and still adequate for the purposes of c-Si PV) [329]. These materials also have wide band gaps > 3 eV and importantly can be contacted directly with transparent conductive overlayers such as ITO. This has allowed them to be applied on the sunward side of the solar cell, already demonstrating a J_{sc} improvement of up to 1.3 mA/cm^2 when replacing doped silicon layers [357].

MoO_3, V_2O_5 and WO_3 have all shown passivation when directly applied to c-Si [355, 356, 361, 379], as shown in Figure 5.12b, although the more successful trials have utilized dedicated passivating interlayers, which greatly reduces the surface recombination. There has been significant discussion on the underlying mechanism for the extraction of holes (and blocking of electrons) using a high work function n-type material, further complicated by common addition of an a-Si:H passivating interlayer [313, 394, 395]. Whilst there is no universal consensus, most agree that a slightly sub-stoichiometric form is required for efficient hole extraction. The most successful candidate within this category is MoO_x, which has been demonstrated in multiple cells above 22% [357, 381]. The poor thermal and moisture stability of MoO_x, however, are significant detractors and have prompted the continued exploration for an alternative metal compound interlayer for hole collection.

5.3.2.4 Dedicated passivating interlayers

In many cases, the difficult task of passivating the c-Si absorber can be performed with a thin (1–6 nm) dedicated passivating interlayer, which itself does not provide a strong preference to electron or hole collection. To date, the two most common passivating interlayers are SiO_x and hydrogenated amorphous silicon (a-Si:H).

In many cases, a thin SiO_x interlayer (nm scale) is unavoidable on the c-Si surface, particularly for deposition/growth processes at elevated temperatures or metal oxides with a higher position in the Ellingham diagram than the Si–SiO_2 system. In these cases, a thin SiO_x layer (typically 4–20 Å thick) grows during deposition or post-deposition thermal treatments. In other circumstances, a SiO_2 layer is

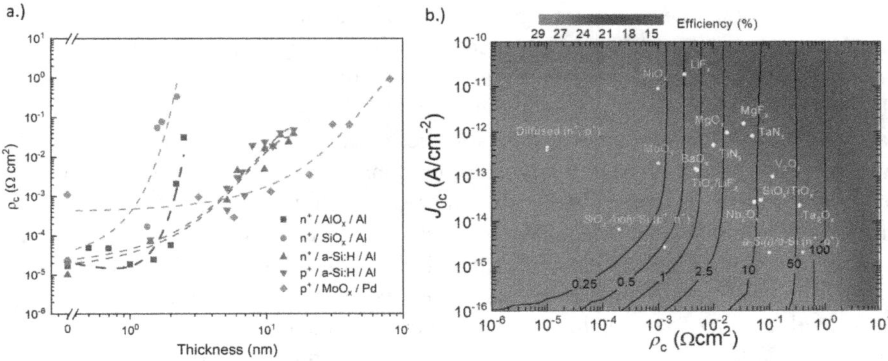

Figure 5.13 *(a) Contact resistivity for AlO_x, SiO_x, a-Si:H and MoO_x interlayers on heavily diffused c-Si surfaces showing a large variation in thickness dependence due to different transport mechanisms. (b) Compilation of metal compound interlayers* J_{0c}/ρ_c *behaviour in comparison to directly metalized heavily diffused 'diffused (n^+, p^+)', polysilicon 'SiO_x/poly-Si (n^+,p^+)' and SHJ 'a-Si(i)/a-Si(n^+,p^+)' contacts. The background contour plot shows the optimum contact fraction (black lines) and resultant idealized efficiency (coloured contours) for each pair of* J_{0c} *and* ρ_c.

intentionally introduced under controlled conditions, such as a short dry oxidation > 600 °C or a nitric acid chemical oxidation, to achieve a higher level of surface passivation with J_0 values as low as 10's of fA/cm² [90, 323]. The thickness of these SiO_2 passivating interlayers is limited to <2 nm, a constraint imposed by the need for direct tunnelling conduction, which places a limitation on the lowest attainable passivation. Furthermore, the incorporation of a SiO_2 interlayer can greatly improve the thermal stability of contacts [323].

As shown in Table 5.10, plasma-enhanced chemical vapour deposited a-Si:H is commonly integrated as a passivating interlayer for metal compound layers in some of the more successful cell architectures. Unlike SiO_x, conduction through a-Si:H is not as strongly dependent on thickness. This is shown in Figure 5.13a, which provides the ρ_c behaviour of SiO_x, a-Si:H, AlO_x and MoO_x interlayers as a function of thickness on diffused c-Si surfaces [355, 396, 397]. Diffused surfaces are included in this example to reduce the influence of c-Si surface barriers on the ρ_c measurements (i.e. ρ_c is dependent mainly on the transport through the interlayer). It is unsurprising to see that wide gap interlayers with larger band offsets, such as SiO_2 and Al_2O_3, have a significantly stronger thickness dependence compared to a-Si:H owing to different conduction mechanisms. As such a-Si:H passivating films are typically thicker, on the order of 4–6 nm thick, a range which achieves exceptional surface passivation (<5 fA/cm²) whilst still allowing appreciable electron or hole flow. A downside of utilizing amorphous silicon is its temperature instability, which typically starts to effect device performance when annealed above 250 °C. Furthermore, in some cases the release of hydrogen from a-Si:H is said to accelerate the degradation of metal compound interlayers such as MoO_x in a-Si:H/MoO_x stacks.

To conclude this sub-section, Figure 5.13b compiles a range of experimentally determined J_{0c} and ρ_c values for different carrier selectivity systems, including contacts with metal compound interlayers, directly metalized heavily doped contacts and doped polycrystalline and a-Si:H-based contacts. These are superimposed on a 'carrier selectivity' plot, which simulates the optimum contact fraction (black lines) and resultant idealized efficiency limit (coloured contour plot) for a given ρ_c and J_{0c} combination. In this plot, it can be generally seen that the directly metalized dopant-diffused contacts are characterized by low ρ_c and high J_{0c} requiring very small contact fractions to achieve optimum efficiency. Conversely, the doped silicon-based passivated contacts (i.e. doped polySi and doped a-Si:H) are characterized by low J_{0c} and moderate/high ρ_c. The metal compound interlayers occupy a region of lower performance compared to the doped-silicon approaches but still offer advantages in terms of transparency and possibly ease of deposition. We also note that Figure 5.13b does not include any results for metal compound interlayers with passivating underlayers (i.e. a-Si:H), which have very similar characteristics to the doped a-Si:H approaches [329]. Finally, this material space continues to grow rapidly and several less developed alternatives, which are not included in Tables 5.8–5.10, warrant mention. These include novel metal compound contact interlayers YbSi, ScO_x and In_2S_3 [398–400], and those borrowed from other solar cell absorber families such as CdS and SnO_x [401, 402].

5.3.3 Deposition techniques for metal compound interlayers

Given the early stage of this technology family, most reported interlayers have been deposited using research scale tools, though their potential for scaling is well known – particularly via spatial ALD. In the research environment, metal compound interlayers are most commonly deposited using thermal/electron beam PVD, sputter-based PVD or batch-type ALD, although a range of other techniques (such as solution-based deposition methods) has been demonstrated to a lesser degree.

5.3.3.1 Thermal and electron beam PVD

These deposition approaches are performed in a vacuum chamber at low pressure (typically 10^{-6} mbar, preferably 10^{-7} mbar scale), by heating a pure source material either directly from a crucible/boat via the Joule effect (thermal evaporation), or by locally bombarding the source with a controlled electron beam (electron beam evaporation). The elevated temperature will result in the sublimation of atoms or clusters of atoms from the source which, provided a low enough pressure is maintained, will travel without collision and condense on the substrate. For example, in the case of LiF_x, the sublimation vapour consists of a mixture of mainly monomer (LiF), dimer (Li_2F_2) and trimer (Li_3F_3) species [403]. This deposition of atoms/clusters is largely directional, meaning that the film will be deposited almost exclusively on the side of the substrate facing the source, a useful characteristic in the fabrication of solar cells.

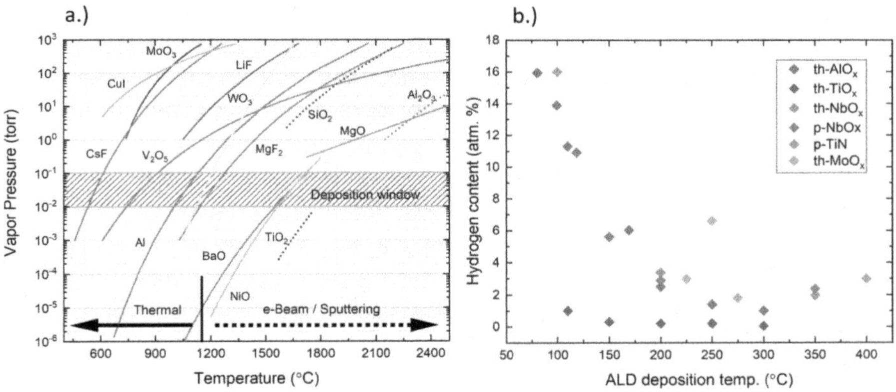

Figure 5.14 *(a) Compilation of vapour pressure behaviour of various metal compound films as a function of temperature. Lines at the bottom of the plot indicate typical regions in which thermal-, electron beam- and sputter-based PVD are preferred [404–409]. (b) Measured hydrogen content in atomic percent for different films deposited by either thermal ALD (th-ALD) or plasma-assisted ALD (p-ALD) [410–414].*

As shown in Figure 5.14a, thermal and electron beam PVD are somewhat complimentary, providing access to materials with different vapour pressures. This plot shows the vapour pressure behaviour of common metal compound interlayers as a function of temperature. For the application of interlayers in c-Si solar cells, thermal evaporation is generally used for materials with high to moderate vapour pressures, which require temperatures up to ~1 000 °C to get appreciable sublimation. Materials with lower vapour pressures, e.g. ceramics, requiring higher temperatures to sublimate, can be deposited by electron beam PVD, which can achieve higher source temperatures. Electron beam PVD also typically provides greater film purity, as the directly heated zone can be small and limited to the source, at the expense of generation of higher energy species which may damage the c-Si surface [415] – e.g. a non-zero percentage of the incoming electron kinetic energy is converted into X-rays.

Most demonstrations of successful interlayers have utilized high purity (i.e. typically with an atomic percentage >99.99) sources of the same stoichiometry as the desired films. In almost all cases, the resultant films are sub-stoichiometric due to partial decomposition of the source. The tendency of a metal compound to decompose under heating can be very roughly estimated by looking at the position of the parent metal in the reactivity series – with the compounds of less reactive metals tending to decompose more readily. In some cases, decomposition is desired, such as the use of metal carbonate sources, which lose CO_2, for the deposition of metal oxides. In others, decomposition is counteracted by post-deposition oxidation. Similarly, oxide films have also been grown by the oxidation of a deposited thin metal layer such as Mg or Ti, yielding excellent surface passivation [326, 333]. The use of advanced strategies to control stoichiometry, such as adding a partial pressure of oxygen or water vapour during the evaporation or co-deposition from multiple sources has not been thoroughly explored for this application.

Inline evaporator systems do exist and have been explored to a limited extent in the context of industrial c-Si PV. For example, the deposition of thick Al films at a rate of ~12 wafers/min (Al thickness of 2 μm) can be achieved through inline thermal evaporation without exposing cells to temperatures in excess of ~350 °C [416]. However, the adaption of this process to metal compound films requiring nm-scale precision is unlikely to be the simplest and most cost-effective pathway towards industrial adoption.

5.3.3.2 Sputtered PVD

In sputter PVD, a plasma is generated within a vacuum chamber, and high energy species within this plasma (commonly Ar^+ ions) are directed to bombard a high purity source to physically sputter atoms from the surface. Similar to the evaporation processes discussed above, once liberated, these atoms can travel to and condense onto the intended substrate surface. Commonly, a magnetron is positioned at the source to confine the plasma plume and improve the sputtering efficiency. Sputtering is an excellent option for low vapour pressure materials particularly those which decompose upon heating (and hence cannot be deposited effectively via electron beam evaporation). Reactive sputtering is a common variant in which reactive gases are added to the input gas mix whilst sputtering from a pure metal target to deposit a metal compound (e.g. introducing O_2 whilst sputtering Ti to deposit TiO_x). Such a process can be used to afford a reasonable level of control over the film stoichiometry. The plasma generation can be achieved by a variety of power sources such as DC, suited for deposition from conductive sources, and RF sources, which can be used to deposit from insulating sources. While deposition can be performed at room temperature, more commonly it is conducted at elevated temperatures to tune characteristics of the deposited film. A commonly described drawback of conventional sputter PVD is the kinetic/thermal damage that the substrate is subjected to during deposition, which can reduce the level of surface passivation and damage thin passivating layers. This damage can be mimimized at the expense of deposition rate by reducing the exposure of the substrate to the plasma plume, and with the addition of a hydrogen source (e.g. via H_2O or H_2 gases) surface passivation can even be achieved using sputtering directly on c-Si [417–419].

Sputtering is also already in use by the PV industry to deposit transparent conductive oxides (TCOs) in SHJ architecture production lines (see Section 5.4). However, these tools primarily employ DC MS and are used to deposit thick (>50 nm) layers. Most demonstrations for nm-scale metal compound interlayers have been conducted using RF MS and hence the translation to an industrial environment is not without significant challenges.

5.3.3.3 Atomic layer deposition

The material versatility, angstrom-scale thickness control and ability to provide surface passivation have also made batch-type ALD a popular choice for the development of metal compound interlayers for c-Si solar cells. This approach involves the exposure of the target surface to sequential atmospheres of metal precursor and reactant, separated by purge cycles, typically at temperatures 50–350 °C. The materials versatility of ALD stems from the broad range of potential precursors/processes available, see, e.g. the extensive database at atomiclimits.com [420]. Relevant to the application of c-Si PV, the incorporation of hydrogen, into films from precursor moieties, is beneficial in the achievement of surface passivation. Figure 5.14b compiles measured hydrogen content for a range of ALD films commonly employed in selective contacts. The atomic concertation of hydrogen in the films is strongly dependent on temperature and can be as high as 20% at low deposition temperatures (<100 °C) but is typically ~5% when depositing at temperatures around 200 °C [421]. During post-deposition annealing at temperatures in the 200–500 °C range, the mobile hydrogen in the films can migrate to the silicon interface to assist with reducing the density of interface traps N_{it}, sometimes reducing the density by an order of magnitude. In some cases, such as TiO_x, excellent passivation can be achieved in the as deposited state (i.e. without a post-deposition anneal). Relatedly, ALD imposes relatively mild conditions on the substrate and any underlying films, a contrast to other deposition techniques that commonly damage thin interlayers.

The angstrom-scale thickness control of continuous/conformal films is also a crucial aspect in the development of metal compound interlayers, which rely on transport mechanisms with extreme thickness sensitivity, such as direct tunnelling which requires thicknesses <2 nm. Several studies have shown a delicate tradeoff between J_{0c} and ρ_c as a function of film thickness reinforcing the necessity of this control [396]. This is also seen in Figure 5.13a introduced earlier, especially for films such as SiO_x and AlO_x. A final advantage of ALD is the potential for composition control, e.g. the deposition of bimetallic oxides or extrinsically doped films using 'super-cycle' ALD processes. This provides a further handle by which to control contact characteristics or transparency.

Due to the above advantages, ALD is commonly cited as the most viable industrial tool for the deposition of metal compound contact interlayers. Both batch and "spatial" ALD (s-ALD) techniques are already widely used in the c-Si PV industry, for the deposition of thin AlO_x layers for the industry dominating PERC architecture. The s-ALD tools spatially separate precursor vapours into different chambers, isolated by inert gas curtains, which removes the need for a vacuum environment. These tools have already advanced to a level where they can process up to three wafers/s at a cost of 0.01–0.03 $/wafer [422]. Further it has been shown that the surface passivation and film quality of spatial ALD are equivalent to that achieved in research scale batch ALD systems.

5.3.4 Remaining challenges

As can be seen in Table 5.10, the >23% efficiency demonstrated on multiple research-scale prototype solar cells employing metal compound interlayers is already in line with the estimated efficiency limit of industrial PERC cells. However, several significant factors need to be overcome before metal compound contact interlayers are seriously considered in the c-Si PV industry.

Thermal and humidity stability remains one of the most commonly ascribed shortcomings of the metal compound interlayer approach. Back-end processing and screen-printing metallization require temperatures of at least 200 °C and preferably higher. This is an issue for MoO_x-based hole contacts, which begin to degrade in performance when annealed in the 150–200 °C range, with some dependence on pre-treatments [423]. MoO_x is also susceptible to moisture, although 1 000-h 85 °C/85% RH DH tests have shown that it can be effectively protected using a TCO overlayer [331]. Electron contacts, such as those fabricated with TiO_x, have shown higher thermal stability, but still undergo performance degradation at temperatures above 350 °C [321, 323]. As such improved stability, particularly thermal stability for the hole contact, would be required before metal compound layers were transferred to industry.

The metal compound interlayer concept must also provide a sufficient improvement in cost-to-performance ratio relative those being transferred to large scale production today, i.e., the SHJ and poly-Si technologies [230]. This would require properly leveraging the optical advantages of metal compound interlayers to realise +25% efficiency, ideally whilst removing the current dependency on PECVD deposited a-Si:H passivating layers.

Finally, as suggested above, an industry applicable fabrication scheme must be developed preferably from tools already commonplace within the c-Si PV industry such as spatial ALD.

5.4 Production tools for evaporation and sputtering TCOs

As already described in the previous chapters, the PVD technology consists in depositing TF and coatings by a process in which a material is vapourized, transported and condensed on to a substrate. PVD processes include sputtering, electron beam evaporation and thermal evaporation [424, 425]. This chapter is reviewing the deposition equipment more suitable for mass production: production tools for evaporation and sputtering of either metals, metal oxides, transparent conductive oxides (TCO). Noting that PVD technology is well adapted to inorganic materials but may also apply to some organic materials (not treated in this chapter).

The following paragraphs will review production process considerations, tool design, throughput, elements influencing the overall cost of a technology.

5.4.1 Prerequisites for mass-volume industrial equipment

The right industrial techniques for deposition are depending on multifactorial parameters, whose principal requirements are enumerated below:

- Characteristics of the material to be deposited and its own constraints. In particular, it depends on the material source easiest form available: solid, liquid, gas. For the common case where the material is a compound, the existence of a stable compound allows simplified equipment design and cost (no need of complex co-evaporation, multiple sources, etc.). Similarly, brittle materials are difficult if not impossible to be shaped into large sputter targets preventing their introduction into mass-volume usage.
- The substrate/device that will receive the deposit. Its sensitivity or incompatibility to the physical/chemical conditions (plasma no plasma, heat, ion bombardment, etching, etc.). This is well known that, e.g. heterojunction cell requires maximum deposition temperature of around 200 °C. New technologies, such as PK cells and tandem cells, have even tighter temperature constrains often <120 °C.
- The deposited material characteristics, properties, thicknesses that are expected to be obtained voluntary but also side effects proper to each technique (level of purity, optical properties, electrical properties, hardness, density, corrosion/environment resistance, step coverage/layer thickness conformity, roughness control for light scattering and/or adhesion concern, etc.)
- Since production tools are of consideration in this chapter, the ability to scale up both the process and equipment from R&D to mass production equipment is determinant in selecting the techniques/technology. Usually, two main approaches are implemented for higher throughput: (a) ability to treat larger surface at the same time (ideally keeping tool footprint reasonable) either on horizontal/vertical flat tray or on batch cassette and (b) increasing the number of wafers per hour thanks to shorter process time and fast handling/automation operations per seconds.
- In addition, the tool operability requires innovative designs: highly automatized equipment including consumable automatic feed-in (gas/liquid/solid source materials) without stopping the equipment, largest possible interval between periodic maintenances (tool production availability).

Fortunately, often there is not only one well-adapted suitable technique. But the industry is prone to select the wider-spread lowest cost technique to start volume production to cope with the cost figure of merit of the PV world, the cost per watt, even if the performance (for the device) is not yet the highest demonstrated in R&D. Then progressively, innovation will be introduced to improve efficiency, or an alternative deposition techniques will decrease its costs to become competitive with reference techniques.

5.4.2 Comparison of several deposition techniques and technologies

While the detailed material properties of the metals, metal oxides, TCOs and their applications have been largely described in the previous chapters, we will come back shortly on the main techniques for depositing these materials. Indeed, many techniques can be used influencing the properties of the layers differently but each one with specific capability to scalability for mass production.

Globally, two generic TF deposition techniques [426] are employed: CVD and PVD.

5.4.2.1 Chemical vapour deposition (CVD)

CVD is a process used in many domains. It is often used in the SC and flat display industries. For typical CVD process, the substrate is maintained at a fixed temperature while exposed to one or more precursors in the gas phase, which react and/or decompose on the surface of the substrate to generate the desired film deposit. Several variant techniques of industrial CVD reactor exist among: MOCVD, atmospheric pressure CVD, low pressure CVD, ALD.

The principle of ALD is a sequential process. The process includes deposit phases, which consist of supplying a small amount of precursors, and purge phases to remove all unreacted precursors. These self-limiting alternating phases make up the deposition cycle, which is repeated to achieve the desired film thickness. A cycle deposits a few angstroms with duration of typically 10 s in order for the precursors to react/decompose on the surface of the substrate. This technique allows extremely precise control of the deposited thickness and the growth of the desired material. Its intrinsic merits allow for the highly reproducible growth of pinhole-free conformal TF with unparalleled control in TF deposition at the nanoscale. The downside is that it results deposition time, which can reach several hours, in particular for thicker films. Additionally, because of this weak speed of growth, a large percentage of impurities may be incorporated into the film. The process of ALD is often performed at lower temperatures than other techniques, which is beneficial when working with substrates that are fragile, and some thermally unstable precursors can still be employed with ALD if their decomposition rate is slow. It is worth noting that although the surface smoothness and quality of ALD films are significantly better than sputtering, the low nucleation rate and low growth rate of ALD ITO have restricted applications in silicon solar cells when there is need for thicker films (several tens nanometers) of TCO. Therefore, these restrictions have limited the applications of ALD-grown films (TCOs, metal oxides) in silicon PV [427, 428] at the remarkable exemption of the success of Al_2O_3 by ALD in commercial silicon solar cells that has shown that ALD can be cost-effective in high-volume manufacturing.

5.4.2.2 Physical vapour deposition

PVD processes [429–431] are atomistic deposition processes in which material vapourized from a source is transported in the form of a vapour through a vacuum

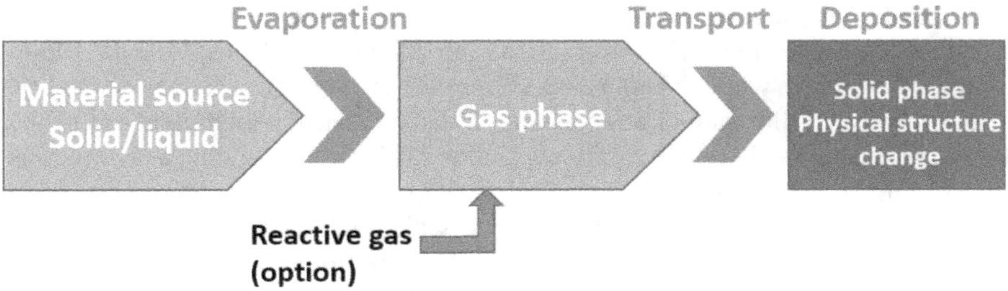

Figure 5.15 PVD basic principles

or low-pressure gaseous environment to the substrate, where it condenses and film growth takes place. The deposited layers generally present a microstructure, which is widely different from the bulk material evaporated. PVD processes can be used to deposit films of compound materials by the reaction of depositing material with the ambient gas environment or with a co-deposited material. The most common PVD processes are sputtering and evaporation, this includes several techniques [432] like vacuum evaporation, sputter deposition, arc vapour deposition, PLD, ion beam sputtering and ion plating. Vacuum evaporation uses thermal vapourization as a source of depositing atoms; sputter deposition uses physical sputtering as the vapourizing source; arc vapour deposition uses a high-current, low-voltage arc for vapourization and ion plating uses concurrent or periodic energetic particle bombardment to modify the film growth (see Figure 5.15).

(a) Thermal vacuum evaporation

Vacuum evaporation or vacuum deposition is the process where the material used for coating is thermally vapourized and then proceeds by potential differences to the substrate with little or no collisions with gas molecules. By running this process under a vacuum, contamination by other gases is greatly reduced. Normal vacuum levels are in the medium to high vacuum range of 10^{-5} to 10^{-9} Torr. Typical deposition rate is of 10–100 nm/min. Depending on the gas contamination of the vacuum chamber, the deposition rate can be much higher when compared to the other methods. Thermal evaporation is normally done by resistive coils or a high energy electron beam. The substrates have a larger distant from the source in order to reduce radiant heat transfer.

(b) Cathodic sputtering

Cathodic sputtering or MS is a technique that synthesizes a material on a substrate from a solid source named a target. The application of a potential difference between the target and the walls of the reactor in an atmosphere rarefied allows the creation of a cold plasma. Under the effect of the electric field, positive ions from the plasma are attracted to the cathode (target) and collide with the target. Then the positive ions communicate their momentum, thus causing the atomization of the target atoms in the form of neutral particles, which are ejected then condense on the substrate to form the film.

The growth rate being higher than the ALD technique, the incorporation of impurities is less. Sputtering process is often preferred to thermal evaporation, because of better step coverage, and wider choice of possible materials (thanks to the ability to use RF plasma or DC plasma) and often higher deposition rates (up to a few 100 nm/min), while forming dense films with an appropriate adhesion quality. The as-deposited film quality often allows to reduce annealing temperatures of the layer post-treatment. Reactive sputtering can be carried out by adding a reactive gas (like O_2, H_2, H_2O, etc.). MS is the standard deposition technique for transparent front contacts in industrial production. It is easy to scale up and substrates up to around 20 m^2 can be coated in large coaters designed for deposition of architectural glass.

(c) Developments of sputtering equipment for production

Intensive work has been carried out by equipment manufacturers on the optimization of high productivity sputtering equipment. Nowadays, production lines use rotatable targets for their higher material utilization rate and, therefore, higher throughput than planar targets. These aspects are developed in the next paragraph.

In the case of PV, wafers are transported either on horizontal tray or on vertical tray. This latter by construction allows reduced equipment footprint and cleaner condition than the former (less dust/particle contamination falling on wafer surface). Despite these advantages, horizontal tray coaters are mainly encountered in production principally because wafer loading/unloading on the tray is simplified in the horizontal version. Furthermore, horizontal tray with hollow pattern where the wafer is supported only by its perimeter has been developed. This tray design allows concurrent deposition from both top and bottom. This feature is particularly adapted to heterojunction devices with TCO deposited on both sides of the cell. Two main advantages are resulting:

- Simplified handling system since the wafer does not need to be flipped anymore.
- By design, the electrical edge isolation is achieved thanks to the shadowing of tray at the wafer edge, while the deposition from bottom target is running. Width of the wafer edge isolation is adjustable but in practice it is preferred to be minimized while keeping enough surface to support the wafer to prevent it to fall into the process chamber.

Problems with back-side deposition may occur if the tray motion has vibration. In particular, it is exacerbated for thinner wafers and if the edge isolation becomes too thin.

Finally, industrial high-volume equipment incorporate automatic tray return system for continuous un-manned operation of the tools. Actually, two version exists, return of empty tray or return of tray with deposited wafers depending if the output of the production is located at the end of the processes or both loading/unloading of the wafer is located at the beginning of the equipment.

(d) Ion plating

Ion-assisted deposition or ion vapour deposition otherwise called ion plating or RPD is a deposition process that combines periodic or concurrent bombardment of the depositing material in order to control or modify the properties of the film. 'In ion plating the energy, flux, and mass of the bombarding species along with the ratio of bombarding particles to depositing particles are important processing variables' [433]. The material may be vapourized by chemical vapour decomposition, sputtering or evaporation. This process can be done where ions for bombardment are received from a separate ion gun or from the plasma itself. The separate ion gun process is sometimes known as IBAD.

The RPD deposit technique was developed for TCO deposits on OLED technologies [434]. This technique has been shown to be relevant for making TCOs without damaging the organic layers. Some heterojunction labs and industries do use this technology, presumably to preserve the hydrogenated amorphous layers. This system operates under vacuum and consists mainly of a gun and a plasma controller. The gun works by pressure gradient, it generates the plasma (argon based). A magnetic field guides the electrons between the cathode (gun) and the heart (target). The plasma is thus located along the electron beam and is routed to the heart where the TCO source is located. This very dense plasma near the TCO generates, through a sublimation process, a strongly ionized vapour. This localized and high-density plasma makes it possible to combine a low-temperature process, denser films and better adhesion. Uniform deposits over large areas seem to be possible by adjusting the positioning (or scanning) of the beam from the plasma controller (deflection of the magnetic coils) and/or by putting in several guns.

(e) Comparison of RPD versus MS

The first point of comparison is the energy aspect of the process. With heterojunction PV technology, we are faced with the problems of degradation of the passivation layers, especially when they are thin. By sputtering, the kinetic energy of the sputtered elements is of the order of 3–5 eV. By ion plating, the kinetic energy of the vapourized elements is 10–20 eV (see Figure 5.16). Therefore, it could be believed that sputtering is less aggressive except if not considering that there is an additional contribution of Ar^+ ions and O^- ions, which implant at energies above 150 eV, and it is this population that can damage surfaces.

The material properties of the ITO film made by ion plating seem equivalent to conventional ITO sputtering and even better in terms of resistivity and roughness. These better characteristics could be explained by a denser microstructure and a more homogeneous crystalline orientation in the (111) direction.

A second point of comparison is the equipment maintainability related to source/target wear. By sputtering, the problem of erosion of the target and especially the formation of accumulations onto the process chamber walls leads to the problems of defects (particles, openings, cleaning). The RPD system is using surface sublimation of the TCO along with constant supply of the level of the source material to extend operation times.

In summary, the ion plating deposition technique allows to obtain an efficient ITO while being low damage (low temperature, low energy) and adaptable to large surfaces. Integrating this solution on a PV line is, therefore, feasible on the other hand no hindsight on cost issues are stated (plasma gun maintenance, source feedthrough maintenance, shielding clean frequency, etc.), causing still limited expansion for this technology.

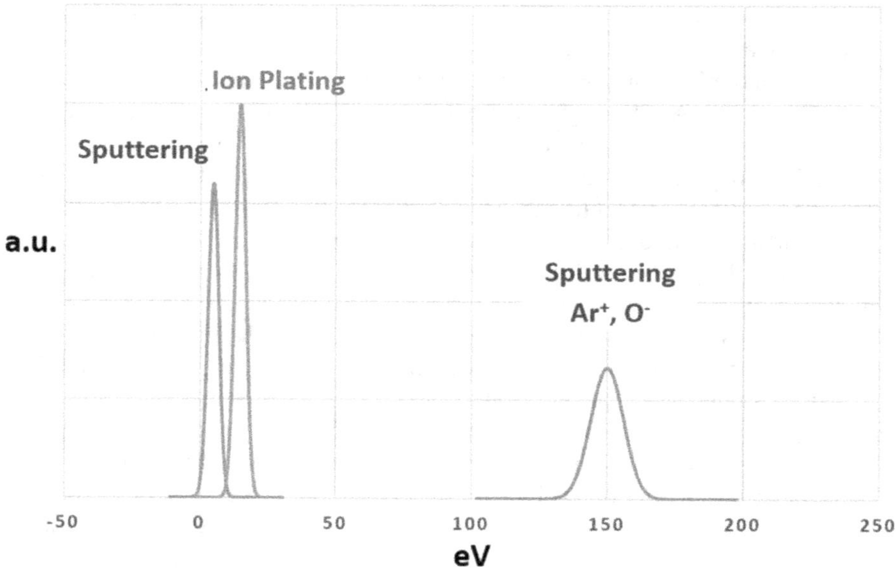

Figure 5.16 *Kinetic energies (eV) of plasma species by sputtering and by ion plating*

5.4.3 Essential metrics for a volume production equipment

Beyond the process considerations (the fulfilment necessary for the final aim of the product), a production equipment has to maximize several key parameter indicators (called KPI) for the sake of productivity. In particular, the equipment productivity is essential to calculate the cost of ownership (COO) of an operation and total COO of a production line. These criteria are fundamentals for the bankability and competitiveness of a technology.

Among the numerous parameters, KPI of main interest are commonly (see Figure 5.17):

- throughput (mechanical, process, net)
- equipment availability for production (uptime/downtime)
- equipment/product yield.

These KPI combined with following parameters will lead to the COO of an operation:

- selling price of the equipment
- maintenance cost of the equipment
- cost of consumables for the continuous operation of the equipment
- equipment footprint (particularly important if operated in a clean room environment)..

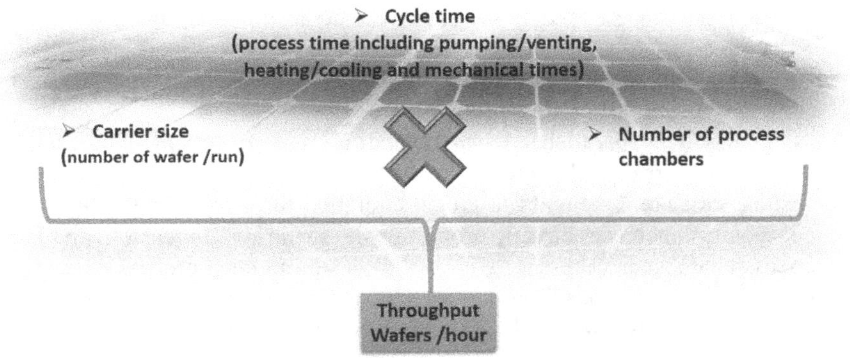

Figure 5.17 Equipment production throughput depends essentially of three parameters: cycle time, carrier size and number of process chambers

The SC industry has well defined and standardized all these parameters over 45 years ago, thankfully it applies directly to the PV equipment even though the PV industry has not yet fully converged towards standardization and domain roadmap.

SEMI standards [435] are voluntary technical agreements originally for the SC, then more broadly for the high-tech field: from FPD, micro-electromechanical systems, PV and high-brightness LED industries.

The SEMI International Standards Program is a service offered by SEMI for the benefit of the worldwide electronics design and manufacturing industries. The program, started over 45 years ago in North America, was expanded in 1985 to include programs in Europe and Japan, and now also has technical committees in China, Korea and Taiwan.

The objective of this paragraph is not to describe exhaustively the KPI but provide elements of reflection and sensitivity to select the best equipment to deposit evaporated contacts of metals, metal oxides or transparent conductive oxides for high volume 24-h/7-day operations. It is understood that these parameters are usually of less interest for institutes or R&D labs developing technologies but are the criteria of merit for economic viability of an equipment and eventually of a technology.

(a) Throughput

The throughput of a deposition equipment is commonly defined as the number of wafers per hour processed and exiting the equipment.

The throughput is the resultant of the combination of tool cycle time (CT), which itself combine the mechanical and process times:

$$\mathrm{CT} = \frac{\text{net production time}}{\text{number of units produced}} \tag{5.3}$$

It takes into account the process time but also the mechanical times for wafer handling, pumping time, heating/cooling time, etc.

In order to increase the productivity of the tool, the equipment manufacturers are acting on:

- Reducing the process time by increasing the deposition rate (nm/s).
- Reducing all handling times.
- Designing small volume vacuum chambers to reduce pumping/purge times in addition to using fast pumping tool.
- Using massively parallelized wafer handling and processing. For example, larger wafer carrier with increased number of wafers (generally three types are existing: cassette, boat and tray). One limitation for horizontal tray is that the tool footprint increases directly proportionally to the tray dimensions and larger process chamber is necessary with skilful design to keep process control uniform and homogeneous film characteristics across the tray. This is partially relaxed for the cassette- and boat carrier-based technologies.

It is understood that there are limits to decrease the mechanical/automation time and process time. Shorter process and automation times can potentially increase the stress of the wafer, respectively, due to thermal chocks and mechanical rough handling if accelerations/decelerations are abrupt. Furthermore, going to larger size wafers from M2 to M12 is, for instance, more challenging in terms of handling due to the inherent mechanical deformations (Young modulus) of the wafer amplified by wafer thickness decrease. One can demonstrate that a M12 wafer at 150 μm is equivalent of a M2 wafer with <90 μm.

The evolution of the throughput of the PV equipment in recent past years is outstanding. It has been accelerated by progress in the manufacturing of process equipment, mainly driven by China but not exclusively. It is widely commented in several conference and articles in particular [436] as illustrated in Figure 5.18. The

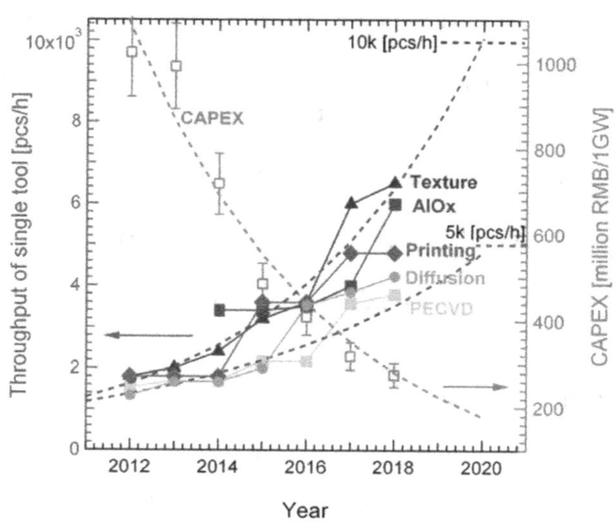

Figure 5.18 Evolution of equipment throughput and capital expenditure over the past decade. From . [436].

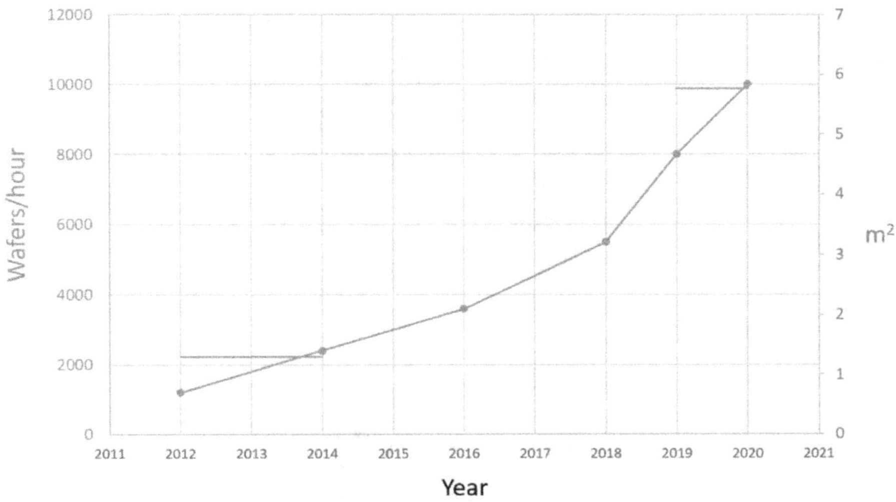

Figure 5.19 Blue: evolution of the throughput of PVD MS equipment dedicated to PV. Orange: typical tray surfaces. (b)

fast decrease in capital expenditure for more productivity is allowing the PV industry to become more financially sustainable, needing less cash to grow to GW level.

A similar trend is observed for the PVD sputtering as shown in Figure 5.19 with new equipment arising with 10 000 wafers per hour. To be noted that in Figure 5.18 the throughput for the equipment over 5 000 wafers per hour is for wafers with M6 dimension (166 × 166 mm). Meaning that the overall tool capacity is further increased compared to tool running lower throughput <4 000 wafers per hour and in addition with smaller M2/M4 wafer dimensions. In consequence, largest throughput equipment makes possible to have only two PVD sputtering per one gigawatt yearly capacity.

A. Uptime

The operation time is the sum of the scheduled downtime (preventive maintenance activities), the unscheduled downtime (corrective maintenance activities to recover from failures), the engineering time (engineering activities such as testing, etc.), the standby time (waiting for incoming products, facilities or operator) and the productive time (production activities). Only stops of more than 6 min will be considerate as failure and the time to recover from each failure will be considerate as corrective maintenance activities (Figure 5.20).

The equipment-related uptime (in %) is the ratio of the time when the entire line is available for production and engineering (sum of engineering time, standby time and productive time) divided by the operation time and multiplied by 100.

The equipment-related downtime (in %) is the ratio of the time when the line is not 100% available for production and engineering, due to related equipment problems and scheduled maintenance, divided by the operation time and multiplied by 100.

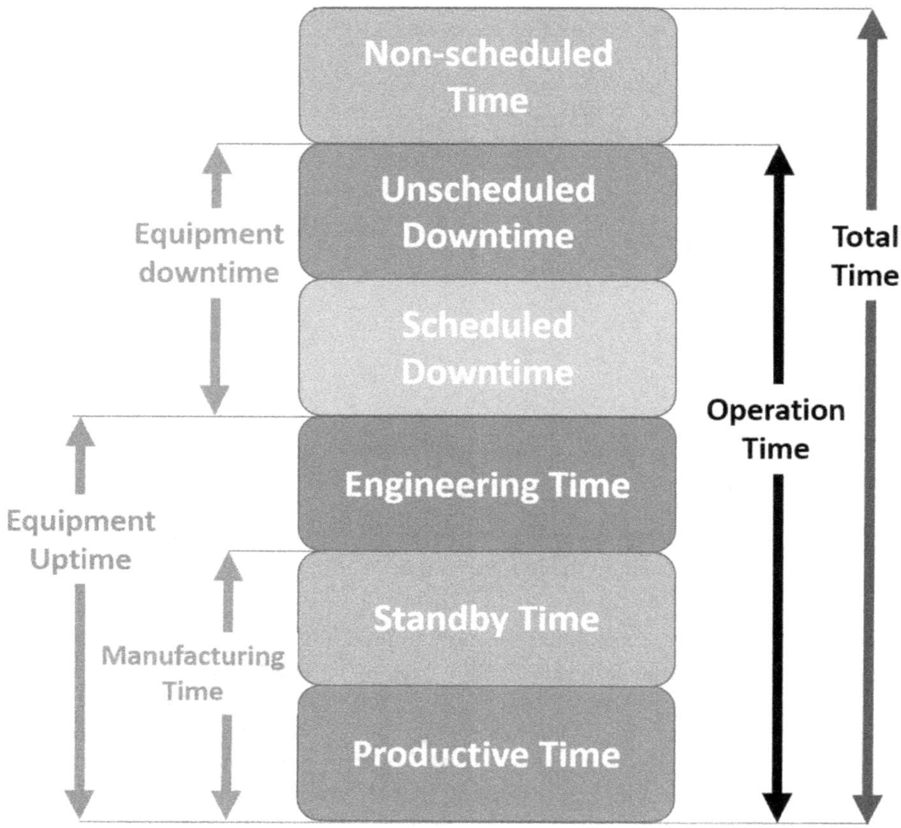

Figure 5.20 Equipment states chart based on SEMI standards

The equipment-related scheduled downtime (in %) is the ratio of the time when the line is not 100% available for production and engineering, due to the scheduled maintenance activities on the equipment, divided by the operation time and multiplied by 100.

Note that if a line requalification is needed after a failure, the requalification time will part of the unscheduled downtime.

(b) Mean time between failure (MTBF)

MTBF = uptime in hours/(number of failures + 1) on several rolling weeks.

A machine micro-stop exists when an operator intervention is needed to restart the machine after a machine failure (alarm), an operator request or a wafer none-quality (breakage or other defect).

A micro-stop is counted as an assist if it lasts less than a few minutes. Above this time, it will count as a failure and will be used to calculate the MTBF. As soon as the machine leaves the automatic mode, it is considered as a micro-stop. For the calculation of the MTBF the following causes are ignored:

- stop for scheduled maintenance
- voluntary stop of the operator
- failure of any facility (air, electricity, vacuum, exhaust, etc.)
- lack of incoming parts
- outgoing product not evacuated.

The MTBF can be calculated according to the information recorded in the log files and any external logging system. This parameter will also be checked together during the performance tests of an equipment and then followed on a regular basis afterwards. A correct estimation of this parameter can only be done when the machine is fully loaded with incoming products, therefore, it will be monitored at least during several consecutive weeks of full production. The MTBF should be consistently 'above' a required value.

(c) Mean time to repair

MTTR = repairing time/number of failures.

Repairing time is the time to repair the system trouble, often excluding the temperature up/down times and the process back-up time.

A machine micro-stop exists when an operator intervention is needed to restart the machine after a machine failure (alarm), an operator request or a wafer none-quality (breakage or other defect). A micro-stop is counted as an assist if it lasts less than a few minutes. Above this time, it will count as a failure and will be used to calculate the MTTR. As soon as the machine leaves the automatic mode, it is considered as a micro-stop.

For the calculation of the MTTR, the following causes are ignored:

- stop for scheduled maintenance
- voluntary stop of the operator
- failure of any facility (air, electricity, vacuum, exhaust, etc.)
- lack of incoming parts
- outgoing product not evacuated.

The MTTR will be calculated according to the information recorded in the log files and any external logging system.

This parameter will be checked together during the performance tests and then followed on a regular basis afterwards.

A correct estimation of this parameter can only be done when the machine is fully loaded with incoming products, therefore, it will be monitored at least during several consecutive weeks of full production. The MTTR should be consistently 'below' the required value.

(d) Product quality issues/non-quality

The different causes of non-quality issues or low quality are classified in two groups.

The first group takes into account the following reasons:

- breakage (i.e. wafer broken in a minimum two pieces, wafer piece missing, etc.)

- macro-cracks (i.e. wafer not broken but exhibiting cracks going through the whole thickness, etc.)
- micro-cracks (only at the exit of the system, the micro cracks detected at the entrance are not accounted for this calculation).

The second group takes into account the following reasons:

- traces
- defect or visual aspects
- processed wafers out of specifications based on colour, lifetime photoconductance, photoluminescence or electroluminescence criteria, for example
- contamination.

Additional quality reject criteria accounted as 'electrical non-quality' measured at the final IV testing are also counted like: low shunt resistance, Joe out of specification, etc.

To prevent from group one defects, the automation systems and process equipment should be designed to take into account the following:

- Wafers are indeed thin and relatively fragile. Therefore, shock, bending or any mechanical stress should be avoided.
- For the same, reason-induced thermal stress should be kept to a minimum in order to avoid any damage to the wafers.

For the handling of the individual wafers by the automation, the gentle manipulation of the wafers becomes essential for highly passivated devices (high V_{oc} well above 710 mV) such as passivated contact cells or heterojunction cells in order to preserve to the maximum the degradation of the surface either by generating surface damages or by contamination of the surface. In summary:

- Si wafer handling causes localized lack of passivation.
- Handling systems can scratch the wafer surfaces, or leave contaminants.
- Wafer handling is expected to become more and more aggressive as production throughputs increase.

The use of damage-free vacuum grippers and special transport belts is necessary. The ideal situation would be handling system without direct contact on the wafer surface but limited to the edge only. It is worth notice that the SC industry has already faced and solved these challenges, but often it cannot be directly transposed to the PV industry due to the very high throughputs considered for this latter.

On the other hand, it is observed and understandable that the surface damages or contamination defects are in principle less critical for the process steps involving PVD steps such as TCO deposition or metal oxide deposition or metal deposition as soon as this process step is distant from the core silicon bulk or silicon surface that are the most sensitive to defects.

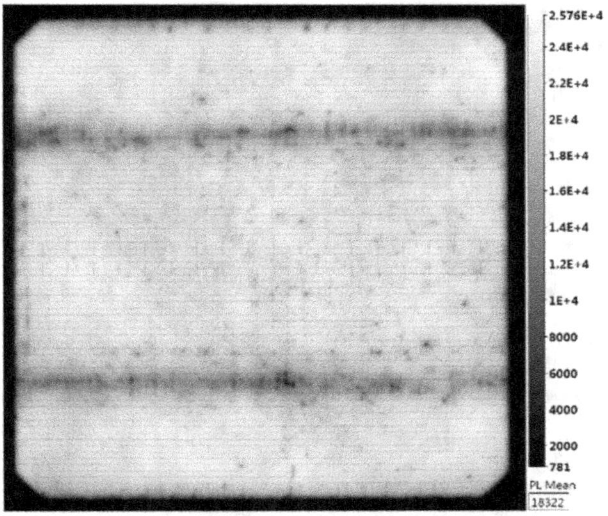

Figure 5.21 *Example of photoluminescence (PL) image of complete cell (busbarless metallization) exhibiting various types of defects (belts marks, localized defect at the periphery, grey zones at the centre, scratches, etc.)*

Several approaches have been explored for identification and quantification of defects caused by hardware. An interesting method [436] is based on photoluminescence image (Figure 5.21) processing software developed to quantify and analyze the defectivity present on SHJ cells. Several defectivity metrics were defined to characterize precisely how the local lack of passivation affects the cell performance. Furthermore, the ability to detect any changes in the behaviour of the production line before their effect on device performance become critical is essential to keep the productivity at high level. This statement is illustrated in Figure 5.22, where the cell efficiency η is plotted as a function of the global defectivity metric (MD) for two batches of cells produced before and after cleaning the automatic transport system (automation: belts, vacuum pickers).

(e) Target material utilization rate

One major improvement that occurred in the year 2000 is the deployment of cylindrical rotary target in the combination of stronger and stronger magnetic field. It did importantly improve the target material utilization rate defined as

$$U_T = \frac{\text{Eroded target volume}}{\text{Nominal target volume}} \tag{5.4}$$

Nowadays, target utilization rate of >80% is common (see Table 5.11). Magnet bars composed of many elementary magnets are finely balanced in order to achieve nearly flat erosion profiles. Strongest magnet fare employed to generate high magnetic field is required to improve plasma density and low target voltage to prevent

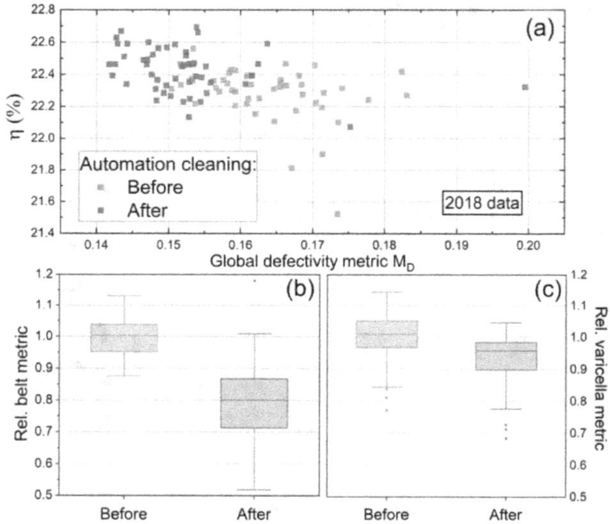

Figure 5.22 *(a) Cell performance versus the global defectivity metric (MD) before and after automation cleaning. (b) Relative belt marks metric before/after cleaning the automation. (c) Relative varicella (punctual, random defects) metric before/after cleaning the automation [437].*

(or reduce) the sputter damage of the cell caused by heavy ions including oxygen radical in the case of metal oxide or TCO deposition were oxygen is always present.

The main advantages of cylindrical rotary targets are:

- Extended period between target change preventive maintenances. It has a direct impact on tool availability and COO.
- Also, tighter thickness non-uniformity across the tray, while deposition trays are exhibiting larger surfaces. Different equipment supplier claiming thickness non-uniformity of <4% and even 3% for the largest tools with 10 000 wafers per hour [438, 439].

(f) Time for the format change (or set up time)

In case the equipment is used with different wafer dimensions, the time for the format change is the time between the exit of the last part of the series in progress (with one wafer format) and the exit of the first good part of the following series (with another wafer format). All necessary reconditioning of the machine such as handling test, tray coating, chamber cleaning, qualification runs will be included in this time. The format change performed on the machine should create the shortest possible production stop. The machine must be able to process equally all the formats considered.

Table 5.11 Effect of a magnet on process parameters and film characteristics. All the samples have been done with adjusted power to reach the same thickness than samples made with magnet 1. Example: for a same power magnet two gives higher voltage (~+14% higher) and lower current 86% than reference magnet 1 with 100% power.

	Magnet 1	**Magnet 2**		**Magnet 3**	
	With standard power (%)	**With standard power (%)**	**With adjusted power (%)**	**With standard power (%)**	**With adjusted power (%)**
Power	100	100	85	100	95
Voltage	100	114	112	95	95
Current	100	86	75	105	100
Thickness	100	120	100	10	100

Table 5.12 Typical KPI values encountered in the industry for mass production tools

KPI	Typical values
Uptime	>90%, best: >93%.
Number of rolling weeks	≥13 weeks
Micro stop duration	<6 min, best <3 min
MTBF	Week, best: weeks
Breakage rate	≤0.3%, best:≤0.25%

In summary, all above parameters and elements presented in this paragraph have a strong impact on the equipment capability to sustain high-volume production at the lowest cost. Therefore, the choice of the type of production equipment is led by the technology that meet the optimum values among all these KPI in addition to the expected material and cell performances. Indicative typical KPI are presented in Table 5.12.

5.4.4 Conclusions

Globally, the PV industry is maturing like SC and flat display industries have experienced in the past. Maybe quicker than previous fields in the search of greater financial margins, indeed they are ones of the smallest margins in high-tech industries. Larger returns on investment are necessary to re-invest in R&D and engineering. Therefore, in order to chase further continuous improvement in mass production, actors in this industry have developed high-throughput equipment (10 000 wafers per hour) and move to so-called Industry 4.0 with fully automated giant fabs where human intervention is re-centred to process optimization, quality controls and maintenance tasks. Automatic guided vehicles are transporting and delivering the wafers to the equipment. Manufacturing equipment systems similarly to SC are managing the production GW fabs with massive real-time data collection and analysis. For this latter, a standardization of data/information format protocol to SEC/GEM PV2 norm based on SEMI E30 for instance would ensure state-of-the-art giant PV fab production management with high rate communication between equipment. It is the path currently followed by key players to renew with competitiveness, enlarge financial margin and cope with the world challenge to produce massively the cells and modules for generating sustainable electricity thanks to the Sun's electromagnetic radiations.

Acknowledgement

The author would like to thank S. W. Glunz, L. Tutsch, B. Macco and H. Nagel for proofreading, S. Dubois and T. Desrues for valuable discussions, B. Stannowski and M. Bivour for updated data, and M. Tomassini, M. Vandenbossche and R. Varache for valuable discussions and informations.

References

[1] Haase F., Hollemann C., Schäfer S., *et al.* 'Laser contact openings for local poly-Si-metal contacts enabling 26.1%-efficient POLO-IBC solar cells'. *Solar Energy Materials and Solar Cells*. 2018;186(6):184–93.

[2] Richter A., Benick J., Feldmann F. 'Both sides contacted silicon solar cells: pptions for approaching 26% efficiency'. Proceedings of the 36th European Photovoltaic Solar Energy Conference and Exhibition, Marseille, September 2019; Munich: WIP Renewable Energies; 2019. pp. 90–5.

[3] Richter A., Benick J., Feldmann F., Fell A., Hermle M., Glunz S.W. 'N-type Si solar cells with passivating electron contact: identifying sources for efficiency limitations by wafer thickness and resistivity variation'. *Solar Energy Materials and Solar Cells*. 2017;173(6):96–105.

[4] Schindler F., Fell A., Müller R., *et al.* 'Towards the efficiency limits of multicrystalline silicon solar cells'. *Solar Energy Materials and Solar Cells*. 2018;185(5):198–204.

[5] Schubert M.C., Schindler F., Benick J., *et al.* 'The potential of cast silicon'. *Solar Energy Materials and Solar Cells*. 2021;219(3):110789.

[6] Zhao J., Wang A., Green M.A. '24·5% efficiency silicon PERT cells on MCZ substrates and 24·7% efficiency PERL cells on FZ substrates'. *Progress in Photovoltaics: Research and Applications*. 1999;7(6):471–74.

[7] Nekarda J., Reinwand D., Hartmann P., Preu R. 'Industrial inline PVD metallization for silicon solar cells with laser fired contacts leading to 21.8 % efficiency'. *Proceedings of the 2nd Workshop on Metallization for Crystalline Silicon Solar Cells, April 2010, Constance, Germany*. Constance: KOPS; 2010. pp. 52–8.

[8] Chapman B. *Glow Discharge Processes*. New York: Wiley; 1980.

[9] ' [Murarka S.P. 'Metallization' in Sze S.M. (ed.). *VLSI Technology*. 2nd ed. New York, NY: McGraw-Hill; 1988.

[10] Vossen J.L., Kern W. (eds.). *Thin Film Processes II*. Boston: Academic Press; 1991.

[11] Kurt j. Lesker Company. *Material deposition chart [online]*. 2021. Available from https://en.lesker.com/newweb/deposition_materials/materialdepositi onchart.cfm?pgid=0 [Accessed Oct 2020].

[12] Mayo S., Galloway K.F., Leedy T.F. 'Radiation dose due to electron-gun metallization systems'. *IEEE Transactions on Nuclear Science*. 1976;23(6):1875–1880.

[13] Depla D., Mahieu S., Greene J.E. 'Sputter deposition processes' in Martin P.M. (ed.). *Handbook of Deposition Technologies for Films and Coatings*. 3rd ed. Burlington, USA: Elsevier; 2010. pp. 253–296.

[14] Wolke W. Kathodenzerstäubung zur Beschichtung von kristallinen Silizium-Solarzellen [PhD thesis].Freiburg, Germany, Albert-Ludwigs-Universität; 2005.

[15] Hamasaki M. 'Radiation effects on thin-oxide mos capacitors caused by electron beam evaporation of aluminum'. *Solid-State Electronics.* 1983;26(4):299–303.

[16] Ning T.H. 'Electron trapping in SiO_2 due to electron-beam deposition of aluminum'. *Journal of Applied Physics.* 1978;49(7):4077–4082.

[17] Serikawa T., Yachi T. 'Electrical damage in mos devices by RF diode sputtered aluminum metallization'. *IEEE Transactions on Electron Devices.* 1981;28(7):882–885.

[18] Vossen J.L., O'Neill J.J., Hughes G.W., Taft F.A., Snedeker R. 'Source of radiation damage to mos devices during S-Gun R metallization'. *The Journal of Vacuum Science and Technology.* 1980;178888(1):400–2.

[19] Blakers A.W., Green M.A., Szpitalak T. 'Surface damage caused by electron-beam metallization of high open-circuit voltage solar cells'. *IEEE Electron Device Letters: A Publication of the IEEE Electron Devices Society.* 1984;5(7):246–47.

[20] Kopp J., Warta W., Aberle A., Glunz S., Knobloch J. 'Impact of metallization techniques on 20% efficient silicon solar cells'. *The Conference Record of the Twenty-Second IEEE Photovoltaic Specialists Conference;*Las Vegas (NV), Oct 1991. Piscataway, NJ: IEEE; 1991. pp. 278–83.

[21] O'Sullivan B.J., Bearda T., Nadupalli S., *et al.* 'Process-induced degradation of SiO_2 and a-Si:H passivation layers for photovoltaic applications'. *IEEE Journal of Photovoltaics.* 2014;4(5):1197–1203.

[22] Schiele Y., Hahn G., Terheiden B. 'Investigation of radiation damage to the Al_2O_3/Si wafer interface during electron beam evaporation by means of C-V and lifetime measurements'. *Proceedings of the 26th European Photovoltaic Solar Energy Conference and Exhibition;* Hamburg, Sep 2011; Munich, WIP Renewable Energies; 2011. pp. 1068–72.

[23] Glunz S.W., Feldmann F. 'SiO_2 surface passivation layers – a key technology for silicon solar cells'. *Solar Energy Materials and Solar Cells.* 2018;185:260–69.

[24] Dingemans G., van de Sanden M.C.M., Kessels W.M.M. 'Influence of the deposition temperature on the c-Si surface passivation by Al_2O_3 films synthesized by ALD and PECVD'. *Electrochemical and Solid-State Letters.* 2010;13(3):H76–79.

[25] Dingemans G., Beyer W., van de Sanden M.C.M., Kessels W.M.M. 'Hydrogen induced passivation of Si interfaces by Al_2O_3 films and SiO_2/Al_2O_3 stacks'. *Applied Physics Letters.* 2010;97(15):152106.

[26] Mette A. 'New concepts for front side metallization of industrial silicon solar cells' [PhD thesis].Freiburg, Germany, Albert-Ludwigs-Universität; 2007.

[27] Reinwand D., Nekarda J., Grohe A., Hartmann P., Trassl R., Preu R. 'Industrial sputtering metallization technology for crystalline silicon solar cells'. *Proceedings of the 24th European Photovoltaic Solar Energy Conference,* Hamburg, 21–25 September 2009; Munich, WIP Renewable Energies; 2009. pp. 1479–82.

[28] Katkhouda K., Feistritzer G., Buechel P., Geppert T. 'Overcoming the damage of passivation layers caused by DC-sputtered metals'. *Proceedings of the 27th European Photovoltaic Solar Energy Conference and Exhibition,* Frankfurt, Sept 2012; Munich, WIP Renewable Energies; 2012. pp. 1747–50.

[29] Dingemans G., Kessels W.M.M. 'Status and prospects of Al_2O_3-based surface passivation schemes for silicon solar cells'. *Journal of Vacuum Science & Technology A: Vacuum, Surfaces, and Films*. 2012;30(4):040802.

[30] Richter A., Benick J., Hermle M. 'Boron emitter passivation with Al_2O_3 and Al_2O_3/SiN_x stacks using ALD Al_2O_3'. *IEEE Journal of Photovoltaics*. 2013;3(1):236–245.

[31] Feldmann F., Bivour M., Reichel C., Hermle M., Glunz S.W. 'Passivated rear contacts for high-efficiency n-type Si solar cells providing high interface passivation quality and excellent transport characteristics'. *Solar Energy Materials and Solar Cells*. 2014;120:270–274.

[32] Mayer J.W., Lau S.S. *Electronic Material Science*. New York: Macmillan; 1990.

[33] Schwartz G.C., Srikrishnan K.V. *Handbook of Semiconductor Interconnection Technology*. 2nd edn. Boca Raton, FL: CRC/Taylor & Francis; 2006.

[34] Campbell R.B., Rohatgi A. 'Investigation of contact metallization systems for solar cells'. *Journal of the Electrochemical Society*. 1980;127(12):2702–2704.

[35] Fischer H., Gereth R. 'Electrochemically passivated contacts for silicon solar cells'. *IEEE Transactions on Electron Devices*. 1971;18(8):459–464.

[36] Gereth R., Fischer H., Link E., Mattes S., Pschunder W. 'Contribution to silicon solar cell technology'. *Energy Conversion*. 1972;12(3):103–107.

[37] Lowe V.E., Day A.C. 'High-temperature-stable contact metallization for silicon solar cells'. *IEEE Transactions on Electron Devices*. 1984;31(5):626–629.

[38] Jiang Y., Pillai S., Green M.A. 'Realistic silver optical constants for Plasmonics'. *Scientific Reports*. 2016;6:30605.

[39] McPeak K.M., Jayanti S.V., Kress S.J.P., *et al*. 'Plasmonic films can easily be better: rules and recipes'. *ACS Photonics*. 2015;2(3):326–333.

[40] Olmon R.L., Slovick B., Johnson T.W. 'Optical dielectric function of gold'. *Physical Review. B*. 2012;86(23):235147.

[41] Palik E.D., Ghosh G. *Handbook of Optical Constants of Solids*. San Diego: Academic Press; 1998.

[42] Palm K.J., Murray J.B., Narayan T.C., Munday J.N. 'Dynamic optical properties of metal hydrides'. *ACS Photonics*. 2018;5(11):4677–4686.

[43] Werner W.S.M., Glantschnig K., Ambrosch-Draxl C. 'Optical constants and inelastic electron-scattering data for 17 elemental metals'. *Journal of Physical and Chemical Reference Data*. 2009;38(4):1013–1092.

[44] Rhoderick E.H., Williams R.H. *Metal-semiconductor Contacts*. 2nd edn. Oxford, UK, New York: Clarendon Press; Oxford University Press; 1988.

[45] Schroder D.K., Meier D.L. 'Solar cell contact resistance—a review'. *IEEE Transactions on Electron Devices*. 1984;31(5):637–647.

[46] Fisher D.J. *Diffusion in Silicon – 10 Years of Research*. Zurich: Trans Tech Publications, Limited; 1998.

[47] Jones S.M. *Diffusion in Silicon*. Georgetown, MA: IC Knowledge LLC; 2008.

[48] Shacham-Diamond Y., Dedhia A., Hoffstetter D., Oldham W.G. 'Reliability of copper metallization on silicon-dioxide'. *Proceedings of Eighth International IEEE VLSI Multilevel Interconnection Conference,* Santa Clara (CA), June 1991; Piscataway, IEEE; 1991. pp. 109–15.

[49] Sze S.M., Irvin J.C. 'Resistivity, mobility and impurity levels in GaAs, Ge, and Si at 300°K'. *Solid-state Electronics*. 1968;11(6):599–602.

[50] Dubois S., Palais O., Pasquinelli M., Martinuzzi S., Jaussaud C. 'Influence of substitutional metallic impurities on the performances of P-type crystalline silicon solar cells: the case of gold'. *Journal of Applied Physics*. 2006;100(12):123502.

[51] Coletti G., Bronsveld P.C.P., Hahn G., *et al.* 'Impact of metal contamination in silicon solar cells'. *Advanced Functional Materials*. 2011;21(5):879–890.

[52] Bartsch J., Mondon A., Bayer K., Schetter C., Hörteis M., Glunz S.W. 'Quick determination of copper-metallization long-term impact on silicon solar cells'. *Journal of the Electrochemical Society*. 2010;157(10):H942–H946.

[53] Fong K.C., Kho T.C., Fell A., *et al.* 'Contact resistivity of evaporated Al contacts for silicon solar cells'. *IEEE Journal of Photovoltaics*. 2015;5(5):1304–1309.

[54] Vaidya S. 'Pit formations at AL/(100)Si contact windows'. *Journal of Materials*. 1981;10(2):337–346.

[55] Couderc R., Amara M., Lemiti M. 'Improvement of back surface metallization in a silicon interdigitated back contacts solar cell'. *Energy Procedia*. 2013;38:684–690.

[56] Gambino J.P., Colgan E.G. 'Silicides and ohmic contacts'. *Materials Chemistry and Physics*. 1998;52(2):99–146.

[57] Maex K., van Rossum M. *Properties of Metal Silicides*. London: INSPEC/IEE; 1995.

[58] Yu A.Y.C. 'Electron tunneling and contact resistance of metal-silicon contact barriers'. *Solid-State Electronics*. 1970;13(2):239–247.

[59] Richter A., unpublished. 'Contact resistivity of evaporated Al (300 nm) and Ti/Pd/Ag (50 nm / 50 nm / 200 nm) contacts was measured with the TLM technique on different $POCl_3$ diffused n+ and BBr_3 diffused p+ surfaces on planar Si wafers. Direct before the evaporation, the samples were dipped in diluted HF. The TLM contacts were structured by means of photolithography and the lift-off technique. After deposition, some of the samples were annealed on a hotplate for 1 min at 350°C or annealed under forming gas (FGA) at 425°C for 25 min. The surface doping concentration was measured via ECV'.

[60] Schroder D.K. *Semiconductor Material and Device Characterization*. 3rd edn. Hoboken: Wiley; 2006.

[61] Mader C., Müller J., Eidelloth S., Brendel R. 'Local rear contacts to silicon solar cells by in-line high-rate evaporation of aluminum'. *Solar Energy Materials and Solar Cells*. 2012;107:272–82.

[62] Benick J., Hoex B., van de Sanden M.C.M., Kessels W.M.M., Schultz O., Glunz S.W. 'High efficiency N-type Si solar cells on Al_2O_3-passivated boron emitters'. *Applied Physics Letters*. 2008;92(25):253504.

[63] Glunz S.W., Knobloch J., Biro D., Wettling W. 'Optimized high-efficiency silicon solar cells with Jsc=42 mA/cm_2 and η=23.3%'. *Fourteenth European Photovoltaic Solar Energy Conference. Proceedings of the International Conference; Barcelona, 1997.*; Bedford, H. S. Stephens & Associates; 1997. pp. 392–5.

[64] Knobloch J., Aberle A., Voss B. 'Cost effective processes for silicon solar cells with high performance'. *Ninth E.C. Photovoltaic Solar Energy Conference. Proceedings of the International Conference,* Freiburg, 1989; Dordrecht, Kluwer Academic Publishers Group; 1989. pp. 777–80.

[65] Zhao J., Wang A., Altermatt P., Wenham S., Green M. '24% efficient perl silicon solar cell: recent improvements in high efficiency silicon cell research'. *Solar Energy Materials and Solar Cells*. 1996;41:87–99.

[66] Steinhauser B., Kamp M., Brand A.A., *et al.* 'High-efficiency n-type silicon solar cells: advances in PassDop technology and NiCu plating on boron emitter'. *IEEE Journal of Photovoltaics*. 2016;6(2):419–425.

[67] Franklin E., Fong K., McIntosh K., *et al.* 'Design, fabrication and characterisation of a 24.4% efficient interdigitated back contact solar cell'. *Progress Photovoltaics: Research and Applications*. 2016;24(4):411–427.

[68] Hendrichs M., Keding R., Spribille A. 'Optimization of multi-layer metallization design for large-area back-contact back-junction solar cells'. *Proceedings of the 29th European Photovoltaic Solar Energy Conference and Exhibition,* Amsterdam, Sep 2014; Munich, WIP Renewable Energies; 2014. pp. 412–6.

[69] Reichel C., Granek F., Hermle M., Glunz S.W. 'Back-contacted back-junction N-type silicon solar cells featuring an insulating thin film for decoupling charge carrier collection and metallization geometry'. *Progress in Photovoltaics: Research and Applications*. 2012;31(5):1063–1076.

[70] Green M.A., Blakers A.W., Zhao J., Milne A.M., Wang A., Dai X. 'Characterization of 23-percent efficient silicon solar cells'. *IEEE Transactions on Electron Devices*. 1990;37(2):331–336.

[71] Richter A. *Aluminum Oxide for the Surface Passivation of High Efficiency Silicon Solar Cells – Technology and Advanced Characterization*. Stuttgart: Fraunhofer Verlag; 2015.

[72] Aberle A.G. 'Surface passivation of crystalline silicon solar cells: a review'. *Progress Photovoltaics: Research and Applications*. 2000;8(5):473–487.

[73] Katkhouda K., Martinez-Limia A., Bornschein L., *et al.* 'Aluminum-based rear-side PVD metallization for nPERT silicon solar cells'. *IEEE Journal of Photovoltaics*. 2014;4(1):160–167.

[74] Green M.A., Blakers A.W., Shi J., Keller E.M., Wenham S.R. 'High-efficiency silicon solar cells'. *IEEE Transactions on Electron Devices*. 1984;31(5):679–683.

[75] Kaminski A., Vandelle B., Fave A., *et al.* 'Aluminium BSF in silicon solar cells'. *Solar Energy Materials and Solar Cells*. 2002;72(1-4):373–379.

[76] Hartley O.N., Russell R., Heasman K.C., Mason N.B., Bruton T.M. 'Investigation of thin aluminium films on the rear of monocrystalline silicon solar cells for back surface field formation'. *Conference Record of the Twenty-Ninth IEEE Photovoltaic Specialists Conference,* New Orleans (LA), May 2002; Piscataway, NJ, IEEE; 2002. pp. 118–21.

[77] Chen J., Du Z., Hoex B., Tey Z.H.J., Aberle A.G. 'Investigation of evaporated rear contacts for Al-LBSF silicon wafer solar cells'. *Energy Procedia*. 2012;25:10–18.

[78] Uruena A., Hernández J.L., John J., Poortmans J., Mertens R. 'Controlling the depth of the local Al BSF in PERC crystalline solar cells using alternative back side metallization'. *Proceedings of the 25th European Photovoltaic Solar Energy Conference and Exhibition and of the 5th World Conference on Photovoltaic Energy Conversion;* Valencia, Sep 2010; Munich, WIP Renewable Energies; 2010. pp. 2562–4.

[79] Schneiderlöchner E., Preu R., Lüdemann R., Glunz S.W. 'Laser-fired rear contacts for crystalline silicon solar cells'. *Progress Photovoltaics: Research and Applications*. 2002;10(1):29–34.

[80] Hezel R. 'Silicon nitride for the improvement of silicon inversion layer solar cells'. *Solid-State Electronics*. 1981;24(9):863–868.

[81] Stang J.-C., Hendrichs M.-S., Merkle A., *et al.* 'ITO-free metallization for interdigitated back contact silicon heterojunction solar cells'. *Energy Procedia*. 2017;124:379–383.

[82] Reichel C., Müller R., Feldmann F., Richter A., Hermle M., Glunz S.W. 'Influence of the transition region between P- and N-type polycrystalline silicon passivating contacts on the performance of interdigitated back contact silicon solar cells'. *Journal of Applied Physics*. 2017;122(18):184502.

[83] Rienacker M., Bossmeyer M., Merkle A., *et al.* 'Junction resistivity of carrier-selective polysilicon on oxide junctions and its impact on solar cell performance'. *IEEE Journal of Photovoltaics*. 2017;7(1):11–18.

[84] Ingenito A., Limodio G., Procel P., *et al.* 'Silicon solar cell architecture with front selective and rear full area ion-implanted passivating contacts'. *Solar RRL*. 2017;1(7):1700040.

[85] Bullock J., Hettick M., Geissbühler J., *et al.* 'Efficient silicon solar cells with dopant-free asymmetric heterocontacts'. *Nature energy*. 2016;1(3):15031.

[86] Zhang Y., Cui W., Zhu Y., *et al.* 'High efficiency hybrid PEDOT:PSS/nanostructured silicon Schottky junction solar cells by doping-free rear contact'. *Energy & Environmental Science*. 2015;8(1):297–302.

[87] Wan Y., Samundsett C., Bullock J., *et al.* 'Conductive and stable magnesium oxide electron-selective contacts for efficient silicon solar cells'. *Advanced Energy Materials*. 2017;7(5):1601863.

[88] Yoon W., Moore J.E., Cho E., *et al.* 'Hole-selective molybdenum oxide as a full-area rear contact to crystalline p-type Si solar cells'. *Japanese Journal of Applied Physics*. 2017;56(8S2):08MB18.

[89] Wan Y., Samundsett C., Bullock J., *et al.* 'Magnesium fluoride electron-selective contacts for crystalline silicon solar cells'. *ACS Applied Materials & Interfaces*. 2016;8(238):14671–7.

[90] Macco B., Black L.E., Melskens J., *et al.* 'Atomic-layer deposited Nb2O5 as transparent passivating electron contact for c-Si solar cells'. *Solar Energy Materials and Solar Cells*. 2018;184:98–104.

[91] Yang X., Aydin E., Xu H., *et al.* 'Tantalum nitride Electron-Selective contact for crystalline silicon solar cells'. *Advanced Energy Materials*. 2018;8(20):1800608.

[92] Yang X., Bi Q., Ali H., Davis K., Schoenfeld W.V., Weber K. 'High-performance TiO_2-based electron-selective contacts for crystalline silicon solar cells'. *Advanced Materials*. 2016;28(28):5891–5897.

[93] Plagwitz H., Nerding M., Ott N., Strunk H.P., Brendel R. 'Low-temperature formation of local Al contacts to a-Si: H-passivated Si wafers'. *Progress Photovoltaics: Research and Applications*. 2004;12:47–54.

[94] Reichel C., Würfel U., Winkler K., *et al.* 'Electron-selective contacts via ultra-thin organic interface dipoles for silicon organic heterojunction solar cells'. *Journal of Applied Physics*. 2018;123(2):24505.

[95] Nagel H., Gledhill S., Kroyer T. 'Solderable PVD Al back contacts for the module integration of high-efficiency c-Si solar cells'. *Proceedings of the 37th European Photovoltaic Solar Energy Conference and Exhibition,* Sep 2020, [online]; Munich, WIP Renewable Energies; 2020–861–6.

[96] Nagel H., Eberlein D., Hoffmann S. 'Ultrasonically tinned PVD al rear contacts on high-efficiency crystalline silicon solar cells for module integration'. *Proceedings of the 33rd European Photovoltaic Solar Energy Conference and Exhibition,* Amsterdam, Sep 2017; Munich, WIP Renewable Energies; 2017. pp. 295–9.

[97] Nagel H., Kamp M., Eberlein D. 'Enabling solderability of PVD Al rear contacts on high-efficiency crystalline silicon solar cells by wet chemical treatment'. *Proceedings of the 32nd European Photovoltaic Solar Energy Conference and Exhibition,* Munich, June 2016; Munich, WIP Renewable Energies; 2016. pp. 48–52.

[98] Jung V., Köntges M. 'Al/Ni:V/Ag metal stacks as rear-side metallization for crystalline silicon solar cells'. *Progress in Photovoltaics: Research and Applications*. 2012;21(5):876–883.

[99] Kumm J., Samadi H., Hartmann P., Nold S., Wolf A., Wolke W. Development of Temperature-Stable, Solderable PVD Rear Metallization for Industrial Silicon Solar Cells. *Proceedings of the 28th European Photovoltaic Solar*

Energy Conference and Exhibition; Paris, September-October 2013; Munich, WIP Renewable Energies; 2013. pp. 1287–90 pp..

[100] Kumm J., Chacko R.V., Samadi H., Hartmann P., Eberlein D., Wolf A. 'Long-term and annealing stable, solderable PVD metallization with optimized Al diffusion barrier'. *Energy Procedia.* 2015;77:374–81.

[101] Stein W., Schwabedissen A., Dimer M. 'A new back metallization conceptfor PERC cells'. *IEEE 7th World Conference on Photovoltaic Energy Conversion (WCPEC) (A Joint Conference of 45th IEEE PVSC, 28th PVSEC & 34th EU PVSEC),* Waikoloa (HI), June 2018; Piscataway, NJ, IEEE; 2018. pp. 1058–61.

[102] Ting C.Y., Crowder B.L. 'Electrical properties of Al/Ti contact metallurgy for VLSI application'. *Journal of the Electrochemical Society.* 1982;129(11):2590–4.

[103] Bivour M., Meinhardt C., Pysch D. 'N-type silicon solar cells with amorphous/crystalline silicon heterojunction rear emitter'. *35th IEEE Photovoltaic Specialists Conference Proceedings;* Honolulu (HI), June 2010; Piscataway (NJ), IEEE; 2010. pp. 1304–8.

[104] Blakers A.W., Wang A., Milne A.M., Zhao J., Green M.A. '22.8% efficient silicon solar cell'. *Applied Physics Letters.* 1989;55(13):1363–5.

[105] Zhao J., Wang A., Green M.A. '23.5% efficient silicon solar cell'. *Progress Photovoltaics: Research and Applications.* 1994;2(3):227–30.

[106] Glunz S.W., Feldmann F., Richter A. 'The irresistible charm of a simple current flow pattern – 25% with a solar cell featuring a full-area back contact'. *Proceedings of the 31st European Photovoltaic Solar Energy Conference and Exhibition; Hamburg, Sep 2015*; Munich, WIP Renewable Energies; 2015. pp. 259–63.

[107] Song N., Romer U., Gentle A., *et al.* 'Metallization method for interdigitated back-contact silicon solar cells employing an insulating resin layer and a Ti/Ag/Cu metal stack'. *IEEE Journal of Photovoltaics.* 2018;8(4):916–22.

[108] van der Vossen R., Feldmann F., Moldovan A., Hermle M. 'Comparative study of differently grown tunnel oxides for p-type passivating contacts'. *Energy Procedia.* 2017;124:448–54.

[109] Nicolai M., Zanuccoli M., Feldmann F., Hermle M., Fiegna C. 'Analysis of silicon solar cells with poly-Si/SiO_x carrier-selective base and emitter contacts'. *IEEE Journal of Photovoltaics.* 2018;8(1):103–9.

[110] Kökbudak G., Müller R., Feldmann F. 'On the determination of the contact resistivity for passivating contacts using 3D simulations'. *Proceedings of the 33rd European Photovoltaic Solar Energy Conference and Exhibition;* Amsterdam, Sep 2017; Munich, WIP Renewable Energies; 2017. pp. 242–6.

[111] Luderer C., Messmer C., Hermle M., Bivour M. 'Transport losses at the TCO/a-Si:H/c-Si heterojunction: influence of different layers and annealing'. *IEEE Journal of Photovoltaics.* 2020;10(4):952–8.

[112] Bivour M., Temmler J., Steinkemper H., Hermle M. 'Molybdenum and tungsten oxide: high work function wide band gap contact materials for hole

selective contacts of silicon solar cells'. *Solar Energy Materials and Solar Cells*. 2015;142:34–41.

[113] Rohatgi A., Rounsaville B., Ok Y.-W., *et al.* 'Fabrication and modeling of high-efficiency front junction N-type silicon solar cells with tunnel oxide passivating back contact'. *IEEE Journal of Photovoltaics*. 2017;7(5):1236–43.

[114] Holman Z.C., Descoeudres A., Wolf Sde., Ballif C. 'Record infrared internal quantum efficiency in silicon heterojunction solar cells with dielectric/metal rear reflectors'. *IEEE Journal of Photovoltaics*. 2013;3(4):1243–9.

[115] Cariou R., Benick J., Feldmann F., *et al.* 'III–V-on-silicon solar cells reaching 33% photoconversion efficiency in two-terminal configuration'. *Nature Energy*. 2018;3(4):326–33.

[116] Hauser H., Mühlbach K., Höhn O., *et al.* 'Tailored disorder: a self-organized photonic contact for light trapping in silicon-based tandem solar cells'. *Optics Express*. 2020;28(8):10909–18.

[117] Feldmann F., Simon M., Bivour M., Reichel C., Hermle M., Glunz S.W. 'Carrier-selective contacts for Si solar cells'. *Applied Physics Letters*. 2014;104(18):181105.

[118] Ingenito A., Nogay G., Jeangros Q., *et al.* 'A passivating contact for silicon solar cells formed during a single firing thermal annealing'. *Nature Energy*. 2018;3(9):800–8.

[119] Bivour M., Schröer S., Hermle M., Glunz S.W. 'Silicon heterojunction rear emitter solar cells: less restrictions on the optoelectrical properties of front side TCOs'. *Solar Energy Materials and Solar Cells*. 2014;122:120–9.

[120] Descoeudres A., Allebé C., Badel N., *et al.* 'Low-temperature processes for passivation and metallization of high-efficiency crystalline silicon solar cells'. *Solar Energy*. 2018;175:54–9.

[121] Schulze P.S.C., Bett A.J., Bivour M., *et al.* '25.1% high-efficiency monolithic perovskite silicon tandem solar cell with a high bandgap perovskite absorber'. *Solar RRL*. 2020;4(7):2000152.

[122] Halbich M.-U., Zielke D., Gogolin R., Sauer-Stieglitz R., Lövenich W., Schmidt J. 'Improved surface passivation and reduced parasitic absorption in PEDOT:PSS/c-Si heterojunction solar cells through the admixture of sorbitol'. *Scientific Reports*. 2019;9(1):9775.

[123] Yang X., Xu H., Liu W., *et al.* 'Atomic layer deposition of vanadium oxide as hole-selective contact for crystalline silicon solar cells'. *Advanced Electronic Materials*. 2020;6(8):2000467.

[124] Allen T.G., Bullock J., Jeangros Q., *et al.* 'A low resistance calcium/reduced titania passivated contact for high efficiency crystalline silicon solar cells'. *Advanced Energy Materials*. 2017;7(12):1602606.

[125] Bower R.W., Mayer J.W. 'Growth kinetics observed in the formation of metal silicides on silicon'. *Applied Physics Letters*. 1972;20(9):359–61.

[126] Canali C., Campisano S.U., Lau S.S., Liau Z.L., Mayer J.W. 'Solid-phase epitaxial growth of Si through palladium silicide layers'. *Journal of Applied Physics*. 1975;46(7):2831–6.

[127] Zhang X., Wan Y., Bullock J., Allen T., Cuevas A. 'Low resistance Ohmic contact to P-type crystalline silicon via nitrogen-doped copper oxide films'. *Applied Physics Letters*. 2016;109(5):52102.

[128] Bullock J., Cuevas A., Allen T., Battaglia C. 'Molybdenum oxide MoO_x : A versatile hole contact for silicon solar cells'. *Applied Physics Letters*. 2014;105(23):232109.

[129] Zollner S., Gregory R.B., Kottke M.L. 'Metrology of silicide contacts for future CMOs'. *AIP Conference Proceedings*. 2007;931:337–46.

[130] Zhu S., Van Meirhaeghe R.L., Forment S., Ru G., Li B. 'Effects of the annealing temperature on Ni silicide/n-Si(100) Schottky contacts'. *Solid-State Electronics*. 2004;48(1):29–35.

[131] Xu B., Littau K., Zesch J., Fork D. 'Front side metallization of crystalline silicon solar cells using selectively laser drilled contact openings'. *Proceedings of the 34th IEEE Photovoltaic Specialists Conference,* Philadelphia (PA), June 2009; Piscataway (NJ), IEEE; 2009. pp. 517–22.

[132] Wan Y., Samundsett C., Yan D., *et al.* 'A magnesium/amorphous silicon passivating contact for *n*-type crystalline silicon solar cells'. *Applied Physics Letters*. 2016;109(11):113901.

[133] Pan G., Chen J., Ge K., *et al.* 'Zn(O,S)-based electron-selective contacts with tunable band structure for silicon heterojunction solar cells'. *Journal of Materials Chemistry C*. 2019;7(15):4449–58.

[134] MasmitjG., Ortega P., Puigdollers J., *et al.* 'Interdigitated back-contacted crystalline silicon solar cells with low-temperature dopant-free selective contacts'. *Journal of Materials Chemistry C*. 2018;6(9):3977–85.

[135] Godfrey R.B., Green M.A. '655 mV open-circuit voltage, 17.6% efficient silicon MIS solar cells'. *Applied Physics Letters*. 1979;34(11):790–3.

[136] Hoppe H., Sariciftci N.S. 'Organic solar cells: an overview'. *Journal of Materials Research*. 2004;19(7):1924–45.

[137] Allen T.G., Bullock J., Zheng P., *et al.* 'Calcium contacts to n-type crystalline silicon solar cells'. *Progress Photovoltaics: Research and Applications*. 2017;25(7):636–44.

[138] Hauser H., Tucher N., Tokai K., *et al.* 'Development of nanoimprint processes for photovoltaic applications'. *Journal of Micro/Nanolithography, MEMS, and MOEMS*. 2015;14(3):031210.

[139] Stüwe D., Mager D., Biro D., Korvink J.G. 'Inkjet technology for crystalline silicon photovoltaics'. *Advanced Materials*. 2015;27(4):599–626.

[140] Tomasi A., Paviet-Salomon B., Jeangros Q., *et al.* 'Simple processing of back-contacted silicon heterojunction solar cells using selective-area crystalline growth'. *Nature energy*. 2017;2(5):17062.

[141] Zieliński B., O'Sullivan B.J., Singh S., *et al.* 'Process simplification for 15.6×15.6 cm_2 interdigitated back contact silicon solar cells by laser doping'. *Solar Energy Materials and Solar Cells*. 2017;163:66–72.

[142] Chen Y., Yang Y., Zhang X., *et al.* 'Forming aluminum electrodes by screen printing and electron-beam evaporation for high performance interdigitated back contact solar cells'. *Solar Energy Materials and Solar Cells*. 2015;143:205–11.

[143] Kamp M., Maywald A., Bartsch J., *et al.* 'Electrochemical contact separation for PVD aluminum back contact solar cells'. *Energy Procedia.* 2015;67:70–5.

[144] Kamp M., Efinger R., Gensowski K., Bechmann S., Bartsch J., Glatthaar M. 'Structuring of metal layers by electrochemical screen printing for back-contact solar cells'. *IEEE Journal of Photovoltaics.* 2018:1–7.

[145] Gensowski K., Kamp M., Efinger R. Selective seed layer patterning of PVD-metal stacks by electrochemical screen printing for solar cell applications. *Proceedings of the 36th European Photovoltaic Solar Energy Conference and Exhibition;* Sep 2019, Marseille; Munich, WIP Renewable Energies; 2019. pp. 266–71.

[146] Chopra K.L., Major S., Pandya D.K. 'Transparent conductors—a status review'. *Thin Solid Films.* 1983;102(1):1–46.

[147] Minami T. 'Transparent conducting oxide semiconductors for transparent electrodes'. *Semiconductor Science and Technology.* 2005;20(4):S35–44.

[148] Exarhos G.J., Zhou X.-D. 'Discovery-based design of transparent conducting oxide films'. *Thin Solid Films.* 2007;515(18):7025–52.

[149] Granqvist C.G. 'Transparent conductors as solar energy materials: a panoramic review'. *Solar Energy Materials and Solar Cells.* 2007;91(17):1529–98.

[150] Pasquarelli R.M., Ginley D.S., O'Hayre R. 'Solution processing of transparent conductors: from flask to film'. *Chemical Society Reviews.* 2011;40(11):5406–41.

[151] Dsy J., Peiris N. 'Overview on transparent conducting oxides and state of the art of low-cost doped ZnO systems'. *Science Forecast of Material and Chemical Engineering.* 2018;1:1004.

[152] Edwards P.P., Porch A., Jones M.O., Morgan D.V., Perks R.M. 'Basic materials physics of transparent conducting oxides'. *Dalton Transactions.* 2004;60(19):2995–3002.

[153] Bright C.I. 'Transparent conductive thin films' in Piegari A., Flory F. (eds.). *Optical Thin Films and Coatings – From Materials to Applications.* Duxford: Woodhead Publishing; 2018. pp. 741–88.

[154] Kawazoe H., Yanagi H., Ueda K., Hosono H. 'Transparent *p*-type conducting oxides: design and fabrication of *p-n* heterojunctions'. *MRS Bulletin.* 2000;25(8):28–36.

[155] Herodotou S., Treharne R.E., Durose K., Tatlock G.J., Potter R.J. 'The effects of Zr doping on the optical, electrical and microstructural properties of thin ZnO films deposited by atomic layer deposition'. *Materials.* 2015;8(10):7230–40.

[156] Kobayashi E., Watabe Y. 'Rear-emitter silicon heterojunction solar cells with cerium oxide-doped indium oxide'. *Proceedings of the 29th European Photovoltaic Solar Energy Conference and Exhibition,* Amsterdam, Sep 2014; Munich, WIP Renewable Energies; 2014. pp. 472–4.

[157] Bädeker K. 'Über die elektrische Leitfähigkeit und die thermoelektrische Kraft einiger Schwermetallverbindungen'. *Annalen der Physik.* 1907;327(4):749–66.

[158] Bauer G. 'Elektrisches und optisches Verhalten von Halbleitern. XIII Messungen an Cd-, Tl- und Sn-Oxyden'. *Annalen der Physik.* 1937;422(5):433–45.

[159] Rupprecht G. 'Untersuchungen Der elektrischen und lichtelektrischen Leitfähigkeit dünner Indiumoxydschichten'. *Zeitschrift für Physik.* 1954;139(5):504–17.

[160] Du J., Chen X.-liang., Liu C.-chi., *et al.* 'Highly transparent and conductive indium tin oxide thin films for solar cells grown by reactive thermal evaporation at low temperature'. *Applied Physics A*. 2014;117(2):815–22.

[161] McMaster H.A. 'Inventor; conductive coating for glass and method of application'. *USA patent US2429420A*. 1947.

[162] Mochel J.M. 'Inventor; coated resistance'. *USA patent US2564706A*. 1951.

[163] Ginley D.S., Hosono H., Paine D.C. *Handbook of Transparent Conductors*. New York: Springer; 2010.

[164] Dates H.F., Davis J.K. 'Inventors; Iridizing method'. *USA patent US3331702A*. 1967.

[165] Grove W.R. 'Vii. on the electro-chemical polarity of gases'. *Philosophical Transactions*. 1852;142:87–101.

[166] Chapin J.S. 'Inventor; sputtering process and apparatus'. *USA patent US4166018A*. 1978.

[167] Han Y., Kim D., Cho J.-S., Koh S.-K., Song Y.S. 'Tin-doped indium oxide (Ito) film deposition by ion beam sputtering'. *Solar Energy Materials and Solar Cells*. 2001;65(1):211–18.

[168] Horstmann F., Sittinger V., Szyszka B. 'Heat treatable indium tin oxide films deposited with high power pulse magnetron sputtering'. *Thin Solid Films*. 2009;517(10):3178–82.

[169] Liu C., Mihara T., Matsutani T., Asanuma T., Kiuchi M. 'Preparation and characterization of indium tin oxide films formed by oxygen ion beam assisted deposition'. *Solid State Communications*. 2003;126(9):509–13.

[170] Suzuki Y., Niino F., Katoh K. 'Low-resistivity Ito films by DC arc discharge ion plating for high duty LCDs'. *Journal of Non-Crystalline Solids*. 1997;218:30–4.

[171] Adurodija F.O., Izumi H., Ishihara T., Yoshioka H., Motoyama M. 'Effect of Sn doping on the electronic transport mechanism of indium–tin–oxide films grown by pulsed laser deposition coupled with substrate irradiation'. *Journal of Applied Physics*. 2000;88(7):4175–80.

[172] Suzuki A., Matsushita T., Aoki T., Yoneyama Y., Okuda M. 'Pulsed laser deposition of transparent conducting indium tin oxide films in magnetic field perpendicular to plume'. *Japanese Journal of Applied Physics*. 2001;40(4B):L401–3.

[173] Schneller T., Waser R., Kosec M., Payne D. *Chemical Solution Deposition of Functional Oxide Thin Films*. Wien: Springer-Verlag; 2013.

[174] Hamberg I., Svensson J.S., Eriksson T.S., Granqvist C.G., Arrenius P., Norin F. 'Radiative cooling and frost formation on surfaces with different thermal emittance: theoretical analysis and practical experience'. *Applied Optics*. 1987;26(11):2131–6.

[175] Gordon R.G. 'Criteria for choosing transparent conductors'. *MRS Bulletin*. 2000;25(8):52–7.
[176] Frank G., Kauer E., Köstlin H. 'Transparent heat-reflecting coatings based on highly doped semiconductors'. *Thin Solid Films*. 1981;77(1-3):107–18.
[177] Market Study Reports. 'Global (United States, European Union and China) transparent electronics market research report 2019-2025'. Selbyville: market study report LLC; 2019. Report No.: MSR2130710'.
[178] Morales-Masis M., De Wolf S., Woods-Robinson R., Ager J.W., Ballif C. 'Transparent electrodes for efficient optoelectronics'. *Advanced Electronic Materials*. 2017;3(5):1600529.
[179] Choi Y.-J., Gong S.C., Johnson D.C., Golledge S., Yeom G.Y., Park H.-H. 'Characteristics of the electromagnetic interference shielding effectiveness of Al-doped ZnO thin films deposited by atomic layer deposition'. *Applied Surface Science*. 2013;269:92–7.
[180] Garcia P.F., McLean R.S., Reilly M.H., Nunes Jr G. 'Transparent ZnO thin-film transistor fabricated by RF magnetron sputtering'. *Applied Physics Letters*. 2003;82(7):1117–9.
[181] Koida T., Ueno Y., Shibata H. 'In_2O_3-Based transparent conducting oxide films with high electron mobility fabricated at low process temperatures'. *Physica Status Solidi*. 2018;215(7):1700506.
[182] Valla A., Carroy P., Ozanne F., Muñoz D. 'Understanding the role of mobility of Ito films for silicon heterojunction solar cell applications'. *Solar Energy Materials and Solar Cells*. 2016;157(99):874–80.
[183] Cullity B.D. *Elements of X-Ray Diffraction*. Reading, MA: Addison-Wesley Publishing Company; 1956.
[184] XY L., HJ L., Wang Z.J., Xia H., Xiong Z.Y., Wang J.X. 'Effect of substrate temperature on the structural and optical properties of ZnO and Al-doped ZnO thin films prepared by DC magnetron sputtering'. *Optics Communications*. 2009;282(2):247–52.
[185] Terzini E., Thilakan P., Minarini C. 'Properties of Ito thin films deposited by rf magnetron sputtering at elevated substrate temperature'. *Materials Science and Engineering: B*. 2000;77(1):110–14.
[186] Ohta H., Orita M., Hirano M., Tanji H., Kawazoe H., Hosono H. 'Highly electrically conductive indium–tin–oxide thin films epitaxially grown on yttria-stabilized zirconia (100) by pulsed-laser deposition'. *Applied Physics Letters*. 2000;76(19):2740–2.
[187] Izumi H., Adurodija F.O., Kaneyoshi T., Ishihara T., Yoshioka H., Motoyama M. 'Electrical and structural properties of indium tin oxide films prepared by pulsed laser deposition'. *Journal of Applied Physics*. 2002;91(3):1213–18.
[188] Mott N. *Metal-Insulator Transitions*. London: Taylor & Francis; 1990.
[189] Gupta L., Mansingh A., Srivastava P.K. 'Band gap narrowing and the band structure of tin-doped indium oxide films'. *Thin Solid Films*. 1989;176(1):33–44.
[190] Fraser D.B., Cook H.D. 'Highly conductive, transparent films of sputtered In2-xSnx O3-y'. *Journal of the Electrochemical Society*. 1972;119(10):1368–74.

[191] Yoshida Y., Gessert T.A., Perkins C.L., Coutts T.J. 'Development of radio-frequency magnetron sputtered indium molybdenum oxide'. *Journal of Vacuum Science & Technology A: Vacuum, Surfaces, and Films*. 2003;21(4):1092–7.

[192] Miao W.-na., Li X.-feng., Zhang Q., *et al.* 'Transparent conductive In_2O_3:Mo thin films prepared by reactive direct current magnetron sputtering at room temperature'. *Thin Solid Films*. 2006;500(1-2):70–3.

[193] Keshmiri S.H., Rezaee-Roknabadi M., Ashok S. 'A novel technique for increasing electron mobility of indium-tin-oxide transparent conducting films'. *Thin Solid Films*. 2002;413(1-2):167–70.

[194] Koida T., Fujiwara H., Kondo M. 'Hydrogen-doped In_2O_3 as high-mobility transparent conductive oxide'. *Japanese Journal of Applied Physics*. 2007;46(2):L685–7.

[195] Manifacier J.-C., Szepessy L., Bresse J.F., Perotin M., Stuck R. 'In_2O_3 : (Sn) and SnO_2 : (F) films – application to solar energy conversion part II – Electrical and optical properties'. *Materials Research Bulletin*. 1979;14(2):163–75.

[196] Koida T., Kondo M. 'Comparative studies of transparent conductive Ti-, Zr-, and Sn-doped $In2O_3$ using a combinatorial approach'. *Journal of applied physics*. 2007;101(6):0637136):1:.

[197] Holman Z.C., Filipič M., Descoeudres A., *et al.* 'Infrared light management in high-efficiency silicon heterojunction and rear-passivated solar cells'. *Journal of Applied Physics*. 2013;113(1):013107.

[198] Calnan S., Tiwari A.N. 'High mobility transparent conducting oxides for thin film solar cells'. *Thin Solid Films*. 2010;518(7):1839–49.

[199] Calnan S., Uphadhyaya H.M., Buecheler S., *et al.* 'Application of high mobility transparent conductors to enhance long wavelength transparency of the intermediate solar cell in multi-junction solar cells'. *Thin Solid Films*. 2009;517(7):2340–3.

[200] Brooks H. 'Theory of the electrical properties of germanium and silicon'. *Advances in Electronics and Electron Physics*. 1955;7:85–182.

[201] Dingle R.B. 'XCIV. scattering of electrons and holes by charged donors and acceptors in semiconductors'. *The London, Edinburgh, and Dublin Philosophical Magazine and Journal of Science*. 1955;46(379):831–40.

[202] Pisarkiewicz T., Zakrzewska K., Leja E. 'Scattering of charge carriers in transparent and conducting thin oxide films with a non-parabolic conduction band'. *Thin Solid Films*. 1989;174(1):217–23.

[203] Kim W.M., Kim I.H., Ko J.H., *et al.* 'Density-of-state effective mass and non-parabolicity parameter of impurity doped ZnO thin films'. *Journal of Physics D: Applied Physics*. 2008;41(19):195409.

[204] Young D.L., Coutts T.J., Kaydanov V.I., Gilmore A.S., Mulligan W.P. 'Direct measurement of density-of-states effective mass and scattering parameter in transparent conducting oxides using second-order transport phenomena'. *Journal of Vacuum Science & Technology A: Vacuum, Surfaces, and Films*. 2000;18(6):2978–85.

[205] Look D.C., Leedy K.D., Vines L., *et al.* 'Self-compensation in semiconductors: the Zn vacancy in Ga-doped ZnO'. *Physical Review B*. 2011;84(11):115202.

[206] Frank G., Köstlin H. 'Electrical properties and defect model of tin-doped indium oxide layers'. *Applied Physics A Solids and Surfaces*. 1982;27(4):197–206.

[207] Erginsoy C. 'Neutral impurity scattering in semiconductors'. *Physical Review*. 1950;79(6):1013–14.

[208] Minami T. 'New *n*-type transparent conducting oxides'. *MRS Bulletin*. 2000;25(8):38–44.

[209] Seto J.Y.W. 'The electrical properties of polycrystalline silicon films'. *Journal of Applied Physics*. 1975;46(12):5247–54.

[210] Stratton R. 'Volt-current characteristics for tunneling through insulating films'. *Journal of Physics and Chemistry of Solids*. 1962;23(9):1177–90.

[211] Simmons J.G. 'Generalized formula for the electric tunnel effect between similar electrodes separated by a thin insulating film'. *Journal of Applied Physics*. 1963;34(6):1793–803.

[212] Seager C.H., Pike G.E. 'Electron tunneling through GaAs grain boundaries'. *Applied Physics Letters*. 1982;40(6):471–4.

[213] Sommer N., Hüpkes J., Rau U. 'Field emission at grain boundaries: modeling the conductivity in highly doped polycrystalline semiconductors'. *Physical Review Applied*. 2016;5(2):024009.

[214] Kim H.-S., Kang S.D., Tang Y., Hanus R., Jeffrey Snyder G. 'Dislocation strain as the mechanism of phonon scattering at grain boundaries'. *Materials Horizons*. 2016;3(3):234–40.

[215] Nguyen V.H., Gottlieb U., Valla A., Muñoz D., Bellet D., Muñoz-Rojas D. 'Electron tunneling through grain boundaries in transparent conductive oxides and implications for electrical conductivity: the case of ZnO:Al thin films'. *Materials Horizons*. 2018;5(4):715–26.

[216] Rey G., Ternon C., Modreanu M., Mescot X., Consonni V., Bellet D. 'Electron scattering mechanisms in fluorine-doped SnO_2 thin films'. *Journal of Applied Physics*. 2013;114(18):183713.

[217] Ellmer K., Mientus R. 'Carrier transport in polycrystalline transparent conductive oxides: a comparative study of zinc oxide and indium oxide'. *Thin Solid Films*. 2008;516(14):4620–7.

[218] Preissler N., Bierwager O., Ramu A.T., Speck J.S. 'Electrical transport, electrothermal transport, and effective electron mass in single-crystalline'. *Physical Review. B*. 2013;88(8):085305.

[219] Macco B., Knoops H.C.M., Kessels W.M.M. 'Electron scattering and doping mechanisms in solid-phase-crystallized In_2O_3:H prepared by atomic layer deposition'. *ACS Applied Materials & Interfaces*. 2015;7(30):16723–9.

[220] Koida T., Sai H., Kondo M. 'Application of hydrogen-doped In_2O_3 transparent conductive oxide to thin-film microcrystalline Si solar cells'. *Thin Solid Films*. 2010;518(11):2930–3.

[221] Shigesato Y., Hayashi Y., Haranoh T. 'Doping mechanisms of tin-doped indium oxide films'. *Applied Physics Letters*. 1992;61(1):73–5.

[222] Hartnagel H., Dawar A.L., Jain A.K., Jagadish C. *Semiconducting Transparent Thin Films*. 1st ed. Philadelphia: Institute of Physics Publishing; 1995.

[223] Meng F., Shi J., Liu Z., Cui Y., Lu Z., Feng Z. 'High mobility transparent conductive W-doped In_2O_3 thin films prepared at low substrate temperature and its application to solar cells'. *Solar Energy Materials and Solar Cells*. 2014;122:70–4.

[224] van Hest M.F.A.M., Dabney M.S., Perkins J.D., Ginley D.S., Taylor M.P. 'Titanium-doped indium oxide: a high-mobility transparent conductor'. *Applied Physics Letters*. 2005;87(3):032111.

[225] Koida T., Kondo M. 'Improved near-infrared transparency in sputtered In_2O_3-based transparent conductive oxide thin films by Zr-doping'. *Journal of Applied Physics*. 2007;101(6):063705.

[226] Wen S.J., Campet G., Portier J., Couturier G., Goodenough J.B. 'Correlations between the electronic properties of doped indium oxide ceramics and the nature of the doping element'. *Materials Science and Engineering: B*. 1992;14(1):115–19.

[227] Bender M., Seelig W., Daube C., Frankenberger H., Ocker B., Stollenwerk J. 'Dependence of oxygen flow on optical and electrical properties of DC-magnetron sputtered Ito films'. *Thin Solid Films*. 1998;326(1-2):72–7.

[228] Gervais F. 'Optical conductivity of oxides'. *Materials Science and Engineering: R: Reports*. 2002;39(2-3):29–92.

[229] Balestrieri M., Pysch D., Becker J.-P., Hermle M., Warta W., Glunz S.W. 'Characterization and optimization of indium tin oxide films for heterojunction solar cells'. *Solar Energy Materials and Solar Cells*. 2011;95(8):2390–9.

[230] Trube J. *International Technology Roadmap for Photovoltaic – 2019 Results*. Frankfurt-am-Main: VDMA, Photovoltaic Equipment; 2020.

[231] Fortunato E., Ginley D., Hosono H., Paine D.C. 'Transparent conducting oxides for photovoltaics'. *MRS Bulletin*. 2007;32(3):242–7.

[232] Bruhat E., Desrues T., Grange B., Lignier H., Blanc-Pelissier D., Dubois S. 'Tco contacts for high efficiency c-Si solar cells: influence of different annealing steps on the Si substrates and tCo layers properties'. *Energy Procedia*. 2017;124:829–33.

[233] Park H., Lee Y.-J., Park J., *et al*. 'Front and back tCo research review of a-Si/c-Si Heterojunction with intrinsic thin layer (HIT) solar cell'. *Transactions on Electrical and Electronic Materials*. 2018;19(3):165–72.

[234] Richter A., Benick J., Feldmann F., Fell A., Hermle M., Glunz S.W. 'N-Type Si solar cells with passivating electron contact: identifying sources for efficiency limitations by wafer thickness and resistivity variation'. *Solar Energy Materials and Solar Cells*. 2017;173(6):96–105.

[235] Feldmann F., Simon M., Bivour M., Reichel C., Hermle M., Glunz S.W. 'Efficient carrier-selective p- and n-contacts for Si solar cells'. *Solar Energy Materials and Solar Cells*. 2014;131(Part A):100–4.

[236] Tutsch L., Feldmann F., Polzin J., *et al.* 'Implementing transparent conducting oxides by DC sputtering on ultrathin SiOx/poly-Si passivating contacts'. *Solar Energy Materials and Solar Cells*. 2019;200:109960.

[237] Lippens P., Büchel M., Chiu D., Szepesi C. 'Indium–tin-oxide coatings for applications in photovoltaics and displays deposited using rotary ceramic targets: recent insights regarding process stability and doping level'. *Thin Solid Films*. 2013;532:94–7.

[238] Lippens P., Frei C. 'What is the ideal Ito composition for touch sensor depositions on flexible substrate made with rotary cathodes?' *SID's Digest of Technical Papers*. 2014;45(1):1328–31.

[239] Kobayashi E., Watabe Y., Yamamoto T., Yamada Y. 'Cerium oxide and hydrogen co-doped indium oxide films for high-efficiency silicon heterojunction solar cells'. *Solar Energy Materials and Solar Cells*. 2016;149(1):75–80.

[240] Morales-Masis M., Rucavado E., Monnard R., *et al.* 'Highly conductive and broadband transparent Zr-Doped In_2O_3 as front electrode for solar cells'. *IEEE Journal of Photovoltaics*. 2018;8(5):1202–7.

[241] Koida T. 'Amorphous and crystalline $In2O_3$-based transparent conducting films for photovoltaics'. *Physica Status Solidi A*. 2017;214(2):1600464-15):1:.

[242] Carroy P., Muñoz D., Métral F., Fournier H., Veirman J., Lerat J.F. 'Transparent conductive oxides for silicon heterojunction solar cells'. Presented at the International Symposium on Transparent Conductive Materials. Chania, Greece; 2018.

[243] Erfurt D., Cruz A., Morales-Vilches A.B., Wang E.C., Schlatmann R., Stannowski B. 'Aiming for an industrially relevant high-performance TCO for silicon heterojunction solar cells'. Presented at the 37th European Photovoltaic Solar Energy Conference and Exhibition, Sep 2020, [online]; 2020.

[244] Morales-Vilches A.B., Cruz A., Pingel S., *et al.* 'ITO-free silicon heterojunction solar cells with ZnO:Al/SiO_2front electrodes reaching a conversion efficiency of 23%'. *IEEE Journal of Photovoltaics*. 2019;9(1):34–9.

[245] Favier A., Muñoz D., Martín de Nicolás S., Ribeyron P.-J. 'Boron-doped zinc oxide layers grown by metal-organic CVD for silicon heterojunction solar cells applications'. *Solar Energy Materials and Solar Cells*. 2011;95(4):1057–61.

[246] Korte L., Conrad E., Angermann H., Stangl R., Schmidt M. 'Advances in a-Si:H/c-Si heterojunction solar cell fabrication and characterization'. *Solar Energy Materials and Solar Cells*. 2009;93(6-7):905–10.

[247] Niemelä J.-P., Macco B., Barraud L., *et al.* 'Rear-emitter silicon heterojunction solar cells with atomic layer deposited ZnO:Al serving as an alternative transparent conducting oxide to In_2O_3:Sn'. *Solar Energy Materials and Solar Cells*. 2019;200:109953.

[248] Ellmer K. 'Past achievements and future challenges in the development of optically transparent electrodes'. *Nature Photonics*. 2012;6(12):809–17.

[249] Cruz A., Wang E.-C., Morales-Vilches A.B., *et al.* 'Effect of front TCO on the performance of rear-junction silicon heterojunction solar cells: insights

from simulations and experiments'. *Solar Energy Materials and Solar Cells*. 2019;195(5):339–45.

[250] Morales-Vilches A.B., Larionova Y., Wietler T., Cruz A., Korte L., Peibst R. 'ZnO:Al/a-SiOx front contact for polycrystalline-silicon-on-oxide (POLO) solar cells'. *AIP Conference Proceedings*. 1999;2018(1):040016.

[251] Nguyen V.H., Bellet D., Masenelli B., Muñoz-Rojas D. 'Increasing the electron mobility of ZnO-based transparent conductive films deposited by open-air methods for enhanced sensing performance'. *ACS Applied Nano Materials*. 2018;1(12):6922–31.

[252] Haschke J., Dupré O., Boccard M., Ballif C. 'Silicon heterojunction solar cells: Recent technological development and practical aspects – from lab to industry'. *Solar Energy Materials and Solar Cells*. 2018;187(6):140–53.

[253] Ru X., Qu M., Wang J., *et al*. '25.11% efficiency silicon heterojunction solar cell with low deposition rate intrinsic amorphous silicon buffer layers'. *Solar Energy Materials and Solar Cells*. 2020;215:110643.

[254] Adachi D., Hernández J.L., Yamamoto K. 'Impact of carrier recombination on fill factor for large area heterojunction crystalline silicon solar cell with 25.1% efficiency'. *Applied Physics Letters*. 2015;107(23):233506.

[255] De Wofl S., Descoeudres A., Holman Z.C., Ballif C. 'High-efficiency silicon heterojunction solar cells: a review'. *Green*. 2012;2(1):7–24.

[256] Favre W., Sciuto M., Danel A., Varache R., Ragonesi A., Di Matteo A. '25% efficient large area silicon solar cell: paving the way for premium PV manufacturing in Europe'. *Proceedings of the 37th European Photovoltaic Solar Energy Conference and Exhibition,* Sep 2020; Munich, WIP Renewable Energies; 2020.

[257] Taguchi M., Yano A., Tohoda S., *et al*. '24.7% record efficiency HIT solar cell on thin silicon wafer'. *IEEE Journal of Photovoltaics*. 2014;4(1):96–9.

[258] Muñoz D., Roux C. 'The race for high efficiency in production: why heterojunction in now ready for market'. *Proceedings of the 36th European Photovoltaic Solar Energy Conference and Exhibition [on;one];* Marseille, Sep 2019; Munich, WIP Renewable Energies; 2019.

[259] Lachowicz A., Wyss P., Geissbühler J., Faes A., Champliaud J., Badel N. 'Review on plating processes for silicon heterojunction cells'. Presented at the 8th Workshop on Metallization and Interconnection for Crystalline Silicon Solar Cells; Constanz (Germany); 2019.

[260] Boccard M., Antognini L., Paratte V., *et al*. 'Hole-selective front contact stack enabling 24.1%-efficient silicon heterojunction solar cells'. *IEEE Journal of Photovoltaics*. 2020;11(1):9–15.

[261] Meyer Burger Technology AG. *Meyer Burger announces record HJT cells with efficiencies over 24% at PV CellTech [online]*. 2019. Available from https://www.meyerburger.com/en/newsroom/news/meyer-burger-announces-record-hjt-cells-with-efficiencies-over-24-at-pv-celltech/ [Accessed 8 Oct 2020].

[262] Watahiki T., Furuhata T., Matsuura T., *et al.* 'Rear-emitter Si heterojunction solar cells with over 23% efficiency'. *Applied Physics Express*. 2015;8(2):021402.

[263] Moldovan A., Pingel S., Roder S., Tutsch L., Temmler J., Bodlak L. 'Improved layer properties combined with light soaking enabling for 23% silicon heterojunction solar cells'. *Proceedings of the 37th European Photovoltaic Solar Energy Conference and Exhibition;* Sep 2020; Munich, WIP Renewable Energies; 2020. pp. 218–22.

[264] Bätzner D.L., Papet P., Legradic B., Lachenal D., Kramer R., Kössler T. 'Alleviating performance and cost constraints in silicon heterojunction cells with HJT 2.0'. *Proceedings of the 46th IEEE Photovoltaic Specialists Conference;* Chicago (IL), June 2019; Piscataway, NJ, IEEE; 2019. pp. 1471–4.

[265] Feldmann F., Simon M., Bivour M., Reichel C., Hermle M., Glunz S.W. 'Carrier-selective contacts for Si solar cells'. *Applied Physics Letters*. 2014;104(18):181105.

[266] Kim M., Kim J., Kim H., Park Y.C., Ryu K., Yi J. 'Emitter controlled SiNx-free crystalline Si solar cells with a transparent conducting oxide film'. *Materials Letters*. 2012;79:284–7.

[267] Nogay G., Ballif C., Ingenito A., Rucavado E., *et al.* 'Crystalline silicon solar cells with coannealed electron- and hole-selective SiC_x passivating contacts'. *IEEE Journal of Photovoltaics*. 2018;8(6):1478–85.

[268] Peibst R., Larionova Y., Reiter S., *et al.* 'Building blocks for industrial, screen-printed double-side contacted POLO cells with highly transparent ZnO:Al layers'. *IEEE Journal of Photovoltaics*. 2018;8(3):1–7.

[269] Larionova Y., Schulte-Huxel H., Min B., *et al.* 'Ultra-thin poly-Si layers: Passivation quality, utilization of charge carriers generated in the Poly-Si and application on screen-printed double-side contacted polycrystalline Si on oxide cells'. *Solar RRL*. 2020;4(10):2000177.

[270] Desrues T., Oliveau C., Seron C., Dubois S., Borvon G., Torregrosa F. 'Doping and hydrogenation processes for passivating contact solar cells using plasma immersion ion implantation (PIII)'. *Proceedings of the 37th European Photovoltaic Solar Energy Conference and Exhibition;* Sep 2020; 2020. pp. 173–5.

[271] Schube J., Tutsch L., Fellmeth T., *et al.* 'Intense pulsed light in back end processing of solar cells with passivating contacts based on amorphous or polycrystalline silicon layers'. *Solar Energy Materials and Solar Cells*. 2020;216:110711.

[272] Han C., Yang G., Montes A., *et al.* 'Realizing the potential of RF-sputtered hydrogenated fluorine-doped indium oxide as an electrode material for ultrathin SiO_x/Poly-Si passivating contacts'. *ACS Applied Energy Materials*. 2020;3(9):8606–18.

[273] Yan D., Cuevas A. 'Empirical determination of the energy band gap narrowing in p^+ silicon heavily doped with boron'. *Journal of Applied Physics*. 2014;116(19):194505.

[274] Yan D., Cuevas A. 'Empirical determination of the energy band gap narrowing in highly doped n^+ silicon'. *Journal of Applied Physics*. 2013;114(4):044508.
[275] Hollemann C., Haase F., Schäfer S., Krügener J., Brendel R., Peibst R. '26.1%-efficient POLO-IBC cells: quantification of electrical and optical loss mechanisms'. *Progress in Photovoltaics: Research and Applications*. 2019;27(11):950–958.
[276] Yoshikawa K., Kawasaki H., Yoshida W., *et al*. 'Silicon heterojunction solar cell with interdigitated back contacts for a photoconversion efficiency over 26%'. *Nature Energy*. 2017;2(5):17032.
[277] Bullock J2016Advanced Contacts For Crystalline Silicon Solar Cells [PhD Thesis].CanberraThe Australian National University
[278] Ng K.K., Card H.C. 'A comparison of majority- and minority-carrier silicon MIS solar cells'. *IEEE transactions on electron devices*. 1980;27(4):716–24.
[279] Kobayashi H., Mori H., Ishida T., Nakato Y. 'Zinc oxide/*n*-Si junction solar cells produced by spray-pyrolysis method'. *Journal of Applied Physics*. 1995;77(3):1301–7.
[280] Ghosh A.K., Fishman C., Feng T. 'SnO_2/Si solar cells—heterostructure or Schottky-barrier or MIS-type device'. *Journal of Applied Physics*. 1978;49(6):3490–8.
[281] Green M.A., Godfrey R.B. 'Mis solar cell—General theory and new experimental results for silicon'. *Applied Physics Letters*. 1976;29(9):610–2.
[282] Godfrey R.B., Green M.A. '655 mV open-circuit voltage, 17.6% efficient silicon MIS solar cells'. *Applied Physics Letters*. 1979;34(11):790–3.
[283] Ponpon J.P., Siffert P. 'Open-circuit voltage of MIS silicon solar cells'. *Journal of Applied Physics*. 1976;47(7):3248–51.
[284] Pulfrey D.L. 'Mis solar cells: a review'. *IEEE Transactions on Electron Devices*. 1978;25(11):1308–17.
[285] Shewchun J., Burk D., Spitzer M.B. 'Mi and SIS solar cells'. *IEEE Transactions on Electron Devices*. 1980;27(4):705–16.
[286] De Wolf S., Kondo M. 'Abruptness of a-Si:H/c-Si interface revealed by carrier lifetime measurements'. *Applied Physics Letters*. 2007;90(4):042111.
[287] Agostinelli G., Delabie A., Vitanov P., *et al*. 'Very low surface recombination velocities on P-type silicon wafers passivated with a dielectric with fixed negative charge'. *Solar Energy Materials and Solar Cells*. 2006;90(18–19):3438–43.
[288] Kerr M.J., Cuevas A. 'Very low bulk and surface recombination in oxidized silicon wafers'. *Semiconductor Science and Technology*. 2002;17(1):35–8.
[289] Wan Y., McIntosh K.R., Thomson A.F., Cuevas A. 'Low surface recombination velocity by low-absorption silicon nitride on c-Si'. *IEEE Journal of Photovoltaics*. 2013;3(1):554–9.
[290] Chen L.-M., Xu Z., Hong Z., Yang Y. 'Interface investigation and engineering – achieving high performance polymer photovoltaic devices'. *Journal of Materials Chemistry*. 2010;20(13):2575–98.
[291] Werner F., Larionova Y., Zielke D., Ohrdes T., Schmidt J. 'Aluminum-oxide-based inversion layer solar cells on *n*-type crystalline silicon:

Fundamental properties and efficiency potential'. *Journal of Applied Physics*. 2014;115(7):073702.

[292] Kuhlmann B., Aberle A.G., Hezel R., Heiser G. 'Simulation and optimization of metal-insulator-semiconductor inversion-layer silicon solar cells'. *IEEE Transactions on Electron Devices*. 2000;47(11):2167–78.

[293] Dauwe S. Low Temperature Surface Passivation of Crystalline Silicon and Its Application to the Rear Side of Solar Cells [PhD thesis].Hannover, Leibniz Universität; 2004.

[294] Sze S.M., Ng K.K. *Physics of Semiconductor Devices*. Hoboken, NJ: John Wiley & Sons; 2006.

[295] Dingemans G., Terlinden N.M., Pierreux D., Profijt H.B., van de Sanden M.C.M., Kessels W.M.M. 'Influence of the oxidant on the chemical and field-effect passivation of Si by ALD Al2O3'. *Electrochemical and Solid-State Letters*. 2011;14(1):H1–4.

[296] Aberle A.G., Glunz S.W., Stephens A.W., Green M.A. 'High-eficiency silicon solar cells: Si/SiO_2, interface parameters and their impact on device performance'. *Progress in Photovoltaics: Research and Applications*. 1994;2(4):265–73.

[297] Schroder D.K., Meier D.L. 'Solar cell contact resistance—A review'. *IEEE Transactions on Electron Devices*. 1984;31(5):637–47.

[298] Mönch W. 'On the alleviation of fermi-level pinning by ultrathin insulator layers in Schottky contacts'. *Journal of Applied Physics*. 2012;111(7):073706.

[299] Tung R.T. 'The physics and chemistry of the Schottky barrier height'. *Applied physics reviews*. 2014;1(1):011304.

[300] Liu Z., Kobayashi M., Paul B.C., Bao Z., Nishi Y. 'Fermi level depinning at metal-organic semiconductor interface for low-resistance ohmic contacts'. *Proceedings of IEEE International Electron Devices Meeting;* Baltimore (MD), Dec 2009; Piscataway, NJ, IEEE; 2009. pp. 1–4.

[301] Connelly D., Faulkner C., Clifton P.A., Grupp D.E. 'Fermi-level depinning for low-barrier Schottky source/drain transistors'. *Applied Physics Letters*. 2006;88(1):012105.

[302] Islam R., Shine G., Saraswat K.C. 'Schottky barrier height reduction for holes by Fermi level depinning using metal/nickel oxide/silicon contacts'. *Applied Physics Letters*. 2014;105(18):182103.

[303] Coss B.E., Loh W.-Y., Carlo Floresca H., *et al.* 'Dielectric dipole mitigated Schottky barrier height tuning using atomic layer deposited aluminum oxide for contact resistance reduction'. *Applied Physics Letters*. 2011;99(10):102108.

[304] Schenk A., Heiser G. 'Modeling and simulation of tunneling through ultrathin gate dielectrics'. *Journal of Applied Physics*. 1997;81(12):7900–8.

[305] Yeo Y.-C., King T.-J., Hu C. 'Metal-dielectric band alignment and its implications for metal gate complementary metal-oxide-semiconductor technology'. *Journal of Applied Physics*. 2002;92(12):7266–71.

[306] Heng J.B., Fu J., Kong B., *et al.* '>23% high-efficiency tunnel oxide junction bifacial solar cell with electroplated Cu gridlines'. *IEEE Journal of Photovoltaics*. 2014;5(1):82–86.
[307] Feldmann F., Bivour M., Reichel C., Hermle M., Glunz S.W. 'Passivated rear contacts for high-efficiency N-type Si solar cells providing high interface passivation quality and excellent transport characteristics'. *Solar Energy Materials and Solar Cells*. 2014;120:270–4.
[308] Robertson J., Offsets B. 'Band offsets, Schottky barrier heights, and their effects on electronic devices'. *Journal of Vacuum Science & Technology A: Vacuum, Surfaces, and Films*. 2013;31(5):050821.
[309] Aberle A.G. *Crystalline Silicon Solar Cells: Advanced Surface Passivation and Analysis*. Sidney: UNSW Photovoltaics; 1999.
[310] Ji W., Zhao Y., Fahad H.M., *et al.* 'Dip coating passivation of crystalline silicon by Lewis acids'. *ACS Nano*. 2019;13(3):3723–9.
[311] Geissbühler J., Werner J., Martin de Nicolas S., *et al.* '22.5% efficient silicon heterojunction solar cell with molybdenum oxide hole collector'. *Applied Physics Letters*. 2015;107(8):081601.
[312] Chiu F.-C. 'A review on conduction mechanisms in dielectric films'. *Advances in Materials Science and Engineering*. 2014;2014(7):1–18.
[313] Messmer C., Bivour M., Schön J., Hermle M. 'Requirements for efficient hole extraction in transition metal oxide-based silicon heterojunction solar cells'. *Journal of Applied Physics*. 2018;124(8):085702.
[314] Wurfel U., Cuevas A., Wurfel P. 'Charge carrier separation in solar cells'. *IEEE Journal of Photovoltaics*. 2015;5(1):461–9.
[315] Smit S., Garcia-Alonso D., Bordihn S., Hanssen M.S., Kessels W.M.M. 'Metal-oxide-based hole-selective tunneling contacts for crystalline silicon solar cells'. *Solar Energy Materials and Solar Cells*. 2014;120:376–82.
[316] Liu Y., Stradins P., Deng H., Luo J., Wei S.-H. 'Suppress carrier recombination by introducing defects: the case of Si solar cell'. *Applied Physics Letters*. 2016;108(2):022101.
[317] Grauvogl M., Hezel R. 'The truncated-pyramid MIS inversion-layer solar cell: a comprehensive analysis'. *Progress in Photovoltaics: Research and Applications*. 1998;6(1):15–24.
[318] Hezel R., Meyer R., Metz A. 'A new generation of crystalline silicon solar cells: simple processing and record efficiencies for industrial-size devices'. *Solar Energy Materials and Solar Cells*. 2001;65(1–4):311–16.
[319] Green M.A., Blakers A.W., Willison Z.R. 'The MINP solar cell - a new high voltage, high efficiency silicon solar cell'. *Conference Record of the 15th IEEE Photovoltaic Specialists Conference,* Kissimmee (FL), May 1981; New York, IEEE; 1981. pp. 1405–8.
[320] Green M.A. 'The path to 25% silicon solar cell efficiency: History of silicon cell evolution'. *Progress in Photovoltaics: Research and Applications*. 2009;17(3):183–9.

[321] Bullock J., Wan Y., Hettick M., *et al.* 'Dopant-free partial rear contacts enabling 23% silicon solar cells'. *Advanced Energy Materials.* 2019;9(0):1803367.

[322] Allen T.G., Bullock J., Jeangros Q., *et al.* 'A low resistance calcium/reduced titania passivated contact for high efficiency crystalline silicon solar cells'. *Advanced Energy Materials.* 2017;7(12):1602606.

[323] Yang X., Bi Q., Ali H., Davis K., Schoenfeld W.V., Weber K. 'High-performance TiO_2-based electron-selective contacts for crystalline silicon solar cells'. *Advanced Materials.* 2016;28(28):5891–7.

[324] Cui J., Allen T., Wan Y., *et al.* 'Titanium oxide: a re-emerging optical and passivating material for silicon solar cells'. *Solar Energy Materials and Solar Cells.* 2016;158(4):115–21.

[325] Yang X., Zheng P., Bi Q., Weber K. 'Silicon heterojunction solar cells with electron selective TiOx contact'. *Solar Energy Materials and Solar Cells.* 2016;150:32–8.

[326] Ling Z., He J., He X., *et al.* 'Excellent passivation of silicon surfaces by thin films of electron-beam-processed titanium dioxide'. *IEEE Journal of Photovoltaics.* 2017;7(6):1551–5.

[327] Bullock J., Zheng P., Jeangros Q., *et al.* 'Lithium fluoride based electron contacts for high efficiency N-type crystalline silicon solar cells'. *Advanced Energy Materials.* 2016;6(14):1600241.

[328] Li H.H. 'Refractive index of alkali halides and its wavelength and temperature derivatives'. *Journal of physical and chemical reference data.* 1976;5(2):329–528.

[329] Bullock J., Hettick M., Geissbühler J., *et al.* 'Efficient silicon solar cells with dopant-free asymmetric heterocontacts'. *Nature Energy.* 2016;1(1):15031.

[330] Lin W., Boccard M., Zhong S., *et al.* 'Degradation mechanism and stability improvement of dopant-free ZnO/LiF_x/Al electron nanocontacts in silicon heterojunction solar cells'. *ACS Applied Nano Materials.* 2020;3(11):11391–8.

[331] Bullock J., Wan Y., Xu Z., *et al.* 'Stable dopant-free asymmetric heterocontact silicon solar cells with efficiencies above 20%'. *ACS Energy Letters.* 2018;3:508–13.

[332] Wan Y., Samundsett C., Bullock J., *et al.* 'Conductive and stable magnesium oxide electron-selective contacts for efficient silicon solar cells'. *Advanced Energy Materials.* 2017;7(5):1601863.

[333] Yu J., Liao M., Yan D., *et al.* 'Activating and optimizing evaporation-processed magnesium oxide passivating contact for silicon solar cells'. *Nano Energy.* 2019;62:181–8.

[334] Chistiakova G., Macco B., Korte L. 'Low-Temperature atomic layer deposited magnesium oxide as a passivating electron contact for c-Si-based solar cells'. *IEEE Journal of Photovoltaics.* 2020;10(2):398–406.

[335] Wan Y., Karuturi S.K., Samundsett C., *et al.* 'Tantalum oxide electron-selective heterocontacts for silicon photovoltaics and photoelectrochemical water reduction'. *ACS Energy Letters.* 2018;3(1):125–31.

[336] Yang X., Aydin E., Xu H., *et al.* 'Tantalum nitride electron-selective contact for crystalline silicon solar cells'. *Advanced Energy Materials.* 2018;8(8):1800608.

[337] Yang X., Liu W., De Bastiani M., *et al.* 'Dual-function electron-conductive, hole-blocking titanium nitride contacts for efficient silicon solar cells'. *Joule.* 2019;3(5):1314–27.

[338] Yu J., Phang P., Samundsett C., Basnet R., Neupan G., Yang X., et al. 'Titanium nitride electron-conductive contact for silicon solar cells by radio frequency sputtering from a tin target'. *ACS applied materials & interfaces.* 2020;12(23):26177–83.

[339] Yang X., Lin Y., Liu J., *et al.* 'A highly conductive titanium oxynitride electron-selective contact for efficient photovoltaic devices'. *Advanced Materials.* 2020;32(32):2002608.

[340] Wan Y., Bullock J., Hettick M., *et al.* 'Temperature and humidity stable alkali/alkaline-earth metal carbonates as electron heterocontacts for silicon photovoltaics'. *Advanced Energy Materials.* 2018;8(22):1800743.

[341] Bullock J., Wan Y., Hettick M., *et al.* 'Survey of dopant-free carrier-selective contacts for silicon solar cells'. *2016 IEEE 43rd Photovoltaic Specialists Conference (PVSC)*; 2016. pp. 210–14.

[342] Zhang Y., Cui W., Zhu Y., *et al.* 'High efficiency hybrid PEDOT:PSS/nanostructured silicon Schottky junction solar cells by doping-free rear contact'. *Energy & Environmental Science.* 2015;8(1):297–302.

[343] Wan Y., Samundsett C., Bullock J., *et al.* 'Magnesium fluoride electron-selective contacts for crystalline silicon solar cells'. *ACS Applied Materials & Interfaces.* 2016;8(23):14671–7.

[344] Rodríguez-de Marcos L.V., Larruquert J.I., Méndez J.A., Aznárez J.A. 'Self-consistent optical constants of MgF_2, LaF_3, and CeF_3 films'. *Optical Materials Express.* 2017;7(3):989.

[345] Kim S.-H., Jung J.-Y., Wehrspohn R.B., Lee J.-H. 'All-room-temperature processed 17.25%-crystalline silicon solar cell'. *ACS Applied Energy Materials.* 2020;3(4):3180–5.

[346] Robertson J. 'Band offsets of high dielectric constant gate oxides on silicon'. *Journal of Non-Crystalline Solids.* 2002;303(1):94–100.

[347] Wang W., He J., Cai L., *et al.* 'Solution-processed electron-selective contacts enabling 21.8% efficiency crystalline silicon solar cells'. *Solar RRL.* 2020;4(12):2000569.

[348] Wang Z., Yang Y., Zhang L., *et al.* 'Modulation-doped ZnO as high performance electron-selective layer for efficient silicon heterojunction solar cells'. *Nano Energy.* 2018;54:99–105.

[349] van de Loo B.W.H., Macco B., Melskens J., Beyer W., Kessels W.M.M. 'Silicon surface passivation by transparent conductive zinc oxide'. *Journal of Applied Physics.* 2019;125(10):105305.

[350] Hussain B., Ebong A., Ferguson I. 'Zinc oxide as an active n-layer and antireflection coating for silicon based heterojunction solar cell'. *Solar Energy Materials and Solar Cells.* 2015;139:95–100.

[351] He J., Wang W., Cai L., *et al.* 'Stable electron-selective contacts for crystalline silicon solar cells enabling efficiency over 21.6%'. *Advanced Functional Materials*. 2020;30(50):2005554.

[352] Yao Z., Cai L., Meng L., *et al.* 'High-performance and stable dopant-free silicon solar cells with magnesium acetylacetonate electron-selective contacts'. *Physica Status Solidi (RRL) – Rapid Research Letters*. 2020;14(6):2000103.

[353] Yao L., Ying Z., Wang W., *et al.* 'Solution-processed and annealing-free zirconium acetylacetonate electron-selective contacts for efficient crystalline silicon solar cells'. *Solar Energy*. 2021;215(11):410–15.

[354] Wang W., Cai L., Meng L., *et al.* 'Cerous fluoride Dopant-Free Electron-Selective contact for crystalline silicon solar cells'. *Physica Status Solidi RRL – Rapid Research Letters*. 2021;2100135.

[355] Bullock J., Cuevas A., Allen T., Battaglia C. 'Molybdenum oxide MoO_x : A versatile hole contact for silicon solar cells'. *Applied Physics Letters*. 2014;105(23):232109.

[356] Gerling L.G., Mahato S., Morales-Vilches A., *et al.* 'Transition metal oxides as hole-selective contacts in silicon heterojunctions solar cells'. *Solar Energy Materials and Solar Cells*. 2016;145:109–15.

[357] Dréon J., Jeangros Q., Cattin J., *et al.* '23.5%-efficient silicon heterojunction silicon solar cell using molybdenum oxide as hole-selective contact'. *Nano Energy*. 2020;70(1):104495.

[358] Bivour M., Macco B., Temmler J., Kessels W.M.M. (Erwin)., Hermle M. 'Atomic layer deposited molybdenum oxide for the hole-selective contact of silicon solar cells'. *Energy Procedia*. 2016;92:443–9.

[359] Gregory G., Feit C., Gao Z., Banerjee P., Jurca T., Davis K.O. 'Improving the passivation of molybdenum oxide hole-selective contacts with 1 nm hydrogenated aluminum oxide films for silicon solar cells'. *Physica Status Solidi*. 2020;217(15):2000093.

[360] Tong J., Wan Y., Cui J., Lim S., Song N., Lennon A. 'Solution-processed molybdenum oxide for hole-selective contacts on crystalline silicon solar cells'. *Applied Surface Science*. 2017;423:139–46.

[361] Yang X., Xu H., Liu W., *et al.* 'Atomic layer deposition of vanadium oxide as hole-selective contact for crystalline silicon solar cells'. *Advanced Electronic Materials*. 2020;6(8):2000467.

[362] Bivour M., Temmler J., Steinkemper H., Hermle M. 'Molybdenum and tungsten oxide: high work function wide band gap contact materials for hole selective contacts of silicon solar cells'. *Solar Energy Materials and Solar Cells*. 2015;142:34–41.

[363] Ali H., Koul S., Gregory G., *et al.* 'Thermal stability of hole-selective tungsten oxide: in situ transmission electron microscopy study'. *Scientific Reports*. 2018;8(1):12651.

[364] Lee C.-Y., Aziz M.I.A., Wenham S., Hoex B. 'Characterisation of thermal annealed WO_x on p-type silicon for hole-selective contacts'. *Japanese Journal of Applied Physics*. 2017;56(8S2):08MA08.

[365] Bivour M., Temmler J., Zähringer F., Glunz S., Hermle M. 'High work function metal oxides for the hole contact of silicon solar cells'. IEEE 43rd Photovoltaic Specialists Conference (PVSC); 2016. pp. 0215–20.

[366] Xue M., Islam R., Chen Y., *et al.* 'Carrier-selective interlayer materials for silicon solar cell contacts'. *Journal of Applied Physics*. 2018;123(14):143101.

[367] Nayak M., Mandal S., Pandey A., Mudgal S., Singh S., Komarala V.K. 'Nickel oxide hole-selective heterocontact for silicon solar cells: role of SiO_x interlayer on device performance'. *Solar RRL*. 2019;3(11):1900261.

[368] Zhao J., Ho-Baillie A., Bremner S.P. 'Pulsed laser deposition nickel oxide on crystalline silicon as hole selective contacts'. *Journal of Vacuum Science & Technology B*. 2020;38(1):014013.

[369] Woods-Robinson R., Fioretti A.N., Haschke J., Boccard M., Persson K.A., Ballif C. Linking simulation and synthesis of nickel oxide hole-selective contacts for silicon heterojunction solar cells. *47th IEEE Photovoltaic Specialists Conference (PVSC)*; 2020. pp. 0569–73.

[370] Zhang T., Hossain M.A., Lee C.-Y., Zakaria Y., Abdallah A.A., Hoex B. 'Atomic layer deposited $Zn_x Ni_{1-x}O$: A thermally stable hole selective contact for silicon solar cells'. *Applied Physics Letters*. 2018;113(26):262102.

[371] Ravindra P., Mukherjee R., Avasthi S. 'Hole-selective electron-blocking copper oxide contact for silicon solar cells'. *IEEE Journal of Photovoltaics*. 2017;7(5):1278–83.

[372] K Markose K., Shaji M., Bhatia S., *et al.* 'Novel boron-doped p-type Cu_2O thin films as a hole-selective contact in c-Si solar cells'. *ACS Applied Materials & Interfaces*. 2020;12(11):12972–81.

[373] Zhang X., Wan Y., Bullock J., Allen T., Cuevas A. 'Low resistance ohmic contact to P-type crystalline silicon via nitrogen-doped copper oxide films'. *Applied Physics Letters*. 2016;109(5):052102.

[374] Lin W., Wu W., Xie Q., *et al.* 'Conductive cuprous iodide hole-selective contacts with thermal and ambient stability for silicon solar cells'. *ACS Applied Materials & Interfaces*. 2018;10(50):43699–706.

[375] Lin W., Wu W., Liu Z., *et al.* 'Chromium trioxide hole-selective heterocontacts for silicon solar cells'. *ACS Applied Materials & Interfaces*. 2018;10(16):13645–51.

[376] Lin W., Wu W., Bao J., *et al.* 'Novel hole selective CrO_x contact for dopant-free back contact silicon solar cells'. *Materials Research Bulletin*. 2018;103(12):77–82.

[377] Li J., Pan T., Wang J., *et al.* 'Bilayer MoO_X/CrO_X passivating contact targeting highly stable silicon heterojunction solar cells'. *ACS Applied Materials & Interfaces*. 2020;12(32):36778–86.

[378] Bullock J., Samundsett C., Cuevas A., Yan D., Wan Y., Allen T. 'Proof-of-concept P-type silicon solar cells with molybdenum oxide local rear contacts'. *IEEE Journal of Photovoltaics*. 2015;5(6):1591–4.

[379] Bivour M., Temmler J., Zahringer F., Glunz S., Hermle M. High work function metal oxides for the hole contact of silicon solar cells. 2016 IEEE 43rd Photovoltaic Specialists Conference (PVSC); 2016. pp. 0215–20.

[380] Wang W., He J., Yan D., *et al.* '21.3%-efficient n-type silicon solar cell with a full area rear TiO_x/LiF/Al electron-selective contact'. *Solar Energy Materials and Solar Cells*. 2020;206:110291.

[381] Wu W., Lin W., Zhong S., *et al.* 'Dopant-free back-contacted silicon solar cells with an efficiency of 22.1%'. *Physica Status Solidi (RRL) – Rapid Research Letters*. 2020;14(4):1900688.

[382] Lin W., Dréon J., Zhong S., *et al.* 'Dopant-free bifacial silicon solar cells'. *Solar RRL*. 2021;5(5):2000771.

[383] Zhang T., Iqbal S., Zhang X.-Y., Wu W., Su D., Zhou H.-L. 'Recent advances in highly efficient organic-silicon hybrid solar cells'. *Solar Energy Materials and Solar Cells*. 2020;204(80):110245.

[384] Wirth-Lima A.J., Alves-Sousa P.P., Bezerra-Fraga W. 'Graphene/silicon and 2D-MoS_2/silicon solar cells: a review'. *Applied Physics A*. 2019;125(4):241.

[385] Allen T.G., Bullock J., Zheng P., *et al.* 'Calcium contacts to N-type crystalline silicon solar cells'. *Progress in Photovoltaics: Research and Applications*. 2017;25(7):636–44.

[386] Matsui T., Bivour M., Hermle M., Sai H. 'Atomic-layer-deposited TiO_x nanolayers function as efficient hole-selective passivating contacts in silicon solar cells'. *ACS Applied Materials & Interfaces*. 2020;12(44):49777–85.

[387] Cho J., Melskens J., Payo M.R., *et al.* 'Performance and thermal stability of an a-Si:H/TiO_x/Yb stack as an electron-selective contact in silicon heterojunction solar cells'. *ACS Applied Energy Materials*. 2019;2(2):1393–404.

[388] Bullock J., Wan Y., Hettick M., *et al.* 'Dopant-free partial rear contacts enabling 23% silicon solar cells'. *Advanced Energy Materials*. 2019;9(9):1803367.

[389] Greiner M.T., Chai L., Helander M.G., Tang W.-M., Lu Z.-H. 'Transition metal oxide work functions: the influence of cation oxidation state and oxygen vacancies'. *Advanced Functional Materials*. 2012;22(21):4557–68.

[390] Allen T.G., Cuevas A. 'Plasma enhanced atomic layer deposition of gallium oxide on crystalline silicon: demonstration of surface passivation and negative interfacial charge'. *Physica Status Solidi (RRL) – Rapid Research Letters*. 2015;9(4):220–4.

[391] Lin F., Hoex B., Koh Y.H., Lin J.J., Aberle A.G. 'Low-temperature surface passivation of moderately doped crystalline silicon by atomic-layer-deposited hafnium oxide films'. *Energy Procedia*. 2012;15:84–90.

[392] Wan Y., Bullock J., Hettick M., *et al.* 'Zirconium oxide surface passivation of crystalline silicon'. *Applied Physics Letters*. 2018;112(20):201604.

[393] MasmitjG., Gerling L.G., Ortega P., *et al.* '$V2O_x$-basedhole-selectivecontactsforc-Siinterdigitatedback-contactedsolarcells'. *Journal of materials chemistry. A*. 2017;5(19):9182–9.

[394] Bhatia S., Antony A., Nair P.R. 'Unraveling the hole-selective nature of Si/MoO_x heterojunction'. *IEEE Journal of Photovoltaics*. 2020;10(6):1566–73.

[395] Vijayan R.A., Essig S., De Wolf S., *et al.* 'Hole-collection mechanism in passivating metal-oxide contacts on Si solar cells: insights from numerical simulations'. *IEEE Journal of Photovoltaics*. 2018;8(2):473–82.

[396] Bullock J., Yan D., Cuevas A. 'Passivation of aluminium N+ silicon contacts for solar cells by ultrathin Al_2O_3 and SiO_2 dielectric layers'. *Physica Status Solidi (RRL) – Rapid Research Letters*. 2013;7(11):946–9.

[397] Bullock J., Yan D., Wan Y., *et al.* 'Amorphous silicon passivated contacts for diffused junction silicon solar cells'. *Journal of Applied Physics*. 2014;115(16):163703.

[398] Cho J., Sivaramakrishnan Radhakrishnan H., Recaman Payo M., *et al.* 'Low work function ytterbium silicide contact for doping-free silicon solar cells'. *ACS Applied Energy Materials*. 2020;3(4):3826–34.

[399] Quan C., Tong H., Yang Z., *et al.* 'Electron-selective scandium–tunnel oxide passivated contact for N-type silicon solar cells'. *Solar RRL*. 2018;2(8):1800071.

[400] Meng L., Yao Z., Cai L., *et al.* 'Indium sulfide-based electron-selective contact and dopant-free heterojunction silicon solar cells'. *Solar Energy*. 2020;211:759–66.

[401] Cai L., Wang W., Jin L., *et al.* '12.29% low temperature–processed dopant-free CdS/p-Si heterojunction solar cells'. *Advanced Materials Interfaces*. 2019;6(12):1900367.

[402] Zheng Y., Jiang B., Gao Z., *et al.* 'Optimization of SnO_2-based electron-selective contacts for Si/PEDOT:PSS heterojunction solar cells'. *Solar Energy*. 2019;193(3):502–6.

[403] Yamawaki M., Hirai M., Yasumoto M., Kanno M. 'Mass spectrometric study of vaporization of lithium fluoride'. *Journal of Nuclear Science and Technology*. 1982;19(7):563–70.

[404] Yaws C.L. *The Yaws Handbook of Vapor Pressure*. 2nd edn. Waltham, MA: Elsevier; 2015.

[405] Luxel Corporation. *Vapor Pressure Chart [online]*. 2013. Available from https://luxel.com/wp-content/uploads/2013/04/Luxel-Vapor-Pressure-Chart.pdf [Accessed 18 May 2021].

[406] Johnston H.L., Marshall A.L. 'Vapor pressures of nickel and of nickel oxide'. *Journal of the American Chemical Society*. 1940;62(6):1382–90.

[407] Groves W.O., Hoch M., Johnston H.L. 'Vapor–solid equilibria in the titanium–oxygen system'. *The Journal of Physical Chemistry*. 1955;59(2):127–31.

[408] Stull D.R. 'Inorganic compounds'. *Industrial & Engineering Chemistry*. 1947;39(1947):540–50.

[409] Blewett J.P., Liebhafsky H.A., Hennelly E.F. 'The vapor pressure and rate of evaporation of barium oxide'. *The Journal of Chemical Physics*. 1939;7(7):478–84.

[410] Groner M.D., Fabreguette F.H., Elam J.W., George S.M. 'Low-temperature Al_2O_3 atomic layer deposition'. *Chemistry of Materials: A Publication of the American Chemical Society*. 2004;16(4):639–45.

[411] Ylivaara O.M.E., Kilpi L., Liu X., *et al.* 'Aluminum oxide/titanium dioxide nanolaminates grown by atomic layer deposition: growth and mechanical properties'. *Journal of Vacuum Science & Technology A: Vacuum, Surfaces, and Films*. 2017;35(1):01B105.

[412] Basuvalingam S.B., Macco B., Knoops H.C.M., Melskens J., (Erwin) Kessels W.M.M., Bol A.A. 'Comparison of thermal and plasma-enhanced atomic layer deposition of niobium oxide thin films'. *Journal of Vacuum Science & Technology A*. 2018;36(4):041503.

[413] Heil S.B.S. Plasma-Assisted atomic layer deposition of metal oxides and nitrides. PhD thesis. Eindhoven, Technische Universiteit Eindhoven. 2008.

[414] Mattinen M., King P.J., Khriachtchev L., *et al.* 'Atomic layer deposition of crystalline molybdenum oxide thin films and phase control by post-deposition annealing'. *Materials Today Chemistry*. 2018;9:17–27.

[415] Terheiden B., Hahn G., Schiele Y. 'Investigation of radiation damage to the Al_2O_3 /Si Wafer interface during electron beam evaporation by means of C-V and lifetime measurements'. *Proceedings of the 26th European Photovoltaic Solar Energy Conference and Exhibition;* Hamburg, Sep 2011; Munich, WIP Renewable Energies; 2011. pp. 1068–72.

[416] Mader C., Müller J., Eidelloth S., Brendel R. 'Local rear contacts to silicon solar cells by in-line high-rate evaporation of aluminum'. *Solar Energy Materials and Solar Cells*. 2012;107(3):272–82.

[417] Zhang X., Cuevas A., Demaurex B., De Wolf S. 'Sputtered hydrogenated amorphous silicon for silicon heterojunction solar cell fabrication'. *Energy Procedia*. 2014;55:865–72.

[418] Li T.-T., Cuevas A. 'Effective surface passivation of crystalline silicon by rf sputtered aluminum oxide'. *Physica Status Solidi (RRL) – Rapid Research Letters*. 2009;3(5):160–2.

[419] Krugel G., Sharma A., Wolke W., Rentsch J., Preu R. 'Study of hydrogenated AlN as an anti-reflective coating and for the effective surface passivation of silicon'. *Physica Status Solidi (RRL) – Rapid Research Letters*. 2013;7(7):457–60.

[420] Buiter J., Kessels E. *Database of ALD processes* [online]. Atomic Limits. 2019. Available from https://www.atomiclimits.com/alddatabase/ [Accessed 18 May 2021].

[421] Guerra-Nuñez C., Döbeli M., Michler J., Utke I. 'Reaction and growth mechanisms in Al_2O_3 deposited via atomic layer deposition: elucidating the hydrogen source'. *Chemistry of Materials*. 2017;29(20):8690–703.

[422] Macco B. *What can atomic layer deposition do for solar cells?* [online]. Atomic Limits. 2019. Available from https://www.atomiclimits.com/2019/10/15/what-can-atomic-layer-deposition-do-for-solar-cells-a-few-after-thoughts-from-my-ald-conference-tutorial/ [Accessed 18 May 2021].

[423] Essig S., Dréon J., Rucavado E., *et al.* 'Towards annealing-stable molybdenum-oxide-based hole-selective contacts for Si photovoltaics'. *Solidi Rapid Research Letters*. 2018;2(4):1700227.

[424] Vossen J.L., Kern W. (eds.). *Thin film processes II*. Boston: Academic Press; 1991.

[425] Mattox D. *Handbook of Physical Vapor Deposition (PVD) Processing: Film Formation, Adhesion, Surface Preparation, and Contamination Control*. New York: Noyes Data Corporation/Noyes Publishing; 1998.

[426] Oluwatosin Abegunde O., Titilayo Akinlabi E., Philip Oladijo O., Akinlabi S., Uchenna Ude AOverview of thin film deposition techniques'. *AIMS Materials Science*. 2019;6(2):174–99.

[427] Hossain M.A., Khoo K.T., Cui X., *et al*. 'Atomic layer deposition enabling higher efficiency solar cells: a review'. *Nano Materials Science*. 2020;2(3):204–26.

[428] Macco B., Wu Y., Vanhemel D., Kessels W.M.M. 'High mobility In_2O_3 :H transparent conductive oxides prepared by atomic layer deposition and solid phase crystallization'. *Physica Status Solidi (RRL) – Rapid Research Letters*. 2014;8(12):987–90.

[429] Mattox D. *The Foundations of Vacuum Coating Technology*. 1st edn. Norwich, NY: William Andrew Publishing; 2003.

[430] Mattox D. *Handbook of Physical Vapor Deposition (PVD) Processing Book*. 2nd edn. Norwich, NY: Elsevier; 2010.

[431] Wasa K., Kanno I., Kotera H. *Handbook of Sputter Deposition Technology Book*. 2nd edn. Elsevier Inc; 2012.

[432] SVC. *Society of vacuum coaters* [online]. 2021. Available from https://SVC.org [Accessed 18 May 2021].

[433] Mattox D.M. *The Foundations of Vacuum Coating Technology Book*. 2nd edn. Norwich, NY: Elsevier; 2018.

[434] Masaru T., Hiroyuki M., Ryouji C., Toshiyuki S., Kiyoshi A. 'Application of the ion plating process utilized high stable plasma to the deposition technology'. *Journal of the Vacuum Society of Japan*. 2001;44(4):435–9.

[435] SEMI. *SEMI Standards* [online]. 2021. Available from https://www.semi.org/en/products-services/standards [Accessed 18 May 2021].

[436] Chen Y., Altermatt P.P., Chen D., *et al*. 'From laboratory to production: learning models of efficiency and manufacturing cost of industrial crystalline silicon and thin-film photovoltaic technologies'. *IEEE Journal of Photovoltaics*. 2018;8(6):1531–8.

[437] Varache R., Dahan J., Hotel J., Favre W., Danel A., Roux C. Mastering the defectivity: prerequisite for high efficiency silicon heterojunction solar cell. *Proceedings of the 37th European Photovoltaic Solar Energy Conference and Exhibition*, 7-11 September 2020; Munich: WIP Renewable Energies; 2020. pp. 252–5.

[438] Von Ardenne. *Equipment for high-efficiency solar concepts – HISS/XEAnova5.5/XEAnovaL* [online]. 2020. Available from https://www.vonardenne.biz/fileadmin/user_upload/VAOP/druckschriften/www/pdf/High-Efficiency_Solar_Equipment_PVCSI_eng.pdf [Accessed 18 May 2021].

[439] Singulus. *Generis PVD online brochure* [online]. 2020. Available from https://www.singulus.com/fileadmin/user_upload/singulus2020/Generis_PVD_070720.pdf [Accessed 18 May 2021].

Chapter 6
Screen-printed contacts

Sebastian Tepner[1], Matthias Hoerteis[2], Linda Ney[1], Marwan Dhamrin[3], Tsuji Kosuke[3], Jan Lossen[4], Marco Galiazzo[5], Tom Falcon[6], Helge Wolter[7], Jaap Hoornstra, and Thorsten Dullweber[8]

6.1 Silver pastes

6.1.1 Introduction to silver pastes

Industrially fabricated solar cells are primarily metallized by printed contacts. The reason for this is based on the simplicity and process speed of printing systems. The time frame for the contact formation of a solar cell is currently <1 s per wafer with the trend to even shorter processing times. Therefore, metallization processes using photolithography, as it is commonly used in microelectronics, is too time-consuming and cannot be used to process solar cells. Screen printing of thick film pastes has been and still is the leading technology in solar cell production.

Before thick film technology was applied to solar cells, it had been historically used in microelectronics and offered an alternative to printed circuit board (PCB) technology. Even today, thick film paste is widely applied in electronics (e.g. conductor paste, via paste, dielectric paste and conductive adhesives). It is not surprising that the first pastes used for solar cells derived from these early existing products addressing similar requirements, which are high electrical conductivity, good mechanical adhesion and superior solderability.

As silicon (Si) solar cells are more demanding substrates compared with PCBs or ceramic substrates, the pastes must fulfil further requirements such as fine line

[1]Fraunhofer Institute for Solar Energy Systems (ISE), Germany
[2]Heraeus Holding GmbH, Germany
[3]Toyo Aluminium K.K., Japan
[4]International Solar Energy Research Center (ISC), Germany
[5]Applied Materials, Italy
[6]ASM Assembly Systems Ltd, UK
[7]TÜV NORD EnSys GmbH & Co., Germany
[8]Institute for Solar Energy Research Hamelin (ISFH), Germany

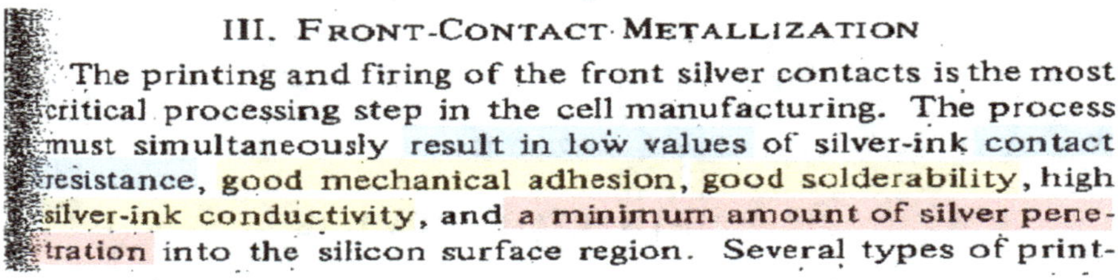

III. FRONT-CONTACT METALLIZATION

The printing and firing of the front silver contacts is the most critical processing step in the cell manufacturing. The process must simultaneously result in low values of silver-ink contact resistance, good mechanical adhesion, good solderability, high silver-ink conductivity, and a minimum amount of silver penetration into the silicon surface region. Several types of print-

Figure 6.1 *Gary C. Cheek et al. 'Thick film metallization for solar cell application', IEEE Trans. Electron Devices, 1984*

printability, low contact resistance to the silicon emitter and most importantly, a controlled contact formation to minimize the damage of the semiconductor. These properties were summarized in an early paper by Gary C. Cheek [1] (see Figure 6.1).

From today's point of view, it is not surprising that these early electronic pastes worked in principle on silicon solar cells. The composition of the paste, manufacturing process and even some application processes are quite similar. The very basic paste composition has not been changed since those early times in the 1980s. The main components are a conductive metal powder, typically silver, accompanied by a glass frit, originally based on lead oxide, with both dispersed in an organic mixture, often called vehicle. Every single component has been intensively studied, developed and optimized with the resulting metallization paste of today being a highly innovated product. In addition, pastes are individually adopted for different solar cell architectures and/or process technologies. Due to an extreme price pressure in the photovoltaics (PV) market and an excessive competition among paste manufacturers, current PV pastes far outperform electronic pastes and have contributed to improvements in solar cell efficiencies by more than 1% absolute. With further understanding of contact formation, paste rheology and printing simulations, efficiency gains of up to 0.05% year over year can be expected from paste improvements.

While most paste improvements in the past could be primarily attributed to new glass formulations, most paste improvements today result from a collaborative approach of polymer scientists, silver experts, glass developers and experienced paste formulators. Due to the right combination of silver particles and glass chemistry, the contact resistivity even at rather low doping concentration of below 1×10^{20} cm^{-2} could be reduced to $\rho_c < 1$ mΩ cm^{-2}. Printability, used as an expression to summarize many aspects of screen printing (line width, line homogeneity, printing speed and printing yield) has progressed impressively. Today, industrially common printing frequencies are below 1 s per wafer, and line width improved from 300 µm to below <30 µm-wide fingers structures in a period of less than 15 years and is still progressing.

These steady improvements are only possible due to the close collaboration between paste manufacturers and solar cell companies, and the fact that each paste iteration can be immediately tested on a large scale. In order to reliably differentiate one paste from another with only 0.02% efficiency difference, more than 200 wafers are required. Quite often paste experiments include printing and firing variations, resulting in studies for which several 1 000 wafers are required. Therefore, statistics and access to an industrial backend line are necessary.

Despite evolutionary step-by-step improvements, it is worth highlighting that game-changing paste developments have had an immediate impact on solar cell performance:

- Ferro was the first to replace silver flakes with highly mono-dispersed crystalline spherical silver particles: →Efficiency improvement of 0.2%
- Heraeus introduced a highly effective thixotropic agent and changed the finger shape from gaussian to nearly rectangular: Efficiency improvement of 0.3%
- Giga Solar Materials used silicone oils to enable fine line printing through openings of less than 40 μm: Efficiency improvements of 0.3%
- DuPont partially replaced lead oxide by tellurium oxide in the glass frit, which led to a significant improvement in the contact formation on lowly doped emitters: Efficiency improvements of 0.3%.

6.1.2 Manufacturing process and composition

Silver paste is typically manufactured in a batch process which is similar to the process of making dough. Initially, both organic and inorganic ingredients are mixed in a large vessel and subsequently homogenized on a three-roll mill. During the milling process, larger agglomerates will break up into their primary particles and the surrounding solvents and surfactants prevent re-agglomeration to form a stable dispersion. After viscosity adjustments, the paste is typically packaged in 2 kg or 4 kg jars.

Silver frontside pastes typically have the following composition, separated by organic and inorganic ingredients:

- 86–94% Silver powder (inorganic): after sintering or curing, the highly conductive metal transports the charge carriers from the semiconductor via fingers and busbars into the interconnectors.
- 0–4% Glass frit (inorganic): responsible for contact formation and adhesion and can promote or suppress silver particle sintering.
- 5–10% Vehicle (organic): all inorganic materials are homogenously mixed and dispersed in an organic medium of solvents, resins and polymers. The vehicle is critical for the printing behaviour. Meanwhile, vehicle formulations are as IP sensitive as glass formulations.
- 0–1% Additives (organic or inorganic): additives are often used to optimize paste sintering properties or printing behaviour to certain screen layouts or specific printing parameters.

Two ingredients are controversially discussed: lead as part of the glass frit, and silver as the main conductive material. While a lead-free glass is a question of regulatory (both health and environmental) and is technically feasible, silver-free pastes will remain a vision, at least for air-fired high-temperate applications. Despite its high cost compared to copper, silver is by far the most successful contact material for silicon solar cells. Solar cells have been metallized using silver for decades, mainly due to its excellent conductivity, corrosion resistance and compatibility to silicon.

Nonetheless, 'every cloud has a silver lining' and in the future we will see Cu-pastes for low-temperature applications like heterojunction, perovskite, thin film and organic solar cells.

Until that time, we will continue to use silver for most solar cells for the following reasons:

- It is cost-effective to precipitate and produce spherical mono-dispersed silver particles from dissolved silver salts.
- The glass frit, responsible for contact formation, is optimized for dissolving silver.
- Silver has the highest conductivity amongst all metals. Only Cu has a similar conductivity but has two main disadvantages; (1) Cu diffuses into the solar cell forming a rather active defect centre at the middle of the bandgap resulting in a strong performance degradation already at operating temperatures of a module and (2) Cu is easily oxidized in air at high temperature.
- Silver contacts have demonstrated a longer history of superior performance with regard to both efficiency and degradation.

6.1.3 Market

The solar paste market used to be dominated by traditional thick film companies such as Ferro, DuPont, Namics and Noritake. Around 2006–2008, when mass production of solar cells started, Giga Solar Materials, Samsung and Heraeus entered the market, too. Today, companies like DKEM (China), Fusion (China) or iSilver (China) and many others are entering the competition for the market share of large Chinese PV manufacturers.

The market is influenced by two major trends: (1) the paste consumption which itself is dependent on the price of silver and (2) the number of solar cells manufactured. In times of high silver price, the increase in the overall PV market was highly saturated by new innovations and the pressure to reduce the amount of silver per cell. The consumed silver per solar cell was reduced from approximately 300 mg/wafer in 2010 to less than 70 mg/wafer today. This reduction is due to smarter grid designs, optimized screens and improved pastes. Further reduction in silver consumption is expected to progress with new technologies like multi-busbar and smart-wire connections. This can be summarized in a general trend, which is shifting the metallization from cell to module level and therefore from silver to copper. Frontside pastes are exclusively based on silver, while most solar cell interconnectors use copper.

Nonetheless, the frontside silver paste market is still attractive, as it is expected to see a rather constant growth in volume due to a more self-sustained and market-driven PV industry independent of feed-in tariffs. The current frontside silver paste market reached about 2 000 tons/year (2020) and is expected to grow with a CAGR of 6%.

6.1.4 Metallization and firing process

The end of screen printing technology has been forecasted for more than ten years due to limitations in line opening, line definition or processing speed. Despite these prophecies of doom, screen printing is still the only technology with significant market share to metallize solar cells today. Due to its relatively simplistic and robust process, broad market acceptance and competition among multiple market players are driving continuous innovations and improvements in screen printing technology.

Over the past decade several alternative printing technologies have failed to enter the market including inkjet printing, laser transfer, laser opening and plating, dispensing, pad printing and even stencil printing. Due to the demand for taller and finer lines, and larger more fragile wafers, faster and potentially contactless printing technologies such as dispensing or transfer printing might gain the market share in the future. The need for speed might also open doors for roll-to-roll technologies.

Regardless of the metallization concepts, they all rely on similar inks and pastes, and the real magic happens during the firing process when three functionalities designed in one paste form a gateway for the photogenerated electrons. A low ohmic contact between semiconductor and metal is formed to extract the electrons from the emitter of a solar cell into the bulk silver of the contact structure. The extremely low resistivity of silver and the ability of individual Ag particles to sinter together is key for a low-loss lateral transport. Lastly, the contacts must remain both mechanically and electrically stable for the lifetime of the module.

All silicon solar cells metallized with a high-temperature paste are exposed to temperatures of approximately 800 °C. A typical firing profile for a silicon solar cell processed in a fast firing furnace is shown in Figure 6.2.

According to Cheek *et al.* [1], ‘the firing of the front silver contacts is the most critical processing step in the cell manufacturing’. During the firing process, a lot of physics and chemistry is happening during a short period of time at rather high temperatures on a sensitive semiconductor substrate. The frontside contact is formed by firing through the anti-reflection or passivation layer. The rear-side contacts are formed, together with a local aluminium back surface field (BSF), which is a highly doped p+ region with the function of reducing the surface recombination at the rear side. Hydrogen from the front side passivation layer is released to increase the bulk carrier lifetime and finally some long-term recombination effects are suppressed.

These processes all happen within a few seconds. It starts with a ‘burn out’ phase, where all organic paste ingredients which are not volatile are thermally decomposed at temperatures of up to 500 °C. Above the temperature of 250 °C the silver particles will begin to neck, which will continue with increasing temperature until the peak of about 800 °C when all silver particles are sintered in one densified

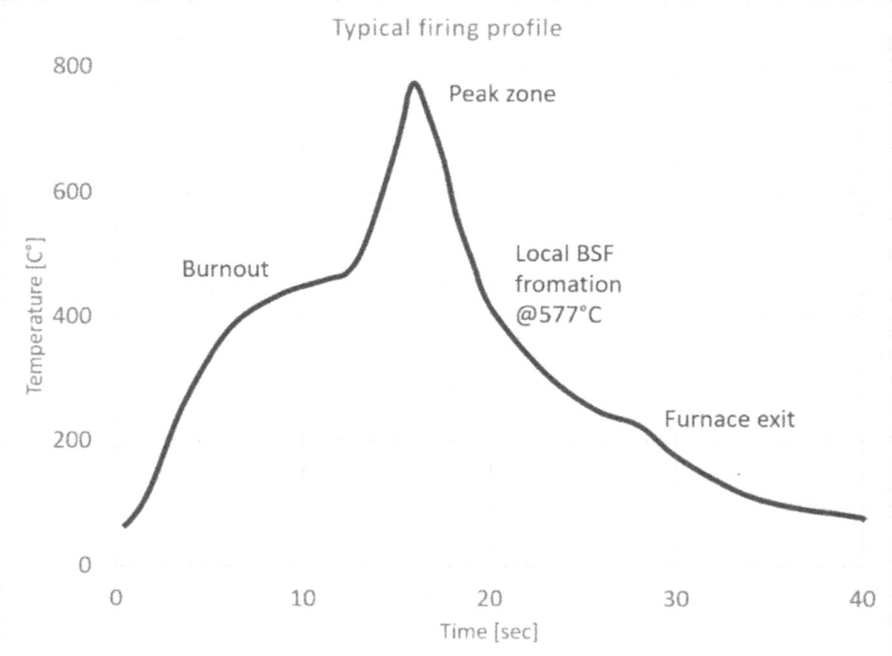

Figure 6.2 A typical temperature profile for a printed silicon solar cell during the firing step

matrix. Depending on the glass chemistry, the glass particle starts to soften between 300 °C and 450 °C, and as soon as a temperature of 400–600 °C is reached, the glass viscosity is low enough for capillary forces to pull the liquid glass to the silicon-silver interface. With increasing temperatures, the glass wets the silicon surface, covering every part of the contact structure. During the peak zone, where the wafer is heated within seconds from 500 °C to 800 °C and is then quickly cooled down to room temperature, the magic of the actual contact formation occurs.

6.1.5 Review of the current models on contact formation

The details of contact formation are one of the most exciting areas of solar cell research. Traditionally, paste development had been reserved for thick film companies, while new glass, vehicle and paste formulations were kept as closely guarded secrets, comparable to the Coca-Cola recipe. Even today, the paste and the unique formulations are highly valued goods, and IP is sensitive information.

Luckily, several scientists, sometimes in cooperation with a solar cell manufacturer or paste manufacturer, have started to reveal the secrets and believed mechanisms of contact formation. Centres for this research were/are Georgia Tech, The University of Konstanz, Fraunhofer ISE, University of Korea and NREL.

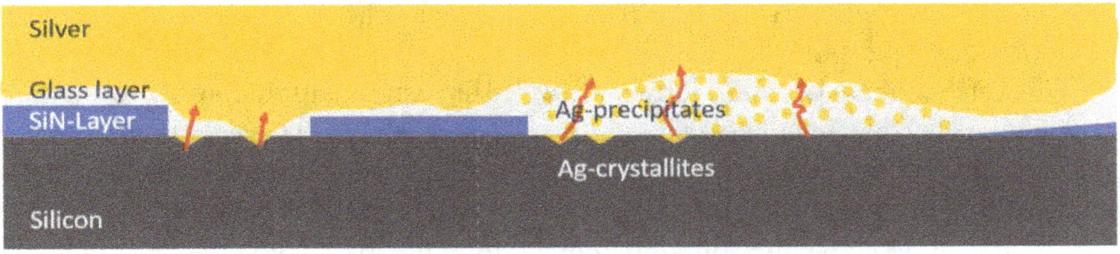

Figure 6.3 Schematic contact cross section

The individual paste components are known or can be analysed, and the fired contact structures on a finished solar cell are also accessible. Detailed microscopic investigations of the interface between silver contact and silicon surface were performed by several groups, revealing that the bulk silver is separated from the silicon surface by a rather inhomogeneous layer of molten glass (see Figure 6.2). Covered by glass and grown into the silicon surface are micron-sized, rather geometric imprints of pure silver, called Ag crystallites (see Figure 6.3). The glass layer itself is infused by a high number of nanoscopic metallic precipitates.

Scanning electron microscopy (SEM), transmission electron microscopy (TEM) and atomic force microscopy (AFM) images together with electrical measurements of solar cells were used to explain the role of the glass, the growth of silver crystallites and silver precipitates, and provided enough latitude for speculations on possible current flow mechanisms.

Clear correlations between microstructure and electrical solar cell measurements could be found. It is also known that firing conditions such as temperature, time and atmosphere can heavily influence the contact quality. It became rather clear that Ag crystallites were playing an important role, and their density correlates with firing temperature and oxygen concentration. It is also strongly dependent on the doping level of the solar cell emitter. However, it is not always beneficial to aim for a high number of contact points from an electrical standpoint, as each of them increases the recombination and therefore lowers the open circuit voltage. In addition, for an optimum contact the number of crystallites should not be too low or too high. The goal is a low contact resistivity while inducing a minimum damage to the solar cell. Therefore, for each paste formulation and solar cell type, the optimum processing parameters need to be adjusted.

However, static images and solar cell data were not enough to explain the formation of this surprisingly complex microstructure.

Two questions around the described structures have been academically discussed: how are these structures formed, and how important are they for the current flow?

The ongoing discussion is whether Ag crystallites or nano-precipitates are needed for low ohmic contact. Do the electrons flow from the emitter via an Ag crystal into the bulk silver, or do they preferably cross the glass layer via percolated tunnelling?

Improved understanding of the contact physics or chemistry is needed in order to answer these questions.

Gunnar Schubert [2] and Hörteis *et al.* [3] investigated the contact formation by studying the reactions between single paste components (silver, glass and silicon) in different combinations, with the aim to understand the reaction mechanisms. Schubert proposed one of the first contact models, described as follows.

Upon heating above the softening temperature, the glass frit begins wetting and reacting with the silicon wafer surface. A redox reaction between lead oxide as part of the glass frit and the silicon nitride (SiN) layer or Si surface occurs, forming a liquid Pb phase. The silicon surface is preferably 'etched' on defects like dislocations, dopant inclusions, edges and tips of surface texture, and the formation of inverted pyramidal pits starts.

$$2\,PbO + Si \rightarrow SiO_2 + 2\,Pb \tag{6.1}$$

$$2PbO + SiN_x \rightarrow 2Pb + SiO_2 + \frac{x}{2}N_2 \tag{6.2}$$

Silica formed during the reactions (6.1) and (6.2) is absorbed by the surrounding glass layer. Lead is partly oxidized, and the reaction continues, and it alloys with silver particles to form a Pb-Ag phase. Upon cooling, phase separation of the liquid Pb-Ag phase causes Ag to precipitate at the inverted pyramidal pits. The remaining Pb-Ag phase is recrystallized in the form of metallic nano-precipitates and lead is oxidized and incorporated into the glass layer.

The theory, however, is not flawless and there are a few facts which couldn't be explained or ignored. For example, the perfect, almost epitaxial growth of the Ag pyramids which is not expected for fast cooling and recrystallizing Pb-Ag alloy. In addition, no Pb can be found in Ag crystallites but pure silver. Furthermore, Ag crystallites are covered by a glass and not an Ag-Pb alloy and metallic lead was only found within the glass layer when the contacts were fired under inert conditions.

The theory is not flawless, and a few facts cannot be explained or ignored. First, the empirical evidence shows perfect, almost epitaxial growth of the Ag pyramids, which is not expected for fast cooling and recrystallizing Pb-Ag alloy. In addition, no Pb can be found within the Ag crystallites. Furthermore, Ag crystallites are covered by a glass and not an Ag-Pb alloy and metallic lead was only found within the glass layer and only when the contacts were fired under inert conditions.

Therefore, another theory was provided. Kyoung-Kook Hong *et al.* [4] proposed a new model, explaining the existence and growth of Ag crystallites from dissolved Ag ions without the need of a liquid metal phase.

With increasing temperature, the glass frit fluidizes, wets the Si surface and dissolves some of the Ag atoms. Some of the Ag atoms dissolved in the glass oxidize to Ag^+ due to the presence of oxygen in the ambient atmosphere:

$$4Ag + O_2 \rightarrow 4Ag^{+}_{(glass)} + 2O^{-2}_{(glass)} \tag{6.3}$$

The silver and oxygen ions in the fluidized glass diffuse towards the glass/Si interface and react with the Si wafer:

$$4Ag^{+}_{(glass)} + 2O^{-2}_{(glass)} + Si \rightarrow 4Ag^{0} + SiO_{2(glass)} \tag{6.4}$$

As long as the temperature is high enough and the glass viscosity low, Ag^+ ions constantly diffuse towards the Si surface, oxidize Si to SiO_2 and deposit metallic Ag on the emitter surface according to reaction (6.4) . At the same time, the glass layer is enriched by silica which increases glass viscosity and limits Ag solubility. During cooling, the solubility of Ag in the melt decreases and nano-precipitates are formed within the glass layer. The glass viscosity and its ability to dissolve Ag are the key for good contact formation.

For both theories, the geometric shape of the Ag pyramids is explained by the lattice structure and preferred etching. On {1,1,1}-Si, the crystallites are quite shallow and lens-shaped, whereas on {1,0,0}-Si the crystallites are pyramid-shaped. This is explained by the fact that {1,1,1}-Si oxidizes three times faster than {1,0,0}-Si.

For quite some time, the academic world of contact formation was divided into two fields with supporting papers and arguments on both sides.

In 2016, Fields *et al.* [5] managed to please both parties when he postulated the latest contact model. Using advanced in situ X-ray diffraction in combination with rapid thermal annealing, he was able to decode the contact formation process.

At temperatures between 500 °C and 650 °C PbO in the frit reacts with and penetrates the SiN_x layer according to Schubert's model. Pb formed during this reaction alloys with Ag and assists silver sintering. However, the Pb-Ag phase is not involved in the formation of Ag crystallites or precipitates. They are formed at above 650 °C when Ag dissolves into the frit and diffuses towards the emitter surface and deposits as Ag crystallites, according to Hong's model (Figure 6.4).

6.1.6 *Possible current paths on printed contacts*

Another controversially discussed topic is the current transport mechanism, or rather, the pathway on which electrons/holes leave the emitter and enter the contact finger. The contact barrier between a metal and a semiconductor is described by the Schottky theory.

The theoretical resistivity value for a Schottky contact between silver and n-type silicon is dependent on the doping concentration and ranges from $\rho_c{\sim}1\times10^{-7}$ to 1×10^{-8} Ω cm² for doping concentrations above $N_d{>}1\times10^{20}$ cm⁻³. These doping concentrations are typical for industrially fabricated solar cells. The measured values for ρ_c however are in the range of $\rho_c{\sim}1\times10^{-3}$ Ω cm². In other words, only 0.01% of the contact area needs to be covered by Ag crystallites to explain the measured values. Roughly 1% of the contact area of a real contact is covered by Ag crystallites which leads to the assumption that there is still a lot of room to improve the contact quality, mainly by reducing the number of crystallites and increasing the conductivity of the glass layer. In the end, the current path needs to be understood.

For the current transport between the emitter and the frontside contact, multiple connections are possible via single Ag crystallites; a single crystallite can either be in direct contact with the bulk silver or be covered by a glass layer which is thin enough to allow the charge carriers to tunnel through. It is also possible that current flows between the emitter and the bulk silver on contact areas where the anti-reflectance coating (ARC) is opened and the bulk silver is in direct or indirect

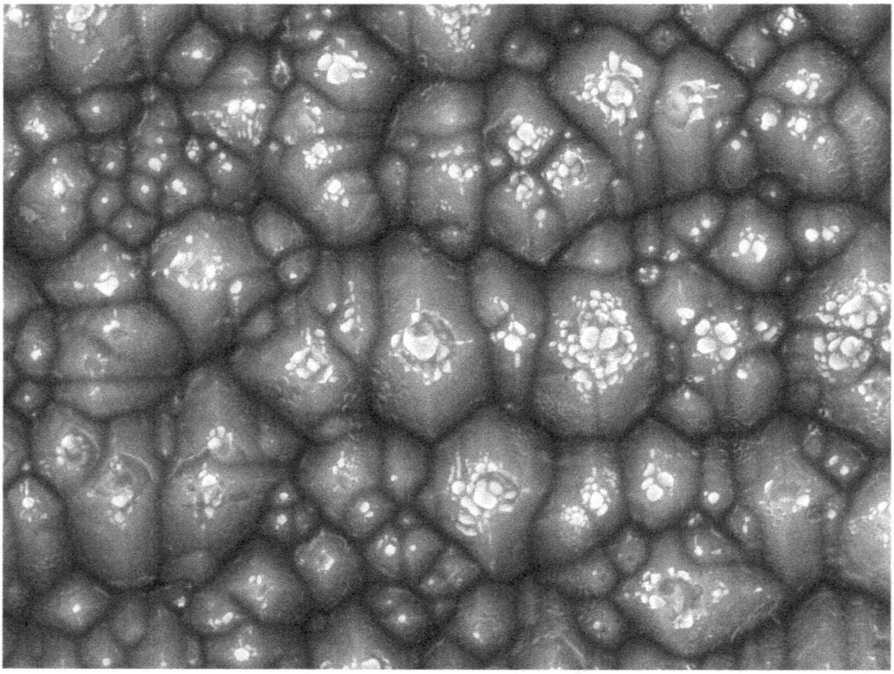

Figure 6.4 Scanning electron microscopy (SEM) top view of a silicon metal interface after the fired contact finger has been chemically removed. The pyramidal surface texture partly covered by Ag crystallites (white spots) is visible.

contact with the emitter, and again, a direct or a tunnelling contact can be imagined. All contact pathways can be combined with percolated tunnelling between silver precipitates distributed throughout the glass layer.

Even if most current paths are related to Ag crystallites grown into the emitter and the theory of a high-quality contact formation due to the epitaxial growth of the Ag crystallites is generally accepted, it is also known that damage of the Si surface by an Ag crystallite increases the recombination and should be avoided. Ideally the opening of the passivating SiN layer is balanced between a low ohmic contact and minimum surface damage. Furthermore, the growth of Ag crystallites should be limited and ideally completely avoided.

This 'smooth' contact formation would have the advantage that the p-n junction is not damaged by the penetration of metal impurities. In this case, the proposed current path is defined by percolating tunnelling via silver precipitates between the silicon surface and the bulk silver.

Such a 'smooth' contact, promoted from time to time, would be theoretically ideal and could increase volatile oxide compound (VOC) and FF values, if the glass used to open the passivation layer exhibited passivating properties, too.

Unfortunately, its existence has yet to be empirically proven. Nearly all SEM images of the contact interface from present solar cells show the typical Ag crystallites in intimate contact with the silicon surface.

6.1.7 The importance of electrons

Hee-Soo Kim *et al.* [6] took another approach when he described and explained the contact formation from an electrochemical point of view. He observed a defect phenomenon using electroluminescent analysis which he called 'gray finger'. Under certain conditions, some areas of a solar cell appeared grey in electroluminescence images, indicating a high contact resistance. He could show that the strongly reduced number of Ag crystallites underneath the gray fingers is correlated with nearby local short cuts. He concluded that the firing reactions in Ag thick film contact formation could be interpreted in terms of the mixed potential theory of corrosion, described as follows.

The electrochemical reactions involved in fire-through Ag contact formation establish a potential difference between the sintered Ag grid lines and the Si emitter separated by molten glass. The molten glass acts as an electrolyte containing mobile Ag^+ and O^{2-} ions during contact firing. Therefore, the short circuiting between the sintered Ag grid lines and Si emitter produces a galvanic cell during contact firing, which inhibits Ag crystallite formation at the contact interface along the grid lines in a short circuit and produces the 'gray fingers'.

In other words, electrons are essential for the reduction of Ag^+ to Ag^0 and without them, no contact can be formed.

Inspired by these findings, Matthias Hoerteis [7] connected a DC power source to a solar cell to provide or remove electrons during the contact formation. At a negative current, no contact was formed as electrons were withdrawn from the front side. Conversely, a positive current where electrons were actively provided during contact formation resulted in a very low ohmic contact. The corresponding microstructure clearly revealed the difference; in the case of an electron-supported contact firing, the silicon surface was almost completely covered by silver crystallites, whereas no single crystallite could be found if electrons were actively suppressed during firing. It could also be shown that contact formation was possible at much lower temperatures when supported by electrons.

Interestingly, the opening or 'etching' of the SiN layer seems not to be affected by the injected current, which further supports Field's model where the SiN opening and Ag crystallite formation are independent, subsequent reactions. Nonetheless, Ag crystallites in the silicon surface could still not be avoided. So, the search for a damage-free 'smooth' contact remains.

Eve Krassowski and Andreas Henning [8] presented a method, called laser enhanced contact optimization (LECO) where they used a laser to scan the surface of a poorly contacted solar cell. The laser is used to excite charge carriers and force them through the metal-semiconductor interface, which results in high current flow through local, very small conductive areas.

With this method, abundant electrons are provided and the reduction of Ag^+ to Ag^0 as described by Kim for a molten electrolyte occurs in a solid state, paving the road for 'smooth' contact formation.

Ryan Mayberry [9] used the LECO method with an optimized glass frit, which did not contact the emitter after regular contact firing and demonstrated the advantages of a low-damage contact. In his SEM images, he did not find any Ag crystallites or damage on the silicon surface and he could give evidence of the existence of a 'smooth' contact.

This new, two-step contact method is enabling new silver paste developments and will facilitate effective contact formation on well-passivated solar cell types like passivated emitter rear contact ($PERC^+$), tunnel oxide passivated contact (TOPCon) and n-type. Even the simultaneous contact formation of n- and p-doped surfaces, as needed for IBC solar cells, should be possible.

With these new findings, the science of contact formation is and remains one of the most needed and exciting topics in solar cell research.

6.2 Aluminium pastes

This sub-chapter introduces aluminium (Al) paste as the key metallization material in conventional silicon solar cell technology. Section 6.2.1 describes the art of paste making, its composition and history. In section 6.2.2, the process of formation of Al-doped p^+ silicon by alloying Al paste to the Si wafer is introduced and the alloying aspects are discussed in addition to the dopant control via the addition of boron to the Al paste. The applications of Al paste in conventional full area BSF, PERC and passivated emitter rear totally diffused (PERT), and passivated contact solar cells is covered in section 6.2.3 with detailed evaluation of carrier recombination impact at the Al/Si interface for advanced narrow Al fingers used in future solar cell contacts.

6.2.1 Aluminium paste composition

The early concept of conductive pastes can be traced back to the early 1930s when General Electric patented what they called conducting varnish that consisting of a mixture of about 50% oil varnish, about 40% of lithopone and about 10% soot or metallic dust [10]. A decade later, further developments were made to form metallization pastes containing metal powders or flacks, an organic vehicle and an inorganic flux, to be later called as glass frit, a composition useful for forming metallic layers on quartz, ceramic material, glass and other materials that are highly adherent solder joints [11]. The organic vehicle was and still is of great importance as it decides the printing properties of the paste, its shelf life and the overall electrical performance of the fired paste. Early works of metallization paste makers address the issue of selecting particular vehicle system properties to match with the paste's industrial need [12].

The usual aluminium paste used to form the backside electrode of a solar cell generally comprises aluminium powder, glass frit and an organic vehicle. The ratio of the vehicle to metal powder plays a key role in the paste design. Heavy metals like silver can be in the range of 90%–95% whereas aluminium must be below 85% due to its high volume that requires more organic vehicle to allow for better printing. For aluminium pastes, the organic vehicle is added in an amount of about 15%–40% by weight in order to improve the applicability or printability of the paste composition when the composition is applied by screen printing. Aluminium powder with a diameter of 1–20 μm is added in an amount of about 60%–85% by mass. Glass frit is added in an amount of about 1%–5% by mass [13–16]. The role of the organic vehicle for the Al paste is to maintain the paste body, which is a viscoelastic substance mainly composed of Al particles, for screen printing, and to impart adhesion between the substrate and the Al coating film after drying. In addition to the Al particle size and particle size distribution, the viscoelasticity of the Al paste can be changed according to the desired printability by the combination of the molecular weight of the polymer and type of the resin and also the solvent that dissolves the resin. Furthermore, in the firing process of metallization, reducing the amount of carbon left from resin, and burning out before the glass transition point or softening point of the glass frit are preferable to avoid the restriction of the sintering reactivity between Al particles and Si wafer. For the solvent choice process, a high boiling point solvent is mainly used to avoid the paste drying during the screen printing process. Additionally, in terms of compatibility, the solvent should have similar polarity and Hildebrand solubility parameter with the main resin used in the vehicle. Finally, for the purpose of improving the wettability between -OH on the Al particle surface and the resin, additives such as fatty acids can be included in the paste to control the solid concentration and to prevent the paste from sedimentation inside the jars.

The organic vehicle is prepared by dissolving one or more of known resins, for example, ethyl cellulose, nitro cellulose, polyvinyl butryal, acrylic resin, alkyd resin, polyimide resin, in one or more solvents such as ether-based glycol, diethylene glycol monobutylether, diethylene glycol monobutylether acetate and dipropylene glycol monomethylether. Additionally, the organic may contain various additives and agents with certain functions such as an antioxidant, a corrosion inhibitor, an antifoaming agent, a thickener, a coupling agent, an electrostatic imparting agent, a polymerization inhibitor, a thixotropic agent, an anti-settling agent, that are important to stabilize the paste and increase its shelf life, for example, polyethylene glycol ester compound, polyethylene glycol ether compound, polyoxyethylene sorbitan ester compound, sorbitan alkyl ester compound, aliphatic polycarboxylic acid compound, phosphate ester compound, amide amine salt of polyester acid, polyethylene oxide series compounds, fatty acid amide waxes and many others that are selected carefully to match with the vehicle system and the surface of the metal powder [13–16].

Conventionally, the design of the Al paste for full area BSF, full area rear junction PERT and PERC solar cells is easier because it needs only to consider few viscoelastic properties for printing a blanket-like Al paste film. As can be seen in Figure 6.5, the screen printing process will start with flooding the paste to cover the whole printing pattern with openings on the emulsion/mesh layer which will be discussed in the sub-chapter 6.6., using a metal scraper blade that moves with speeds

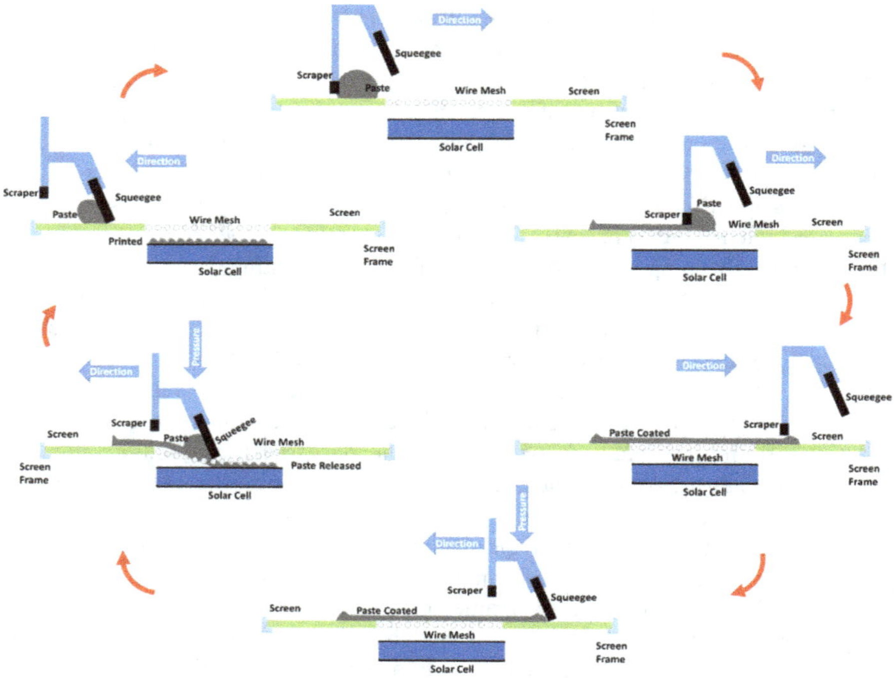

Figure 6.5 Movement sequences during paste screen printing

exceeding 300 mm s^{-1} forming a paste film with uniformly distributed thickness. Therefore, the paste design should ensure that there is no screen sagging so the paste does not seep out via the opened pattern during flooding and should avoid a high tackiness level to ease the release of paste when squeezing it via the full printed area pattern. On the other hand, new cell designs such as bifacial PERC and rear junction PERT with Al fingers will need an Al paste with higher viscosity and thixotropic properties than the conventional Al paste for full-area printing to realize high aspect ratio fingers printed using 100–200 μm. The sheer rate of the Al paste during printing changes when squeegeeing the Al paste in the opening of the screen mask and separating with a substrate and screen mask. The sheer rate will be over 1 000 s^{-1} when the squeegee starts to print the Al paste on the wafer. At such a high sheer rate, the paste viscosity will decrease and, subsequently, the screen mask pushed by the squeegee will seep out from the substrate due to mask tension, and then the paste is released from stress and the shear rate decreased to 0.1–1 s^{-1} or less. Figure 6.6 shows the time-dependent viscosity of the controlled shear rate test to represent paste deformation during the screen printing process. This measurement protocol comprises three sections with different deformation properties. A sheer rate of 0.1 s^{-1} was applied in region I and III to represent the behaviour during preprint and at rest, after the paste was pushed through the finger opening. In region II, we applied a sheer rate of 1 000 s^{-1} to simulate paste transfer through the finger opening. It is

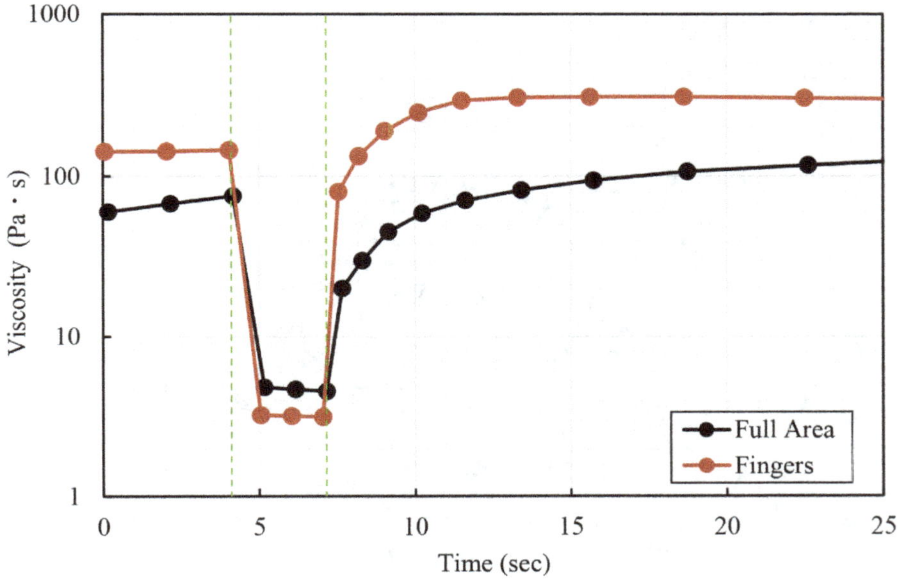

Figure 6.6 *The viscosity of time-dependent controlled sheer rate test*

necessary to have a higher paste yield point and short recovery time of viscosity in order to avoid the paste spreading at the levelling process to keep the finger shape as best as possible.

The glass frits are of great importance as they melt without being volatilized at peak temperature during firing and re-solidify during cooling, providing excellent adhesion. Additionally, the glass frits increase the reliability and electric performance of the contacts. The Al paste for solar cell metallization uses glass frit systems that include metal oxides such as PbO-B_2O_3-SiO_2 and environment-friendly lead-free Bi_2O_3-B_2O_3-SiO_2 [16].

The role of the glass frit is to improve the adhesive strength between the Al-sintered layer and the Si substrate as an inorganic binder. It controls the sinterability of Al particles, improves Al and Si reactivity and suppresses deterioration of characteristics by long-term reliability. For external appearance properties, the glass frit improves the surface roughness and suppresses the bumps on the surface of the final contacts to eliminate any cell breakage during solar module fabrication. As for the glass frit selection criteria, softening point, melt viscosity, wettability with Al particles and the standard Gibbs energy of glass oxides related to the redox reaction are taken into consideration. In particular, in order to make the glass frit function as an inorganic binder, a glass frit that softens and flows during the solar cell contact preheat is selected and glass frits with softening points between 300 °C and 600 °C are suitable. For instance, tellurium glass is used for Ag pastes [17] because of its low softening point and excellent work function that help in decreasing the contact resistance between Ag fingers and n^+ doped silicon. In Al paste, lead or bismuth

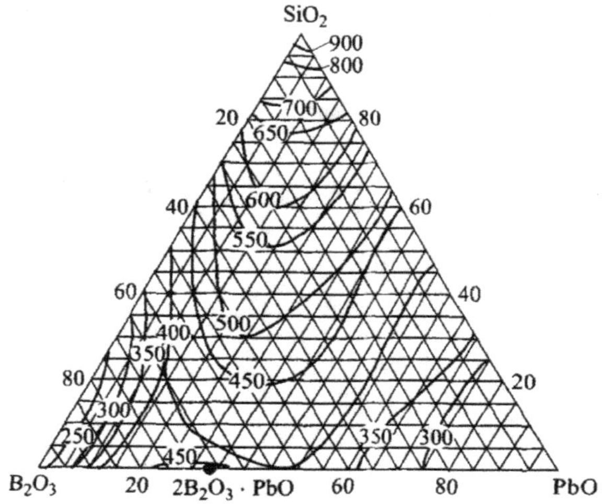

Figure 6.7 *Glass frit T_g taken from Trubnikov [18] in lead borosilicate glasses as a function of its molar composition*

glass frit systems are used. There are two main constituents of glass; the network formers that build up the bulk of the glass by a cross-linked network of chemical bonds and the network modifiers. The common network-forming constitute in glass frit industry is silicon oxide. Lead glass SiO_2-PbO-B_2O_3 is a typical example of low-melting-point glass. If the SiO_2 content, which is the network former component in the glass frit, is increased, the glass softening point increases, while if it is too low, the chemical resistance and water resistance of the glass frit decrease [18] as can be seen in Figure 6.7. B_2O_3 is a component that is highly effective in lowering the softening point and melt viscosity, but if the content is too high, the water resistance may decrease. PbO can lower the softening point or adjust the coefficient of thermal expansion. However, since lead is regulated by the restriction of the use of certain hazardous substances (RoHS) directive, etc., other low-melting-point lead-free glasses are required to reduce the environmental impact, and bismuth-based Bi_2O_3-B_2O_3-SiO_2 and the like are preferred. The average glass frit particle size is in the range of 0.1–5 μm, which increases the number of contact points with the Al particles and makes it easier to get wet with the Al surface and improve its reactivity with Si. In terms of reliability, the lead glass in the formed Al electrode is corroded by moisture and the electrode performance is deteriorated. In order to improve the wettability and reactivity with aluminium particles to suppress the formation of aluminium hydroxide, oxide glass containing vanadium oxides and phosphorus oxides can improve the long-term reliability of Al pastes if used [19].

The selection criteria for glass frit for Al paste that will be applied to PERC solar cells are more complicated. The Al paste requires a glass frit that does not fire through the passivation stacks made of aluminium oxide and capped by SiN_x layers and at the same time ensures not only a great bonding between the Al particles in

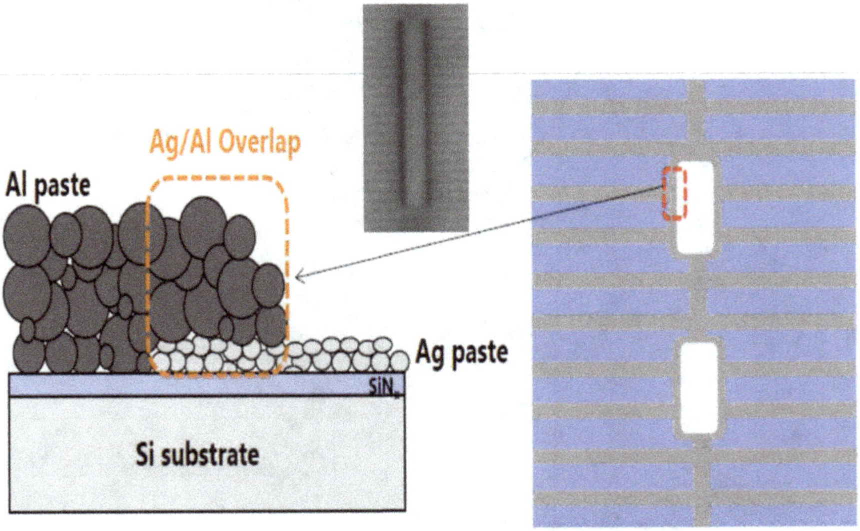

Figure 6.8 *The impact of the PbO glass frit in the Al paste on the quality of Ag/Al interfaces for bifacial passivated emitter rear contact (PERC) solar cells*

the paste matrix but also a great adhesion of the fired electrode to the SiN_x. In conventional full area and bifacial PERC, two paste systems are used. The first is the Al paste for the local BSF and the second is the rear tapping Ag paste for forming solderable pads for module string assembly. Al paste containing non-optimized lead glass will fire through SiN_x at the overlap area at the Ag/Al interface of the back silver pads as can be seen in Figure 6.8; this will reduce the open circuit voltage unless the fire-through features of lead glasses is optimized to avoid firing through SiN_x or lead-free glass frit system in the Al paste is used.

6.2.2 Forming p+ silicon by the aluminium alloying process

The solar cell metallization process starts with printing a thick film of metal pastes on both surfaces followed by a high-temperature curing or sintering step. Depending on the cell design, the front Ag grid printing steps might vary from a single printing step up to three printing steps as a combination of dual and double printing. The Al contact printing at the rear side of conventional full area BSF, PERC and PERT solar cells will include at least two printing steps where the first step involves screen printing of low-solid Ag paste for rear solder pads followed by a screen printing step of Al paste. The typical sintering profile is illustrated in Figure 6.9, on the top left, with a cell peak temperature of 760 °C. Since the solar cells move at a speed of >4 000 mm/min in the belt furnace, the whole firing process may take less than 60 s depending on the length of the furnace. The thick metal pastes, illustrated in Figure 6.9, left, contain about 10%–30% of organic vehicle which will require a drying process by

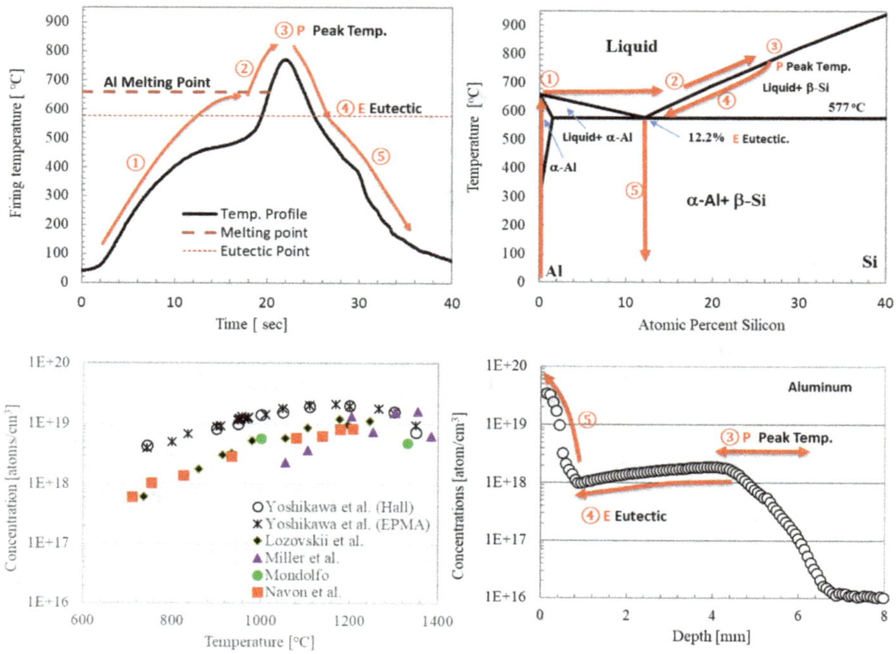

Figure 6.9 *Top left: Solar cell firing profile. Top right: Al-Si phase diagram redrawn after Roberts and Wilkinson [20]. Bottom left: Al concentration, max solubility, in Si as a function of temperature [21–23]. Bottom right: Secondary-ion mass spectrometry (SIMS) profile of Al in epitaxially regrown p+ Si.*

a conveyor belt furnace at a temperature between 400 °C and 500 °C, step 1 in Figure 6.9 top left, to evaporate the solvents used in the Al paste vehicle. The actual contact formation will take place during a higher temperature sintering step above 660 °C and below 900 °C for a few seconds. By the combination of belt speed and adjustable zone temperature settings, the firing profile can be adjusted to have a preheat area as a long plateau allowing slow evaporation of organic residue and fast ramp-up of temperature when the solar cells go through the temperature peak zone, usually set to 700–900 °C, at the end of the furnace.

When the temperature of the Al paste in contact with Si crosses 660 °C, the Al particles will melt inside the high-tension aluminium oxide shells and start wetting and dissolving the Si atoms at the surface via oxide shell cracks. Parts of these dissolved Si will diffuse to the Al particles via the shell cracks replacing and ejecting Al outside the shells. The ejected Al will form liquid Al beneath the Al paste particle matrix as illustrated in Figure 6.10, centre. Both the melted Al particles in the paste matrix and the Al liquid on the Si surface are saturated with Si, forming a melted Al-Si liquid phase, according to the phase diagram [20], at point *P* as shown in Figure 6.9, top right. The solution of melted Al-Si will be the source for

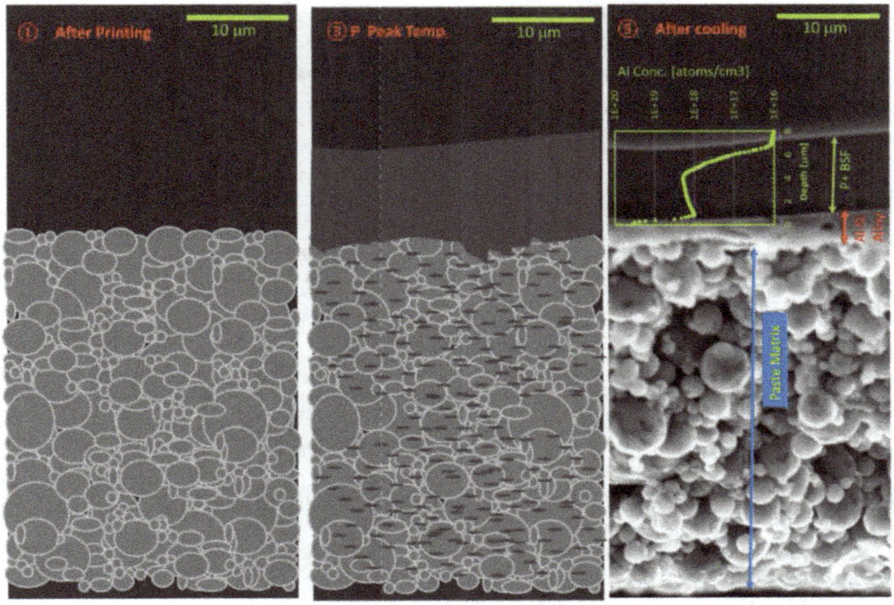

Figure 6.10 *Left: Al particle matrix after printing. Centre: At maximum firing temperature. Right: A scanning electron microscopy (SEM) image of the cross section of conventional full area Al back surface field (BSF) solar cell. The green curve is the secondary-ion mass spectrometry (SIMS) profile of Al in epitaxially regrown p^+ Si.*

the liquid-phase epitaxial process [20, 24]. As the firing temperature decreases, the silicon is rejected from the Si-rich Al-Si liquid phase according to the liquidus line in the phase diagram and deposited at the undissolved Si side with Al concentrations following the solid solubility of Al in Si, as can be seen in Figure 6.9, bottom left [21–23], forming a crystalline Al-doped p+ Si layer. When the temperature reaches the eutectic point *E*, the remaining Al-Si solution solidifies with high Al concentrations at the top surface of the Si wafer forming an excellent silicide contact.

The final doping profile measured by secondary-ion mass spectrometry (SIMS) is shown in Figure 6.9, bottom right, and also in Figure 6.10, right, to give the reader a better picture of the doping profile obtained by the liquid phase epitaxy process. In principle, the thick p^+ layer formation is essential to limit the carrier recombination at the backside of the classic full area Al BSF solar cells. Assuming that the printed Al paste is distributed uniformly across and is in contact with the whole silicon surface, the thickness of the p^+ layer can be calculated using Alamo *et al.* [25]:

$$T_p^+ = \frac{w_{\mathrm{Al}}}{A\rho_{\mathrm{Si}}}\left(\frac{P}{100-p} - \frac{E}{100-E}\right) \tag{6.5}$$

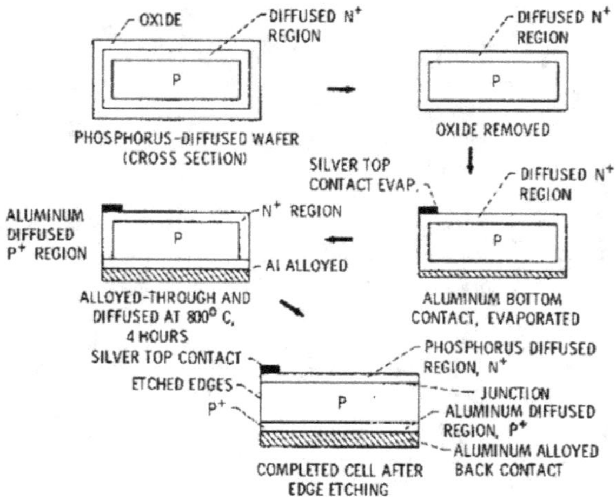

Figure 6.11 *Cell process steps used in the work of Mandelkorn and Lamneck [27]*

where w_{Al} is the weight of effective deposited Al in the paste, A is the printed area, ρ_{Si} is the silicon density, P is the Si weight percentage at firing peak temperature and E is the Si weight percentage at eutectic temperature. From the above equation, such thick p^+ formation will require a combination of high deposition of Al paste and higher firing temperature.

6.2.3 Applications of Al paste in conventional and advanced solar cells

Though the introduction of BSF as a concept, but not wording, goes back to as early as 1953 [26] where it was used in power rectifiers to reduce the contact series resistance and consequently decrease the reverse leakage current, it took almost 10 years to be introduced to the solar cell process. It was in 1972 when Mandelkorn and Lamneck [27] introduced BSF as a concept and used Al which can, miraculously, provide the p^+ BSF layer, which is an excellent contact and overall a great back reflector of near IR light. The cells were very simple (Figure 6.11), featuring frontside phosphorus diffusion followed by oxide removal and evaporation of frontside Ag and few micrometres of Al covering the entire back surface of the cell. The cells were then annealed for four hours in ambient argon dissolving the silicon beneath it and alloying with it forming about 1 mm p^+ layer which was described as diffused Al to the bulk [27], which we know, with our current knowledge, is a rather complicated process to describe simply as Al diffusion; we will shed more light on the process and its physics later in this chapter.

Since the evaporation process requires expensive vacuum and time consumption equipment, not to mention the low throughput and long time required for alloying, it was inevitable to look for other possibilities that adapt fast throughput, lower cost and excellent reproducibility. By then the screen printing technology was widely used in ceramic printed electronics and the attempt to use in solar cell fabrication was only possible after its successful introduction by E.L. Ralph from Spectrolab Inc. in 1975 [28]. The paper for the first time described a process flow to be used for decades to come as the main process flow of the conventional process of full area Al BSF solar cells. As can be seen in Figure 6.12, taken from the same paper, the process includes screen printing of Al followed by a high-temperature step to form p^+ BSF by alloying to the silicon wafer. By that time, the fire-through Ag paste was not yet available and the process therefore included another sintering step to form the front and rear Ag pad contacts. Other cell processes at that time used opening of thermal oxides by masking processes to allow contacting the cells which required another hard-to-do alignment precision.

It was the introduction of another great concept by Haigh [29], who named it fired-through contact, another step towards what is considered today as the state-of-the-art fire-through process. He demonstrated for the first time the feasibility of silver fired-through printed frontside contacts on thermally grown SiO_2 films. This process was then developed further with improvement in printing, paste chemistry and different ARCs such as TiO_2 [30]. By the end of 1970s and early 1980s, Japanese companies started the production of fully screen printing silicon solar cells featuring a full area Al BSF and front Ag that fired-through the SiN_x ARC layers formed by plasma-enhanced chemical vapour deposition (PECVD) [31], a process to be the state-of-the-art technology that dominated the mainstream of solar cell industry for decades to come. Lolgen *et al.* [32] reported surface recombination velocities between 130 cm/s and 170 cm/s in Al-doped p^+ silicon layers formed by screen-printed Al. That was a good indication of the possibilities one can realize using screen-printed Al metallization to form BSF for industrial solar cells. It was obvious that, to reduce the recombination at the Al alloyed BSF, we have to increase its thickness or to increase its electron shielding property by increasing the Al dopant level and further bend the band gap at the p^+ layer. Both will require a combination of thick Al paste deposition and even higher firing temperature. The high deposition of Al will cause a severe bowing of the solar cell that will increase the breakage of cells during module lamination. On the other hand, the high-temperature approach is usually avoided to improve the passivation quality and the overall silicon solar cell performance. Another approach besides the thick p^+ layer is to increase the hole dopant concentration of a thin p^+ layer by increasing the p^+ dopants. Though the Al concentration is limited by its low solid solubility in silicon, the industry looked for an alternative dopant of group III such as boron to be added to the pastes due to its higher solubility in silicon than Al [33]. Boron might be added in several forms including, but not limited to, metallic boron, boron oxide and Al-B alloys [15, 32, 34, 35]. Figure 6.13 shows the typical boron and aluminium concentrations measured by SIMS at the epitaxially regrown p^+ Si formed by screen printing of Al paste that contains boron as additive. As can be seen, the BSF layer has higher boron doping levels than Al itself and therefore, one might call it an Al-assisted boron BSF. Additionally, by adjusting the contents of the paste and doping it with high concentration of phosphorus, one can even change the polarity of the epitaxially

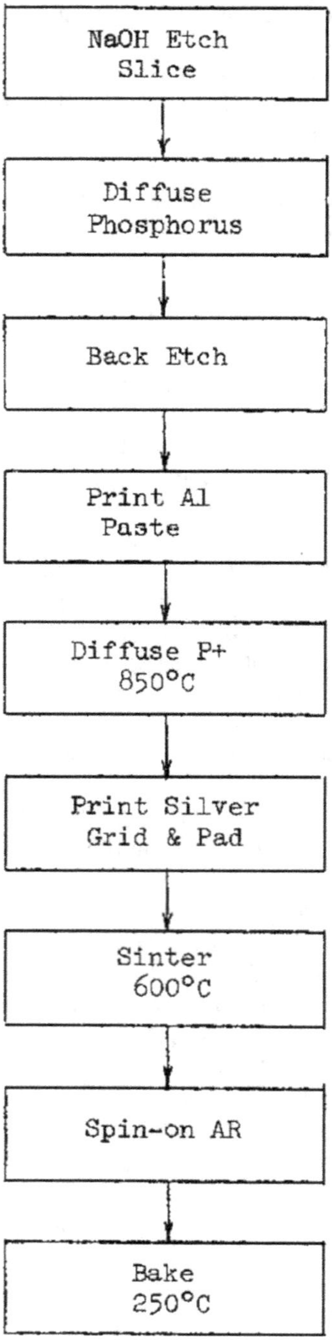

Figure 6.12 The process flow of solar cells developed by Spectrolab Inc. [28]

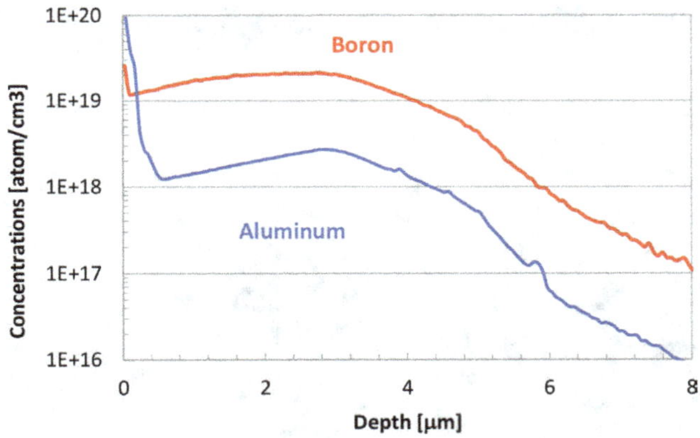

Figure 6.13 *Typical boron and aluminium concentrations measured by secondary-ion mass spectrometry (SIMS) at the epitaxially regrown p^+ Si formed by screen printing of Al paste that contains boron as additive*

regrown layer from p^+ to n^+ as the solid solubility of phosphorus is far higher than Al and boron in Si [13]. Great work has been done by Rauer *et al.* [36] in addressing the electrical properties of BSF formation using Al and boron in the Al screen printing paste.

At the beginning of 2010, PERC solar cells started to gain more attention replacing the full area BSF solar cells due to its rear passivation quality and local Al contacts. Since Al is interacting with Si at the local points, the phenomenon of Si diffusion out to the aluminium paste via a laser local contact opening (LCO) is rather complicated as can be seen in Figure 6.14.

So far, the use of Al paste was limited to fabrication of state-of-the-art industrial p-type full area BSF solar cells. Several researchers started to implement the formation of a p^+ layer by alloying Al to Si as a rear alloyed emitter in n-type rear junction and interdigitated back contact (IBC) solar cells [37–41]. The attempts were successful on the cell level and even efficiencies close to 20% were realized [37]. However, since the whole surface had to be alloyed with Al, it was not possible to use the combination of Ag pads with Al at the back of the solar cells for soldering in a similar way to p-type solar cells. Several attempts to solve the issue by utilizing low-temperature Ag/Al pads on Al [42] or solder Al directly using ultrasonic bonding [43] or to print Ag or tin layers on the alloyed Al [44] were introduced. The introduction of Al pastes that contact the Si locally via the passivation layer, for example in PERC solar cells, makes it possible from the concept point of view to use Al and Ag pads combination in n-type solar cells. Comparotto *et al.* [45], Tous *et al.* [46] and Dullweber *et al.* [47] reported bifacial n-PERT rear passivated junction (RJ) solar cells fabricated with screen-printed Al fingers aligned to LCOs ablated by laser with efficiencies exceeding 22%. A similar process is commonly used in bifacial PERC. In addition, to easily making the switch to n-PERT RJ solar cells, Peng *et al.* [48] and Tous *et al.* [46]

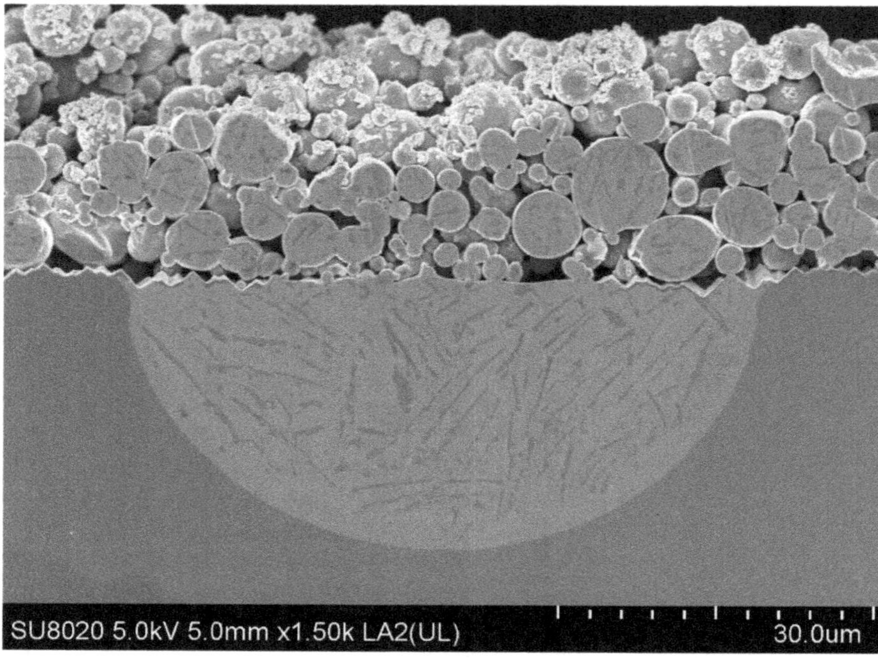

Figure 6.14 Scanning electron microscopy (SEM) image of local aluminium contacts to passivated emitter rear contact (PERC) solar cell

presented n-PERT RJ solar cells with full area Al screen printing on the rear side. However, applying Al metallization as a contact concept on normal front junction n-PERT was always an avoided concept due to Al fingers shading which significantly reduces the short circuit current (J_{sc}) of the solar cell if no proper narrow Al grid printing and LCO alignment are applied. Even with the advancement in Al paste chemistry that allowed less than 100 μm wide Al fingers to be printed, as seen in Figure 6.15, the quality of the contacts of such 100 μm wide fingers as line resistance and recombination at the contact points needs to be optimized. The combination of different LCO patterns with different Al finger widths was investigated [49]. A reduction of the metal finger width below 150 μm significantly increases the metal-silicon saturation current density ($J_{0\text{-met}}$) for line LCO geometry. For finger widths below 100 μm, point contacts with a dot-to-dot distance of ~80 μm to 140 μm were clearly superior as can be seen in Figure 6.16. Such a combination of narrow Al fingers with optimized LCO can be used in passivated contact solar cells. For instance, p-type back junction solar cells featuring frontside screen printing Al fingers and passivating polysilicon at the rear side were successfully fabricated and showed efficiencies above 22.4% with an open circuit voltage of 716 mV [49]. Screen-printed Al contacts on n^+ doped silicon was demonstrated giving the possibility to fabricate Ag-free solar cells using Al to contact both n^+ and p^+ Si in advance cell concepts [50]. Another future possibility is the interesting approach to fabricate silicon-germanium

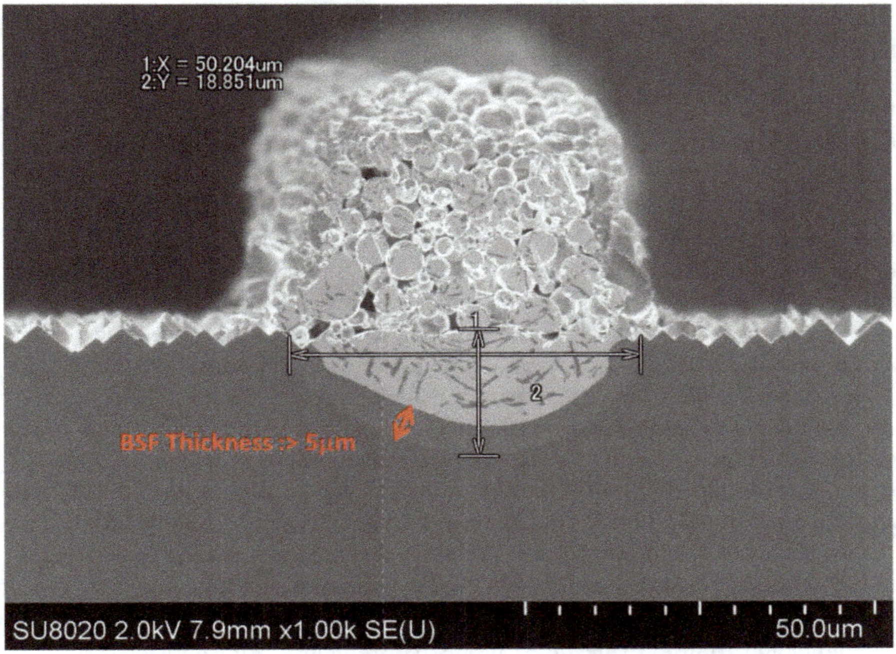

Figure 6.15 *Scanning electron microscopy (SEM) image of screen-printed aluminium fingers on n-type frontside* p^+ *layer*

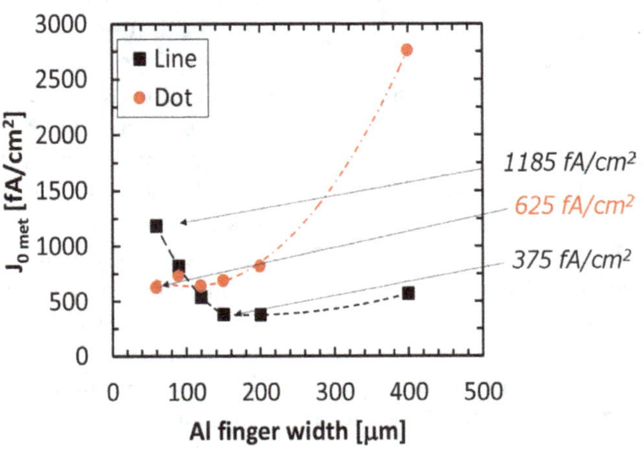

Figure 6.16 *Recombination current Jomet at the aluminum contact in dependence of the aluminum finger widths for line and dot-shaped laser contact openings*

layers by screen printing and alloying Al paste with germanium contents to Si, an approach to fabricate low-cost III–V solar cells on Si [50].

6.3 Contacting of boron emitters with Ag and Ag-Al pastes

A promising alternative to the industrially predominant solar cell types Al-BSF and PERC, which use p-type wafers are solar cells on n-type-doped wafer with boron-doped p^+ emitter. Although some processing steps might be technologically more demanding, a higher conversion efficiency can reward the effort.

The use of n-type wafers is attractive as these typically feature a higher lifetime and in most cases also a higher minority carrier diffusion length compared with p-type wafers produced from the same feedstock, as a result of asymmetric capture cross section for a series of common metallic impurities [51]. Further boron-doped p-type wafers are prone to light induced degradation (LID) through formation of boron oxygen complexes. It is possible to deactivate the defects or to avoid them altogether by use of gallium as doping species, but both require further effort.

Two main technological advantages which favoured p-type solar cells in industrial production so far are an easy-to-handle formation of emitters in $POCl_3$ diffusion, and the very cost-effective formation of base contacts with relatively low recombination from alloying of Al pastes. Nevertheless, with implementation of passivating contacts technology, n-type solar cells, such as TOPCon or IBC, might become the easier-to-produce cell types for the next generation of cells.

The most obvious choice for the creation of emitters in n-type cells is doping by boron atoms. Boron doping can be achieved by diffusion, by implantation, by epitaxial growth, or through deposition of boron-doped Si layers. This sub-chapter focuses on the contact formation from high-temperature thick-film silver pastes to diffused boron emitters, typically prepared in a BBr_3 quartz tube diffusion process. However, boron emitters diffused from APCVD or PECVD source layers and implanted boron emitters generally feature similar profiles. In the BBr_3 diffusion process first a boron silicate glass layer is grown on the silicon surface, from which the boron is subsequently diffused into the silicon at a temperature between 850 °C and 1 050 °C under inert and oxidizing atmospheres. As the solubility of boron in SiO_2 is higher than in Si, the boron concentration typically decreases towards the surface.

In the early 2000s screen-printed Al-BSF p-type solar cells with fired-through-SiN_x front side metallization from Ag paste became the main stream process for the industrial production of solar cells. In the quest for increasing conversion efficiencies, several groups worked on screen-printed n-type Si solar cells with boron emitter, such as the company Shell [52], ECN Institute [53] and the University of Konstanz [54, 55]. But poor contacting of boron emitters from Ag paste posed a problem.

Kopecek *et al.* used Ag-Al paste designed for solder pads on the rear side of Al-BSF cells for contacting of the boron emitter and demonstrated a significantly reduced contact resistance compared to standard Ag paste [55]. Even though the line resistance of the paste was high, the solder pads paste resulted in the best solar cells of their study.

Kerp *et al.* [56] from Ferro and Instituto de Tecnología Microelectrónica (TIM) systematically investigated the addition of Al particles to a commercial Ag front-side screen printing paste. Contact resistance was found to decrease with increasing Al content, which they correlated to an increased density of contact islands on the emitter found in SEM micrographs. However, best conversion efficiencies were achieved for cells with a relatively low Al content, as for higher Al content the cells were severely shunted.

It was demonstrated in the following, that the addition of silicon to the paste is an efficient measure to increase shunt resistance [57]. The high diffusivity of silicon into aluminium leads for pastes containing only Al particles to the creation of large etch pits in the substrate during the heating-up process, which are then at higher temperature filled with liquid metal, resulting in pyramidal metal spikes deep enough for penetrating the emitter. SEM micrographs of Ag-Al contacts to flat 110 silicon wafers showed a high number of up to 5 μm deep contact spots of pyramidal shape. In contrast an Ag-Al paste with approx. A6% silicon content relative to the Al content featured no etching pits in SEM micrographs and resulted in a strongly increased shunt resistance. When silicon is added to the paste, less silicon from the substrate gets dissolved in the Al phase during heating up, similar to pure Al pastes.

In 2010 industrial production of n-PERT solar cells was started by Yingli [58], and around the same time by PVGS in Japan. This raised the interest of more paste companies in this cell type and triggered a first generation of commercially available Ag-Al pastes from different suppliers.

6.3.1 Model of contact formation

Deeper insight into the contact formation process was gained through detailed SEM studies conducted on the interface between paste and emitter. Riegel *et al.* created as series of samples with increasing firing temperature, analysed the changes of the interface in top view and cross-sectional images and formulated a first model for contact formation (see Figure 6.5) [59]. A number of similar investigations followed, which improved the understanding. It was shown that creation of Al/Ag phase facilitates the transport of Al towards the interface, where the reaction with Silicon leads to formation of Al/Ag/Si contact spots [60]. Further it was shown that for emitters coated with a dielectric layer the shape of the surface is maintained by the contact spots [61], and that the resulting contact spots contain mostly Ag and only a small fraction of Al (Figure 6.17) [63, 64].

An improved model explaining these phenomena was published by Fritz *et al.* [62, 65]. According to this model, the evolution of contact formation during the firing process can be described as following, compare Figure 6.18.

At temperatures above 500 °C: The paste densifies and glass accumulates around the aluminium particles. The oxide shells covering the Al particles are starting to get dissolved and the aluminium enters the glass. Glass at the silicon surface starts to etch the dielectric layer.

Figure 6.17 *Scanning electron microscopy (SEM) micrograph of contact spots in top view and cross-sectional view, reprinted from PhD thesis of Susanne Fritz [62]. Contact spots form in clusters below Al-containing regions. In cross-sectional view, it can be seen that contact spots build (111) surfaces with the silicon.*

At temperatures between 660 °C and peak firing temperature: The aluminium particles melt and their oxide shells are cracked. In aluminium-containing regions, an inhomogeneous microstructure develops, consisting of aluminium-containing glass and the silver-aluminium phase. Where the aluminium-containing glass is getting in contact with the wafers surface, the etching of the dielectric layer interestingly is less pronounced, while in other parts it is completely etched by the glass. The preserved dielectric layer acts as mould and keeps the surface structure. Where the silver-aluminium phase gets in direct contact with the wafer through holes in the dielectric layer, silicon is dissolved into the liquid silver-aluminium phase, and silver-aluminium fills the mould, defined by the dielectric layer, which is further undercut. As a result of the slower dissolution of <111> planes, on <100>-oriented silicon wafers, imprints in the shape of inverted pyramids grow into the silicon surface. On textured surfaces domains parallel to the pyramid surface grow into the surface. In other parts, where Ag paste without aluminium gets in contact with the silicon the surface is corroded by the glass(Figure 6.18).

During cool-down: When the solubility limit of silicon in the silver-aluminium-silicon phase is exceeded, silicon precipitates are formed. Upon solidification of the silver-aluminium-silicon phase, silver crystals with small contents of aluminium and silicon are formed.

In conclusion, the mechanism of contact formation of Ag-Al pastes is quite different from that of pure Ag pastes. In pure Ag pastes, silver and silicon are dissolved by the glass frit, which makes a certain glass coverage of the contact spots almost inevitable. In Ag-Al pastes, silver and silicon are dissolved in liquid Al. According to the phase diagram, Ag-Al alloys with up to 95% Ag stay liquid at a typical peak firing temperature [66]. The high solubility of silicon in aluminium is the reason for the much bigger contact spots of Ag-Al pastes. Upon solidification, a direct metallic connection remains between contact spots and the bulk metal.

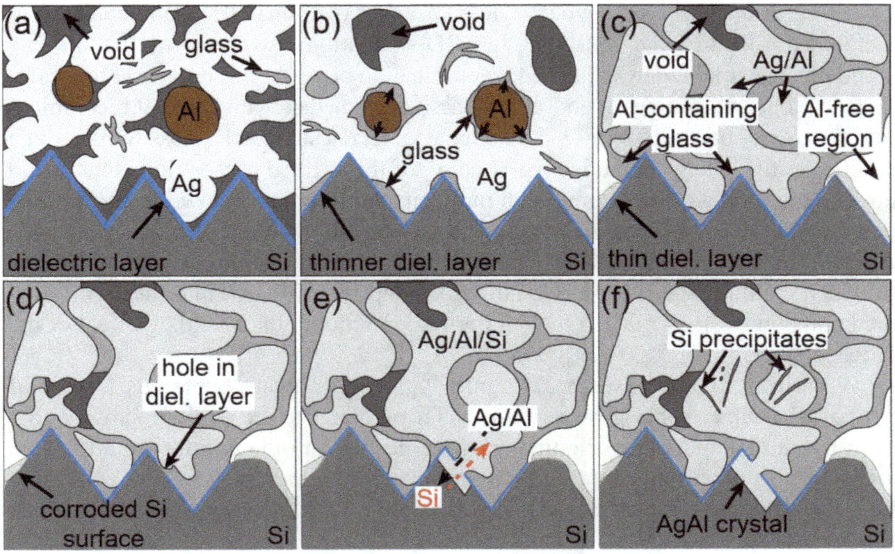

Figure 6.18 *Schematic illustration of phenomenological model for contact formation showing cross-sections of screen-printed Ag/Al contacts at different steps of contact formation. Reprinted from PhD thesis of Susanne Fritz [62]: (a) Contact after drying. (b) Contact densifies. Glass accumulates around Al and starts etching oxide shell around Al particles and dielectric layer. Al diffuses through oxide shell into surrounding glass. (c) Al particles melt. Where the oxide shell is opened, Al dissolves Ag, forming an Ag/Al phase. (d) Below Al-containing regions, glass etches openings in dielectric layer. (e) Where the Ag/Al phase gets in contact with the Si surface, Si is dissolved into the Ag-Al phase. Pyramidal imprints are etched in Si and the Ag/Al/Si phase is formed. Material transport occurs through openings in the dielectric layer. (f) During cool-down, Si precipitates in the Ag/Al/Si phase and the melt solidifies in Ag crystal structure. Contact spots show the shape of inverted Si pyramids.*

6.3.2 *Recombination of Ag-Al contacts*

Shunting, as represented in an equivalent circuit by low ohmic resistance in parallel to the diode was mainly an issue for very early versions of Ag-Al pastes. But increased recombination both with diode factor $n = 1$ and with higher diode factor were found to be a problem for Ag-Al pastes. An increased recombination factor $n = 1$ can be imagined as recombination at the surface or in the bulk of the device. And increased diode current with factor $n = 2$ or higher is associated with recombination in the space charge region, e.g. by metal spikes reaching into the junction.

Edler *et al.* observed a voltage loss of approx. 40 mV from unmetallized to metallized n-PERT solar cells [67]. In a 2D simulation, it was shown that this loss could not be explained by enhanced recombination beneath the contacts at the wafer surface alone. Assuming an entirely unpassivated surface with thermal recombination, a velocity of $S = 10^7$ cm s^{-1} should have yielded a smaller V_{OC} loss for the given emitter profile. They concluded that etching of the surface passivation alone could not explain the loss. By assuming an unpassivated interface below the surface, the model could display the loss more accurately. In follow-up studies J_{0met} values for Ag-Al pastes on different emitter profiles were determined, and it was simulated to which etching depths these would correspond [68, 69]. An interface with high recombination velocity inside the emitter region reduces the shielding effect of the diffused layer.

Wöhrle *et al.* composed a 3D model with the Software Sentaurus Device of pyramidal contact spots growing into a flat emitter surface [70]. It showed that from pyramidal spots only the tips reaching deeper into the emitter would have a major influence on the saturation current value while the flanks would contribute little. Recombination increases over-exponentially with depth of the contact spots. However, a spot density of 10% had to be assumed in their simulations in order to match the results with the measured J_{0met} values for different profiles, which did not match well with an observed density of approx. 1%.

A model with localized planar contact spots was presented by Kiefer *et al.* [66] It considers that on textured surfaces, the <111> orientated interfaces of the contact spots are in parallel to the pyramidal surface structure. Also in this model the emitter saturation current density strongly increases with increasing depth of the assumed spikes, especially when the remaining emitter thickness becomes less than 200 nm. Deeper emitter profiles provide a better shielding. The recombination of contacts from two generations of Ag-Al pastes were analysed. For both pastes, sets of structural parameters could be found, for which a fair agreement between measured and simulated recombination of the contacts for different emitter profiles could be obtained. For the newer paste an area fraction of 1%, same as it was found in a SEM micrograph analysis, and a depth of 500 nm explained well J_{0met} values of 1 400–840 fA cm^{-2} for different emitters. For the older paste, an area fraction of 4% and a depth of 650 nm had to be assumed to explain the J_{0met} values of up to 3 420 fA–cm^{-2}. This did not fully match with the area fraction found in SEM micrographs of 1.2% for this paste, but recombination for deeper spikes could not be solved numerically.

Heinz *et al.* characterized contact spots with microscopical imaging methods, such as Mirco-Raman Spectroscopy (μRS), Mirco-Photoluminescence (μPLS) and Micro-Light Beam Induced Current (μLBIC) [64]. They found that deep spikes punching through the emitter result in high recombination, and correlated loss of pseudo fill factor and V_{OC} to increased number and size of contact spots. They assume from their results the formation of a Schottky contact from deep contact spots to n-type bulk. However, for newer pastes, a certain discrepancy remains: cells which are not shunted feature in SEM analysis sometimes contact spots which seem deeper than the emitter. It might be, that the potential barrier between metal and lowly doped silicon is too wide for a significant current flow over a direct contact [66].

6.3.3 Contact resistance

Contact resistance of Ag-Al pastes to boron emitter was experimentally found to decrease with maximum doping concentration and with depth of the boron emitter profile. The number of contact spots increases only moderately and cannot explain this decrease. Lohmüller *et al.* [71] and Fritz [62] calculated contact resistance analytically according to the Schottky theory of current transport over metal-semiconductor interfaces, following a methodology proposed by Schubert and others for Ag paste contacts [2, 72].

Contacts are assumed as inverted pyramids grown into a (100) silicon surface, in direct contact with the bulk of the Ag paste and the silicon surface. The pyramid surface is divided into slices, in contact with the emitter at different depths. For each slice the contact resistance towards silicon is calculated according to the Schottky theory of current transport over metal-semiconductor interfaces, considering the mechanisms of thermionic emission, thermionic field emission and field emission depending on the local doping concentration of the emitter. The total contact resistance of the pyramid is given as a parallel connection of the contact surfaces of all slices. And macroscopic contact resistance can be calculated by dividing the contact resistance of a single pyramid by the observed area fraction of contacts. However, several parameters in the equations are not precisely known, e.g. the effective tunnelling mass and the Schottky barrier [62].

Absolute values of ρ_c calculated by Lohmüller *et al.* are about two orders of magnitude lower compared with measured ρ_c values for two different pastes, suggesting that either only a small fraction of crystallites is in contact with bulk and silicon, or that input parameters are strongly different than assumed [71]. Relative differences between the contact resistances measured on three emitters of same surface concentration but different depths were in accordance with the model, suggesting that contact resistance is dominated by contacts with deeper penetration depths of several hundred µm.

Fritz also found for emitters with the same surface concentration, that the contact resistance decreases in calculated data in the same relative range as in the measured data. For emitters with increasing maximum doping concentration up to 8.6 e19 cm^{-3}, the calculated contact resistance decreases significantly stronger than the measured contact resistance. It is hypothesized that additional current paths could exist, reducing the total contact resistance for lowly doped emitters [62].

6.3.4 Contacting of boron emitters with Al-free Ag pastes

A first example for low-ohmic contacting of boron emitters from Al-free Ag pastes was given by Richter *et al.* in 2010. They deposited seed layers by aerosol jet printing from an ink containing Ag particles and glass frit on a boron emitter. The laydown of this printing method was very low. After fast firing in a belt furnace the seed layers were electroplated for reaching sufficiently low line conductivity. Contact resistivity ρ_c down to 1 mΩ cm^2 was measured and 2 × 2 cm^2 solar cells reached fill factors above 80%. SEM micrographs of the contact interface showed formation of silver crystallites with size increasing from ≈0.1 μm at firing temperature of 700 °C to ≈ 0.5 μm at 840 °C. However, best contact was achieved at 720 °C suggesting that not the size of the crystallites but the formation of insulation glass layer determines the contact resistance.

Also for inkjet printed metallization it was shown by Jesswein *et al.* that the same Ag ink was capable of contacting phosphorous and boron emitters alike [73]. They further showed that inhomogeneous contact formation over full area solar cells could be suppressed by a print pattern with segmented busbars.

Successful contacting of boron emitter with screen-printed Al-free Ag pastes was reported by Engelhardt *et al.* First they found that emitters diffused from a PECVD-deposited BSG glass [74] could be contacted with Ag paste, but later the same was also proven for BBr_3 diffused emitters [75]. They showed that Ag pastes containing tellurium glass frit can achieve lower contact resistances than those based on Pb glass. Contact resistivity as low as 2 mΩ cm^2 on a 130 Ω/sq boron emitter of 0.3 μm depth, and 1 mΩ cm^2 on a 50 Ω/sq boron emitter of 0.55 μm depth were shown over a broad range of firing temperatures. However, solar cells did not surpass 75% FF, and resulted in lower efficiencies as reference cells metallized with Ag-Al paste for the 50 Ω/sq emitter. For the shallow 130 Ω/sq emitter it was at least demonstrated that cells metallized with Al-free Ag pastes featured considerably lower J_{02} values and higher FF, compared to the cell with Ag-Al paste, indicating that contact formation of pure Ag paste is more shallow than contact formation of Ag-Al pastes.

Insight why contact formation with pure Ag paste to boron emitters is difficult was provided by a series of studies by Korea University of Seoul. Kim *et al.* emphasized the electrochemical nature of the redox reactions during contact firing from Ag pastes, with glass acting as electrolyte [6]. While on the Ag electrode at the paste side the dissolution of Ag is accompanied by the reduction of oxygen (6.1), the Si dissolution on the Si electrode at the wafer side is accompanied by the reduction of Ag+ ions (or Pb2+ ions).

$$4Ag_{(s)} + O_{2(g)} \rightarrow 4Ag^{+}_{(in\ glass)} + 2\ O^{2-}_{(in\ glass)} \tag{6.6}$$

$$Si_{(s)} + 4Ag^{+}_{(in\ glass)} \rightarrow Si4^{+}_{(in\ glass)} + Ag^{0}_{(in\ glass)} \tag{6.7}$$

Reaction (6.2) is seen responsible for etching of pyramidal pits into the silicon surface and formation of Ag crystallites.

They observed that inkjet-printed cells with phosphorous emitters with fully connected H-Grid design had bad contact formation in some area, while solar cells with segmented busbars had homogeneously low contact resistance all over the wafer surface. They speculated that the Ag-Si contacts are not formed simultaneously over the whole wafer during firing. As soon as the first contact spots are formed, the electrodes of the galvanic cell are short circuited, and electrons created during oxidation of silicon can flow over the contact grid to the Ag electrode. Instead of the reactions described above, another set of reactions occurs, in which no

Ag^+ ions are reduced at the surface, and hence no Ag crystallites are formed. They called it the 'short-circuit effect'.

A further study focused on contact formation from Ag pastes to boron emitters [76]. It was explained, that due to the reduced presence of electrons in highly boron-doped silicon emitters, the etching of the Si surface by reduction of Ag^+ ion is strongly reduced. Same is also known for etching of silicon by hydroxide ions in KOH solutions. boron-doped silicon is etched significantly slower compared to phosphorous doped silicon. They presumed that injection of electrons into the emitter via an external circuit would stimulate Ag precipitation at the emitter. This could be verified in a test set-up. Under injection of electrons, the contact resistivity was drastically reduced to values well below 1 mΩ cm^{-2}. SEM micrographs showed an increased number of Ag crystallites. Further it was shown that increased concentration of oxygen in the atmosphere during firing increases the concentration of Ag crystallite, confirming higher dissolution of Ag^+ according to reaction (6.1).

Following these learnings, Chu *et al.* investigated systematically the influence of busbars on contact formation [77]. They compared solar cells fired with an H-grid of fingers and busbars with cells that only featured parallel fingers but no busbars during firing. Lower contact resistances were consistently found for cells fired without busbars for phosphorous as well as for boron emitters of different sheet resistance and on different surface morphologies. The effect was particularly pronounced for surfaces difficult to contact. For an 85 Ω/sq boron emitter of 0.55 μm depth, the contact resistance was found to decrease from a value of >80 mΩ cm^2 to 1 mΩ cm^2, confirming the relevance of the short circuit effect for boron emitters. The achieved contact resistance value was even lower as the contact resistance of the Ag-Al paste on the same emitter amounting to approx. 3 mΩ cm^2. The contact resistance of the Ag-Al paste interestingly did not change significantly between configuration with and without busbars. As for Ag-Al pastes the silicon is not etched by Ag^+ ions, but dissolved due to high solubility into the aluminium containing phase, contact formation is not influenced by the electron concentration.

Further it is interesting to note that metal-induced recombination current density J_{0met} did not change significantly between the firing Ag paste with and without busbars. Compared to the Ag-Al paste fired at the same temperature, only a slight reduction of J_{0met} was found, suggesting that the Ag paste also significantly damages the passivation layer and surface of the emitter.

6.3.5 *Further development of pastes and actual status*

For contacting boron emitters on planar surfaces, such as the rear side of Zebra IBC solar cells, the use of pure Ag pastes has been advantageous. Mihailetchi *et al.* found that the metal induced recombination current density J_{0met} for Al-free Ag paste on planar phosphorous and boron surfaces can be mostly explained by the damaged (etched) dielectric passivation layer [78]. They showed that by increasing the thickness of the dielectric layer, recombination loss can be reduced without significantly increasing the contact resistance, possibly due to incomplete removal of the dielectric layer. For a 140 Ω/sq boron emitter with low surface-doping concentration of around 2 E 19 cm^{-3}, a J_{0met} below 800 fA cm^{-2} at contact resistivity of 3 m$\Omega$$cm^2$ was demonstrated for increased SiN_x thickness. With optimized metallization layout and firing process, actual Zebra IBC solar cells achieve V_{OC} above 695 mV, using Al-free silver pastes [55].

For contacting of textured boron emitters, such as in n-PERT and TOPCon solar cells, to the author's knowledge, both Ag and Ag-Al pastes have been used in industrial productions, with Ag-Al pastes still representing the majority. Similar as Ag frontside pastes, Ag-Al pastes have been improved over the years regarding printability and line definition. Today in 2020 a printed finger width below 40 μm is achievable with state-of-the-art Ag-Al pastes and improved screens.

As an example, a gain in J_{SC} of $\approx$ 0.1 mA/cm^2 was observed for a reduction in front finger width by $\approx$ 4 μm [79]. Due to higher aspect ratio, based on reduced width and increased height, J_{SC} could be improved without adversely impacting the finger line resistance.

One advantage of Ag-Al pastes is that they can provide good contact formation at relatively low fast firing temperatures, as melting of Al particles and dissolution of silicon start at low temperatures. Contact formation from Ag pastes for doped polysilicon layers on the rear side of TOPCon cells, also favours a low firing temperature, in order not to penetrate the poly layer [80].

Limiting the metallization induced recombination current density $J_{0met,}$ remains the challenge for Ag-Al pastes. Improvements have been achieved by control of reaction depth and by use of glasses causing lesser damage to the passivation layer. Padhamnath *et al.* recently compared the performance of two screen-printed and fired-through Ag-Al pastes on a front boron emitter of 80 Ω/sq on large area TOPCon solar cells [79]. For the latest generation Ag-Al paste J_{0met} could be reduced to 806 fA cm^{-2}, compared to 1 743 fA cm^{-2} for an older generation paste, though at the expense of a somewhat increased contact resistivity. SEM images revealed that the etch pits formed by the new paste were significantly smaller and less deep.

Further metal-induced recombination can be reduced by shielding the metallization by a deeper junction, as explained above [66]. This effect is shown quite impressively in a recent work of Bao *et al.* [81] They diffused ultra-deep boron emitter profiles with sheet resistances between 50 Ω/sq and 95 Ω/sq and peak doping concentration as low as 1 e19 cm^{-3} to 2 e19 cm^{-3}, and contacted these with Ag-Al pastes. The depth of these profiles presumably amounted up to 2 μm.

Lowest J_{0met} of 541 fA cm^{-2} was achieved for the deepest profile, with a relatively high surface concentration and the lowest sheet resistance of 50 Ω/sq. But this emitter also featured the highest recombination current in the non-metallized area. Best balance with lowest total recombination from metallized and non-metallized areas was found for a 95 Ω/sq emitter with maximum doping concentration slightly above 1 e19 cm^{-3}, resulting in J_{0pass} of 25 fA cm^{-2}, J_{0met} of 666 fA cm^{-2} and contact resistance of 1.6 mΩ cm^2, allowing for a cell efficiency of 24% at a V_{OC} of 703.5 mV, as champion cell from a larger production batch of 20 000 manufactured pieces. Figure 6.19 compares J_{0met} values reported for commercial Ag-Al pastes over the last decade, showing significant improvement. But as passivating contact technology has drastically reduced the rear-side recombination, the recombination of the

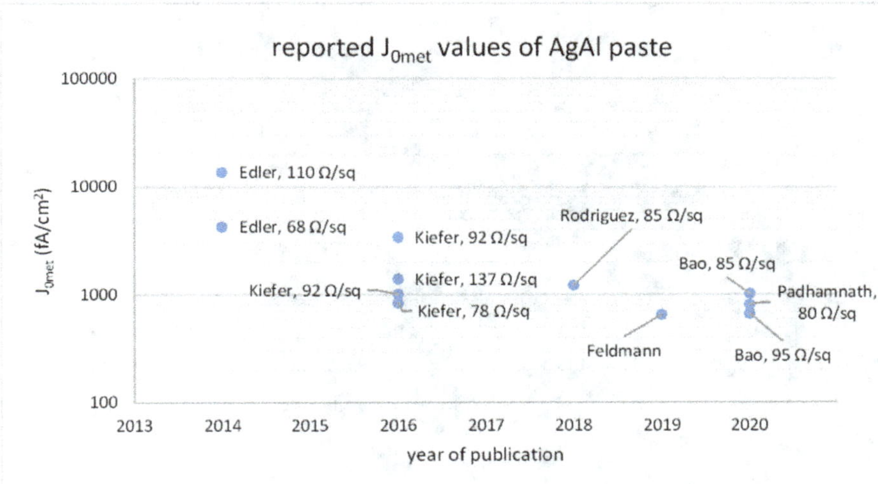

Figure 6.19 *Reported values of metal-related saturation current density J_{0met} for Ag-Al paste on boron emitters. When known, the sheet resistance of the emitter is also presented in the graph. Reduction of J_{0met} of the time can be attributed in part to better pastes and in part to deeper emitter profiles [66, 79, 81–83].*

emitter and the frontside contact are again the dominant limitations for the voltage of TOPCon cells.

Future developments to further reduce the metal-induced recombination could be the introduction of selective boron emitters and localized passivated contacts on the front side. However both options are technologically demanding, so gradual improvements of paste and printing could mark the path to higher efficiencies of industrial TOPCon cells in the next years.

6.4 Screen printer machine technology

Screen printing was invented in China around one thousand years ago and since its introduction it was applied in a variety of fields: art, decoration of goods, textile industry, initially mostly in manual operation. Then, automated or semiautomated screen printing machines have been adopted in several industries with some important milestones.

In 1907 Samuel Simon of Manchester, England, patented the screen printing process for wallpaper decoration. Later the technique was adopted for advertisements. An historical perspective of screen printing is covered by Biegeleisen [84].

In 1960 American entrepreneur, artist and inventor Michael Vasilantone started to use, develop and sell a rotatable multicolour garment screen printing machine.

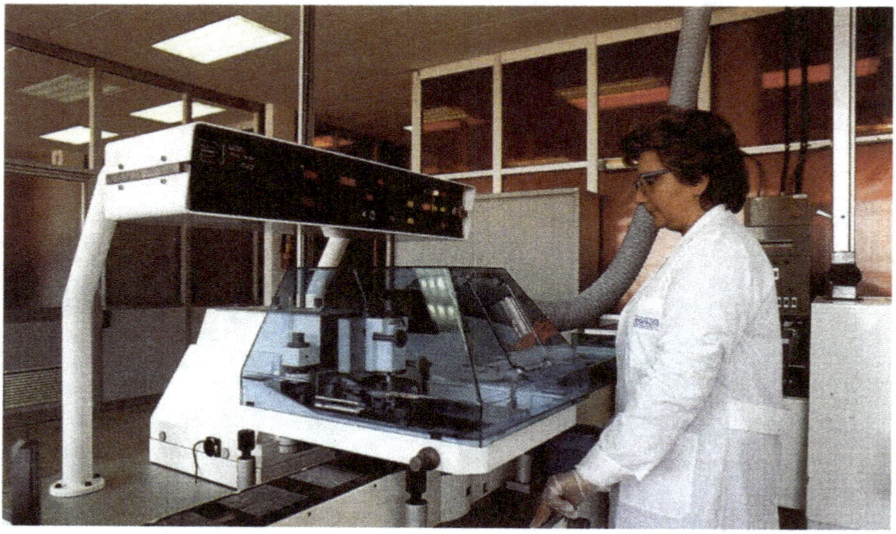

Figure 6.20 *One of the first screen printing semi-automated lines developed in Europe in 1983 by Baccini for Ansaldo (courtesy of Pier Enrico Zani)*

Vasilantone later filed for a patent [85] on his invention in 1967, granted in 1969. It was the first implementation of automated screen printing.

In 1962 artist Andy Warhol realized with serigraphic technique Marilyn Diptych, a famous portrait of Marilyn Monroe in multiple colours.

Screen printing of solar cells was considered for the first time by Ralph in a publication in 1975 [86]. Main motivation was to apply a relatively cheap processing technique for transferring a thick paste deposit on the substrates with a defined pattern. Since then, several companies investigated the applicability of the screen printing in research or pilot manufacturing, with small-scale batch processing with manual loading and unloading of wafers (Figure 6.20).

First implementation of fully automated lines in PV (including loading and unloading of substrates) has been reported in the early 1990s and since then several process improvements have been adopted with the aim to decrease the manufacturing costs and increase the cell efficiency. In general, screen printing has been widely accepted, with an overall share stable at levels higher than 95%. This allowed for constant evolution in terms of consumables and equipment, that ultimately led to a consistent learning curve in both productivity increase and feature size reduction.

Considering the last decade (2010–2020) the increase in productivity was five-fold, with a tact time reduction from 2.5 s per wafer to less than 1 s per wafer, meaning 3 600 wafers per hour productivity in a single-lane configuration. Over the years, the number of lanes per tool have been increased, with dual lane configurations being the standard, and some manufacturers even adopting three lanes. At the same time, minimum printed feature width decreased by 70% from 100–110 μm to

30–35 μm [87], preserving or even increasing feature aspect ratio (ratio between feature height and width). As a direct consequence, total paste deposit per cell was reduced significantly with a major impact on overall cost reduction. Additionally, equipment alignment precision improved during this period, with values of ±10 μm and below achieved routinely in mass production by multiple manufacturers. This aspect enabled several alternatives to traditional cell processes: dual and double printing front metallization and pattern alignment on pre-deposited structures such as selective emitter and laser contact openings for PERC cells.

The future of screen printing will depend on its capability to adapt to emerging industry trends, such as large wafers up to 230 mm in size (doubling the printing area versus the reference size at 156 mm) and different cell architectures: PERC and bifacial-PERC, heterojunction, interdigitated back contacts and passivated contacts. Due to the strong dependency between module interconnection options (full cells or cut cells, ribbon or wire-based connection, busbars or busbar-less, with cell gaps or gapless, mono or bifacial, back contact) and the cell layout, cell and module designers have to work together to fully optimize these concepts in mass production.

6.4.1 Solar cell process details

Three key steps are adopted during the cell metallization process:

Screen printing backside Ag paste: scope is to create the soldering pads for subsequent module assembly. The pattern is defined by multiple pads or sections placed at the same pitch of the busbars along the full cell length. The thickness of this layer is in the range of 10–15 μm depending on the required soldering adhesion. The area dedicated to this printing step is minimized because it has high recombination and is based on Ag paste, with a loading range of 40%–70% depending on each model.

Screen printing backside Al paste: scope is to locally contact the cell backside (in correspondence to the laser opened areas for the PERC cell processing) to form BSF and additionally to carry electrical charges to the backside busbars. Due to the low specific conductivity of Al the usual thickness is in the range of 20–30 μm and the metal loading is ranging from 70% to 85%. The key metrics for these materials are capability to form a thick BSF layer, adhesion to the substrate and reduced cell bowing after firing.

Screen printing frontside Ag paste: scope is to contact the emitter and to carry the current from the fingers to the front busbars. Since the light shading on the frontside is directly correlated to the finger width the key metric is to reduce the minimum feature size without impacting line resistance. At the same time it is important to guarantee good solderability on the busbars. In general, for these pastes the metal loading is approximately 90% and the deposited thickness varies from 10 μm to 25 μm depending on the number of busbars and consequently on the line conductivity.

6.4.2 Screen printing automation design

In this section the basic phases of an automatic printing tool are detailed:

- Wafer loading into the printing nest: from a conveyor system, normally a belt, the wafer is transferred to the printing nest, a planar surface where it sits during the printing process. This operation can be performed by a simple belt-to-belt transfer or by a pickup, for example, a cartesian gantry or a multi-axis robot.
- Wafer chucking (by vacuum) and image capture: when the wafer is deposited on the printing nest it is important to avoid further movements that could lead to alignment errors. A vacuum circuit or chucking system is required, and once the wafer is stabilized a camera system acquires the image to detect its exact position on the calibrated area.
- Nest movement to the printing area: after the image has been acquired and coordinates have been transferred to the multi-axis printing head, able to move precisely in X, Y and theta directions, the wafer is transferred under the screen. The wafer transfer and the printing head positioning operations happen simultaneously.
- Screen printing: this phase is the core of the tool and will be analysed in detail in the next paragraph. During printing, the paste is deposited onto the wafer and after the screen is completely separated from the wafer surface the nest can move.
- Nest movement outside the printing area: in order to unload the wafer it is necessary to move the nest outside the area occupied by the printing head and the screen.
- Automated optical inspection: depending on the tool design it is possible to integrate an image acquisition step after the print process is completed, to check overall quality, pattern placement accuracy and feature dimensions.
- Wafer unloading: the last step is needed to remove the wafer from the printing nest and deposit it on the next section of the tool, with a belt-to-belt transfer or wafer pickup from the wafer rear side.

The standard configuration of the screen printer is based on a rotary table with four or more printing nests, due to process parallelization as shown in Figure 6.21: while nest 1 is performing phases 1–2 (loading and chucking), nest 2 is performing phase 4 (printing) and nest 3 is performing phases 6–7 (inspection and unloading). The nest is normally equipped with a paper roll or a belt able to carry the wafer during the loading and unloading phases and has an embedded vacuum circuit. Additional benefit of having the paper or an alternative material on top of the nest is to limit paste contamination from one wafer to the other (inking process) and avoid paste leaking on the equipment in case of screen breakage.

6.4.3 Machine cycle time

In machine design it is critical to understand the critical path, or the cycle with the longest duration. For a rotary table printer this is the sum of the screen printing time and the table rotation. As an example, table rotation normally accounts for 0.45 s and printing 0.7 s, for a total cycle time of 1.15 s.

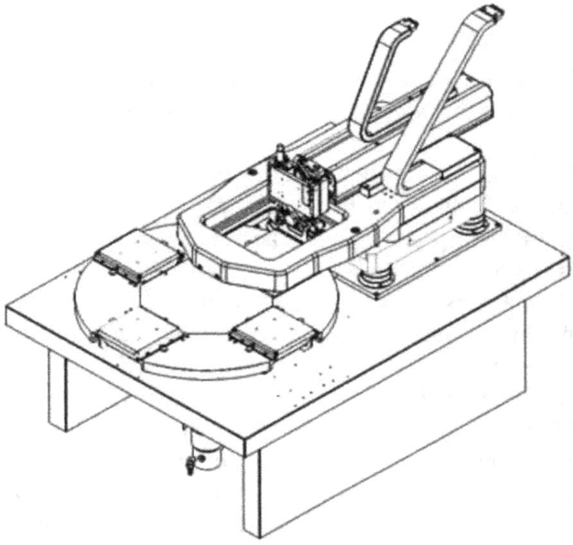

Figure 6.21 *Example of automatic screen printer based on a rotary table automation*

During the last few years the printing speed increase enabled by paste and screen manufacturers has been fundamental to achieve the productivity enhancements experienced by the whole industry. Specifically, in around ten years this value progressively increased from 150–200 mm/s to around 400–450 mm/s. In parallel, the flooding speed was normally set at around three to four times the printing speed.

Several architectures have been proposed as an alternative to the rotary table, with the main scope of reducing the cycle time or increasing the process time available for printing. For example, a tool based on two linear shuttles can achieve a cycle time of 1 s by reducing the nest transfer time to less than 0.3 s.

6.4.4 Dual printing, double printing

The frontside printing process is one of the most delicate steps in cell manufacturing because it should satisfy multiple requirements, especially in terms of overall cost, efficiency and reliability. For this reason, alternative printing processes have been adopted in mass production depending on each cell design, complementing or replacing the usual single printing where busbars and fingers are printed in the same step.

Dual printing is based on two-step metallization of the front side: busbars are normally printed first and fingers are printed in the second step. Between the two steps the paste is dried in a standard oven. This process has the main benefit of saving Ag paste: in fact, busbar paste can be optimized for pull force requirements, even at low deposits, while finger paste can be designed for good contact resistance to the emitter and line conductivity. Additionally, the alignment between the two printing steps could be in the range of ±50 μm or higher depending on the metallization layout tolerances. The efficiency level of dual printing is comparable or slightly higher than the single printing process: in fact, busbar paste does not contact the emitter and this would in turn increase the open circuit tension of the device.

Double printing is required when the aspect ratio of the fingers (height to width ratio) must be higher than 0.3–0.35, versus the usual values of 0.15–0.3 characteristic of a single printed finger. For example, if the number of busbars of the cell is less than five, this technology is adopted in order to increase the fill factor of the cell and consequently the cell efficiency. The technique is based on two-step printing: in the first step only fingers are printed, while in the second step fingers and busbars are printed together. Between the two steps the paste must be dried. If compared to a single printing finger, the line width on the screen can be reduced by 5–10 μm and so the final finger width, with a direct impact in terms of J_{sc} increase for the cell. Clearly, the fingers of the two metallization steps must precisely overlap in order to enable similar or reduced final line width versus single printing, increasing cell current. This process is delicate in several aspects: the screens must be designed for high dimensional stability, the pastes must be modified in terms of contacting and rheological behaviour and the equipment should enable very precise alignment, in the range of ±10 μm and below. Additionally, skilled operators are required to set up and control the process and overall costs can increase versus the single printing process. The benefits of double printing are manifold: reduce or eliminate the risk of finger interruptions, enable higher cell efficiency if high aspect ratio features are required and reduce the paste consumption if compared to single printing, because second printing paste could be designed for higher conductivity at low deposits.

6.4.5 *Printing on large wafers*

With the introduction of larger wafer formats new challenges arise for the printing process. For example, the increase from the usual 156.75 mm or 158.75 mm wafer size to 210 mm (and up to 230 mm) will require an increase in the stroke length in the first place, and a consequent increase of the printing time that will lead to reduced machine throughput in terms of wafer per hour. The screen format will likely scale linearly with the wafer size: the frame will then increase from 15″ in size to up to 20″. In case of high-precision applications such as printing on selective emitter structures or frontside double printing the alignment tolerances of the equipment and the dimensional control of the screens will become more critical, to avoid the risk of having misalignment on the corners. The planarity of the system must be precisely controlled on a larger area in order to guarantee a uniform paste deposit. The wafer transfer steps will be more critical especially if large wafers will have the same thickness of smaller formats, due to increased bending and vibrational movements. All these aspects have to be considered when designing or adapting a tool to process such wafers in a production environment.

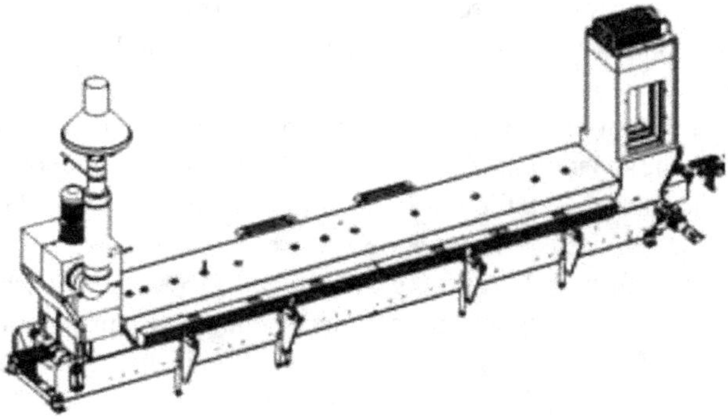

Figure 6.22 *View of the interior chamber of a drying oven, including air inlet, transport belt and exhaust pipe*

6.4.6 Drying ovens

After each printing step, the paste is dried in a belt oven in order to evaporate the solvents and enable further wafer handling steps without paste smearing or removal. Due to improved paste formulations and decreasing deposits, over the years the industry reduced the thermal budget required in the drying, determining the increase of line productivity and simultaneously reducing manufacturing floorspace.

An oven consists of four key elements: a conveyor lane, single or multiple heating sources, an exhaust and ventilation channel and a frame. For each element several options are possible, depending on the time and temperature requirements and the amount of solvent to be removed (Figure 6.22).

Conveyor lane: a metal belt with wafer standoffs is adopted, in order to avoid an extended contact between the wafer surface and the belt, limiting the paste scratching effects when reaching processing temperatures in the range of 300 °C and above. The material and geometry of the belt should be designed for a long lifetime upon continuous operation. Alternatives to the metal belt include a polymeric continuous belt (Teflon or similar material) able to withstand temperatures up to 200–250 °C, a series of ceramic or polymeric rollers and metallic wafer carriers pulled by a chain mechanism.

Heating source: a combination of IR lamps and hot air flow heats the wafers by simultaneous irradiation and convection, with the flexibility to select the IR wavelength range suitable for each paste and substrate. Hot air flow is needed to avoid solvent saturation inside the oven channel, for a fast removal of the volatile components of the paste. An alternative oven design is based on conduction heating by hot plates, with the need to chuck the wafer by vacuum and different handling requirements.

Exhaust: in order to remove the solvents from the chamber a ventilation pipe is provided, and in some implementations a VOC burner is included.

Frame: the outer frame guarantees efficient thermal insulation for safety and energy consumption purposes. Connections to customer facilities should be considered in the frame design.

Due to the different requirements for each printing step, it is important to properly characterize the drying process. For back Ag paste, with deposits in the order of 20–50 mg, the thermal profile could be relatively short and so the oven length. In contrast, for back Al paste processing the weight ranges from 1 g to 1.3 g for full PERC cells and from 0.2 g to 0.5 g for bifacial PERC. In this case the amount of solvents is significantly higher and care has to be taken during solvent management and proper airflow setting, to avoid condensation in cold areas of the oven and residue accumulation. Frontside Ag paste drying is normally adopted when multiple print steps are present. One important requirement is to reduce the handling of the cell backside with deposited Al, that could be softened by the drying process and then removed.

After these steps the wafer is processed in a fast firing furnace at around 850 °C to sinter the paste and simultaneously contact the emitter on the front side and form the local BSF on the cell rear side.

6.5 Screen technology

The following chapter will explain the fabrication and significant parameters of the mesh. Later the emulsion, its attributes and manufacturing process, is addressed. The last chapter will introduce the mathematical modelling of screens and several methods of analysing opening areas for paste transfer. In Figure 6.23, some of the key parameters of a typical screen for solar cell frontside metallisation are shown in a SEM image.

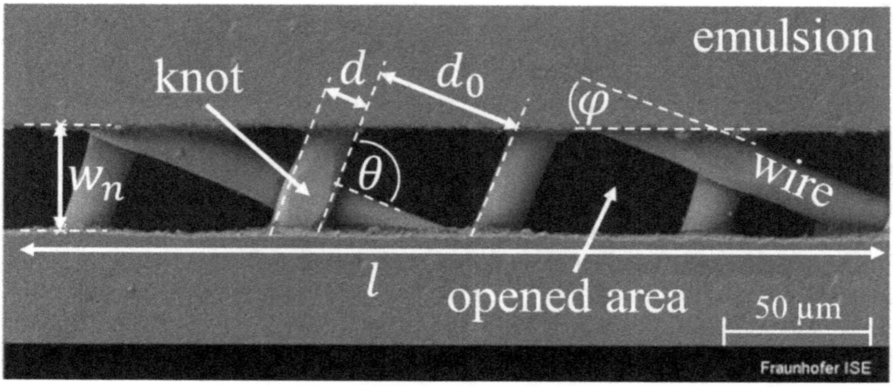

Figure 6.23 Scanning electron microscopy (SEM) image of a screen opening channel and its parameters [88]

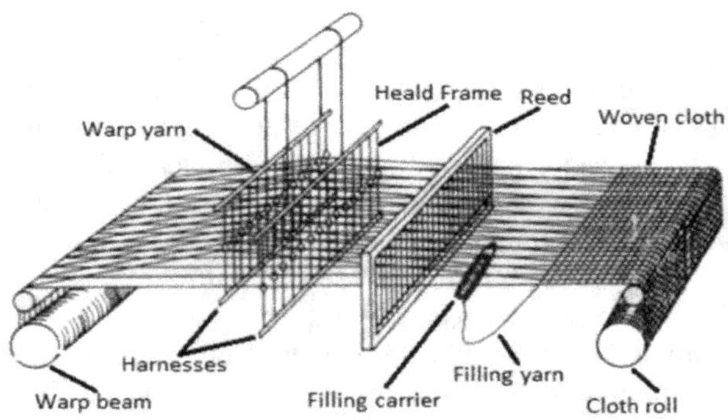

Figure 6.24 Schematic figure of a weaving loom [89]

6.5.1 The mesh

The basis of metallic wire meshes are cold formed wires of different metals and metal alloys. The wire material must have high elasticity so that the wires can be interwoven without tearing apart. They are manufactured on specifically designed weaving machines. The pretensioned warp wires, running parallel to the weaving direction, form the carrier material into which the weft wires are continuously drawn perpendicularly. The weft wire is wound from a spool and passed on to the shuttle, which fills the warp beam. Depending on how the warp and weft wires are combined during the weaving process, different types of weave with different apertures between the wires can be produced. Figure 6.24 schematically illustrates the typical design of a weaving machine for the production of a metal wire cloth.

The first step in the production of a wire mesh is the preparation of the warp, the so-called warping. The wires are placed on spools, ideally with the desired warp length. The spools are individually placed on the warping creel by hand and pulled forward over the deflection roller onto the warp beam as bundles. By rotating the warp beam accordingly, the desired length of warp material is then wound onto the warp beam. In case of the usual weave width of 122 cm, approx. 23 000 warp wires have to be warped manually for a 480 mesh.

After the warp wire has been wound onto the warp beam, it must be drawn-in in a separate production step. First, the warp wire is drawn into the eye, i.e. the opening, of the heald. The heald in the heald frame brings the warp wire into the correct 'up' or 'down' position. Then the warp wire is drawn through the opening of the reed. In the later weaving process, the weaving reed has the task of bringing the weft wire to the correct position in the mesh. Like the warping process, the drawing-in is also completely manual.

Afterwards, the warp beam including heald frame and weaving reed can be installed in the weaving machine and the machine is set up accordingly. In particular, the positions of the individual cylinders, the tension conditions in the warp wires, the time sequence between the up and down movement of the heald frames and the stop of the reed, must be adjusted to the corresponding mesh.

A modification of the mesh count (MC) and the wire diameter in the warp is only possible by producing a completely new warp. For this purpose, all previously mentioned production steps have to be run through again. For each new fabric, a correspondingly adapted weaving reed and adapted healds are required with a new warp beam and a new weaving machine set-up.

In the following, the key mesh parameters are introduced.

The maximum mesh tension, σ_{max} $[N/cm]$

Mesh stability is a crucial parameter in the selection of the appropriate MC and wire diameter configuration. The maximum mesh tension σ_{max} $[\mathrm{N/cm}]$ depends on the MC, the wire cross-section and the material (6.1) [90]. Typically used stainless steel has a tensile strength of σ_{uts} = 800 N/mm^2 [88].

$$\sigma_{max} = \sigma_{uts} \cdot MC \cdot \pi \frac{d^2}{4} \tag{6.8}$$

The yield strength of the material must not be exceeded during the printing process, which is why a sufficient elastic stretch reserve must be taken into account when tensioning the mesh [41]. The elongation of the mesh has already been calculated by Huebner [91] as a function of screen length, fabric tension and squeegee pressure. At high mesh tension, the take-off height can be reduced and the length and position accuracy of the printed image improved [90]. The calendering of the mesh also contributes to stabilization. Typical tension values are around σ_{max} = 20 N/cm [92]. The higher the elasticity of the material, the higher the snap-off must be set, otherwise there will be so-called dragging on the substrate [93]. However, if the snap-off is set too high, the print image will be distorted and the mesh will be subjected to increased stress [92]. The higher the tension and the lower the squeegee speed (related to the throughput), a smaller the snap-off can be chosen. The lifetime of the screen depends on several factors, some of which are difficult to estimate, e.g. number of squeegee passes, friction from the squeegee (squeegee pressure), squeegee hardness, tension of the mesh, emulsion adhesion, number of cleanings, surface roughness and hardness of the substrate (texture), the type of frame and the interaction of all process parameters [94]. Analyses of the squeegee shape were carried out to minimize the stress on the screen by the squeegee [95]. As soon as one wire is torn apart, the emulsion breaks off at one point or the mesh becomes fatigued, the screen must be replaced.

The tensioning process is done by pneumatic tensioning devices, whereby the mesh is stretched to the desired tension. The longer this tension is maintained (relaxation time), the lower the subsequent tension drop. This tension drop usually decreases by 10%–20% of the final tension at the end of the tensioning process and stabilizes after approx. 24 hours [90]. The tensioned fabric is then glued into a steel frame, which is roughened beforehand to improve the adhesion of the adhesive. The lower the MC and the higher the tension, a longer drying time must be kept [92]. Adjustable support profile frames are used to create an angle between the fabric and the frame. Later, this results in the so-called screen angle. The loss of tension

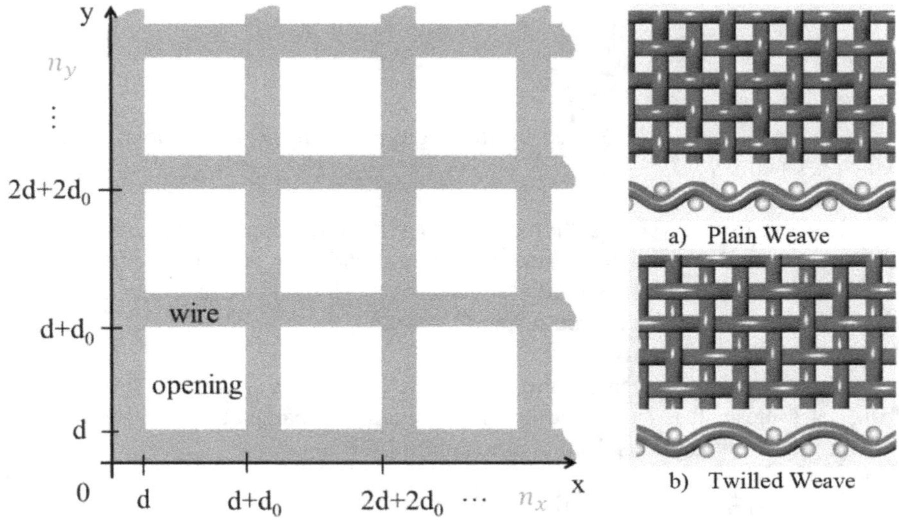

Figure 6.25 Left: Schematic of a mesh with wire diameter d and opening length d_0. Right: Weaving types: plain weave and twilled weave [92]

depends on the tensioning and relaxation time, the tensioning technology, the bending strength of the frame, the temperature and the gluing process. After a constant mesh tension has been established, the application of the emulsion begins (see Subchapter 6.5.3).

Wire diameter, d [μm]

The wire diameter is one of the most important parameters of the mesh and is measured in the non-woven state. Conventional screens for solar cell metallisation consist of monofilament (single core) stainless steel wires. Multi-core threads are made of several twisted single wires and are used more rarely [90]. Wire diameters in the order of 20 μm can have tolerances of approx. ±0.4 μm. Various materials have been analysed by Dubey [96] with focus on the parameters: tensile strength, elasticity, hardness and solvent absorption for screen printing process applications.

Mesh opening, d_0 $[\mu m]$

The mesh opening or aperture d_0 is defined by the free space between the wires. In literature often the letter 'w' is used, but in the scope of solar cell metallisation, this would lead to confusion with the screen parameter 'channel opening width d_0' within the next chapter. Figure 6.25 shows the mesh geometry within a coordinate system.

Mesh count, MC [1/inch]

The MC, defines the number of wires per unit length (usually per inch). Together with the wire diameter the mesh is described e.g. 380/0.014′, which accounts for MC = 380 1/inch and d = 14 μm. Common MCs for frontside metallisation are in the range of 300–400 1/inch. In general, the MC (in 1/μm) can be calculated with the following equation [97].

$$MC = \frac{1}{(d + d_0)} \qquad [1/m] \tag{6.9}$$

Mesh angle, θ [°]

The mesh angle θ is defined by the angle between the warp and weft wires, which is usually 90°.

Mesh thickness, D [μm]

Directly after the webbing, the maximal mesh thickness can be measured as $D=2d$. The mesh can be calendered (flattened) on both sides or on one side by several heatable steel rollers. This slightly reduces the mesh thickness and thus the paste application and increases the mesh stability [90].

Open area, OA [per cent]

The open area [per cent] is the ratio of the opened area [μm²] to the total area [μm²] of a mesh. It is calculated using only the mesh parameters (d and d_0). OA plays a major role during selection of the mesh [90].

$$\mathrm{OA} = \frac{d_0^2}{(d+d_0)^2} \times 100[1/\mu m] \tag{6.10}$$

Open mesh volume, V_o [m³/m²]

The open mesh volume is calculated by the difference of the overall mesh volume and the volume of two wires with the length of $d + d_0$ [97].

$$V_o = (d + d_0)^2 \cdot D - 2\pi\,(d + d_0) \cdot \frac{d^2}{4} \tag{6.11}$$

With low MC, (6.4) should be seen critically, since then the channel width limits the volume. The printed paste volume is much smaller than V_o, since the paste is not transferred completely out of the mesh after printing. The influence of the emulsion height is also not taken into account. Many other parameters such as amount of paste on the screen, adhesion to the substrate, squeegee angle, squeegee velocity and many more influence the paste volume [93]. Therefore (6.4) should only be considered as a guideline.

Weave of the mesh

The weaving of the warp and weft wires can be done in different ways, as shown in Figure 6.13. The plain weave is the most common, since twilled weaves have a larger contact area of the fabric on the substrate [92].

6.5.2 The emulsion on the mesh

The stencil represents the non-printable areas of the mesh. First the mesh is degreased, removing impurities and particles. A liquid emulsion is spread on the mesh and irradiated with UV light through an indirect stencil template. The used liquid consists of a polymer emulsion and UV-reactive resins. During the exposure to UV light, the emulsion hardens at the non-printable areas on the underside of the mesh. Afterwards the still liquid part, which was not exposed to light, is washed out. The hardening of the emulsion is influenced by its water content, spectral sensitivity, the light intensity, the light spectrum of the lamp

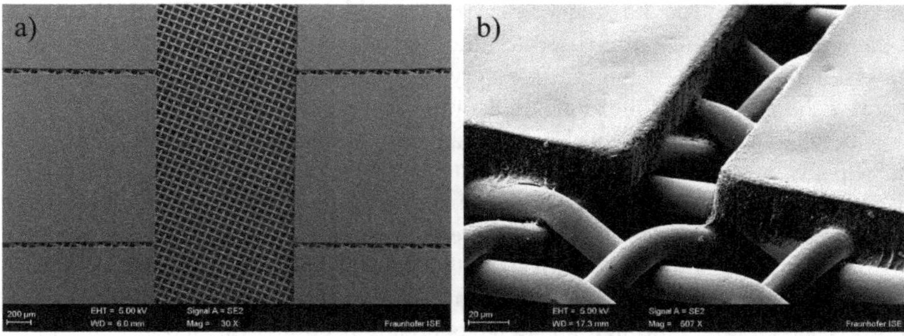

Figure 6.26 *Scanning electron microscopy (SEM) images of a flatbed screen with view on the down-facing side of the screen. (a) Busbar and fingers are seen as uncovered by the emulsion. (b) Close-up view at one cross-point of busbar and finger.*

and the exposure time. These parameters can be used to adjust the thickness of the emulsion. A high emulsion over mesh (EOM) does not always result in a greater finger height. In general, a smaller emulsion thickness should be selected for narrow channels, otherwise the flow resistance of the paste in the channel becomes too high and has a negative effect on the aspect ratio [98, 99]. After drying, the screen is visually analysed for clogged areas or faulty stencil locations. Figure 6.26 shows two SEM images of coated screens for flatbed screen printing.

In the following, the key parameters of the screen are introduced.

Emulsion over mesh, *EOM* [µm]

The emulsion thickness (EOM) determines the volume of paste printed in addition to the mesh thickness. It is adapted to the channel width and varies between 15 µm and 20 µm [100]. During the printing process it comes into contact with the textured substrate. Depending on the surface of the emulsion (or its coating), the paste flows (or slips [101]) more or less smoothly within the opening channels [102]. If it seals the gap between substrate and screen too poorly, the printed finger width will increase and so-called sawtooth corners are formed. If it seals too well, suction can occur during the process of separating the screen from the substrate, which also increases the finger width [90].

Screen angle, φ [°]

Screen frame, finger direction and squeegee direction are all parallel to each other. However, the mesh wires may have an angle to this direction. If the mesh would have a 0° or 90° orientation to the screen frame, individual wires could lie in the middle of a channel opening. In this case, not enough paste flows through the opening and the printed image and hence the solar cell becomes unusable. In order to avoid these angles, an angle of 22.5° (or less often 30°) was originally chosen in microelectronics, as this allows complex, rectangular patterns to be printed [103].

This angle is established as the standard in the screen printing process. Since the H-pattern layout is most often used for the frontside metallisation of solar cells and the channel openings of busbars are very large, it might be advantageous to re-examine the influence of the angle on the printing result.

So-called 'knotless screens' have an angle of $\varphi = 0°$ between the channel and the wires of the mesh, thus when perfectly aligned, they have no wire crossings inside the channel opening. These special screens are manufactured by only a few companies. The advantage of these 0° screens is the extraordinarily high open area OA. Since there are no wire crossovers in the channel, but only short lengths of weft wires, the printed fingers have a particularly high print quality. The effects of the knotless screen on the printed image have already been investigated at Fraunhofer ISE [104]. Printing with a knotless screen increases the aspect ratio and therefore improves the contact resistance. However, the problem lies in the production of these special screens. With today's manufacturing tolerances, a precise definition of the emulsion channels on the mesh to this extent is only possible with great effort. On a cell with about 100 fingers, it is quite possible that an entire wire lies in the channel opening, thus making the entire screen unusable. Nevertheless, in order to be able to produce these screens, there are two possibilities:

1. Individually selected wires, which occur inside the opened channel are pulled out of the mesh, which creates local low MCs and hence a reduction of the screen tension.
2. With the use of a mesh analyser, each mesh produced must be precisely measured for a few minutes. Then the emulsion template is adapted to each mesh and channels are created exactly where no wires are present. Despite the same mesh design, this process must be carried out individually for each mesh, since even small tolerances in MC can cause an unfavourable alignment of the stencil on the mesh.

For both manufacturing options, the cost of the necessary precision cannot yet be justified industrially despite the benefit of the increased aspect ratio.

Channel width, w_n [µm]

The channel width significantly influences the width of the printed contact finger. The actual finger width w_f differs from the channel width w_n by several µm [102, 105]. To reduce the finger width, the channel width must also be reduced. Scheer [90] developed an empirical relationship between the mesh parameters and the theoretically printable finger width $w_{f,th}$ (6.5).

$$w_{f,th} = \frac{\sqrt{2}d\,(2d+d_0)}{d_0} \qquad [\mathrm{m}] \tag{6.12}$$

Channel length, l [µm]

The channel opening is created by washing the emulsion out of the mesh. The channel length of a finger structure is typically about 156 mm [106].

The Moiré pattern

The first explanation of the Moiré pattern goes back to Lord Rayleigh in the year 1874 [107]. He investigated the superposition of two patterns with parallel, evenly distributed lines. Later, Moiré patterns were also discovered by superimposing concentric circles [108]. The mathematical analysis of this effect can be found in the study by Oster *et al.* [109]. Any pattern with regularly recurring geometries can be represented as an oscillation. If this oscillation is sampled with a frequency that is less than twice the frequency of the pattern, the aliasing effect occurs, also known from signal processing. The Moiré pattern is a form of this effect by under-sampling. Moiré patterns often occur by recording screens when the image frequencies overlap unfavourably. Likewise, in screen printing, these interference patterns can also occur in the screen and should be avoided at all costs. The frequency of the usual meshes is the MC and lies in the range of about 350 1 inch^{-1} = 14 1 mm^{-1}. The frequency of the recurring channels of the stencil is about 0.64 lines mm^{-1}. The positioning of the lines is mainly due to electrical and optical optimisation considerations, rather than possible pattern formation in the screen. Angles of 0°, 45° and 90° should be avoided if the alignment of the mesh to the stencil is not fine enough. Due to manufacturing tolerances of MC and mesh distortion during the printing process, Moiré effects may occur locally [90]. In the metallisation of solar cells and the screens used, there has not yet been any specific investigation of Moiré pattern effects.

6.5.3 *Simulation of screens*

Various studies have addressed the question what screen parameters will result in optimal printing performance. However, Figure 6.23 underlines that the geometry of the printed structure heavily depends on the local distribution of mesh wires in the screen opening channel. Even though the opening rate OA answers the question of how much of the opening channel is on average blocked by wires, it lacks any information on the local distribution of these wires. To address this problem, Ney *et al.* developed an approach to simulate any arbitrary screen architecture which allows to study the dependency of the OA on all two-dimensional geometric design parameters of a screen such as the channel width w_n, the screen angle φ, the MC, wire diameter d and channel position on the screen κ [88, 105, 110]. Furthermore, they introduced the novel parameter σOA, which describes the average deviation of the OA across the channel length. Figure 6.27 illustrates the definition of that parameter. Here, the local distribution of the OA across the channel length for an exemplary opening channel with a small screen angle is shown. It becomes clear that, depending on the position, the opening rate varies significantly. The average deviation σ_{OA} quantifies this rate of change. In order to ensure stable printing at narrow line width, the screen should be designed in a way that this deviation becomes minimal.

They further provided simulation results for a broad range of screen parameters as illustrated (Figure 6.28). Here, the average deviation of the opening rate σ_{OA} is simulated for different screen angles and channel widths. The established industry standard for the screen angle of $\varphi = 22.5°$ would dramatically increase the average deviation of the opening rate σ_{OA} when the demand for smaller opening channels is continued. One solution would be to increase the screen angle accordingly as

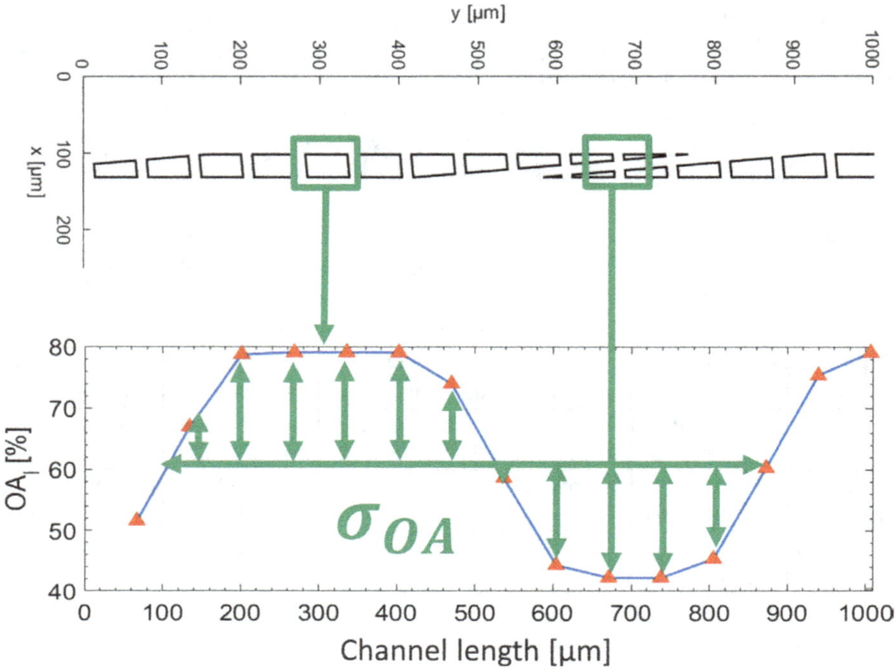

Figure 6.27 Definition of the deviation of the opening rate σ_{OA} [88]

suggested in Figure 6.28. However, the screen manufacturer also must consider production cost when designing the screen architecture. Tepner *et al.* provided a mathematical analysis on the mesh cutting losses depending on the screen angle, showing that cutting of mesh at higher angles will increase waste [111]. This comparison highlights the complex trade-offs in screen design. A full analysis and optimization of all involved parameters cannot be done manually any more at state-of-the-art requirements in the PV industry.

To address this, they continuously advance their algorithm's prediction capabilities. They provide a comprehensive dataset on the shape, size and exact location of all individual opened areas which can occur within a screen opening [88, 112]. The algorithm discovered that only seven different types of shapes can occur no matter what parameters for the MC, wire diameter, screen angle and opening width may be chosen (assuming a straight rectangular channel and quadratic underlying mesh). Figure 6.29 illustrates six of these opening shapes, excluding the full quadratic mesh unit. Throughout a series of publications, they studied the dependency of these individual openings on the macroscopic screen architecture and the impact on the printing result [88, 112, 113].

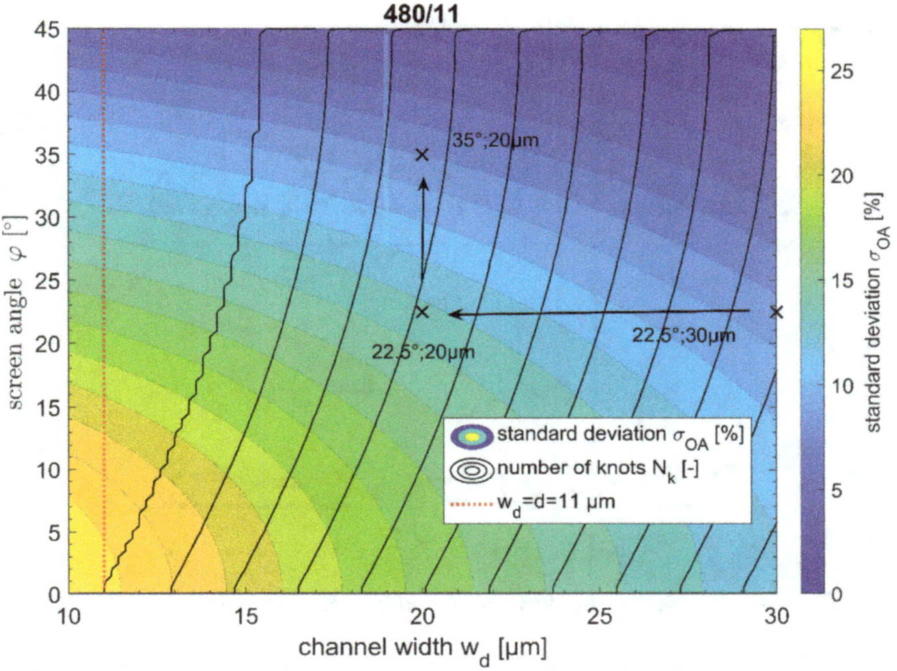

Figure 6.28 *The dependency of the average deviation of the opening rate σ_{OA} and the number of knots (contour lines) on the screen angle and channel width. Here, a mesh with a mesh count of MC = 480 1/inch and wire diameter of d = 11 μm is assumed. The figure has been taken from Clement et al. [110] and is based on Ney et al. [88].*

6.6 Screen printing process mechanics

The basic working mechanics of the screen printing process are illustrated in Figure 6.30. The set-up consists of four main components: A *squeegee* pushes the *printing paste* through a *screen* onto a *substrate*. First, the squeegee, placed at a certain angle below 90° with respect to the screen, pushes the paste over the screen. Due to the angle of the squeegee, a downward velocity component for the bulk of the paste is created, forcing the paste to flow into the opening units of the metal mesh. During

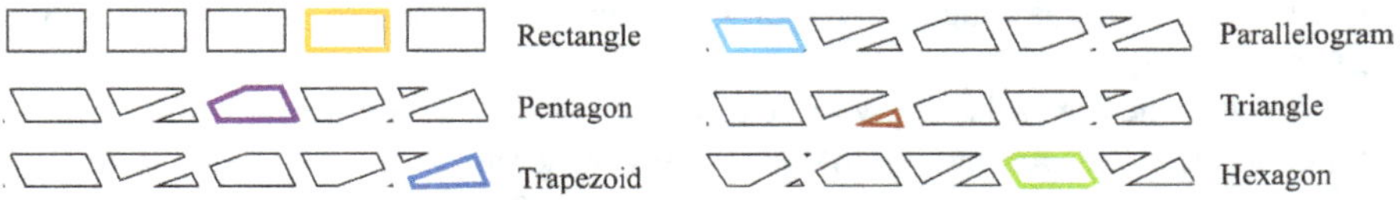

Figure 6.29 *Illustration of different opening shapes which can occur in a screen channel [112]*

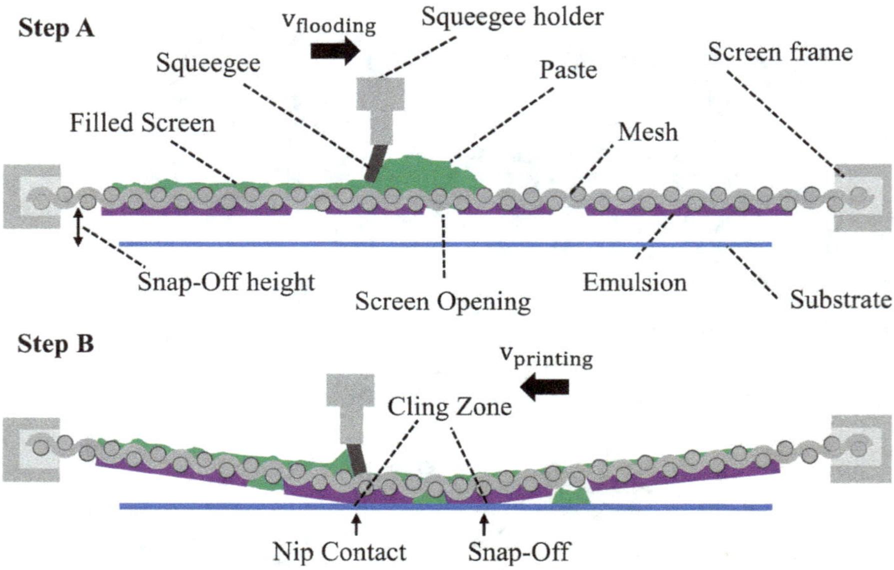

Figure 6.30 Schematic illustration of the screen printing process [101]

that motion the squeegee does not exert any forces onto the screen which means that the screen remains at a certain distance – the so-called snap-off distance – to the substrate. This process step is known as screen flooding and is illustrated in Figure 6.30 in step A. Once the squeegee has fully flooded the mesh, a second squeegee motion is initiated. This time the actual printing occurs. Now, the squeegee applies a certain pressure onto the screen, pushing it downward until contact with the substrate is reached under the squeegee position. This motion further pushes the paste through the opening channel of the emulsion layer onto the substrate. When the squeegee now continues to move across the screen, the pressure vanishes behind the squeegee. Due to the tension of the mesh (see Sub-chapter 6.5), the screen snaps off the substrate. The printed line adheres to the substrate and separates itself from the emulsion channel as long as the adhesion to the substrate is stronger than towards the emulsion channel walls. The area of contact between screen and substrate is called cling zone and its length is a function of the force balance between paste adhesion to the screen and restoring forces of the mesh. The operator of a screen printing machine would like to minimize the length of that cling zone in order to prevent unwanted process instabilities. Therefore, he has to carefully choose printing parameters such as squeegee velocity and snap-off distance as well as screen and paste parameters.

Now the question arises how these different parameters are linked to each other. Tepner and Lorenz summarized in their review on screen printing how modelling of the screen printing process has evolved over the last 20 years and provided a deeper insight on the underlying process mechanics [114]. Riemer was the first to provide a comprehensive theoretical model of the paste flow during squeegee motion

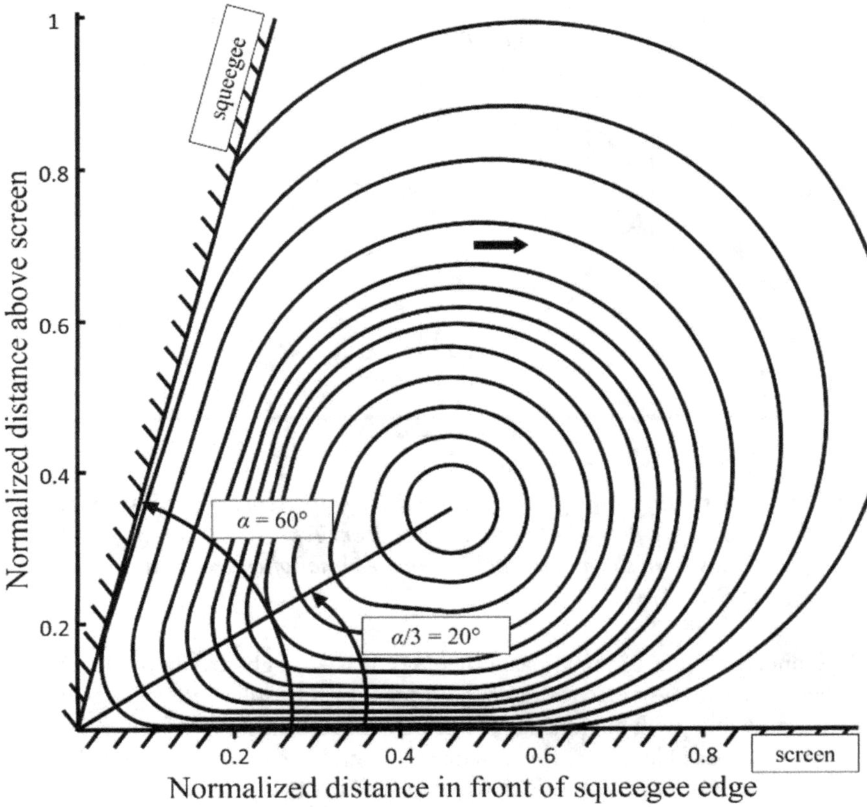

Figure 6.31 *Streamline pattern of a paste in front of the squeegee after Riemer. The figure has been taken from Tepner and Lorenz [114] and has been slightly adapted and simplified from Riemer's original publication [117].*

in screen printing. He mathematically modelled the pressure distribution based on the squeegee orientation [115–118]. Figure 6.31 shows his famous illustration of the streamlines in front of the squeegee tip. He derived this model by applying a hydrodynamic model which was based on Benson [119] and Messerschmitt [120]. Tepner and Lorenz further argued that this model served as the basis for what has also become known as the 'experience of the screen printing operator' in today's industry. Careful adjustment of the squeegee angle will create the downward velocity component of the paste to sufficiently flood the mesh. This relationship makes the screen printing process work and it is closely connected to the rheology of the printing paste. Later Chapman *et al.* [121] also included the shear thinning phenomena of the non-Newtonian power law fluids to the model. These findings initiated hundreds of studies on flatbed screen printing, the impact of paste rheology on the printing outcome and further optimization strategies.

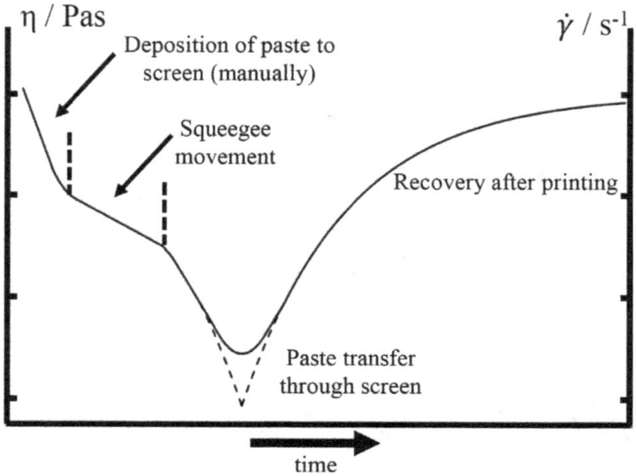

Figure 6.32 *Viscosity of the printing paste during different stages of the screen printing process. Figure is taken from Tepner and Lorenz [114] and has been slightly adapted by the original after Trease and Dietz [141].*

The understanding of metal paste rheology was strongly driven by the industry's need for narrow Ag electrodes on the front side of Si solar cells. The goal was to print narrow lines with homogeneous line height in the most reproducible and fast manner. As discussed in the previous chapter, common Ag pastes for frontside metallization of PERC contain a mass share of up to 95% Ag particles with average particle sizes of 3–4 µm and a more or less spherical shape [114, 122, 123]. Moreover, so-called glass frits are added for contact formation to the substrate [124, 125]. The particle matrix is embedded into an organic vehicle which contains a binder, rheological additives as well as various solvents. This vehicle significantly influences the rheological properties of the paste. Today's Ag pastes can be categorized as non-Newtonian fluids with so-called plastic flow characteristics. This means that a certain shear stress must be applied to the paste volume before it transitions from an elastic – solid-like response to a plastic – fluid-like deformation. Furthermore, the paste viscosity reduces significantly when the bulk of the paste is exposed to significant shearing forces [126]. According to Tepner and Lorenz [114], the impact of this rheological behaviour on the printing outcome has been studied by Hoornstra *et al.* [127], Hsu *et al.* [128], Thibert *et al.* [98, 129], Jewell *et al.* [130, 131], Potts *et al.* [132], Pospischil *et al.* [126, 133–135], Yüce *et al.* [123, 136, 137], Xu *et al.* [102, 138], Tepner *et al.* [101, 139, 140] and couple others. Their research activities can be summarized by the famous illustration of Trease and Dietz [141] in Figure 6.32.

They introduced this simple model in 1970 to describe the different phases of shear rates during screen printing. Because the common paste follows the non-Newtonian plastic flow behaviour, the viscosity drops by multiple orders of magnitude when sheared extensively. This shearing occurs even before the actual process because the operator has to apply the paste manually onto the screen. The high-yield stress of the paste requires him to even stir the paste excessively before application, making it almost impossible to ensure reproducible rheological properties every time. The lowest point of viscosity is reached when the paste enters the metal mesh. Here, a high gradient of velocity at small scale occurs. After the printing sequence the paste begins to recover its internal structure depending on its thixotropic characteristics. Even though this model is a very simplistic representation of the process dynamics, it still serves as the fundamental starting point for many researchers in this field. Tepner and Lorenz summarized research activities in the field in respect to this model by asking the following five questions [114, 140]:

1. Which values on the right y-axes for the shear rate are relevant in today's screen printing process and how are they linked to the process and screen parameters?
2. Which values then result on the left y-axes for the viscosity of the investigated paste and how to measure it in the relevant shear rate regime?
3. How exactly does the presented curve unfold over time during the printing process?
4. How to directly test this model by experiment?
5. What are the limits of this model and what other phenomena may have to be included in order to predict the flow behaviour of the investigated paste during the process steps (e.g. slip effects)?

They argue that these questions summarize the current state of research in the field of screen printing and its relationship to paste rheology. Most studies deal in some sense with at least one of these questions. For example, Hsu *et al.* [128] and Hoornstra *et al.* [127] focused on question 2. They studied the relationship between paste composition and shear rate dependent viscosity and its thixotropic behaviour. Thibert *et al.* [98] focused more the first two questions by looking at different screen printing set-ups (e.g. variations in screen parameters). They found that the screen printing operator can reduce paste consumption and increase process stability by carefully adjusting screen parameters to a certain paste rheology. Tepner *et al.* [139] worked on question 2 by experimentally studying how the dilution of state-of-the-art commercial Ag pastes can improve the paste transfer for fine-line printing below 20 μm. Figure 6.33 presents the typical dependency of the paste viscosity on the shear rate indicating the dramatic change in flow behaviour for fine-line printing. Here, pastes A–C show different rates of dilution of the same paste.

Tepner and Lorenz further categorized the work of Xu *et al.* [102], Potts *et al.* [132] and more recently Abdel Aal *et al.* [142] to address primarily questions 3 and 4 because they utilize a high-speed camera set-up to experimentally study the printing procedure. Xu *et al.* and Abdel Aal *et al.* replaced the substrate with a transparent glass plate in order acquire video images of the printing process from below. Using such an approach allowed them to directly study the spreading of paste right after deposition onto the substrate. One of their main conclusions was that spreading mainly occurs within the first 50 ms after the screen release. Potts *et al.* developed a set-up which allowed them to film the screen snap-off mechanics from a side view.

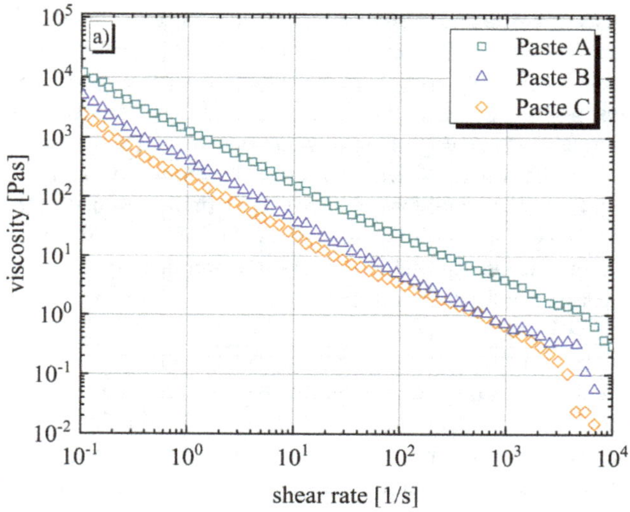

Figure 6.33 Typical relationship of viscosity and shear rate for commercial frontside Ag pastes. Figure has been taken from Tepner et al. [139].

With such an experiment, they studied the relationship between squeegee velocity, snap-off distance and the shape-printed structures, finding a clear correlation in the viscoelastic behaviour of the paste [132].

For the fifth question, Tepner and Lorenz summarized in their review how so-called wall slip behaviour of metal pastes can improve the paste transfer through the screen [114]. This effect is well known in the field of suspension rheology and has been intensively studied by Kalyon *et al.* [143, 144] and Barnes [145, 146]. The latter described the effect in the following way: the reason why a highly filled suspension might slip along a smooth rigid surface is the fact that the particle fraction of the suspension is reduced near the fluid-wall interface because spherical rigid particles can only touch the wall at a specific point [145]. This leads to a small liquid layer between the bulk of the paste and the adjacent wall. The existence of this highly dynamic effect and its magnitude depends on most involved parameters such as particle sizes, their shape, orientation, the total volume fraction of particles, their surface properties and the rheological characteristics of the surrounding liquid phase as well as the macroscopic shear history of the suspension [101, 140, 145]. On top of that, the surface properties and geometry of the adjacent wall also play a major role if slip occurs. Durairaj *et al.* then investigated how this effect influences the stencil printing process. They found that finger interruptions can be significantly reduced when slip at the apparatus surface exist. The results indicate that the operator of a printing line can utilize this effect by reducing the nominal opening width of the apparatus without compromising process stability [147]. For screen printing,

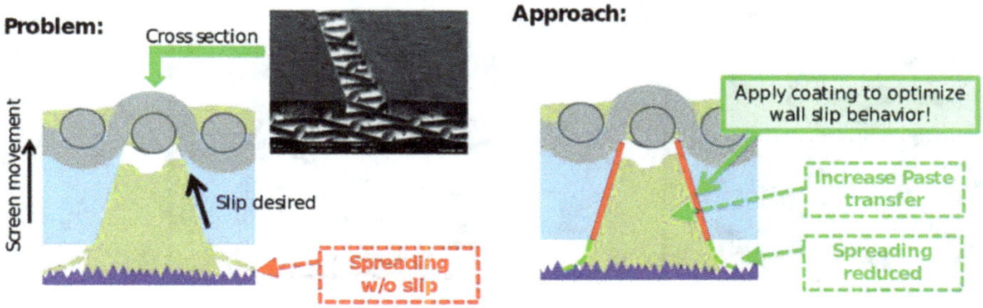

Figure 6.34 *Optimization approach for the paste–screen interaction. The emulsion surface is coated with a functional coating to promote slip behaviour. Data suggest that improve wall slip behaviour increases the paste transfer and reduces spreading on the substrate [101]. Figure has been taken from Clement et al. [110].*

Xu *et al.* [138] and Tepner *et al.* [101] studied the impact of wall slip at the paste emulsion interface. While Xu *et al*. focused on different adaptions of Ag pastes which were specifically desired to be prone to slip, Tepner *et al*. studied the impact of functional coatings on the emulsion surface on the onset for slip and its corresponding magnitude. They achieved that by applying emulsion material/coating onto the parallel plate geometry of a commercial parallel-plate rheometer. Both studies show promising results by reducing achievable finger widths and finger interruptions significantly. Figure 6.34 illustrates the approach of Tepner *et al*. for wall slip optimization in screen printing. At the paste emulsion interface, shearing forces can be significantly reduced by the existence of a small liquid slip layer. This leads to increased paste transfer and reduced spreading.Figure 6.35 presents experimental data for this statement. The application of a functional coating allowed to reduce the printed finger width while at the same time reduce lateral finger resistance, thus increasing the cross-sectional area by increased paste transfer. These results indicate that the flow behaviour of paste can be significantly influenced by the screen properties. Whenever the operator of a screen printing line is looking for an optimized printing sequence, he must consider the interaction between paste and screen as both heavily influence the design choice of each other. This means that there is no perfect printing paste and perfect screen architecture on its own. Both components of the screen printing process have to be optimized to specially suit each other.

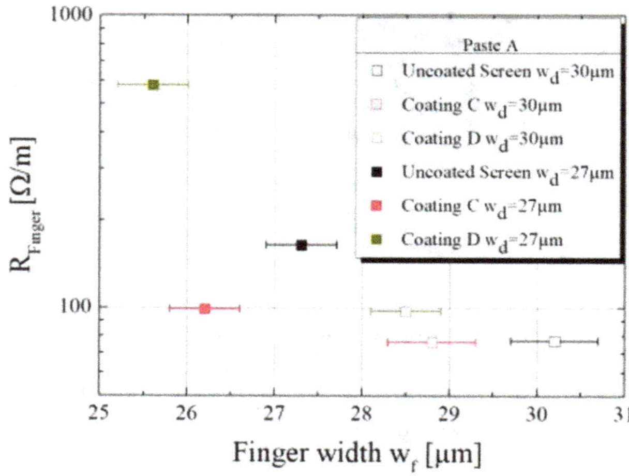

Figure 6.35 *Correlation between the finger width w_f and the corresponding lateral finger resistance R_{Finger}. Data is taken from Tepner et al. [101] and the figure has been taken from Clement et al. [110].*

6.7 Stencil printing

As described in the previous sub-chapters, screen printing is the most widely used and dominant metallization technology in the PV industry since decades due to its continuously improving and cost-effective process delivering a sufficient Ag front grid quality for high-efficiency silicon solar cells. Nevertheless, in the past 25 years significant effort has been made to develop, evaluate and commercialize stencil printing as an alternative front grid printing technology. Stencil printing is widely used in the microelectronics industry to print the solder pads on PCBs. Whereas screen printing applies a wire mesh embedded in an emulsion, stencil printing uses a very thin metal foil that is fully opened at the areas where paste is to be printed as described in section 6.7.1 in detail. As outlined in section 6.7.2, a 100% open area of stencils enables very uniform Ag finger profiles. If the Ag paste rheology is adjusted, in addition stencils obtain very narrow Ag fingers with high aspect ratios. In the past, these advantages led to around $0.3\%_{abs}$ higher conversion efficiencies of silicon solar cells with stencil-printed Ag front grid in comparison to solar cells with screen-printed Ag front grid; see Section 6.7.3 for experimental results. However, despite these excellent technological results, stencil printing is not applied in mass production of solar cells, yet. With the momentum of large-scale application of screen printing and, as mentioned, the ongoing improvements, the threshold to apply stencil printing seems still too high for the industry. As described at the end of this sub-chapter, one objection is that stencil-printed Ag grids sometimes exhibit locally widened Ag fingers probably due to dust particles between the wafer and the stencil. These locally widened Ag fingers do not impact the solar cell efficiency but are detected as local defects in the optical quality sorting test at the end of a solar cell production line and hence would reduce the solar cell production yield. If this issue could be solved in the future, stencil printing would become an attractive solar cell metallization production technology.

6.7.1 Stencil fabrication and Ag finger grid design

Stencils have first been developed in the 1980s for printing solder pads on PCBs with surface-mount technology (SMT). The stencils were made of brass foils and structured by photolithography and chemical etching. In the 1990s, the stencil technology advanced to three different stencil manufacturing approaches resulting in different stencil qualities for different applications:

1. Chemically etched stainless steel for coarse-pitch applications
2. Laser cut stainless steel for moderately fine-pitch applications enabling a cheap and fast stencil manufacturing
3. Electro-formed nickel for very fine-pitch, as used in the highest level of PCB miniaturization, for example, cell phones.

The Ag front grid of silicon solar cells is much more demanding in terms of feature size, line conductivity and production throughput compared to most PCB solder pad applications. Hence, Hoornstra and Heurtault concluded in their work on stencils for solar cells in 2009, that 'since an SMT-board is very much different from a thin, fragile silicon solar cell, and solder paste is far different from silver paste, we can learn from the SMT stencil print application, but not copy their practice' [148].

For PV, first results on stencil print were published in 1996 [148, 149], although earlier studies in Germany on stencil print for solar cells go back to the late 1980s [150]. These stencil printing experiments were executed on a small research production line at Siemens AG in Munich with so-called Hi-mesh type electroformed stencils and double layer etched stencils made by Stork Veco. With a line width of 50 μm, a printed finger width of 100 μm could be achieved, demonstrating poor paste behaviour. The double or dual layer of stencil build-up came from the demand to realize the full two-busbar H-pattern print in one step, as was dictated by cost consideration. The Hi-mesh material production was too complex and discontinued. In the later, and independent studies [148, 149], a double- or dual-layer electroformed stencil approach was developed by Stork Veco; see for manufacturing process details a few further lines below. In the pioneering experiments, results on cell level were limited through non-consistent stencil manufacturing accuracy, printer quality, non-flat at alkaline etched crystalline cell surfaces, unmatched paste rheology or even too large paste particles. Single layer stencils demonstrated improved aspect ratio and print quality through the full opening.

Early work on single layer stencils for solar cell metallization mostly used around 50 μm thin stainless steel foils with around 70 μm narrow finger openings produced by laser structuring [148, 151] similar to the PCB industry. A disadvantage of laser structuring was non-smoothness of the internal walls and material residues, limiting paste release. As of 2012, electroformed nickel stencils were supplied by

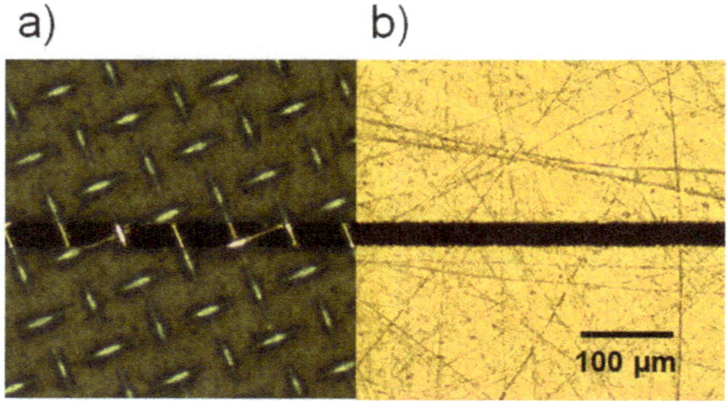

Figure 6.36 *Optical microscopy images of (a) a screen containing wires in the finger opening and (b) a stencil with 100% finger opening. Both are designed with a finger opening width of 40 µm.*

ASM (formerly DEK) and intensively evaluated for PV applications [152–154]. In electroforming, first a photoresist is deposited on a conductive substrate and structured by photolithography which defines the approximately 30 µm narrow finger openings. Afterwards, a 30 µm thin nickel layer is deposited by electroplating on the conductive substrate at areas where the photoresist has been removed. Finally, the nickel foil is separated from the conductive substrate. This stencil manufacturing process allows precise definition of the finger openings including tapering of the openings in the nickel foil for optimum paste release. Figure 6.36 shows microscope images of a nickel stencil in comparison to a conventional mesh screen, both with 40 µm finger openings. The completed nickel foil is mounted in a spring-tensioned framing system which has the advantage of offering more precise and consistent control over the foil tension than a regular mesh mounted stencil.

Whereas screens can support any kind of finger opening layout, the solar-cell-typical H-pattern design of around 100 Ag fingers per wafer and, to date, up to 12 perpendicular busbars is challenging for stencils since a mechanical support in single layer stencils is missing in the H-pattern design. Hence, the so-called dual print approach has been developed where first the busbars are printed with a conventional screen and afterwards the Ag fingers are printed with a stencil [148, 152, 153]. This approach offers the additional benefit of using two different Ag pastes. A non-firing-through Ag paste is printed for the busbars which is optimized for soldering the ribbons for module interconnection. Afterwards, a high-viscosity firing-through Ag paste is used for the stencil-printed Ag fingers which is optimized for fine-line, high aspect ratio printing and contacting lightly doped emitters. The non-firing-through busbar Ag paste reduces the metal contact area to the emitter thereby reducing carrier recombination at the front metal contact which increases the solar cells V_{oc} by around 2 mV [154]. A benefit of multiple busbars, for stencil print, is shorter finger length, especially since cell sizes also gradually increased. Shorter finger length will make the stencil sturdier, both in case of tensioning in the frame, and also during the print process.

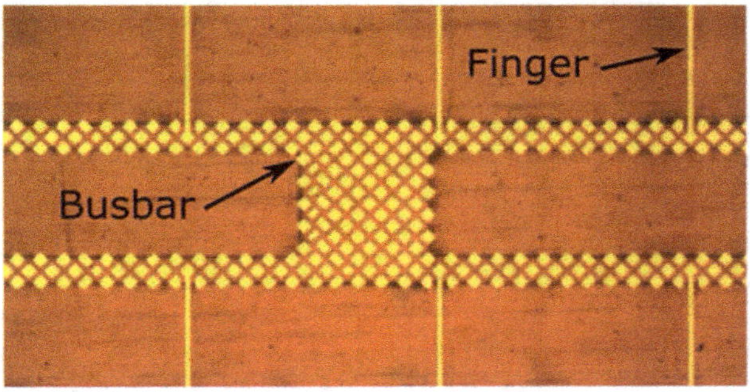

Figure 6.37 *Light microscope image of a single print stencil consisting of two different nickel layers. The top layer of the stencil has a mesh grid over the busbar regions which acts as a bridge. The finger regions are 100% open and integrated in both nickel layers [154].*

However, the dual print process described above requires two printing and drying steps and hence is more costly than a single screen print process. Therefore, the single print stencil shown in Figure 6.37 has been further developed by ASM and allows to print fingers as well as busbars in one printing step. The single print stencil consists of a two-layer nickel foil, with each Ni layer being fabricated by a separate photolithography and nickel electroplating process as described above. As displayed in Figure 6.37, the top (or squeegee side) layer of nickel has a mesh grid over the busbar regions. Both, the top and the bottom foil contain the 100% opened finger regions.

6.7.2 *Stencil printed Ag finger properties*

Early work on stencil printing focussed on reducing the front grid Ag finger width and increasing the aspect ratio compared to screen-printed Ag fingers. High aspect ratio metal fingers are desired for solar cell front grid applications in order to obtain narrow Ag fingers for minimizing shadowing losses and high finger cross-sectional areas for low grid line resistivity, simultaneously. Since single layer stencils feature 100% opened finger lines, Ag pastes with much higher viscosity can be used to minimize paste spreading and hence widening of the fingers. In 2009, Hoornstra and Heurtault applied a 50 μm thin metal stencil with 60 μm finger opening and a high-viscosity Ag paste and obtained about 70 μm narrow and 30 μm high Ag fingers as shown in Figure 6.38 [148]. The stencil-printed Ag finger profile in Figure 6.38 nicely reproduces the stencil opening feature revealing only minimal finger widening. For comparison, Figure 6.38 also shows a screen-printed Ag finger profile using a typical screen printing process at that time resulting in around 110 μm wide and 15 μm high Ag fingers clearly showing the spreading and widening of the screen-printed Ag fingers.

In an attempt to further reduce the stencil-printed Ag finger width by reducing the finger opening widths towards 30 μm, Heurtault and Hoornstra [148] found that

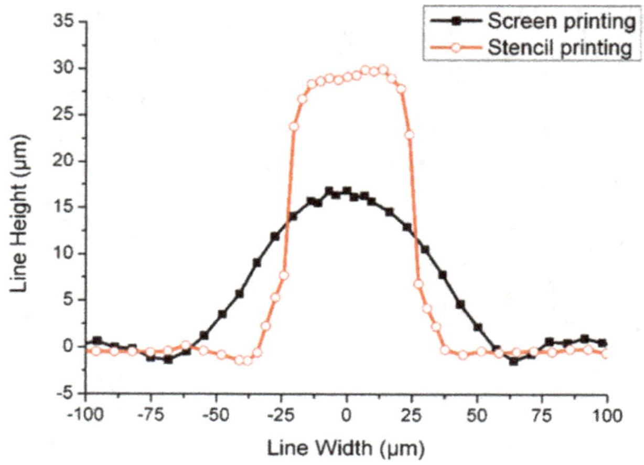

Figure 6.38 *Profilometer images of a stencil-printed Ag finger using a high-viscosity Ag paste in comparison to a screen-printed Ag finger. The high aspect ratio of the stencil-printed finger is desired for solar cell front-grid applications in order to obtain narrow Ag fingers for little shadowing losses and high finger cross-sectional areas for low line resistivity, simultaneously. Reproduced from Hoornstra and Heurtault [148] with permission.*

with a 50 µm thick stencil using only 30 µm opening, the Ag finger cross-sectional area drastically reduced as displayed in Figure 6.39. The reason is, that if the finger opening widths become narrower than the stencil foil thickness, the Ag paste adheres to the stencil sidewalls rather than to the silicon wafer substrate resulting in very bad paste release from the stencil and hence very small finger cross-sectional areas. As shown in Figure 6.39, reducing the stencil thickness to 30 µm enables very good paste transfer even for 30 µm finger-opening widths. Hence, as a rule of thumb, the finger opening widths should be similar or larger than the stencil thickness in order to obtain good finger cross-sectional areas. Accordingly, Falcon *et al.* demonstrated 35 µm narrow stencil-printed Ag fingers in 2012 by using a 30 µm thin stencil foil containing 30 µm narrow finger openings [156]. The screen-printed reference group resulted in 50 µm narrow fingers at that time.

One of the biggest advantages of stencil printing over screen printing is that due to the 100% open finger area stencil printing enables extremely uniform Ag finger grid lines with a finger height variation below ±1.5 µm as shown in Figure 6.40 a [153, 154]. In contrast, when using a mesh screen containing wires in the finger openings as shown in Fig. 6.6.1 a, during printing the Ag paste has to flow below the wires in order to establish a continuous grid line. This leads to reduced Ag paste transfer below the wires and hence a rather strong variation of the screen-printed Ag finger height of ±6 µm as shown in Figure 6.40 b [153, 154]. At areas where two wires form a node in the finger opening, the Ag paste may also stick to the wires and during the release of screen from wafer this extra surface area may pull the

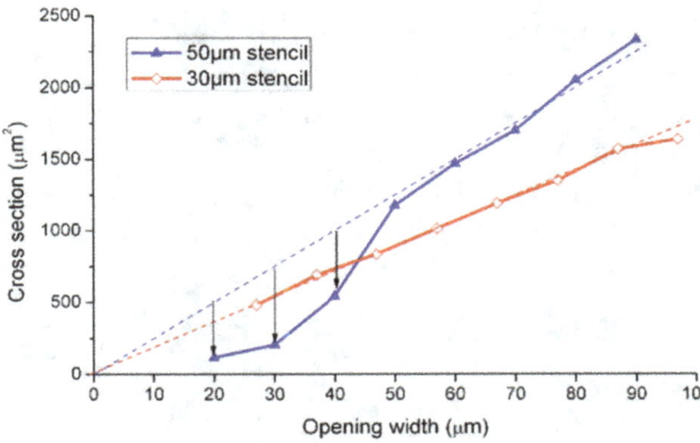

Figure 6.39 *Cross-sectional area of stencil-printed Ag fingers for different finger opening widths and stencil foil thicknesses of 30 µm and 50 µm. As soon as the finger opening width is smaller than the foil thickness, the Ag paste starts sticking to the stencil foil rather than the silicon wafer resulting in strongly reduced Ag finger cross-sectional area due to worse paste release from the stencil. Hence, as a rule of thumb, the finger opening width should be similar or larger than the stencil thickness.Reproduced from Heurtault et al. [155] with permission.*

paste in the finger upwards to form the 'camel humps' visible in Figure 6.40 b. The extremely uniform finger profile of stencil-printed Ag fingers allows to minimize the consumption of expensive Ag paste by adjusting the Ag finger height to the minimum value which is required to obtain the desired grid line resistance. In doing so, Hannebauer *et al.* demonstrated a record low Ag paste consumption of 67.7 mg

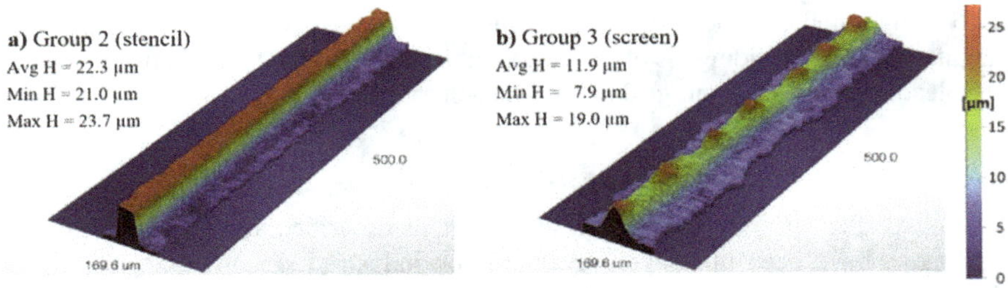

Figure 6.40 *Optical profilometer images of (a) a 36 µm narrow stencil-printed Ag finger with a finger height of 22.3 µm ± 1.5 µm and (b) a 47 µm wide screen-printed Ag finger with a finger height of 11.9 µm ± 6 µm [152]. Whereas the mesh in the screen causes quite a high variation of the screen-printed Ag finger height, the 100% open area of stencils enables very uniform Ag finger grid lines. Reproduced from Hannebauer et al. [152] with permission.*

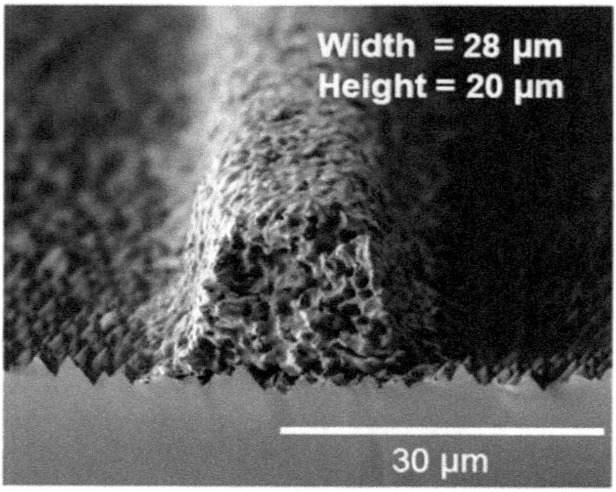

Figure 6.41 *Scanning electron microscopy (SEM) image of a 28 µm narrow and 20 µm high stencil-printed Ag finger on the textured front side of an industrial passivated emitter rear contact (PERC) solar cell demonstrated in 2016 by ASM and ISFH*

for the Ag front grid of a full-sized high-efficiency industrial PERC silicon solar cell in 2013 [157]. It was also demonstrated that in stencil printing the Ag finger height can be systematically adjusted not only by changing the stencil thickness but also by adjusting the shore hardness of the squeegee [158]. The softer the squeegee, the deeper it penetrates into the stencil opening and the shallower is the printed Ag finger, but still with extremely uniform finger height. Figure 6.41 shows a SEM image of a stencil-printed Ag finger of an industrial PERC solar cell processed at ISFH in 2016 using a nickel stencil from ASM. The SEM image reveals the uniform finger height and the steep sidewalls of the 28 µm narrow and 20 µm high Ag finger, which were leading-edge Ag front grid print results at that time.

6.7.3 Stencil-printed silicon solar cells

Due to excellent print results on research scale as described above, stencil-printed Ag fingers have been intensively evaluated on industrial silicon solar cells as an alternative mass production technology to screen printing. Until 2015, the so-called aluminium BSF (Al-BSF) silicon solar cell was the dominating industrial solar cell design applying p-type silicon wafers, a phosphorus-doped and SiN-passivated front surface emitter, and printed Ag front fingers and a printed full-area Al rear contact obtaining conversion efficiencies between 17% and 20%. From 2016 onwards, the PERC design became the industrially preferred solar cell design due to its higher conversion efficiencies, which today are in the range of 22% to 23%. PERC solar cells are similar to Al-BSF cells on the front side but apply an AlOx/SiN surface passivation on the rear side with locally alloyed Al rear contacts. See Sub-chapter 9.1 for more details on Al-BSF and PERC solar cells.

With the print results shown in Figure 6.38, Hoornstra and Heurtault have already demonstrated in 2009 that the 70 µm narrow stencil-printed fingers reduce light reflection, increase the J_{sc} and thereby the conversion efficiency of Al-BSF cells by 0.4%$_{abs}$ to 16.6% in comparison to the 110 µm wide screen-printed Ag fingers [148]. A few years later, with further advances in printing and solar cell technology, Hannebauer *et al.* [152] showed that stencil printing allows to either increase the grid line cross-sectional area obtaining lower grid line resistances or to reduce the consumption of expensive Ag paste compared to screen printing. With comparable Ag paste consumption, the stencil-printed Al-BSF cells obtained a conversion efficiency of 19.1% versus 18.6% of the screen-printed Al-BSF cells mostly due a lower grid line resistance and a lower shadowing loss of the stencil-printed Al-BSF cells [152]. Similar Al-BSF cell results were also demonstrated by SERIS and ASM in 2015 [153] by optimizing the finger opening widths for each print technology. The 0.2%$_{abs}$ efficiency advantage of 30 µm stencil-printed Ag fingers versus 45 µm wide screen-printed fingers was supported by numerical simulations of the impact of the Ag finger properties on the solar cell current-volatge (IV) characteristics [153].

First applications of stencil-printed Ag fingers to industrial-type PERC silicon solar cells led to conversion efficiencies up to 19.8% in 2013 with record low Ag paste consumption of 67.7 mg for stencil-printed front fingers and screen printing of three-segment busbars [157]. In 2016, ASM and ISFH investigated the single print stencil shown in Figure 6.37 and systematically compared it to alternative front grid metallization technologies on high-efficiency industrial five-busbar PERC solar cells with the resulting IV parameters displayed in Figure 6.42 [154]. In addition to the single print stencil group, the single print screen printing serves as industrial reference group. Dual print stencil means that the Ag fingers are printed with a stencil, whereas the busbars are screen-printed separately with a non-firing-through Ag paste as in dual print. The dual print screen group also uses the dual print approach with screen-printed Ag fingers and separately screen-printed busbars using the same non-firing through busbar Ag paste as in the dual print stencil group. As can be seen in Figure 6.42, the stencil groups obtain 0.3%$_{abs}$ higher conversion efficiencies compared to screen printing due to 0.3 mA/cm^2 higher J_{sc} values due to a 10 µm narrower Ag finger widths and hence less shadowing loss. Also, the stencil groups obtain 0.3% higher FF values due to a higher finger aspect ratio and hence lower grid line resistance. The dual print groups obtain 1 mV to 2 mV higher open circuit voltages V_{oc} since the non-firing through busbar Ag paste reduces the metal contact area and hence charge carrier recombination in contrast to the firing-through busbar paste of the single print groups.

In general, as shown by the solar cell results above, the 100% open stencil area enables narrower Ag fingers reducing shadowing losses and at the same time increasing the finger height and thereby the finger cross-sectional area reducing the grid line resistance thereby enabling a significant efficiency increase compared to screen printing. In the past decade, the number of busbars continuously increased from 3 to 5 and nowadays to 9 or 12 busbars per wafer. Thereby, the Ag finger

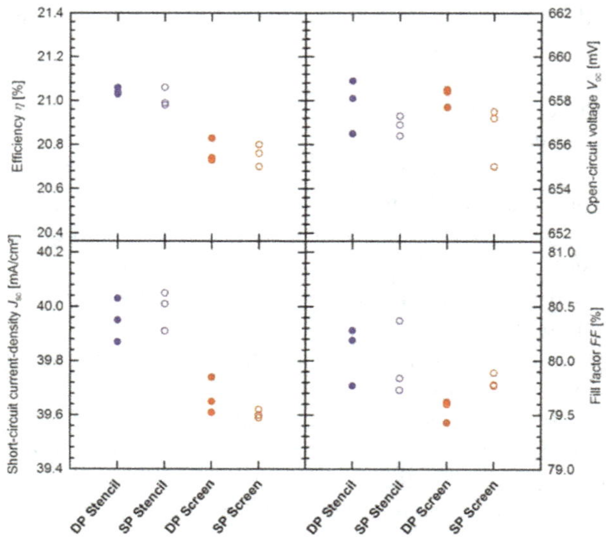

Figure 6.42 *Current-voltage (IV) parameters of industrial PERC solar cells comparing four different Ag front grid metallization technologies. Dual print stencil means that the Ag fingers are printed with a stencil, whereas the busbars are screen-printed separately with a non-firing-through Ag paste as in dual print. The single print stencil group applies the single print stencil in Figure 6.37 thereby printing fingers and busbars simultaneously. The dual print screen group uses the dual print approach with screen-printed Ag fingers and separately screen-printed busbars using the same non-firing-through busbar Ag paste as in the dual print stencil group. Finally, the single print screen group refers to the industry-typical single print screen printing. The stencil groups obtain 0.3%$_{abs}$ higher conversion efficiencies compared to screen printing due to higher J_{sc} values (reduced Ag finger widths and hence less shadowing losses) as well as higher FF values (higher aspect ratio and hence lower grid line resistance). Reproduced from Hannebauer et al. [154] with permission.*

length (half busbar to busbar distance) is shortened from around 26 mm in 2010 to around 6.5 mm today. The reduced finger length drastically relaxes the requirement of the grid line resistance from below 0.5 Ω cm^{-1} in the past to below 3 Ω cm^{-1} now. Accordingly, with a higher number of busbars the finger height can be reduced thereby minimizing the expensive Ag paste consumption which is economically very important. Due to the uniform Ag finger profile as shown in Figure 6.40, stencil printing allows to reduce the finger height to the minimum value required to obtain the desired grid line resistance as it was evaluated in [158]. Another important front

Figure 6.43 *Light microscopy image of a stencil-printed Ag front grid revealing a locally widened Ag finger. Such a local finger broadening appears on one or two spots on almost every stencil-printed solar cell. The likely root cause are dust particles which act as local spacers between the silicon wafer and the stencil allowing the Ag paste to flow beyond the stencil opening. Whereas these locally widened fingers do not impact the solar cell performance, they fail the optical quality sorting test at the end of the solar cell production line thereby strongly reducing the production line yield. This is one of the main reasons why stencil printing has not been applied in mass production of silicon solar cells, yet. Image reproduced from [159] with permission.*

grid parameter is the contact resistance of the Ag fingers to the phosphorus-doped emitter. Since Ag pastes optimized for screenprinting and stencil printing use the same glass frit composition in order to dissolve the SiN layer and alloy with the phosphorus emitter during firing, the contact resistance is independent of the printing technology.

With the PERC solar cell results shown in Figure 6.42, a high-performance and cost-effective single print stencil printing process was available for mass production to the PV industry in 2016. Nevertheless, it was never applied in high-volume production so far. One of the main reasons is, that in pilot production trials of stencil printing quite often optical defects in the front grid pattern were detected by camera inspection systems at the end of the solar cell production line. An example of such an optical defect is shown in Figure 6.43, where an Ag finger is locally widened [159]. Even though these locally widened fingers do not impact the solar cell performance, they fail the optical quality sorting test at the end of the solar cell production line thereby reducing the production line yield below the one set for screen printing. The local finger broadening is likely caused by dust particles which act as local spacers between the silicon wafer and the stencil allowing the Ag paste to flow beyond the stencil opening. So far, no technical solution has been found to eliminate the local finger broadening. However, in case the presumed dust issue will be solved, stencil printing offers an attractive front grid metallization technology due to its superior grid line properties enabling increased solar cell conversion efficiencies with minimal Ag paste consumption.

References

[1] Cheek G.C., Mertens R.P., van Overstraeten R., Frisson L. 'Thick-film metallization for solar cell applications'. *IEEE Transactions on Electron Devices*. 1984;**31**:602–9.

[2] Schubert G. 'Thick Film Metallisation of Crystalline Silicon Solar Cells: Mechanisms. Models and Applications [PhD Thesis]'. Konstanz: Universität Konstanz; 2006.

[3] Hörteis M., Gutberlet T., Reller A., Glunz S.W. 'High-temperature contact formation on N-type silicon: basic reactions and contact model for seed-layer contacts'. *Advanced Functional Materials*. 2010;**20**:476–84.

[4] Hong K.-K., Cho S.-B., Huh J.-Y., Park H.J., Jeong J.-W. 'Role of PbO-based glass frit in Ag thick-film contact formation for crystalline Si solar cells'. *Metals and Materials International*. 2009;**15**:307–12.

[5] Fields J.D., Ahmad M.I., Pool V.L., Yu J., van Campen D.G., Parilla P.A. 'The formation mechanism for printed silver-contacts for silicon solar cells'. *Nature Communications*. 2016;**7**:11143.

[6] Kim H.-S., Cho S.-B., Kim H., *et al.* 'Electrochemical nature of contact firing reactions for screen-printed silicon solar cells: origin of "gray finger" phenomenon'. *Progress in Photovoltaics Research and Applications*. 2016;**24**:1237–50.

[7] Hoerteis M. 'Contact recombination and microstructure evaluation of screen printed and firing-through metallization'. *7th Metallization and Interconnection Workshop*. 23.10.2017.

[8] Krassowski E. '2CO.13.3 laser enhanced contact optimization – a novel technology for metal-semiconductor-contact optimization for crystalline silicon solar cells'. *EU PVSEC 2020*. 2020.

[9] Mayberry R.W. 'Laser enhanced contact optimization (LECO) and LECO-specific pastes – a novel technology for improved cell efficiency'. *EU PVSEC 2019*. 12.09.2019.

[10] General Electric. 'Conductive Varnish'. 1933.

[11] Christensen Carl J., Rigterink Merle D., Treptow Arnold W. 'Metallizing composition'. *Patent US2461878A*.

[12] Gray Robert B., Steele Raymond H. 'Silver paint'.

[13] Dhamrin M., Suzuki S., Kikuchi K., Nakahara M., Morishita N. 'Paste composition and method for forming silicon germanium layer'. *Patent US10916423B2.*

[14] Masahiro Nakahara T.W., Dhamrin M. 'Paste composition'. *WO2013115076.*

[15] Tsuji K., Dhamrin M., Matsubara M., Nakahara M., Wa T. 'Paste consumption and solar battery element'. 2017-09-27.

[16] Haruzo Katoh T.W. 'Aluminum paste composition and solar cell element using the same'.

[17] 'DuPont de Nemours and company'. *The Method for Manufacturing a Solar Cell.* 2013.

[18] Trubnikov I.L. 'Thermal expansion, vitrification temperature, and corrosion behavior of lead-borosilicate glasses'. *Refractories and Industrial Ceramics.* 2000;**41**:169–71.

[19] Takashi N., Aoyagi T., Tachizono S., *et al.* 'Electronic component'. *Conductive Paste for Aluminum Electrode to Be Applied to the Same, and Glass Composition for Aluminum Electrode.*

[20] Roberts F.M., Wilkinson E.L.G. 'The effects of alloying material on regrowth-layer structure in silicon power devices'. *Journal of Materials Science.* 1971;**6**:189–99.

[21] Yoshikawa T., Morita K. 'Solid solubilities and thermodynamic properties of aluminum in solid silicon'. *Journal of the Electrochemical Society.* 2003;**150**:G465.

[22] Miller R.C., Savage A. 'Diffusion of aluminum in single crystal silicon'. *Journal of Applied Physics.* 1956;**27**:1430–2.

[23] Navon D., Chernyshov V. 'Retrograde solubility of aluminum in silicon'. *Journal of Applied Physics.* 1957;**28**:823–4.

[24] Mitsuhata T. 'Phenomenological theory on liquid phase epitaxy'. *Japanese Journal of Applied Physics.* 1970;**9**:90–4.

[25] Del Alamo J., Eguren J., Luque A. 'Operating limits of Al-alloyed high-low junctions for BSF solar cells'. *Solid-State Electronics.* 1981;**24**:415–20.

[26] Lindholm F.A., Chen Y.H. 'Current-voltage characteristic for bipolar P-N junction devices with drift fields, including correlation between carrier lifetimes and shallow-impurity concentration'. *Journal of Applied Physics.* 1982;**53**:8863–6.

[27] Mandelkorn J., Lamneck J.H. 'Simplified fabrication of back surface electric field silicon cells and novel characteristics of such cells'. *Solar Cells.* 1990;**29**:121–30.

[28] Ralph E.L. 'Recent advancements in low cost solar cell processing'. 11th Photovoltaic Specialist Conference (PVSC); 6-8 May; Scottsdale; 1975. pp. 315–16.

[29] Haigh A.D. 'Fired through printed contacts on anti-reflection coated silicon terrestrial solar cells'. *12th IEEE PVSEC, USA.* 1976.

[30] Frisson L., Honoré M., Mertens R., Govaerts R., van Overstraeten R. 'Silicon solar cells with screen printed diffusion and metallization'. IEEE Photovoltaic Specialists Conference (PVSC); 7-10 January; San Diego, CA; 1980. pp. 941–2.

[31] Sasaki T. 'The past, present, and future development of photovoltaic system in Japan'. 1984.

[32] Lolgen P., Leguijt C., Eikelboom J.A., Steeman R.A., Sinke W.C., Verhoef L.A. 'Aluminium back-surface field doping profiles with surface recombination velocities below 200 cm/s'. *Piscataway*. Piscataway, NJ: Institute of Electrical and Electronics Engineers; 1993. pp. 236–42.

[33] Vick G.L., Whittle K.M. 'Solid solubility and diffusion coefficients of boron in silicon'. *Journal of the Electrochemistry Society*. 1969;**116**:1142.

[34] Gee J.M., Bode M.D., Silva B.L. 'Boron-doped back-surface fields using an aluminum-alloy process [Si solar cells]'. Piscataway, NJ: IEEE; 1997. pp. 275–8.

[35] Huster F., Schubert G. 'ECV doping profile measurement of aluminum alloyed back surface fields'. *Proceedings of 20th EUPVSEC*; Barcelona; 2005. pp. 1462–5.

[36] Rauer M., Schmiga C., Tuschinsky A., Glatthaar M., Glunz S.W. 'Investigation of aluminum-boron doping profiles formed by coalloying from screen-printed pastes'. *Energy Procedia*. 2013;**43**:93–9.

[37] Schmiga C., Hermle M., Glunz S.W. 'Towards 20% efficient n-type silicon solar cells with screen-printed aluminium-alloyed rear emitter'. *23rd EUPVSEC*.

[38] Kopecek R., Buck T., Libal J., Rover I., Wambach K., Geerligs L.J. 'Large area screen printed N-type silicon solar cells with rear aluminium emitter: Efficiencies exceeding 16%'. *Piscataway*. Piscataway, NJ: IEEE; 2006. pp. 1044–7.

[39] Schmiga C., Nagel H., Schmidt J. '19% efficient n-type Czochralski silicon solar cells with screen-printed aluminium-alloyed rear emitter'. *Progress in Photovoltaic Research and Applications*. 2006;**14**:533–9.

[40] Gong C., van Kerschaver E., Robbelein J., *et al.* 'Screen-printed aluminum-alloyed P^+ emitter on high-efficiency N-type interdigitated back-contact silicon solar cells'. *IEEE Electron Device Letters*. 2010;**31**:576–8.

[41] Bock R., Schmidt J. 'Brendel R. N-type silicon solar cells with surface-passivated screen-printed aluminium-alloyed rear emitter'. *Physica Status Solidi (RRL)*. 2008;**2**:248–50.

[42] Halm A., Popescu L., Theobald J., Peter K., Kopecek R. 'Low temperature pads on Al-emitter or Al-BSF'. *2009 EU PVSEC*. 2009.

[43] Heimann M., Klaerner P., Luechinger C., Mette A., Mueller J.W., Traeger M. 'Ultrasonic bonding of aluminum ribbons to interconnect high-efficiency crystalline-silicon solar cells'. *Energy Procedia*. 2012;**27**:670–5.

[44] von C.H., Huber S., Meyer S., Reiff S., Vietor J. 'Direct tin-coating of the aluminum rear contact by ultrasonic soldering'. 27th European Photovoltaic Solar Energy Conference and Exhibition; 2012. pp. 1150–3.

[45] Comparotto C., Lossen J., Mihailetchi V.D. 'Bifacial screen-printed N-type passivated emitter rear totally diffused rear junction solar cells'. *AIP conference proceedings*. 1999;**100001**(2018).

[46] Tous L., Chen J., Choulat P., *et al.* 'Large Area Monofacial Screen-Printed Rear-Emitter nPERT Cells Approaching 23% Efficiency'. *35th EUPVSEC*. 2018.

[47] Dullweber T., Wehmeier N., Nowack A., Brendemühl T., Kajari-Schröder S.B.R. 'Industrial bifacial N-type silicon solar cells applying a boron co-diffused rear emitter and an aluminum rear finger grid'. *Physica Status Solidi*. 2016;**213**:3046–52.

[48] Peng Z.-W., Nakahara M., Buck T., Kopecek R. 'Towards 22% efficiency n-PERT rear junction solar cells with screen printed Al point back contact'. 2018:100002.

[49] Min B., Wehmeier N., Brendemuehl T., Haase F., Larionova Y., Nasebandt L. '716 mV Open-circuit voltage with fully screen-printed P -type back junction solar cells featuring an aluminum front grid and a passivating poly-silicon on oxide contact at the rear side'. *Solar RRL*. 2021;**5**:2000703.

[50] Suzuki S., Peng Z.-W., Tsuji K., Nakahara M., Morishita N., Kuroki T. 'Screen-Printed aluminium contacts on n+-Doped silicon'. 37th European Photovoltaic Solar Energy Conference and Exhibition; 2020. pp. 542–4.

[51] Macdonald D., Geerligs L.J. 'Recombination activity of interstitial iron and other transition metal point defects in P- and N-type crystalline silicon'. *Applied Physics Letters*. 2004;**85**:4061–3.

[52] Froitzheim A., Münzer K.E., K.- H., Tölle R., Schlosser R., Winstel M.G. 'N-Type silicon solar cells based on industrial technology'. 20th EU PVSEC Conference Proceedings; 2005. pp. 594–9.

[53] Tool C.J.J., Rieffe H.C., Kinderman R., Kopececk R., Wambach K., Geerligs L.J. 'Solar cells on N-type multicrystalline silicon wafers by industrial processing techniques'. 19th EU PVSEC Conference Proceedings; 2004. p. 970.

[54] Buck T., Kopecek R., Libal J., *et al.* Industrial screen printed n-type silicon solar cells with front boron emitter and efficiencies exceeding 17%. 21st EU-PVSEC; 2006. pp. 1264–7.

[55] Kopecek R., Buck T., Libal J., *et al.* Large area N-type multicrystalline silicon solar cells with B-emitter: efficiencies exceeding 14%. *15th International Photovoltaic Science & Engineering Conference*; 2005. pp. 892–4.

[56] Kerp H., Kim S., Lago R., Recart F., Freire I., Pérez L. 'Development of screenprintable contacts for P+ emitters in bifacial solar cells'. *21th EU PVSEC Conference Proceedings*. 2006:892–4.

[57] Lago R., Pérez L., Kerp H., Freire I., Hoces I., Azkona N. 'Screen printing metallization of boron emitters'. *Progress in Photovoltaics Research and Applications*. 2010;**18**:20–7.

[58] Burgers A.R., Geerligs L.J., Carr A.J., Gutjahr A., Saynova D.S., Li L. '19.5% Efficient n-type Si solar cells made in production'. *26th EU PVSEC*. 2011.

[59] Riegel S., Mutter F., Hahn G., Terheiden B. 'Contact formation in the silver/aluminum thick film firing process – a phenomenological approach'. 2010.

[60] König M., Rudl A., Georg N., Hörteis M. *Advanced model for the contact formation of Al-containing Ag thick film paste on boron-doped emitter of Si solar cells*. Chambery, France; 2013.

[61] Fritz S., Riegel S., Herguth A., König M., Hörteis M., Hahn G. 'Preservation of Si surface structure by Ag/Al contact spots – an explanatory model'. *Energy Procedia*. 2015;**67**:43–8.

[62] Fritz S. 'Contact formation of Ag/Al screen-printing pastes to heavily B-Doped c-Si'. Konstanz; 2016.

[63] Fritz S., Riegel S., Hammud A., Deniz H., Hahn G. 'Crystalline nature of metal spikes and silicon inclusions in Ag/Al screen-printing metallization'. *IEEE Journal of Photovoltaics*. 2016;**6**:79–85.

[64] Heinz F.D., Breitwieser M., Gundel P., *et al*. 'Microscopic origin of the aluminium assisted spiking effects in N-type silicon solar cells'. *Solar Energy Materials and Solar Cells*. 2014;**131**:105–9.

[65] Fritz S., Konig M., Riegel S., Herguth A., Horteis M., Hahn G. 'Formation of Ag/Al screen-printing contacts on B emitters'. *IEEE Journal of Photovoltaics*. 2015;**5**:145–51.

[66] Kiefer F. *Kontaktierung Ionen-implantierter Boremitter in n-Typ Silizium-Solarzellen mittels Silber/Aluminium-Pasten*. 1st edn. München: Verlag Dr. Hut; 2016.

[67] Edler A., Mihailetchi V., Kopecek R., Harney R., Böscke T., Stichtenoth D. 'Improving screen printed metallization for large area industrial solar cells based on N-type material'. *Energy Procedia*. 2011;**8**:493–7.

[68] Koduvelikulathu L.J., Edler A., Mihailetchi vd., *et al*. 'Metallization and firing process impact on VOC – a simulation study'. *27th EU PVSEC*. 2012.

[69] Edler A., Mihailetchi V.D., Koduvelikulathu L.J., Comparotto C., Kopecek R., Harney R. 'Metallization-induced recombination losses of bifacial silicon solar cells'. *Progress in Photovoltaics Research and Applications*. 2015;**23**:620–7.

[70] Wöhrle N., Lohmüller E., Werner S., Greulich J. 'Development, characterization and modelling of doping profile, contact resistance, and metal spiking in diffused and screen-printed boron emitters'. *31nd EU PVSEC*. 2015.

[71] Lohmüller E., Werner S., Hoenig R., Greulich J., Clement F. 'Impact of boron doping profiles on the specific contact resistance of screen printed Ag–Al contacts on silicon'. *Solar Energy Materials & Solar Cells*. 2015;**142**:2–11.

[72] Kontermann S. *Characterization and Modeling of Contacting Crystalline Silicon Solar Cells*. München: Dr. Hut; 2009.

[73] Jesswein R., Fastnacht P., Weiß M., Lossen J., Meyer K., Krokoszinski H.-J. 'Process optimization of single step inkjet printed front contacts for industrially fabricated solar cells leads to an efficiency gain of 0.3 % abs with consumption of less than 60 Mg silver'. *28th EU PVSEC*. 2013.

[74] Engelhardt J., Frey A., Gloger S., Hahn G., Terheiden B. 'Passivating boron silicate glasses for co-diffused high-efficiency N-type silicon solar cell application'. *Applied Physics Letters*. 2015;**107**:42102.

[75] Engelhardt J., Fritz S., Emre E., Hahn G. 'Contact formation on p-doped Si by screen-printing pure Ag pastes for bifacial N-type Si solar cells'. *32nd EU PVSEC*. 2016.

[76] Kim C., Choi J.-W., Choi S., Kim S., Park H., Song H. 'Effects of current-injection firing with Ag paste in a boron emitter'. *Scientific Reports*. 2016;**6**:21553.

[77] Chu H., Preis P., Lossen J., Mojrova B., Buchholz F., Becht G. 'Impact of the presence of busbars during the fast firing process on contact resistances'. *IEEE Journal of Photovoltaics*. 2018;**8**:1–7.

[78] Mihailetchi V.D., Chu H., Kopecek R. 'Insight into metal induced recombination losses and contact resistance in industrial silicon solar cells'. *2018 IEEE 7th World Conference on Photovoltaic Energy Conversion (WCPEC) (A Joint Conference of 45th IEEE PVSC, 28th PVSEC & 34th EU PVSEC); 6/10/2018–6/15/2018; Waikoloa Village, HI*. Piscataway, NJ: Institute of Electrical and Electronics Engineers; 2018. pp. 2673–7.

[79] Padhamnath P., Khanna A., Balaji N., Shanmugam V., Nandakumar N., Wang D. 'Progress in screen-printed metallization of industrial solar cells with SiOx/poly-Si passivating contacts'. *Solar Energy Materials and Solar Cells*. 2020;**218**:110751.

[80] Çiftpınar H.E., Stodolny M.K., Wu Y., Janssen G.J., Löffler J., Schmitz J. 'Study of screen printed metallization for polysilicon based passivating contacts'. *Energy Procedia*. 2017;**124**:851–61.

[81] Bao J., Chen C., Ma L., Huang C., Zhang X., Wang Z. 'Towards 24% efficiency for industrial n-type bifacial passivating-contact solar cells with homogeneous emitter'. *37th European Photovoltaic Solar Energy Conference and Exhibition*. 2020:160–3.

[82] Rodriguez J., Wang E.-C., Chen N., JW H., Li M., Buatis J.K. 'Towards 22% efficient screen-printed bifacial n-type silicon solar cells'. *Solar Energy Materials and Solar Cells*. 2018;**187**:91–6.

[83] Feldmann F., Fellmeth T., Steinhauser B., Nagel H., Ourinson D., Mack S. 'Large area Topcon cells realized by a PECVD tube process'. *36th EU PVSEC*. 2019.

[84] Biegeleisen J. *The Complete Book of Silk Screen Printing Production*. New York: Dover Public; 1975.

[85] Vasilantone M. 'Stencil printing machine'. *Patent US3427964*.

[86] Ralph E.L. 'Recent advancements in low cost solar cell processing'. Conference Record (A76-14727 04-44) New York. 11th Photovoltaic Specialists Conference; Scottsdale; 1975. pp. 315–6.

[87] ITRPV. *ITRPV. International Technology Roadmap for Photovoltaic (ITRPV): Results 2020*. 11th edn; 2020.

[88] Ney L., Tepner S., Linse M., Lorenz A., Bechmann S., Weber R. 'Optimization of fine line screen printing using in-depth screen mesh analysis'. *AIP Conference Proceedings*. 2019;**20006**.

[89] Weaving Machine. Available from https://abadikabeh.blogspot.com/1969/12/weaving-machine-air-jet-loom-diagram.html [Accessed 31 Jan 2021].

[90] Scheer H.G. *Siebdruck-Handbuch*. Lübeck: Verlag der Siebdruck; 1999.
[91] Huebner G. 'Considerations on bulged-out print shoulders due to mesh depression and high emulsion over mesh in screen printing'. *Journal of Print and Media Technology Research*. 2017:4–17.
[92] Sefar A.G. *Siebdruck-Handbuch*. Thal, Schweiz; 2007.
[93] G. BOPP +Co.AG F. BOPP SD. Available from www.bopp.com/ [Accessed 14 Feb 2019].
[94] Buzby D. 'Fine line screen printing of thick film pastes on silicon solar cells'. *SGIA Journal*. 2008.
[95] Horvath E., Harsanyi G. 'Mechanical modelling and life cycle optimisation of screen printing'. *Journal of Theoretical and Applied Mechanics*. 2012:1025–36.
[96] Dubey G.C. 'Screens for screen printing of electronic circuits'. *Microelectronics Reliability*. 1974;**13**:203–7.
[97] Hahne P. *Innovative Drucktechnologien: Siebdruck-Tampondruck*. Lübeck: Verlag der Siebdruck; 2001.
[98] Thibert S., Jourdan J., Bechevet B., Chaussy D., Reverdy-Bruas N., Beneventi D. 'Influence of silver paste rheology and screen parameters on the front side metallization of silicon solar cell'. *Materials Science in Semiconductor Processing*. 2014;**27**:790–9.
[99] Fortu Z., Song L.X. 'Identifying screen properties to optimise solar cell efficiency and cost'. *Energy Procedia*. 2013;**33**:84–90.
[100] Tavares R., Dobie A., Buzby D., Zhang W., Mesh O.S. 'Optimal screen mesh, emulsion chemistry, and emulsion thickness for fine-line front metallization pastes on crystalline silicon solar cells'. 2011.
[101] Tepner S., Wengenmeyr N., Ney L., Linse M., Pospischil M., Clement F. 'Improving wall slip behavior of silver pastes on screen emulsions for fine line screen printing'. *Solar Energy Materials and Solar Cells*. 2019;**200**(6):109969.
[102] Xu C., Willenbacher N. 'How rheological properties affect fine-line screen printing of pastes: a combined rheological and high-speed video imaging study'. *Journal of Coatings Technology and Research*. 2018;**15**(6):1401–12.
[103] Stalnecker, Jr. S.G. 'Stencil screens for fine-line printing'. *Electro Component Science and Technology*. 1980;**7**(1–3):47–53.
[104] Lorenz A., Linse M., Frintrup H., Jeitler M., Mette A., Lehner M. 'Screen printed thick film metallization of silicon solar cells – recent developments and future perspectives'. *35th European Photovoltaic Solar Energy Conference and Exhibition*. 2018:819–24.
[105] Clement F., Linse M., Tepner S., Wengenmeyr N., Ney L., Krieg K. '"Project FINALE" - screen and screen printing process development for ultra-fine-line contacts below 20μm finger width'. *36th EU PVSEC Conference Proceedings*. 2019:259–62.
[106] Equipment V.P. International Technology Roadmap for Photovoltaic (ITRPV). 2018. Available from http://www.itrpv.net/ [Accessed 14 Feb 2019].

[107] Rayleigh LXII. 'On the manufacture and theory of diffraction-gratings'. *The London, Edinburgh, and Dublin Philosophical Magazine and Journal of Science*. 1874;**47**(310):81–93.

[108] Nishijima Y., Oster G. 'Moiré patterns: their application to refractive index and refractive index gradient measurements'. *Journal of the Optical Society of America*. 1964;**54**(1):1.

[109] Oster G., Wasserman M., Zwerling C. 'Theoretical interpretation of moiré patterns'. *Journal of the Optical Society of America*. 1964;**54**(2):169.

[110] Clement F., Tepner S., Linse M., Ney L., Wengenmeyr N., Pospischil M. 'Screen printing process development towards 20 μm front-side Ag-electrodes on Si-solar cells'. *PV International*. 2020;**44**.

[111] Tepner S., Ney L., Singler M., Preu R., Pospischil M., Clement F. 'A model for screen utility to predict the future of printed solar cell metallization'. *Scientific Reports*. 2021;**11**(1).

[112] Tepner S., Ney L., Linse M., *et al.* 'Screen pattern simulation for an improved front-side Ag-electrode metallization of Si-solar cells'. *Progress in Photovoltaic Research and Applications*. 2020;**28**(10):1054–62.

[113] Tepner S., Ney L., Linse M., Lorenz A., Pospischil M., Clement F. 'Studying knotless screen patterns for fine-line screen printing of Si-solar cells'. *IEEE Journal Photovoltaics*. 2020;**10**(2):319–25.

[114] Tepner S., Lorenz A. 'Printing technology and its impact on the pv industry – a comprehensive review'. *Progress in Photovoltaics Resications Applications submitted for publication.*

[115] Riemer D.E. 'Analytical engineering model of the screen printing process'. *Part 1. Solid State Technology*. 1988;**8**:107–11.

[116] Riemer D.E. 'Analytical engineering model of the screen printing process'. *Part 2. Solid State Technology*. 1988;**9**:85–90.

[117] Riemer D.E. 'The theoretical fundamentals of the screen printing process'. *Microelectronics International*. 1989;**6**(1):8–17.

[118] Riemer D.E. 'Ein Beitrag Zur Untersuchung Der physikalisch-technischen Grundlagen des Siebdruckverfahrens'. *Dissertation, University of Berlin*. 1988.

[119] Benson A.M. 'Thick-film screen printing'. *Solid State Technology*. 1969:12.

[120] Messerschmitt E. 'Rheological considerations for screen printing inks'. *Screen Printing*. 1982;**10**:136–9.

[121] Chapman S.J., Fitt A.D., Please C.P. 'Extrusion of power-law shear-thinning fluids with small exponent'. *International Journal of Non-Linear Mechanics*. 1997;**32**(1):187–99.

[122] Li W., Wu T., Jiao R., *et al.* 'Effects of silver nanoparticles on the firing behavior of silver paste on crystalline silicon solar cells'. *Colloids and Surfaces A: Physicochemical and Engineering Aspects*. 2015;**466**:132–7.

[123] Yüce C., Okamoto K., Karpowich L., Adrian A., Willenbacher N. 'Non-volatile free silver paste formulation for front-side metallization of silicon solar cells'. *Solar Energy Materials and Solar Cells*. 2019;**200**(12):110040.

[124] Jeon S.J., Koo S.M., Hwang S.A. 'Optimization of lead- and cadmium-free front contact silver paste formulation to achieve high fill factors for industrial screen-printed Si solar cells'. *Solar Energy Materials and Solar Cells.* 2009;**93**(6–7):1103–9.

[125] Mette A. *New Concepts for Front Side Metallization of Industrial Silicon Solar Cells.* 1st edn. *München*: Dr. Hut; 2007.

[126] Pospischil M. 'A parallel dispensing system for an improved front surface Metallization of silicon solar cells'. *Fraunhofer Verlag*. 2017.

[127] Hoornstra J., Weeber A.W., De Moor H.H.C., Sinke W.C. 'The importance of paste rheology in improving fine line: thick film screen printing of front side metallization'. 14th European Photovoltaic Solar Energy Conference and Exhibition; 1997.

[128] Hsu C.P., Guo R.H., Hua C.C., Shih C.-L., Chen W.-T., Chang T.-I. 'Effect of polymer binders in screen printing technique of silver pastes'. *Journal of Polymer Research.* 2013;**20**(10):277.

[129] Thibert S., Chaussy D., Beneventi D. Silver ink experiments for silicon solar cell metallization by flexographic process. *2012 IEEE 38th Photovoltaic Specialists Conference (PVSC); 6/3/2012–6/8/2012*; Austin, TX, USA. Piscataway, NJ: IEEE; 2012. pp. 2266–70.

[130] Jewell E.H., Claypole T.C., Gethin D.T. 'Viscosity control in the screen printing of ceramic transfers'. *Surface Coatings International Part B: Coatings Transactions.* 2003;**86**(2):155–63.

[131] Jewell E., Hamblyn S., Claypole T., Gethin D. 'Deposition of high conductivity low silver content materials by screen printing'. *Coatings.* 2015;**5**(2):172–85.

[132] Potts S.-J., Phillips C., Jewell E., Clifford B., Lau Y.C., Claypole T. 'High-speed imaging the effect of snap-off distance and squeegee speed on the ink transfer mechanism of screen-printed carbon pastes'. *Journal of Coatings Technology and Research.* 2020;**17**(2):447–59.

[133] Pospischil M., Zengerle K., Specht J., *et al.* 'Investigations of thick-film-paste rheology for dispensing applications'. *Energy Procedia.* 2011;**8**(4):449–54.

[134] Pospischil M., Specht J., Konig M., *et al.* 'Paste rheology correlating with dispensed finger geometry'. *IEEE Journal of Photovoltaics.* 2014;**4**(1):498–503.

[135] Pospischil M., Specht J., Gentischer H., König M., Hörteis M., Mohr C. 'Correlations between finger geometry and dispensing paste rheology'. *27th EU PVSEC*. 2012.

[136] Yüce C., Willenbacher N. 'Challenges in rheological characterization of highly concentrated suspensions – a case study for screen-printing silver pastes'. *Journal of Visualized Experiments*. 2017.

[137] Yüce C., König M., Willenbacher N. 'Rheology and screen-printing performance of model silver pastes for metallization of Si-solar cells'. *Coatings.* 2018;**8**(11):406.

[138] Xu C., Fies M., Willenbacher N. 'Impact of wall slip on screen printing of front-side silver pastes for silicon solar cells'. *IEEE Journal of Photovoltaics*. 2017;**7**(1):129–35.
[139] Tepner S., Wengenmeyr N., Linse M., Lorenz A., Pospischil M., Clement F. 'The link between Ag-paste rheology and screen-printed solar cell metallization'. *Advanced Materials Technologies*. 2020;**5**(10):2000654.
[140] Tepner S. 'Contribution to the Understanding of Screen Printing and its Implication for the PV Industry [Dissertation]'. *Fraunhofer ISE*. 2021.
[141] Trease R.E., Dietz R.L. 'Paste rheology can improve your fine line printing'. Proceedings ISHM; 1970.
[142] Abdel Aal K., Willenbacher N. 'Front side metallization of silicon solar cells – a high-speed video imaging analysis of the screen printing process'. *Solar Energy Materials and Solar Cells*. 2020;**217**:110721.
[143] Kalyon D.M. 'Apparent slip and viscoplasticity of concentrated suspensions'. *Journal of Rheology*. 2005;**49**(3):621–40.
[144] Aral B.K., Kalyon D.M. 'Effects of temperature and surface roughness on time-dependent development of wall slip in steady torsional flow of concentrated suspensions'. *Journal of Rheology*. 1994;**38**(4):957–72.
[145] Barnes H.A. 'A review of the slip (wall depletion) of polymer solutions, emulsions and particle suspensions in viscometers: its cause, character, and cure'. *Journal of Non-Newtonian Fluid Mechanics*. 1995;**56**(3):221–51.
[146] Barnes H.A. 'Rheology: principles, measurements and applications'. *Powder Technology*. 1996;**86**(3):313.
[147] Durairaj R., Man L.W., Ekere N.N., Mallik S. 'The effect of wall-slip formation on the rheological behaviour of lead-free solder pastes'. *Materials & Design*. 2010;**31**(3):1056–62.
[148] Hoornstra J., Heurtault B. 'Stencil print applications and progress for crystalline silicon solar cells'. *Proceedings of 24th European Photovoltaic Solar Energy Conference*. 2009;**992**:989.
[149] de Moor H.H.C. 'Fine-line screen printing for silicon solar cells'. 6th Colorado Silicon PV Workshop, Snowmass; 1996. pp. 154–70.
[150] Münzer K.A. 'Private communication'.
[151] Cotter J.E., Yao G., Eggleston B. 'Laser-formed stencils for printed silicon solar cells'. Proceedings of 31st IEEE Photovoltaic Specialists Conference; 2005.
[152] Hannebauer H., Dullweber T., Falcon T., Brendel R. 'Fineline printing options for high efficiencies and low Ag paste consumption'. *Energy Procedia*. 2013;**38**:725–31.
[153] Shanmugam V., Wong J., Peters I.M., *et al*. 'Analysis of fine-line screen and stencil-printed metal contacts for silicon wafer solar cells'. *IEEE Journal of Photovoltaics*. 2015;**5**(2):525–33.
[154] Hannebauer H., Falcon T., Cunnusamy J., Dullweber T. 'Single print metal stencils for high-efficiency PERC solar cells'. *Energy Procedia*. 2016;**98**:40–5.

[155] Heurtault B., Hoornstra J. 'Towards industrial application of stencil printing for crystalline silicon solar cells'. *Proceedings of 25th European Photovoltaic Solar Energy Conference*. 2010:1912–16.

[156] Falcon T., Anderson R.D., Clark D.A. 'Fine line printing of front side metallizations using ultra fine mesh and stencil technologies'. *Proceedings of 27th European Photovoltaic Solar Energy Conference*. 2012:1739–43.

[157] Hannebauer H., Dullweber T., Falcon T., Chen X., Brendel R. 'Record low Ag paste consumption of 67.7 mg with dual print'. *Energy Procedia*. 2013;**43**:66–71.

[158] Hannebauer H., Schimanke S., Falcon T., Altermatt P.P., Dullweber T. 'Optimized stencil print for low Ag paste consumption and high conversion efficiencies'. *Energy Procedia*. 2015;**67**(8):108–15.

[159] Hannebauer H. 'Optimierung der Vorderseitenmetallisierung und des Emitters für hocheffiziente industrielle Silizium Solarzellen'. *Ph.D. thesis, Leibniz university Hannover*. 2016.

Chapter 7
Alternative printing technologies

Maximilian Pospischil[1], *Andreas Lorenz*[2], *Saskia Kühnhold-Pospischil*[2], *and Adrian Adrian*[3]

7.1 Parallel dispensing

7.1.1 Introduction: working principle of dispensing

Dispensing offers a single-step, contactless printing process, where a functional medium is forced through a small nozzle opening which is moved relatively towards the substrate forming either dots or continuous lines. Although some applications use jetting valves, the typical approach incorporates some kind of pump or valve initiating a continuous flow or a precisely controlled deposition of dots. Similar to inkjet printing or aerosol printing, dispensing belongs to direct writing technologies which means patterns can be designed electronically without processing of a printing form [1]. Opposite to most printing technologies, the substrate only gets in contact with the functional medium itself which enables the use of fragile substrates. Furthermore, an inert handling of the medium inside the system allows for a wide range of industrial applications including deposition of solder paste [2–5], fillings [6], adhesives [7–9], sealings [10–12] and other functional media [13–15], especially in surface mount technology (SMT) and microelectromechanical systems (MEMS) but also in pharma and medicine technology.

Its application in Si PV front-side metallization provides homogeneous high aspect ratio (AR) contact lines at finger widths down to 17 μm and hence could be a future alternative to screen-printing once industrial feasible equipment and processes are available.

7.1.2 Dispensing lines on solar cells

7.1.2.1 Historical outline of research activities in Si-PV

Dispensing metal grid lines on solar cells was patented by Hanoka in 1992 [16] and led to a first industrial application at former Schott Solar AG in Alzenau, Germany in the production of highly fragile EFG (Edge-defined Film-fed Growth) solar cells.

[1]Fraunhofer-Institut für Solare Energiesysteme ISE (until 4/21), HighLine Technology GmbH, Germany
[2]Fraunhofer-Institut für Solare Energiesysteme ISE, Germany
[3]International Solar Energy Research Center (ISC) Konstanz e.V., Germany

However, due to fast-emerging screen-printing technology at that time, the approach was terminated when switching to other, more common solar wafer technologies. At Fraunhofer ISE, Daniel Biro initiated research activities regarding dispensing in his group in the year 2009 with the first dispensed silver grid demonstrated by Specht *et al.* [17] and an application of etching pastes by Spribille *et al.* [18], both presented at the EUPVSEC in 2010. A few months earlier, Chen *et al.* [19] had already showed some dispensed grid lines at the IEEE PVSC conference followed by two more publications in the year 2011 and 2012 [20, 21]. Meanwhile, record cell efficiencies of 20.6% on 125×125 mm² FZ p-type material using dispensing technology on MWT-PERC (metal wrap through – passivated emitter and rear cell) solar cells, featuring a selective emitter structure were presented by Lohmüller *et al.* [22] and clearly demonstrated the potential of the technology. Ever since, the focus was set on the development of an industrial feasible dispensing process [23]. For this reason, the number of nozzles operated in parallel was scaled from 10 [24] to 50 [25] and finally the ability to freely adopt nozzle diameter and pitch over the whole width of a 6″ solar cell. These developments were enabled by precise rheological paste characterization that was then transformed to a CFD model and finally allowed for simulation of respective print head and nozzle geometries [26].

With respect to the solar cell application, homogeneous, fine-line printed front contacts allowed for efficiency increases of around $\Delta\eta/\eta = 1\%$ compared to respective screen-printed reference over the years and the absence of mesh marks reduced the silver (Ag) laydown by 20% without reducing grid conductivity with commercially available screen-printing pastes [27].

Last but not least, high printing speeds are key to ensuring future throughput rates in PV production and a successful demonstration of an intermittent operating parallel print head has been demonstrated in 2017 [28]. Since then, potential solar cell applications were expanded beyond the traditional route of high temperature front-side metallization on aluminium back surface field (Al-BSF) and PERC solar cells. In 2019, a growing market share of bifacial cell concepts and a more mature production of heterojunction (HJT) solar cells, dispensing of Al pastes for bifacial PERC solar cells as well as applications on n-PERT and HJT were demonstrated [29]. The latter involves rheologically challenging, polymer-based low-temperature pastes and has been focused since then due to its variety of applications regarding possible future cell concepts including HJT, Perovskite and even thin film PV, where a metallization of CIGS mini-modules was demonstrated recently by Gensowski *et al.* [30], who was also experimenting with etching pastes for dispensing applications in a previous study [31].

7.1.2.2 Previous industrial approaches to increase throughput rates

An increase of throughput rates is crucial in order to enable industrial production cycles comparable to commercial screen-printing metallization lines. At former Schott Solar AG, continuously operating parallel dispensing units [16] were in use for many years in their production lines . A customized print head containing an array of exchangeable single nozzles was applied for contact fingers and excess paste was re-collected and then applied for both dispensed busbars. During inline dispensing, squared solar wafers were closely driven together in order to allow for a continuous paste extrusion of grid lines from one wafer to the next. Controlled line interruption was then achieved by separating two adjacent wafers again. Since alignment precision was not yet challenging at that time, a pure inline belt transport with mechanical alignment along the wafer edges could be realized including processing on the fly. Interestingly, a special approach was patented by von Campe and Schwirtlich that prevented spreading of high AR contact fingers on the wafer by subsequently spraying a tenside foam that extracts solvents from freshly dispensed contact fingers [32].

A first intermittent grid line extrusion was achieved by Chen *et al.* [19]. They developed a parallel print head applying several n-scrypt smart pumps, each feeding up to three parallel aligned single nozzles. However, the approach was not scaled to a larger number of nozzles and activities were focused more into other challenging precise micro-dispensing applications rather than large-scale PV production.

The co-extrusion approach as patented by Fork *et al.* [33] and applied for solar cell metallization by Beutel *et al.* [34] and Richter *et al.* [35] at former Solar World Innovations GmbH included a parallel print head [36] that enables ultra-fine contact lines of down to $w_f = 23.7$ μm by using a sacrificial medium surrounding and focusing the Ag paste inside its 200–220 μm wide nozzle openings. Here, a production tool was developed employing two print heads in a dual printing lane and reaching a rated throughput of 2 700 wafers/hour. Although, more than 100 000 solar cells have been printed with this tool and even an intermittent printing of pseudo square wafers has been demonstrated [34], several aspects prevented a broad application: due to the laminar sheath flow inside the nozzle, a special Ag paste with adapted rheology had to be developed and maintained. Furthermore, an active distance control between the print head and the wafer limits printing velocity and the grid design is fixed to the configuration of the print head that contains channels and valves for two media at each nozzle. Due to the challenge of keeping up with fast-evolving screen-printing pastes at that time, many research activities were closed down around the year 2014. Finally, one of the relatively mature industrial printing devices (Pelican) was successfully taken over by Fraunhofer ISE in the year 2018 and transformed to serve as a development platform for its conventional dispensing approach with 6″ print heads in the PV-TEC.

7.1.2.3 Parallel dispensing at Fraunhofer ISE and HighLine Technology GmbH

Within the funded industrial research project GECKO, a demonstration of parallel dispensing using a ten-nozzle print head was achieved [24] reaching line widths of 27 μm with Heraeus pastes and successfully applied for solar cell processing [24, 37]. Here, the focus was to achieve a homogeneous distribution to a variable number of dispensing nozzles by using a single paste supply. The main impact for this achievement was an in-depth study of paste rheology including wall slip properties and its transformation to a CFD simulation approach that allowed for an

Figure 7.1 *Intermittent dispensed solar wafer and parallel print head mounted into an industrial screen printer of type XS2 via the cuckoo setup. This process runs with up to v_{disp} = 500 mm/s with low temperature pastes (Source: HighLine-Technology.com).*

optimization of both, print head and nozzle geometries [25, 26]. After this initial project was finished in 2014, knowledge was transferred to a new developed 6″ print head that enabled metallization at 650 mm/s using commercially available pastes, more than three times the printing speed of screen printers at that time. However, with decreasing research funding and the main issue, an intermittent process, not being solved yet, activities were reduced to samplings with the existing equipment. Nevertheless, the ability to apply commercially available silver pastes allowed for remarkable results on PERC solar cells, where screen-printed reference groups were constantly outreached with different pastes [28]. A record geometry of 17 μm in width and a paste laydown of just 35 mg for 150 parallel dispensed grid lines was demonstrated in 2018.

With the disastrous lab fire in PV-TEC in early 2017, the focus was completely set on the development of intermittent parallel print heads that are key for an industrial application of the approach (Figure 7.1). Thanks to its CFD model, the print heads were already designed for an automated priming procedure (i.e. filling up the print head) without the inclusion of air that could be reused for an in situ start and stop of all parallel aligned nozzles with the freedom to customize the arrangement of nozzles within minutes regarding pitch and diameter by exchanging its bottom part, the nozzle-KIT.

In order to finally make the move out of the institute and into the market, it was further decided to launch a start-up to provide dispensing solutions for PV metallization. Since April 2020, Freiburg based HighLine Technology GmbH provides continuous hardware and process development as well as machine integration in close

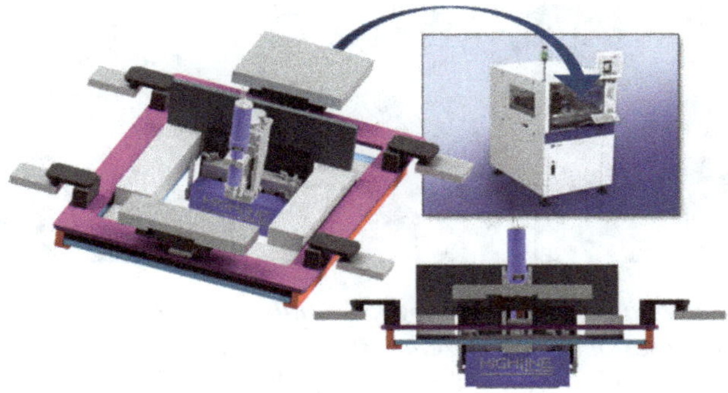

Figure 7.2 *HighLine's cuckoo: An industrial screen printing platform of type XS2 was equipped with a two-axis dispensing stage in exchange for screen and squeegee unit [38]*

cooperation with Fraunhofer ISE. The latest state of industrialization, as presented at the metallization workshop 2020 contains an integration Kit in commercial screen printers like EKRA's XS2 (see Figure 7.2). Here, the squeegee set-up is removed making space for an independent y–z axis system controlling the dispensing print head. Cell handling as well as screen alignment system remains original and in the simplest configuration, communication includes I/O handshakes and a web interface in order to control parallel dispensing units in this industrial approved printing platform [38]. In the meantime, this set-up (the 'cuckoo') was integrated in a screen-printing line at Fraunhofer ISE's PV-TEC.

7.1.3 Improving dispensed contact geometries

7.1.3.1 Influence of paste rheology during dispensing

Due to the non-contact approach (i.e. nozzle and substrate do not touch), dispensing offers the possibility to adapt paste rheology in a much wider range than other thick film printing technologies. The shape of resulting contact fingers (Figure 7.3) on the wafer can be adjusted with respect to an optimum trade-off between mechanical robustness, shading losses and electrical contacting behaviour that often had not been sufficient for these small structures when screen-printed contacts were still far beyond concerning their finger width. The most crucial parameter for the resulting finger AR turned out to be the yield stress τ_y of the paste, hence the stress at the transition between a more liquid state during shear flow and a rather solid appearance, e.g. after deposition on the wafer. This in combination with a minor influence of thixotropic paste properties and wall slip inside the nozzles typically determines the shape of the resulting contact [39, 40].

However, with contact widths decreasing far below 50 μm, surface effects turned out to strongly influence the shape additionally to the yield stress. When this effect was finally accessible quantitatively by the pendant thread approach, a very interesting conclusion could be drawn for dispensing pastes (Figure 7.4):

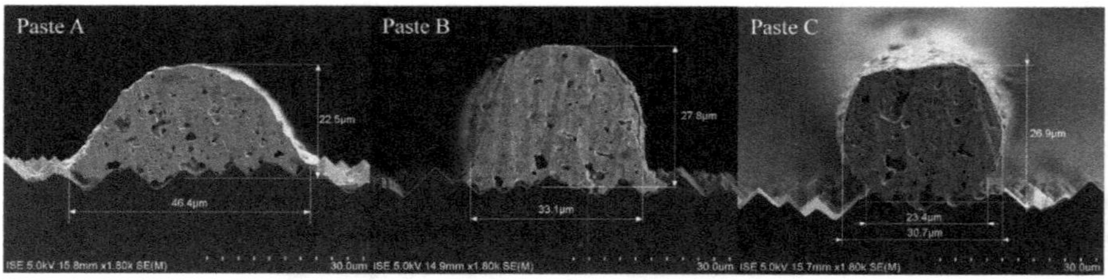

Figure 7.3 *Adaption of paste rheology: Scanning electron microscope (SEM) pictures of three different dispensing pastes (A–C) on texturized Czochralski-grown (Cz) material. With increasing yield stresses τ_y from A to C, the optical aspect ratio (AR_o) increases while the optical finger width w_o decreases from 46 µm at paste A down to 31 µm at paste C. The phenomenon of paste necking becomes relevant at paste C. Here, contact width w_c is even smaller than the optical finger width w_o. All dispensed contact fingers were processed using a conical dispensing nozzle with an outlet diameter of D = 50 µm [39].*

the electrical AR $AR_{el} = A_f/w_f^2$ linearly correlates with the yielding capillary number $Ca_y = 1/2 \cdot \tau_y \cdot D \cdot \Gamma$.

The extrapolated offset value directly influenced by the material properties of the textured wafer surface – hence contact geometries on mc wafers always showed lower ARs than those on Czochralski-grown (Cz) wafer and a decreasing nozzle diameter D requires an increase of either yield stress τ_y or a decrease of surface tension Γ [23].

The ideal AR is always a trade-off between performance of cells and modules on the one hand and contacting behaviour and Ag cost on the other. The effect of reduction of shading however is not just a function of finger width but also influenced by the AR, since a steeper contact shape enhances internal reflection of the silver grid. Expressing this effect by means of an effective width EW factor shows that the actual shaded area of the contact could drop to around 70% for high AR dispensed contacts on cell level and to around 45% on module level where internal reflection with the module glass contributes another share (Figure 7.5).

Today, Ag paste development in screen-printing technology addresses fine-line printing in order to save Ag costs. Effects like wall slip and yield stress were found to also matter there and hence, characteristics of printing pastes resemble more or less those applied for dispensing. Differences may occur in the vehicle that has to prevent agglomeration of particles at elevated pressures and the amount of glass frit that has to suit the desired AR in order to ensure mature contact formation. This brings a huge advantage, since new results and findings from screen-printing paste development can be directly transferred to dispensing technology. In a first step, new paste systems are then evaluated regarding their applicability for dispensing. Here, very homogeneous finger geometries are obtained but typically with medium

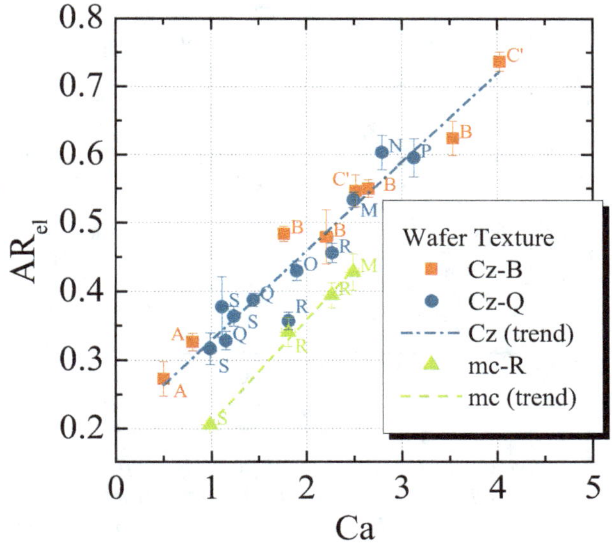

Figure 7.4 *Electrical aspect ratio depending on yielding capillary number for Ag printing pastes, investigated between 2010 and 2016 at Fraunhofer ISE. The influence of the surface texture is clearly visible in the offset value as well as the difference between Cz and mc textured surfaces [23].*

ARs. Once, the evaluation of a novel paste has been successful, paste rheology has to be adapted towards more beneficial finger geometries with high ARs and substantially reduced line widths [39].

7.1.3.2 Other approaches

Around the year 2015, a Korean research group came up with the idea of enhancing dispensing results by the implementation of electric fields that should increase the AR. Interestingly, they were able to also enhance contact resistivity by adding 2% wt. of fluorosurfactants and applying an electric voltage of 0.7 kV at a distance of only 100 μm to their eNano Jet from EnJet Inc. However, a speed of only 7 mm/s at a nozzle diameter of 40 μm is not even close to industrial feasibility [41]. Although small line interruptions typically do not have an impact on cell results, as Lossen *et al.* presented on the 8th Metallization & Interconnection Workshop in 2019 [42], they are, when apparent, more visible than, e.g. for screen-printed structures. Hence, an approach to prevent these interruptions would be to improve elongational characteristics of dispensing pastes, an approach that was actively investigated by Kunpai *et al.* [43] in the year 2017. The addition of graphite nano-fibers to existing pastes formulations, the dispense gap as well as the finger width could be reduced

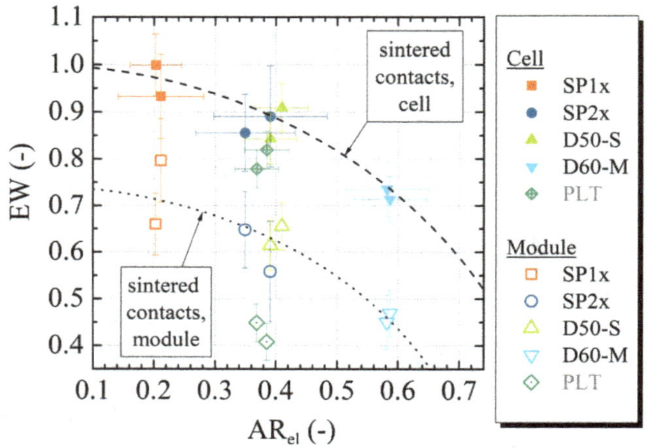

Figure 7.5 Effective contact width (EW) depending on the electrical aspect ratio (AR_{el}) of screen printed and dispensed contacts at cell and module levels. Improved light reflection at steeper finger edges significantly improves contact optics of high AR contact lines. Plated contacts showed even lower values for EW due to their higher surface reflection [23].

drastically. From a rheological point of view, this approach certainly increased the yield stress that can be measured in shear and extensional flow.

7.1.4 Dispensing applications in PV

7.1.4.1 Overview

Dispensing has many applications in industry; parallel fine-line dispensing, however, was specially developed to enhance the front-side grid of Si solar cells. Increasing the AR and hence reducing shading losses without making a trade-off regarding grid conductivity was the main motivation in the past. In these days, other advantages like preventing mesh marks and thus enabling a further reduction of Ag consumption plays a more important role similarly to the prospective of further increasing line throughputs by high speed dispensing as well as cutting down screen cost that certainly increases with finer mesh and larger substrate sizes. All these advantages also apply for the rear side of most future cell concepts including n-PERT, TopCon, hetero junction or possibly perovskite- tandem devices. For bifacial PERC cells (with Al rear side) as well as busbar printing and during module interconnection, the dispensing approach could further enhance and accelerate production lines by potentially substituting all screen-printing steps and enabling new designs of cell transport, etc. A first demonstration of such an approach was given in [29] where an all dispensed bifacial mini module has been produced, containing six consecutive dispensing steps, i.e. front-side and rear-side contact fingers on cell level (busbarless grid design) followed by front and rear busbar edge definition and both side dispensed ECA for interconnection.

7.1.4.2 Dispensing the front-side grid of PERC solar cells

PERC has become the mainstream concept in Si-PV with more than 150 GW installed by now. Regarding parallel dispensing, many successful demonstrations on R&D level have been published in the past. Typically, industrially pre-processed precursors including locally opened rear-side contacts as well as screen-printed Al rear surfaces were divided into several groups comparing screen-printed and dispensed front-side patterns. In [27], the application of two standard screen-printing pastes from different manufacturers was demonstrated leading to an efficiency increase of $\Delta\eta = +0.3\%$ abs. of the dispensed groups with finger widths of $w_f = 27$ µm compared with the screen-printed references. At the same time, the resistivity of the dispensed grid reached the same level as the screen-printed reference although reducing the Ag paste laydown by $\frac{\Delta m}{m} = -20\%$ simply due to the absence of meshmarks.

The last demonstration was conducted in 2019, on a six-busbar layout [29]. Here, four groups with dispensed solar cells of different grid parameters and a screen-printed reference group were processed. For the dispensed solar cells, a fine-line applicable Ag paste was used in order to compensate substantially different contact geometries compared with the screen-printed reference. The grid parameters varied regarding the respective nozzle diameter D and a suitable number of contact fingers N_f according to previously conducted grid simulations [39].

As visible in Figure 7.6 , dispensed solar cells with nozzle diameters of D = 20 µm and D = 25 µm clearly outperformed the screen-printed reference, reaching a top efficiency of $\eta = 21.9\%$. In a typical dispensing experiment, one would expect the increase coming from substantial gains in the cell current due to reduced shading losses. In this case, however, the main reason was found to be the contacting behaviour of this specially designed paste as visible from record Fill Factors(FFs) obtained for those small contact fingers. A specific contact resistivity of ρ_c=2.3 mΩ cm² was reached for the group dispensed with D = 20 µm openings and corresponding finger widths of w_f = 17 µm (see Figure 7.7.). A complete FF analysis (given in Table 7.1) also demonstrates the influence of the quasi ideal homogeneous contact shape that allows for lowest grid resistances R_{Grid} although applying paste laydowns of only $\Delta m_{Ag,wet}$ = 35 mg when dispensing with D = 20 µm nozzle openings.

7.1.4.3 Dispensing of AgAl and Al-FT Pastes on n-PERT and BiPERC

One could think that a perfectly suited paste for the front of a bifacial n-PERT solar cell, such as BiSoN [44], or the rear of a bifacial PERC cell would be a fire-through Al (Al-FT) paste, as introduced by Navarrete-Astorga *et al.* [45]. These pastes can open the passivation layer during fast firing and further form a local highly doped region along their AlSi alloy. As demonstrated by Fellmeth on the rear side of bifacial PERC solar cells [46], the challenge however lies in the prevention of

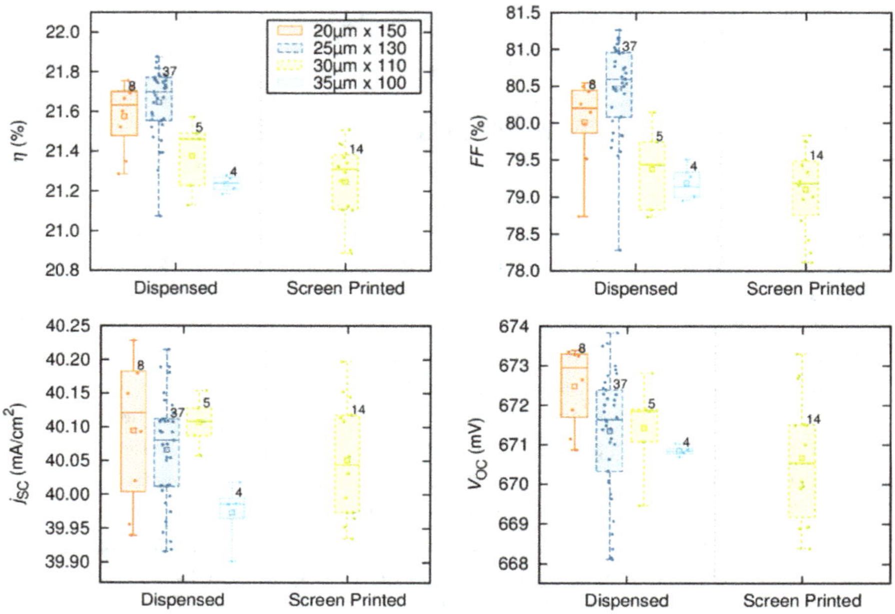

Figure 7.6 Solar cell results on industrial passivated emitter and rear cell (PERC) precursors. Dispensed samplings with four different nozzle diameters were compared with screen printed reference samplings at best furnace settings, respectively. The nozzle (screen) opening as well as the respective number of contact fingers N_f is indicated [29].

parasitic etching at the edges of printed contact lines where the amount of Al is too little to form a decent BSF. Hence, the laser contact opening (LCO) in combination with a non-Al-FT paste is preferred since higher open circuit voltages are reached. Furthermore, the line resistivity of Al-FT pastes is a factor of 20 higher than that of conventional silver pastes, hence an application with dispensed high AR contacts not only provides steeper edges but also sufficient conducting material. All of these challenges also apply for n-PERT solar cells, although the process here competes with a fire-through AgAl paste and small line width additionally should be reached in order to limit shading losses. A trial batch [29] demonstrated the applicability of AgAl and Al-FT pastes by dispensing on n-PERT solar cells. Although dispensed AgAl contacts slightly outperformed the screen-printed contacts, the Al-FT paste in combination with $D = 60$ µm nozzle openings significantly increased shading losses and still could not ensure sufficient grid conductivity.

Nevertheless, the challenge of aligning screen and laser patterns to each other on the rear side of PERC solar cells as well as an improved bifaciality by high AR contacts remain as motivation. Ourinson *et al.* [47] hence further improved the process sequence of BiPERC solar cells with Al-FT pastes on the rear by using multilayer parallel dispensing.

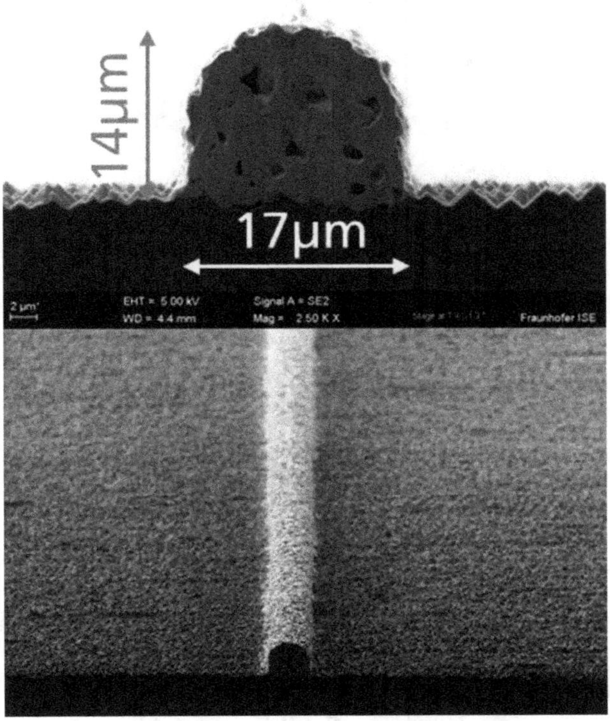

Figure 7.7 *Scanning electron microscope (SEM) images of dispensed contact fingers on a passivated emitter and rear cell (PERC) solar cell. These contact geometries were obtained with nozzle openings of D = 20 µm [29].*

A record AR of 1.76 was reached, with a four-layer wet-in-wet process in the same machine and L-BSF formation and grid conductivity was significantly improved. When fired separately and hence using a plated front side, a two-layer dispensed grid reached a similar performance as a conventional LCO screen-printed reference process. However, in a co-firing approach, it turned out that standard front-side pastes tend to overfire with the applied firing profile for Al-FT pastes that require an increased thermal budget to penetrate the passivation and then form the BSF. Finally, only steep finger edges could ever enable this process sequence, hence a further consideration is not expected before dispensing has successfully entered the market.

Table 7.1 *Fill factor analysis, mean values, sorted by technology and nozzle/screen opening*

Number of contact fingers N_f and Printing Technology	**Opening diameter D(µm)**	**Fill factor FF (%)**	**Series resistance R_s(Ω cm²)**	**Grid res. R_{Finger} (Ω/m)**	**Specific contact resistivity ρ_c (mΩ cm²)**
150 contacts, dispensed	20	80.2	0.56	103	2.3
130 contacts, dispensed	25	80.6	0.45	94.5	2.4
110 contacts, dispensed	30	79.4	0.70	66.5	3.8
100 contacts, dispensed	35	79.1	0.75	46.6	11.2
110 contacts, screen printed	30	79.1	0.66	108	7.4

7.1.4.4 Dispensing low temperature pastes

The last part of this subchapter is dedicated to dispensing of low temperature curing pastes, known for their temperature sensitivity that make them ideal for temperature-controlled closed system applications like dispensing. These pastes can be described as highly filled ECAs and hence contain a larger polymer content with substantially different flow behaviour compared to high temperature pastes. Applicable for both sides on silicon heterojunction (SHJ) cells and possibly on tandem perovskite concepts their future market share should definitely increase. Even CIGS thin-film PV recently considers adding metallized front-side patterns to their modules in order to reduce its relatively large cell-to-module losses [30].

Although Fraunhofer ISE's spin-off HighLine Technology mainly applies these pastes, until today, all published results on solar cells were achieved using the smaller GECKO print head. Similar to the standard PERC approach, a high Ag reduction and throughput potential as well as the aim to reach finest contact lines applies along with a further advantage due to the proven concept of busbar-less cell processing with similar pastes on the front and rear sides.

Despite the fact that fine-line dispensing with conventional high temperature Ag sinter pastes already provides remarkable contact geometries (finger width of 17 μm at 14 μm height), similar results were not yet reached with low temperature curing pastes as necessary on SHJ solar cells [48]. In [29], a first successful demonstration of silver and silver coated copper pastes was achieved, two investigated pastes were suitable for application of a nozzle diameter of $D = 40$ μm and one allowed for stable dispensing with $D = 35$ μm. In the mean time, the minimum nozzle diameter had already reached $D = 25$ μm and mass laydowns reached 0.31 mg per contact finger, as recently described by Gensowski *et al.* [38].

Furthermore, the so-called die swell phenomenon leads to an increase of the thread diameter after exit of the nozzle since polymers tend to regain their initial structure after being forced through small openings. Consequently, the resulting contact width is not yet on the same level as with high temperature pastes but already decreased from of $w_f = 48$ μm in [29] to 34 μm in [38] on cell level. When comparing this with a screen-printed contact with the same paste, the difference in homogeneity between the two printing methods becomes clear (Figure 7.8.).

In a following experiment, Gensowski *et al.* [38] compared three different low-temperature paste systems on industrial SHJ precursor material using dispensed and screen-printed front side contacts, respectively. With finger widths in the range of $w_{shading} = (34 \pm 2)$ μm for a rheological adapted paste , a clear difference between the groups advantage of the dispensed patterns was visible. The screen-printed groups were printed with a screen opening of $w_s = 40$μm and obtained $w_{shading} = (65 \pm 2)$ μm. For two pastes, a clear improvement of cell efficiency of $\Delta\eta = +0.3\%$ abs. compared to the reference groups was demonstrated, reaching top values of $\eta_{max} = 22.5\%$, see Figure 7.9. The main reason for this were substantially reduced shading losses of the dispensed samples even though the grid design was taken from previous PERC experiments and the number of contact lines N_f was increased by 10% and 30% compared to the screen printed references.

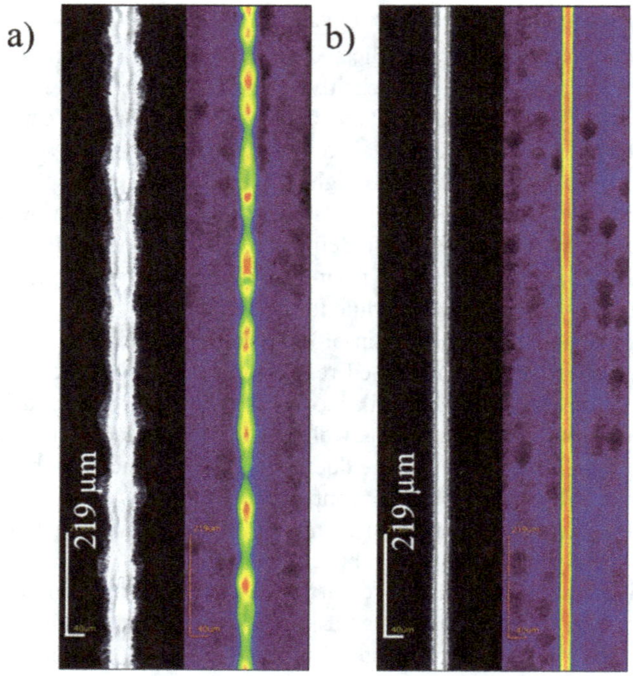

Figure 7.8 *Comparison of screen-printed (left) and dispensed (right) contact line obtained from laser scanning confocal microscopy. With typical meshmarks in the left illustration (a), a clear improvement of homogeneity of the dispensed grid line on the right (b) is visible [38].*

Furthermore, a specific paste laydown of only m_{Ag} = 0.30 mg per contact finger was achieved with this 130 × 25 μm nozzle setup and thus demonstrated its applicability for modern multi-busbar concepts or even busbar-less interconnection.

Finally, parallel dispensing is applicable to different solar cell concepts and its advantages can be beneficial in many ways. Its contact homogeneity allows for a significant reduction of paste laydown without loosing grid conductivity. The throughput potential enables even faster production concepts for the future and with the cuckoo, a system based on an industrial proven printing platform has been demonstrated.

7.2 Rotary printing

7.2.1 Classification of rotary printing methods

Among the different printing and coating technologies, rotary printing methods are the most common technologies for large-volume production of all kinds of printing applications. Rotary printing can be applied for many kinds of printed matter from

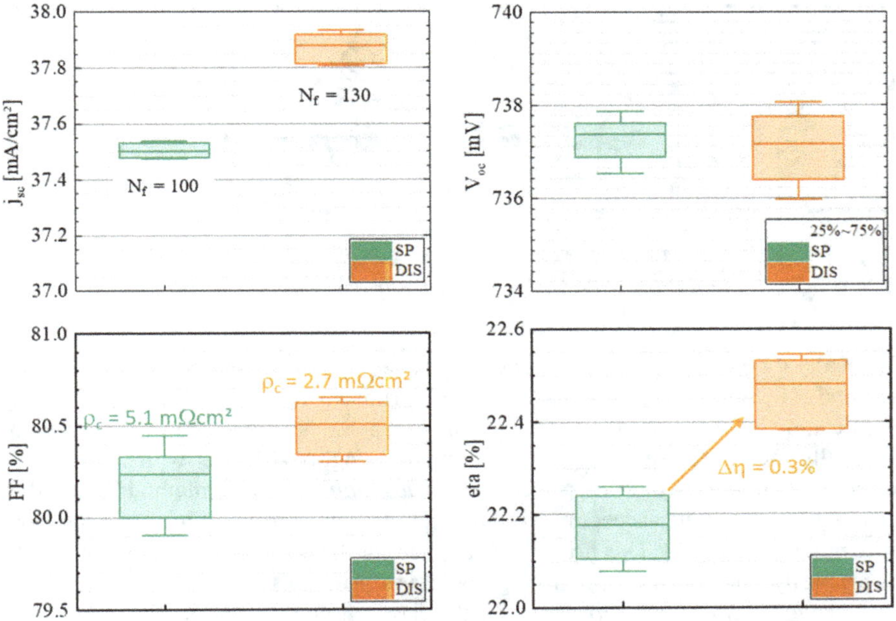

Figure 7.9 *IV results of a benchmark dispensing experiment on industrial heterojunction (SHJ) precursor material. All solar cells were tested by GridTOUCH setup. All rear-side metal grids contained N_f = 100 contact fingers and were screen-printed. The front contact metallization contained N_f = 100 contact fingers, printed with a screen opening of w_s = 40 µm, the dispensed group used parallel arranged nozzles with a diameter D = 25 µm @ N_f = 130 fingers [38].*

large-volume commercial and package printing to high-precision printing and coating steps for the production of technical products like sensors, batteries, fuel cells or organic solar cells. The large majority of rotary printing machines is designed for the production of web-based materials like foil, paper or composite materials using a *roll-to-roll* machine concept. However, printing and coating steps are also performed on non-bendable, single-item substrates and components. Printing processes for such applications require non-impact printing processes (i.e. inkjet, dispensing), *flat-to-flat* printing processes (i.e. flatbed screen-printing or stencil-printing) or *roll-to-flat* printing processes (rotary printing), see Figure 7.10. The metallization step for silicon solar cells represents such a highly sophisticated mass-production printing process on single-item components.

Besides the application concept, printing methods can be distinguished regarding their printing principle. Focusing on form-bound printing methods, four main printing principles can be differentiated (Figure 7.11). *Relief printing methods* are specified by a hard or flexible printing form with an elevated printing layout which

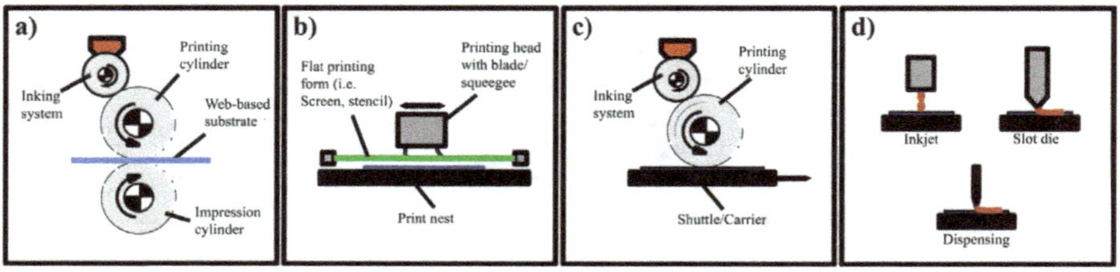

Figure 7.10 Schematic drawing of the application concepts roll-to-roll (a), flat-to-flat (b), roll-to-flat (c) and non-impact printing (d)

is usually directly printed on the substrate. Today, the most common relief printing method is flexographic printing [49] which is primarily applied for package printing applications (see section 0). *Gravure printing* methods (see section 0) work with a usually metal-based printing cylinder with an engraved layout which is printed directly or indirectly onto the substrate. *Planographic printing* methods use a flat printing form with ink-accepting and ink-repelling areas realized by particular surface modification [50]. This principle is applied for the *offset printing* method which is primarily applied in the field of graphic arts printing [50]. Methods following the *through-printing principle* transfer the paste through a flat or cylindric mesh which is partly covered by an emulsion layer. The most common through-printing method is flatbed screen printing (FSP) which is also known as *silk screen printing*. A related printing method is *stencil-printing* which uses a partly open metal foil instead of a mesh-based screen.

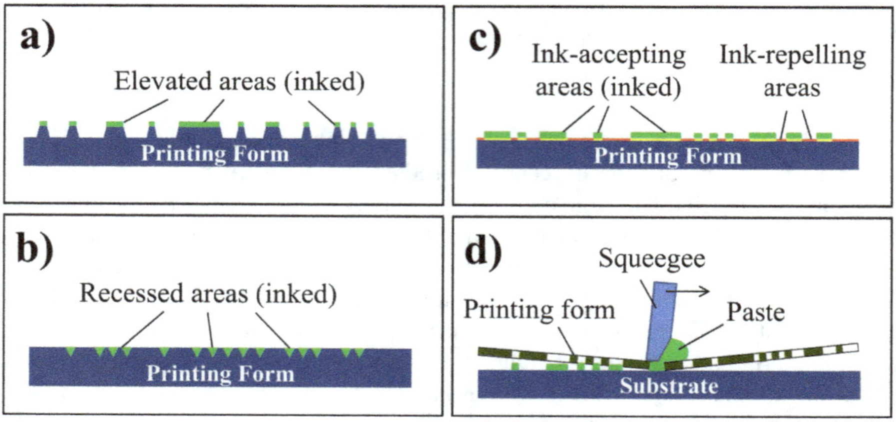

Figure 7.11 Schematic drawing of four major ink/paste transfer principles with focus on form-bound printing methods: relief printing principle (i.e. flexographic printing) (a), gravure printing principle (b), planographic printing principle (i.e. offset printing) (c) and through-printing principle (i.e. screen printing) (d)

7.2.2 Rotary screen printing

7.2.2.1 Historical and technical background

FSP is the commonly known method of choice for the application of thick-film printing structures. Due to the flexibility of this printing method, the ability to transfer all kinds of printable paste-like fluids and the relatively easy handling, FSP has established itself as a standard process in many kinds of graphic arts and technical applications. However, it is rather unknown that the opportunities and benefits of the through-printing principle for the application of thick films can be combined with the throughput potential of a rotary printing method. This is the case for rotary screen printing (RSP), a printing method which is rather exotic and unknown except for specific applications. Primary fields of application for RSP are label and textile printing, usually on roll-to-roll machines using web-based materials such as foil, paper, textile fabrics and cardboard [50]. High-speed printing machines for web-based materials (e.g. label printing machines) enable printing velocities of 160 m/min. (2.7 m/s) and more today [51]. RSP offers high flexibility to transfer various kinds of printable pastes including pastes for graphic arts purposes, varnishes, special effects and all kinds of functional coatings and technical applications. Most paste-like fluids can be printed using RSP, provided that the required rheological properties can be adjusted. This offers a great potential to apply this highly productive printing method in many fields of printing applications. Beside typical application fields like graphic arts, label and textile printing, RSP can also be applied for technical printing applications like flexible organic solar cells [52, 53], RFID antennas [54] or conductive circuits [55]. However, while FSP can be applied with relatively simple manual or semi-automatic machines according to the flat-to-flat concept, RSP requires roll-to-roll or roll-to-flat printing machines which are considerably more complex, large and cost-intensive. Thus, RSP is less attractive for small-scale laboratory applications but rather attractive for large-scale mass production processes with high production volumes.

7.2.2.2 Working principle and influence factors

The working principle of a rotary screen-printing unit for solar cell metallization is shown in Figure 7.12. Like FSP, RSP is a thick-film printing method according to the through-printing principle. The printing form can be a mesh-based screen which is partly covered with an emulsion layer or a stencil-like, electroformed metal foil. While FSP works with a flat printing form, a cylindric screen or stencil-like cylinder is used for RSP. During the printing process, the cylindric printing form rotates while the paste is transferred through the open areas using a squeegee system which is located inside the cylinder. While the printing principle of FSP and RSP are similar, both methods have substantial differences regarding the properties of the printing form, paste rheology requirements, printing sequence and printing result. Thus, RSP should be considered as an independent printing method rather than a closely related derivate of FSP. In the following section, the main components of a rotary screen-printing unit intended for the metallization of silicon solar cells are described in detail.

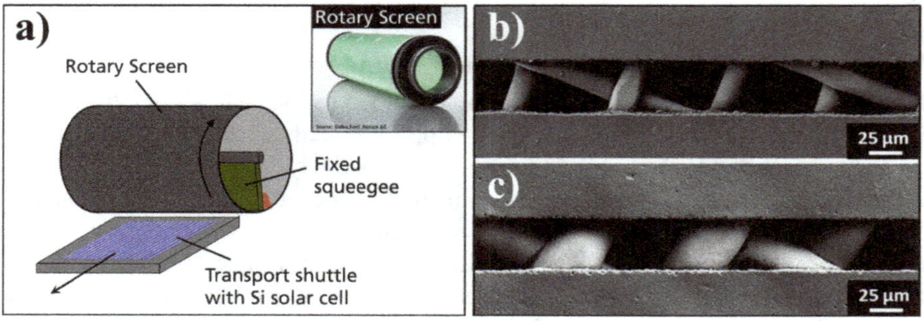

Figure 7.12 *(a) Working principle of a rotary screen printing unit for solar cell metallization: The paste is transferred onto the silicon wafer by a fixed squeegee inside the cylindric, rotating screen. Due to stability reasons, rotary screens require considerably thicker wires of the woven screen mesh compared with screens for flatbed screen printing. The comparison of a fine-line channel for contact finger printing illustrates the substantial difference of a flatbed screen (b) compared with a rotary screen (c).*

7.2.2.3 Rotary screen/stencil cylinder

The cylindric printing form for RSP is either based on a woven mesh of fine wires usually made of stainless steel (in the following referred to as *rotary screen*) or electroformed, structured nickel foil cylinders with a hexagonal honeycomb structure (in the following referred to as *rotary stencil*). In this chapter, rotary screens are primarily considered as this type of printing form is – according to the current state of science – particularly well suited for the metallization of silicon solar cells.

Similar to flatbed screen, the rotary screen is based on a regular, woven mesh of fine wires usually made of stainless steel. The main difference of a rotary screen compared to a flatbed screen is the substantially greater thickness of the wires. Flatbed screens are fixed to the frame with relatively high tension which provides a good stability during the printing process even when using fine meshes with small wire thickness. Thus, flatbed screens with ultra-fine meshes up to a mesh count of 900 and more wires/inch and a wire thickness down to 9 µm and less are applicable for specific applications. Rotary screens – in contrast – require a considerably higher stability of the mesh, as the cylindric screen is stretched laterally in the printing unit but with a significantly lower tension. Thus, considerably thicker wires of the screen mesh in the range of 30–60 µm are required to enable a sufficient stability of the rotary screen during the printing process. The strongly increased wire thickness substantially reduces the percentage open screen surface and thus has a considerable impact on the paste transfer capacity through small channel openings. Figure 7.12 shows a fine-line contact finger channel opening of a rotary screen with a mesh count of 400 wires/inch, Figure 7.12 a fine-line channel opening within a fine mesh flatbed screen. The considerably lower open screen surface of the rotary screen mesh strongly affects the amount of paste which can be transferred through the screen openings. The thick wires also affect the uniformity of the printed patterns, leading to potentially stronger local deviations of the line height and width (mesh marks). However, the necessity to use metallic pastes with lower viscosity compared to FSP metallization pastes counteracts extensive mesh marking by the mesh wires, as the paste flow after the printing process levels such structures to a certain extend.

The stainless steel mesh of rotary screens can be additionally covered with a thin layer of electroplated nickel to increase the stability and enhance the printing properties [56]. The mesh is coated with an UV-reactive photopolymer emulsion layer which defines the printing layout with open (printing) and covered (non-printing) areas. The screen fabrication process of a ready-to-print rotary screen can either be realized by the conventional, analogue method using UV-exposure with a positive film or a digital process using computer-to-screen (CtS) UV laser process [57]. The screen feedstock (reel-based screen mesh with pre-coated, unexposed emulsion) is provided in different variants by material suppliers. Using the conventional screen fabrication process, a sheet of the screen feedstock is cut from the reel web to the required size and punched with register holes. The flat screen sample and a positive film with the desired printing layout are precisely positioned on the exposure table and subsequently exposed by UV radiation under high vacuum followed by a wash-out step with water. The exposed and developed screen sample is briefly dried using hot air. In a following step, the flat screen sheet is mounted to a cylindric form and laterally attached to stencil rings by gluing (Figure 7.12). Using the digital process, the unexposed screen sheet is directly mounted to a CtS device and exposed by a UV laser beam. Subsequently, the exposed sheet is washed out, dried and mounted to the stencil rings. A third method is the selective laser-based ablation of a so-called laser ablation mask system (LAMS) layer which substitutes the film by a laser-ablative mask during the subsequent UV exposure process. The following processing steps are similar to the described process routes.

The choice of the optimal rotary screen depends on the desired printing pattern and resolution (coarse or fine structures) and the required wet layer thickness. Different configurations of meshes and emulsion exist to cover a wide range of possible applications. Solid-area patterns with high layer thickness are typically printed with 'coarse' screens using low mesh counts down to 64 wires/inch and less. Such screens have a larger percentage of open screen surface and can thus transfer more paste per unit area. Depending on the paste properties, a wet layer thickness of more than 250 µm can be obtained using such type of screens. Printing fine patterns requires fine mesh screens up to 400 wires/inch. According to manufacturer specifications, the maximum resolution of such screens is specified down to a line width of w_{min} = 100 µm [58, 59]. Yet, substantially finer lines with good quality have already been demonstrated for solar cell front-side metallization [60]. While fine mesh screens provide a good resolution, the amount of transferable paste per unit area is limited due to considerably lower open screen surface down to ~15% [58]. This limits the ability to obtain a high layer thickness.

7.2.2.4 Squeegee system

The paste inside the rotating cylinder screen is transferred by the squeegee system through the open areas of the screen (Figure 7.12). The squeegee system consists of a squeegee holder with the flexible squeegee bar. The properties and adjustment of the squeegee system obviously has a strong impact on the printing result [61]. The global pressure adjustment of the squeegee is usually controlled pneumatically. The squeegee itself is usually based on polyurethane (PU) with varying Shore hardness. Depending on the printing application and layout, three variations of Shore hardness are typically used: soft squeegees (55°–60° Shore A), medium squeegees (65°–70° Shore A) and hard squeegees (70°–75° Shore A). While soft squeegees are optimal to transfer a higher amount of paste on coarse patterns, hard squeegees are typically used for fine patterns like texts or technical fine-line applications. A sufficient edge sharpness of the PU squeegee is very important to obtain good printing results, as worn-out or locally damaged squeegee bars lead to inhomogeneous printing result. Thus, the edge sharpness of the squeegee has to be regularly controlled and if necessary re-grinded manually or by a specific squeegee grinding device. To mount a squeegee to the holder, the PU squeegee stripe is cut from a roll to the desired length and precisely mounted into the squeegee holder. It is very important to mount the squeegee precisely parallel to the edge of the squeegee holder to avoid an inconsistent pressure distribution in the printing zone. Furthermore, the position of the squeegee system in the direction of printing has to be finely adjusted to avoid blurring or defects along the edges of printed patterns. Further important factors are the pre-adjusted angle of the squeegee to the substrate plane and the length of the metallic support strip in the squeegee holder, which affects the rigidity of the PU squeegee.

7.2.2.5 Further influencing factors

A further important influence factor is the so-called *flow path* [62], which can be described as the available time slot to equalize local deviations in the wet film directly after printing. Typical variations include mesh marks, pinholes and edge kerfs related to the paste transfer through the screen openings. The available effective *flow path* depends on the printing velocity, the rheological properties of the paste and the (geometrical) distance between the rotary screen-printing unit and the subsequent drying unit. The intended uniformity of the printed film defines the required *flow path* to levelize the existing variations. The required flow path should be considered in the planning for an optimal machine/dryer configuration. The requirements with respect to uniformity can be quite different depending on the printing layout and the intended purpose of functionality of the printed patterns. While coarse structures like solid area usually require a longer flow path, fine structures might demand a minimal flow path in order to avoid intense spreading.

7.2.3 Application of RSP for solar cell metallization

The application of RSP for the metallization of silicon solar cells is appealing as RSP generally combines the technological advantages of through-printing methods with regard to application of thick films with the potentially high productivity of a continuous rotary printing process. In comparison to FSP, time-consuming process steps like wafer positioning, flooding and printing step can be avoided or performed inline during the movement of the wafer on a shuttle. This allows substantially higher printing velocities and offers the potential to reduce the cycle time per wafer substantially and thus increase the effective throughput of a production line. The idea to apply RSP for the metallization of silicon solar cells has been – to the authors' knowledge – first proposed by Hahne [63]. However, the usage of nickel-based rotary stencils in these early attempts revealed clear limitations with respect to fine-line front-side contacts [63] and thus prevented further activities. Since 2016, the application of RSP for the metallization of silicon solar cells using mesh-based rotary screens has gained increasing attention [64]. Various studies have been carried out focusing on the rear-side and front-side metallization of Al BSF and PERC solar cells [60, 64–66] and a demonstrator machine for an industrial inline metallization of solar cells using RSP has been developed [67]. Focusing on solar cell metallization, the following influence factors have been found to be particularly important for the printing/metallization results:

- Machine platform: transport concept, alignment, printing units, drying unit
- Paste properties: rheology, particle size
- Rotary screen specification: mesh type (mesh count, wire thickness, material), emulsion
- Rotary screen fabrication: resolution, quality, roughness, angle
- Squeegee: shore hardness, angle, edge sharpness, pressure, position
- Printing parameters: printing velocity, environmental conditions
- Substrate: free surface energy, macroscopic and microscopic topography, tolerances

7.2.3.1 Machine platform and drying equipment

An essential requirement for an industrial metallization of silicon solar cells using RSP is the availability of a sufficient machine platform which can be implemented into existing PV production lines. Such a machine has to meet several specific requirements like a fast, stable and precisely aligned wafer transport, specially designed roll-to-flat printing units and sufficient drying solutions. A demonstrator machine for this purpose has been recently developed and demonstrated by a consortium of industry and research institutes [67] (Figure 7.13).

An optimal machine configuration for RSP solar cell metallization has to consider the required flow path and suitable drying solutions in order to achieve best results. For the rear-side metallization, a sufficiently long flow path depending on the typical printing velocity should be planned as a good uniformity of the printed layer is primarily important. The drying technology should be selected and configured to reliably dry the metallized thick-film layers without extensive length of the drying section. The opposite requirements apply for fine-line front-side contacts. In an ideal case, the fine-line contact geometry (narrow width, adequate height) should

Figure 7.13 *(a) Demonstrator machine for high-throughput rotary screen and flexo printed metallization of silicon solar cells. (b) Rear-side metallization process of a passivated emitter and rear cell (PERC) solar cell using the rotary screen printing unit of the demonstrator machine.*

be 'frozen' directly after printing to avoid further spreading before the metallization is cured or fired. This would require an ultra-short flow path after the metallized cell leaves the printing nip. Such instant drying or 'freezing' might be realized by UV drying [68], intense pulsed light (IPL) [69] or other ultra-fast drying approaches.

7.2.3.2 Properties of the metallization paste

The characteristics of the metallization paste play an important role for the printability, the printing results and the resulting functionality of printed patterns. Contrary to FSP, a pre-filling (flooding) step of the screen, which is required for highly filled, highly viscous pastes cannot be applied using the continuous RSP process. While this is a clear advantage with respect to printing velocity and throughput on the one hand, it limits the printability of highly viscous pastes on the other hand. Thus, metallization pastes for FSP usually cannot be processed using RSP without a sufficient modification of the rheological properties. Experiments have shown that diluting typical FSP metallization pastes with a suitable amount of thinner enables a stable RSP process [60, 65]. However, obtaining optimal results requires a dedicated modification of the paste rheology with respect to the individual conditions of the rotary screen-printing metallization process. This is particularly necessary when printing ultra-fine patterns for the solar cell front-side metallization.

7.2.3.3 Optimal specification of the rotary screen

The rotary screen specification has to be carefully selected with respect to the required layer thickness and geometrical characteristics of the desired metallization pattern. Printing coarse structures like busbars, Ag-pads or full-area Al rear-side metallization requires a screen with a higher open screen surface and thus a medium-to-low mesh count in order to transfer an adequate amount of paste. However, using a screen with very low mesh count for the rear-side metallization leads to a high wet layer thickness above the required optimum, resulting in unnecessary paste consumption and excessive bowing of the solar cells after rapid thermal processing

(RTP). Optimal results for the rear-side metallization of Al BSF and PERC solar cells have been achieved using a mesh count of around ~150 wires/inch, resulting in a layer thickness of approx. 20–25 µm after RTP [65]. Such screen meshes usually have thick wires in the range of 50–60 µm in diameter. However, due to the lower viscosity of the paste, mesh marks in the printed patterns equalize to a certain extent if an adequate flow path is considered. Thus, a comparable uniformity and electric performance of the rotary screen-printed rear-side metallization compared with FSP can be obtained [65].

Printing the fine-line front-side grid or – in case of bifacial solar cells – the rear-side grid of silicon solar cells is considerably more challenging compared with the rear-side metallization. The front-side contact fingers should have a narrow width and an adequate height/cross-sectional area to enable sufficient lateral conductance. Furthermore, interruptions and disruptions of the contact fingers like excessive mesh marking, smearing and lateral spreading should be avoided or minimized. These extremely challenging and partly contradictory requirements for ultra-fine lines represent one of the most sophisticated disciplines in the field of technical printing. Focusing on RSP, printing fine contact fingers requires a suitable screen specification. Such screens have a high mesh count and optimally a small wire diameter to increase the open screen surface and minimize mesh marking. The screen-making process of such screens is particularly challenging, as materials, equipment and processes are intended for structures which usually do not drop below a width of approx. 100 µm. Realizing channel openings below 50 µm width in the emulsion layer thus requires a high-resolution positive film, careful selection of emulsion and very precise handling of all screen fabrication steps. The possibility to realize fine-line channel openings down to 30 µm and less with excellent quality has been demonstrated [70], see Figure 7.14.

A further important parameter is the angle of the screen mesh in relation to the stencil rings or respectively the direction of printing. The screen mesh is usually cut out of the screen feedstock with a certain angle and attached to the stencil rings in order to prolong the lifetime of the screen and enable a good printing result of vertical, inclined and horizontal printing patterns. However, the selected angle of the screen has a substantial impact on the effective open screen surface and the presence

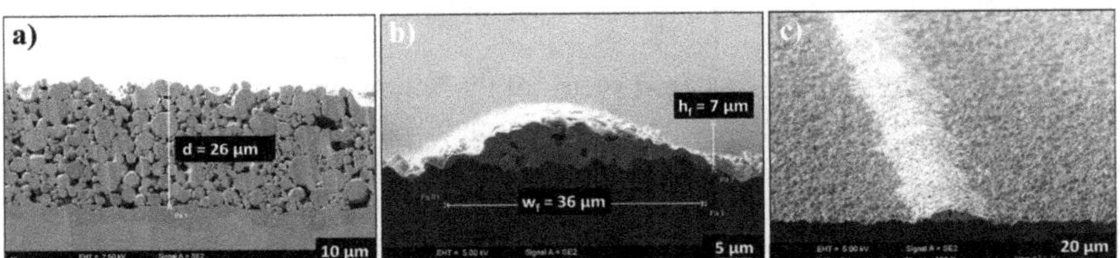

Figure 7.14 (a) Scanning electron microscope (SEM) cross-sectional image of a rotary screen printed aluminium rear-side metallization (b, c) SEM images of rotary screen-printed fine-line front-side contacts (images taken from Lorenz [70])

of wire intersections (knots) in the fine-line channels of the emulsion [71, 72]. This strongly affects the paste transfer capability and the uniformity of the printed fingers. Thus, a screen angle which provides a good balance between screen stability, printability and quality of the printing result should be chosen. Excellent results have been achieved using a screen angle of 15° in relation to the direction of printing (contact fingers aligned in the direction of printing).

7.2.3.4 Selection and adjustment of the squeegee

The choice of a suitable squeegee regarding Shore hardness and the adjustment of the squeegee system within the rotary screen has a considerable impact on the uniformity and quality of the printed patterns. Particularly the sensitive front-side metallization requires an absolute precise adjustment of the squeegee and a regular inspection of the squeegee condition. Furthermore, the squeegee pressure has to be adjusted carefully. A further important process parameter is the printing velocity which affects the time span for equalization on the one hand and the rheological condition of the paste due to increased shearing on the other hand. In an optimal case, the printing velocity should be chosen according to the intended throughput/cycle time per cell and kept constant throughout cell processing.

7.2.4 Rotary screen-printed solar cell metallization – results and future perspectives

RSP is a highly promising candidate to meet the task of tomorrow's challenges for solar cell metallization – combining high productivity with excellent printing quality. RSP captivates through the general benefits of the through-printing principle, easy handling and the ability to use existing materials – i.e. metallization paste – with minor adaptions.

The application of high-quality metallization patterns for the rear side of monofacial solar cell concepts and coarse patterns like busbars is a comparatively easy task for RSP. Such patterns can be printed with comparable quality to FSP by using slightly modified FSP metallization pastes and sufficient screen configurations [65, 70] (see Figure 7.14) a. Applying fine-line patterns like the front-side grid or the rear-side grid of bifacial solar cells is considerably more challenging. The ability of RSP to print fine-line contact fingers down to a width of approx. 40 μm and below using diluted FSP Ag paste has been successfully demonstrated (b, c) [60, 70]. The limitation of RSP with respect to finger height might even be beneficial in the future in order to reduce the silver consumption of front-side metallization. First results indicate that it is possible to print the contacts with a sufficient mean finger height/lateral line resistance for a cell interconnection using six busbars or multiple wires [70]. However, obtaining uniform front-side contacts with high reproducibility at the industrial level requires further understanding of the process and consequent optimization of the material properties. Primarily the rheological properties of the metallization paste and paste transfer characteristics of the rotary screen should be the focus of further research and development.

7.2.5 Flexographic printing

7.2.5.1 Historical and technical background

Flexographic printing goes back to the so-called aniline printing process, which was patented by Carl Holweg in 1907 [73]. Aniline printing – an upstream rotary letterpress printing process for bag production machines – was originally realized with elastic rubber plates, aniline inks and rubber-coated squeegee rollers to control the ink film thickness [74]. The development of engraved and chrome-coated anilox rollers (in reference to aniline printing) in the 1930s was an important milestone towards an improvement of this printing method. The wrong assumption that aniline inks could be toxic led to a long-lasting negative image of the aniline printing process. Thus, an initiative to rename this printing method in 1952 resulted in the new name *flexographic printing* [49]. The development of UV-exposed photopolymer printing plates (also known as clichés) in the 1970s was a further important milestone. This new type of printing plate quickly replaced the rubber plates used up to that time [49]. Further important innovations were the introduction of flexographic printing sleeves (cylinder-shaped printing form on the basis of fiberglass-reinforced plastic) instead of flat printing plates, chambered doctor blade systems instead of knife blades and laser-engraved ceramic anilox rollers. Nowadays, flexographic printing is mainly used in packaging printing on web-based substrates such as film, paper, cardboard and corrugated board. A further relevant and widespread field of application is the application of integrated flexo coating units in offset printing presses. Flexographic printing machines work almost exclusively with the roll-to-roll principle. Flexographic printing presses for package printing consist of up to 16 printing units each for a separate (graphic) ink or varnish. Today, modern flexographic printing systems can achieve printing speeds of up to 1200 m/min [75]. Depending on the type of machine, the printing units can be arranged in series (inline design) or in form of satellite units around a central cylinder (central cylinder design). Applications using non-flexible, single-item substrates according to a roll-to-flat principle are very rare niche applications and thus require a specially designed machine concept. Flexography is also successfully applied for various technical printing purposes. Examples for such applications are flexo-printed conductive circuits [76–78], battery layers [79], antennas [80], organic solar cells [52] or semiconducting layers [81].

7.2.5.2 Working principle and influence factors

Flexographic printing is a letterpress printing process with a flexible printing form (sleeve or plate). The elevated areas of the printing form are printed directly and with physical contact onto the substrate. Figure 7.15 illustrates the working principle of a roll-to-flat flexographic printing unit intended for the metallization of silicon solar cells. The main components and influence factors are introduced in the following.

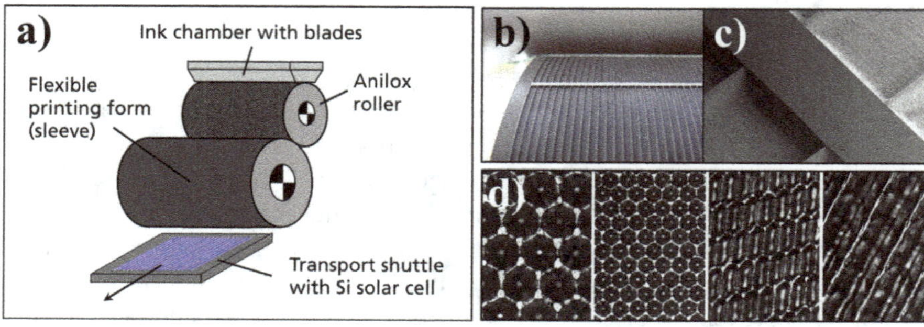

Figure 7.15 *(a) Working principle of a flexographic printing unit intended for the metallization of silicon solar cells: The anilox roller with a finely textured surface is filled with a defined amount of ink from the ink chamber. Excessive ink is removed by blade knifes of the ink chamber system. Elevated areas of the flexo plate/sleeve are inked by the anilox roller and directly printed onto the substrate (silicon wafer). (b) Flexo-printing form with fine-line layout for solar cell front-side metallization. The elevated areas (printing areas) are clearly visible. (c) Scanning electron microscope (SEM) image of a busbar-finger intersection on a flexo-printing plate. (d) Microscopic images of different laser-gravure variants for anilox rollers (from left to right): large-sized hexagonal micro-wells, small-sized hexagonal micro-wells, SteppedHex micro-wells, trihelical gravure (with courtesy of Kurt Zecher GmbH).*

Ink supply system and ink chamber

The ink supply system consists of an ink chamber which is manually or automatically supplied with ink from an ink reservoir (e.g. ink tank) by a pump. The ink chamber continuously fills the micro-structured surface of the anilox roller with a defined amount of ink, while excessive ink is removed by doctor blades arranged on both sides of the ink chamber. The ink chamber can be equipped with additional measuring devices to control, e.g. the ink temperature or pH value.

Anilox oller

The anilox roller is an essential part of a flexo-printing unit and has a vital impact on the printing result. The main purpose of the anilox roller is pre-dosing and uniform distribution of the ink film for transfer onto the elevated areas of the printing form. A steel or aluminium cylinder serves as the basic body of the anilox roller. Traditionally, the anilox roller surface is coated with an electroplated copper layer, electromechanically engraved using a diamond graver and finally finished with an electroplated chromium layer. Today, a ceramic surface is applied by means of plasma deposition and subsequently structured by laser engraving. Various patterns of micro-well arrays or line-shaped structures can be realized depending on the required ink transfer properties (Figure 7.15). The cumulative volume of the micro-wells per unit area is expressed as the theoretical *dip volume* V_d (denoted in [cm³/m²] or [ml/m²]) and defines the theoretical ink transfer capacity. The real ink volume which is actually transferred by the anilox roller onto the printing form is usually considerably smaller than the theoretical dip volume and depends on several factors, namely the rheological properties of the printing ink, the pressure setting in the nip, the printing speed, the surface properties (i.e. free surface energy) of the printing form, the temperature of the rollers and further parameters. The filling and emptying behaviour of the anilox roller is further strongly affected by the three-dimensional shape of the engraved micro-structure. Functional surface coating techniques or a special after-treatment can be applied to optimize the ink transfer properties. The second important parameter of the anilox roller is the *lineatur* l [lines/cm] which defines the amount of wells per unit length, referring to the fact that the micro-wells are engraved in spiral lines with a certain angle. Due to the fabrication process, the lineatur of mechanically engraved chrome rollers is limited to a maximum of approx. 200 lines/cm while laser-engraved ceramic rollers can obtain much higher lineatures of up to 700 lines/cm [50]. With increasing lineatur, the area and volume of the micro-wells and thus the nominal dip volume V_D decreases. Focusing on graphic applications, finely engraved anilox rollers (high lineatur, small micro-wells) are typically used for high-resolution printing layouts while anilox rollers with small lineatur and thus high dip volume are used for solid-area motives. However, this does not necessarily account for technical applications with focus on the functionality of the printed patterns. The most commonly used anilox roller configuration is a hexagonal shape of the micro-wells with an engraving angle of 60°. However, engraving angles of 0°, 30°, 45° and 60° are also possible for special applications [49].

Printing plate cylinder

The printing form cylinder serves to hold the elastic printing form. The manufacture of flexographic printing machines requires highest precision in the production of cylinders, gears and bearings, since even the slightest geometric deviations of the rollers (e.g. concentricity, cylinder radius, surface tolerances) are visible in the printed image. Bearings in modern machines therefore require a total indicated runout (TIR) ≤ 5 µm [49]. In addition to maintaining tight manufacturing tolerances, the avoidance of vibrations during the printing process also plays a major role in ensuring high and reproducible print quality.

The flexographic printing form

The flexographic printing form with the elevated printing relief directly prints the pattern onto the substrate surface. Traditionally, flat printing plates which are mounted on the impression cylinder using compressible foam tape are used as printing form. Today, an elaborate system using cylindric sleeves on the basis of glass fiberglass-reinforced plastic, which is coated with one or more layers of photopolymer or elastomer, is commonly used for highly productive, industrial printing presses. The sleeve is manually pushed onto an air cylinder and fixed in place using compressed

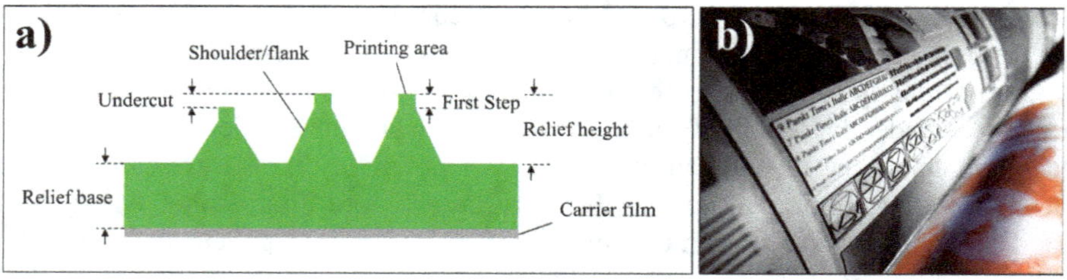

Figure 7.16 (a) Elements on a flexo-printing form (adapted from Lorenz [70]). (b) Laser-engraved ethylen propylen dien rubber (EPDM) flexo-printing form (sleeve) in a roll-to-roll printing press (with courtesy of ContiTech Elastomer Coatings GmbH).

air [50]. This procedure considerably simplifies the mounting of the printing plate and thus reduces machine downtimes. Since flexographic printing is a letterpress process, the imaging elements of the printing plate are elevated. Figure 7.16 shows the schematic cross section of a flexographic printing form including the designation of typical elements. Figure 7.16 depicts a flexographic printing sleeve in a roll-to-roll printing press.

The fabrication of the printing plate is either realized by analogue or digital exposure methods. Using an analogue process, the raw plate is first exposed on the rear side using UV radiation to create the relief base. During this process, monomers within the raw plate solution are cross-linked by UV radiation exposure. In the next step, the front side with the printing layout is structured by selective UV exposure using a negative film to protect the non-printing areas from the UV light. Within the subsequent development process, the unexposed areas are washed out, followed by a drying step and a post-exposure. Nowadays, the digital fabrication process is commonly used. The raw plate or sleeve is covered with an additional, opaque LAMS mask on the front side. This mask is partly ablated by laser, subsequently exposed by UV light and washed out. UV-reactive photopolymer printing plates have excellent printing properties, but also considerable disadvantages with respect to functional printing applications. The chemical resistance against many solvents is critical, resulting in uncontrolled swelling of the printing form. Combined with mechanical stress during the printing process, fine printing elements are easily damaged with negative effects on the printed patterns.

An alternative to UV-reactive photopolymer materials are printing forms based on elastomeric materials like ethylen propylen dien rubber (EPDM). In contrast to photopolymers, EPDM is highly resistive against a large variety of solvents and cleaning agents as well as considerably more stable against mechanical abrasion. EPDM printing forms are structured by selective laser ablation using high-resolution fiber lasers [82, 83]. Modern high-definition (HD) UV exposure systems for photopolymer plates are limited to a (theoretical) resolution of 4 000 dpi [84], which is in reality not achievable due to limitations of the photopolymer material and not required for graphic arts purposes. In contrast, high-precision fiber laser engraving systems for EPDM forms are able to realize structures down to approx. 5 μm [85]. Thus, laser-engraving of EPDM printing forms is clearly the method of choice for high-resolution functional printing applications.

Impression cylinder/transport system

Roll-to-roll flexo presses for printing on web-shaped substrates have an impression cylinder that guides the web-shaped substrate printing unit. This can be either realized by a central drum (central cylinder machines) or by an induvial impression cylinder in each printing unit (inline machines). Printing on rigid, single-item substrates using the roll-to-flat printing principle requires a printing table or – in an inline process – an individual shuttle or belt transport system to carry, align and fix the substrate during the printing process. Due to the sensitivity of flexography regarding local pressure differences, the transport system/printing table has to fulfil high requirements regarding alignment precision, planarity and surface tolerances. If more than one layer should be printed in consecutive, sequential printing units, a high accuracy of the lateral, circumferral and diagonal register has to be assured. While graphic arts or package printing requires a typical register accuracy in the range of a few hundredths of a millimeter [50], technical printing applications can be more sophisticated if individual layers must be positioned exactly in relation to each other or a double printing step is required.

7.2.5.3 Application of flexographic printing for solar cell metallization

The first activities to apply flexography for the metallization of silicon solar cells were started in 2010 [86, 87]. Initially, these activities addressed a seed-and-plate approach using a flexo-printed fine-line seed layer metallization with low-viscous, firing-through silver ink, RTP and subsequent reinforcement using Ag electroplating. Promising first results using this approach on small-sized silicon wafer samples (mini solar cells) were demonstrated by two independent research groups [87–90]. In the following, a successful scale-up on 6″ Cz-Si Al BSF solar cells was demonstrated using a roll-to-flat flexography laboratory printing machine [91, 92]. In order to reduce the Ag consumption on the front side, the focus was shifted towards a seed-and-plate approach using subsequent nickel-copper-silver (Ni-Cu-Ag) electroplating [93]. Within the funded project ‘Rock-Star’ (starting in 2015), the focus shifted towards the development of a process for directly flexo-printed contacts without subsequent plating which was successfully demonstrated in 2016 on Cz-Si Al BSF solar cells with busbar-less front-side grid for multi-wire interconnection [85, 94]. Following research and development activities focused on further optimization of ink properties, printing plate, anilox roller and further important process parameters [70, 85] which resulted in a successful realization of flexo-printed contact fingers down to a width of less than 30 μm [70].

A variety of influence parameters affect the printing result of the flexo-printed front-side metallization. The following aspects have been identified as particularly important for the printing result:

- Machine platform: transport concept, alignment, printing units, drying unit
- Anilox roller: Dip volume, lineatur and shape of the micro-wells
- Printing plate/sleeve: Material properties, compressibility and relief structure
- Ink: Rheology, solvent, binder and particle properties, surface tension
- Printing parameters: pressure adjustment of rollers, (local) printing pressure between plate/sleeve and substrate, printing velocity
- Substrate: free surface energy, macroscopic and microscopic topography, tolerances

7.2.5.3.1 Machine platform and drying equipment

The application of flexographic printing for an industrial metallization of silicon solar cells requires – similar to RSP (see section 0) – a sufficient machine platform which can be implemented into a PV production line. The described general requirements of such a device also largely apply for flexographic printing. However, the requirements regarding planarity and tolerances of the wafer transport shuttles are – in contrast to RSP – considerably more critical as flexographic printing is highly sensitive towards local pressure variations in the printing nip. Thus, a production platform has to fulfil high demands with respect to tolerances of the relevant components like shuttles/printing table, cylinders and sleeves. Such a demonstrator machine has been recently developed and demonstrated [67] and is currently under evaluation. As flexographic printing is primarily suited for the application of fine patterns with limited layer thickness/line height, a sufficient drying technology which enables ultra-short drying or hardening of the printed patterns is favourable. Similar to RSP, drying technologies like UV drying [68], IPL [69] or other ultra-fast drying approaches should be addressed.

7.2.5.3.2 Anilox roller specification

Among the different influence factors, the anilox roller in combination with the rheological properties of the ink has a particularly strong impact on the geometric and electric properties of flexo-printed fine-line contacts. Thus, the choice of an adequate anilox roller specification is essential in order to obtain the desired results. However, the criteria for the selection of an adequate anilox roller for solar cell fine-line metallization have to be clearly distinguished from 'typical' criteria applied for graphic arts printing. For typical flexo-printing applications, i.e. in the field of package printing, images are separated into rasterized images with thousands of small-sized, dot-shaped printing elements on the flexo-printing form. If the anilox roller lineatur (and thus the size of the micro-wells) is too coarse, small dot elements on the printing form dip into the micro-wells, resulting in a complete flooding of these elements. This so-called 'dot-dipping effect' leads to considerable colour shifts in the printed image. To avoid this effect, the lineatur of the anilox roller is usually chosen with a resolution that is three to six times higher than the resolution of the printing image/layout [95]. Various test runs revealed that this effect is not relevant when printing fine-line patterns like solar cell front-side grids. It is assumed that

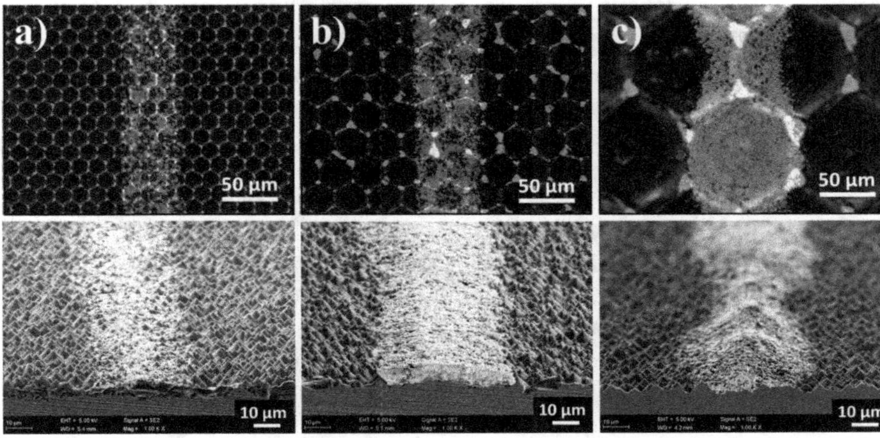

Figure 7.17 *Microscopic overlay images and scanning electron microscope (SEM) images of anilox roller micro-well patterns and the resulting flexo-printed finger geometry: (a) Finely engraved anilox roller with a lineatur of l = 700 l/cm. (b) Medium anilox roller with a lineatur of l = 340 l/cm. (c) Coarse anilox roller with a lineatur of l = 100 l/cm. The impact of the anilox roller dip volume and micro-well pattern on the printed finger geometry is clearly visible in the corresponding SEM images (images modified based on [70]).*

line-shaped structures are supported by the – typically honey-comb shaped – edge walls of the micro-wells which prevents a complete dipping and thus flooding of the structures. Thus, the desired ink transfer capacity and thus layer thickness of the printed patterns is the predominant aspect for the choice of the anilox roller specification. While anilox rollers with high lineatur (small micro-wells) and thus low dip volume V_D are optimal for the application of narrow seed layer contacts with low finger height and thus low silver consumption [96]. Using a finely engraved anilox roller with a high lineatur of 350–700 l/cm in combination with low or moderate viscosity should be chosen for this metallization approach [70]. The small dip volume of such high-lineatur anilox rollers provides seed layer contacts with very small height of approx. $h_f \approx$ 1–3 µm and thus a very low overall silver consumption on the front side [92]. Subsequent reinforcement of the seed layer grid by electroplating usually enables a sufficiently low lateral resistance of the front-side grid [93] (Figure 7.17).

In contrary, directly flexo-printed front-side contacts without subsequent electroplating demands narrow contact fingers with a sufficient finger height to enable a sufficiently low lateral grid resistance with respect to the selected cell interconnection method [70]. Achieving such contact geometries using flexography is considerably more challenging compared to seed layer patterns, as two contradictory effects of the anilox roller have to be considered. The demand to print the contacts with a sufficient finger height requires the choice of anilox rollers with high dip volume and thus low lineatur. Using anilox rollers with a maximal dip volume V_D and a lineatur of around 100 l/cm in combination with high-viscous silver ink results in flexo-printed contact fingers with a medium height of approx. $h_f \approx 8 - 12$ µm [70] which is adequate for an intended interconnection with six busbars or multiple wires. However, such coarse anilox rollers have, on the other hand, a negative impact on the uniformity of the printed fine lines which is clearly visible when analysing the microscopic finger geometry after printing. 'Braid-like' disturbance patterns of the printed lines can be observed depending on the (random) contact position of the line-shaped elements on the printing form with the anilox roller surface (Figure 7.17). These disturbance patterns are suspected to have a considerably negative effect on the global lateral grid resistance [70]. Thus, further research activities will strongly focus on optimizing the anilox roller properties or even developing an alternative approach for the uniform inking of the printing form.

7.2.5.3.3 Flexographic printing plate/sleeve

The flexo-printing form as the carrier of the printing image has obviously a considerable impact on the printing result. Considering the material properties, EPDM printing plates are the material of choice for solar cell metallization, as a high chemical and mechanical stability is required due to typcially used solvents and the abrasive, textured wafer surface [70]. However, the chemical compatibility of the solvents used for the ink formulation should be tested individually on the intended type of printing form using swell tests under standardized conditions. Laser-engraved EPDM printing forms also have clear advantages regarding the resolution of ultra-fine lines and the ability to design the optimal three-dimensional shape of the finger elements.

A major parameter of the printing form is the thickness, Shore hardness of the EPDM material and the compression properties of the plate/sleeve. EPDM printing sleeves and plates are available with several thickness and Shore hardness variations. It is further possible to implement a porous, sponge-like layer in the base layer of the plate/sleeve which allows a considerably better compressibility of the printing form and thus prevents a deformation of the printing elements on the relief. Optimal results for solar cell front-side metallization have been achieved using printing plates with a thickness of 1.7 mm, a Shore hardness of 65° Shore A and an incorporated compressible layer.

Engraving tests have demonstrated that it is possible to create homogeneous and reproducible line elements down to a nominal width of the printing-relevant line ridge down to $w_n = 5$ µm [85]. Such minimum structures are irrelevant for typical graphic arts applications but highly relevant for technical applications like ultra-fine conductive lines. The width of the printed contact fingers obviously shows a linear relation to the nominal width of the finger elements in the digital printing layout. It is thus beneficial to use a very small nominal finger width in order to print narrow contacts [70]. However, such fine elements require a certain stability to prevent inhomogeneous deformation and lateral elevation under pressure. This

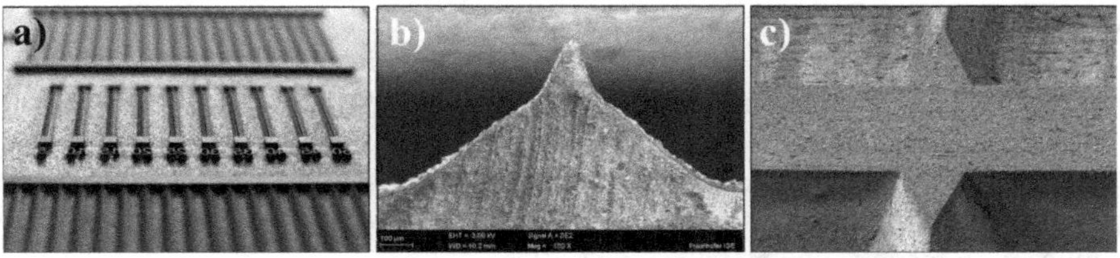

Figure 7.18 *(a) Test layout on laser-engraved ethylen propylen dien rubber (EPDM) printing plate comprising finger elements with different nominal widths from 5 μm to 50 μm. (b) Optimal three-dimensional cross section of a laser-engraved finger element on an EPDM printing form. (c) Tapered busbar-finger intersection to avoid missing areas at the transition (images modified based on Lorenz [70]).*

can be effectively achieved by optimizing the three-dimensional shape of the finger elements. A two-step design with a broad base to stabilize the finger element and a narrow first step with steep flanks and a minimal ridge line have shown the best printing results (Figure 7.18).

A further important aspect is the design of the printing layout. Printing fine-line contact fingers and orthogonally arranged, considerably broader busbars with one printing form is problematic, as the load change when rolling over the busbar-finger intersection leads to missing areas exactly at the transition. This is highly critical for the electric performance of the front-side grid, as it prevents the current flow from fingers into the busbars. This negative effect can be prevented by a tapered transition design between busbar and finger elements (Figure 7.18).

7.2.5.3.4 Ink properties

The printing medium or fluid plays a crucial role in most printing and coating processes. Focusing on solar cell metallization, several important aspects of the ink have to be considered for the flexo-printing process:

- Rheological properties of the ink
- Surface tension and wetting behaviour
- Proportion, size and shape of the metallic particles
- Properties of the solvent and binders (i.e. boiling temperature, chemical compatibility with the flexo-printing form)
- Storage and processing conditions (sedimentation, solvent evaporation)
- Drying conditions

The flexo-printing process entails several special requirements with respect to the ink formulation. Thus, the selection of the ingredients, i.e. solvents, binders and additives has to consider several important aspects regarding storage, material compatibility, process requirements (printing), printability and functionality (i.e. conductivity, geometry of the printed structures). Covering these different demands poses considerable challenges with respect to the ink formulation. While several silver-based conductive flexo inks for printed electronics applications are commercially available, this is currently not the case for solar cell metallization purposes. Thus, flexo-printable Ag inks have been fabricated and optimized to evaluate the applicability for solar cell metallization [70].

An important aspect which has to be considered particularly for particle-based inks is the stability of the ink during storage and processing. Medium-to-low viscous particle-containing inks often tend to agglomerate and sediment quickly in a motionless condition. This has to be considered for the ink formulation, i.e. by adding sufficient stabilizing agents as well as for the storage and processing conditions. Storing of such inks should be realized under constant motion using, i.e. magnetic stirrers or tube rollers. Furthermore, a constant motion of the ink within the ink chamber should be ensured. The selection of the (metallic) particles has to consider the specification and interaction with the anilox roller. Large particles or agglomerates within the ink tend to cant within the small micro-wells of the anilox roller or in between the knife blades of the ink chamber and the anilox roller surface resulting in strip-shaped disruption patterns in the printing image and scratch corruptions on the anilox roller. Thus, the particle size distribution (D95) has to be selected with consideration of the intended size of the micro-wells on the anilox roller. In order to reliably fill the micro-wells during the inking process, particles should be approx. five to ten times smaller than the diameter and the depth of the micro-wells. The micro-wells of a high-lineatur anilox roller with 700 l/cm have a diameter of approx. 14 μm providing a hexagonal gravure. Thus, particles should not exceed a diameter of $d_{p,D95} \approx 1.4 - 3$ μm (D95) when using such finely engraved anilox rollers. Using a coarsely engraved anilox roller with a lineatur of $l = 100$ l/cm would allow a considerably higher particle diameter up to $d_{p,D95} \approx 10–20$ μm. Thus, the specification of the designated anilox roller has to be considered for the ink formulation and vice versa.

A major impact for printability and the printing results are the rheological properties of the flexo ink. Various test runs with different ink formulations have shown that flexo-printing has a relatively large 'printability corridor' regarding the viscosity of the ink. The printability ranges from water-like inks to paste-like inks closely comparable to rotary screen-printing paste. The shear-dependent viscosity strongly depends on the amount of silver particles in the ink. Optimal results for direct metallization of front-side contacts have been achieved with an ink viscosity in the range of 650–750 mPas at a shear rate of $\dot{\gamma} = 1\,000/s$ [70]. For a seed and plate approach, the ink should have a lower viscosity in the range of 300–400 mPas $\dot{\gamma} = 1\,000/s$ [70].

7.2.5.3.5 Printing pressure

Flexography is considerably sensitive to the actual pressure conditions in the printing nip between printing form and substrate. While the global pressure adjustment can be controlled by manual or digital engagement of the printing cylinder, the actual spatially resolved pressure conditions in the printing nip are influenced by

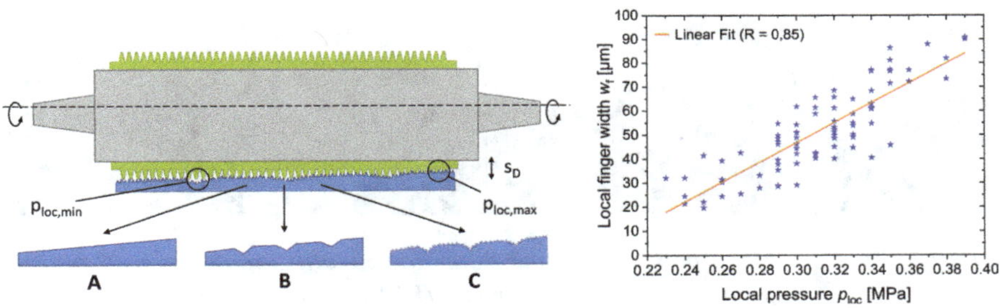

Figure 7.19 *Schematic drawing of wafer geometry deviances and their impact on the local pressure between the wafer surface and the flexo sleeve. The graph illustrates the impact of an increasing local pressure p_{loc} on the resulting printed finger width w_f (images taken from Lorenz [70]).*

a multitude of factors which can lead to significant deviations along the contact line. Causes for such local pressure deviations can be local variations of the wafer geometry (thickness and roughness), the printing form and geometry tolerances, or adjustment deviations of relevant machine components like the printing cylinder and the transport shuttle/printing table [70]. Particularly the wafer material can exhibit considerable geometry deviations of different orders (i.e. wedge form, saw marks and micro-roughness) (Figure 7.19). The average wafer thickness can easily vary between 25 µm and 30 µm within one production batch; variations in a comparable range are possible over the area of one wafer [70]. These variations can be superimposed by further tolerances resulting from the mentioned sources and have a considerable effect on the local pressure conditions and thus the printed local geometry (primarily the finger width w_f) of the fine-line contacts. Experiments have shown that increasing the local pressure by $\Delta p_{loc} = 0.1$ MPa results in an additional broadening of the average printed finger width by $\Delta w_f \approx 40$ µm [85]. This rough assessment which is in reality depending on various influence factors emphasizes the strong impact of the local printing pressure on the resulting finger geometry. Thus, tolerances of all kinds have to be minimized in order to achieve a reproducible and high-quality printing result of the front-side metallization using flexographic printing.

7.2.5.4 Flexo-printed solar cell metallization – results and future perspectives

Flexographic printing is a promising method for the front-side metallization of silicon solar cells. The ability to apply narrow contacts with low silver consumption is attractive for a seed and plate approach with subsequent electroplating of alternative metals to silver (Figure 7.20b). Various studies have demonstrated that solar cells on the same performance level to screen printed cells can be fabricated using

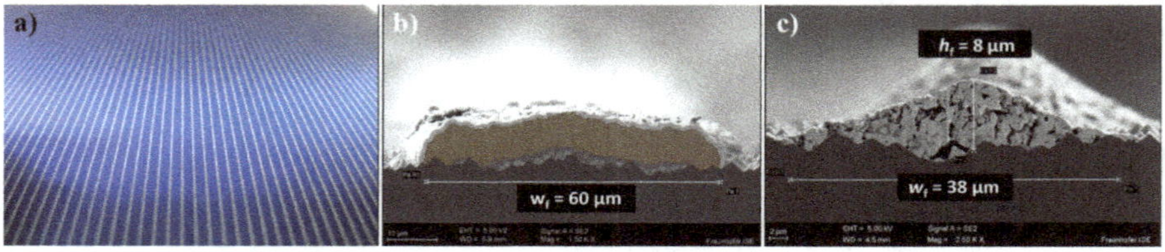

Figure 7.20 *(a) Flexo-printed fine-line front-side metallization of a busbarless passivated emitter and rear cell (PERC) solar cell. (b) Scanning electron microscope (SEM) cross-sectional view of a flexo-printed seed layer contact reinforced with Ni/Cu/Ag electroplating. (c) SEM cross-sectional view of a flexo-printed front-side contact without subsequent plating. (Images taken from Lorenz [70].).*

this approach [90, 93]. However, using a seed and plate approach has become less attractive due to low silver prices and progress with direct metallization approaches.

Flexography has also proven its ability to print a fine-line front-side metallization with adequate finger height to enable an electrically well-performing cell interconnection via multiple busbars or multi-wire interconnection [94] (Figure 7.20a). By consequently optimizing materials and process parameters, a directly flexo-printed reproducible finger width below 40 μm (Figure 7.20c) and even further to a mean width of approx. 30 μm [7] has been demonstrated. However, further research and development is required particularly with regard to an optimization of the uniformity (section 0) of the printed fine lines.

The ongoing trend towards a cell interconnection with multiple busbars or wires makes printing approaches with low silver consumption increasingly attractive. The higher lateral resistance of such contacts does not necessarily limit the fill factor of the cell due to considerably reduced finger segment lengths in between the busbars/wires. Flexography is able to meet this demand and additionally provides the economic potential to increase the throughput of the metallization step significantly. Thus, flexography can be regarded as a potential 'hidden champion' among various competing metallization approaches.

7.2.6 Rotogravure printing

A further promising method to apply the front-side grid of silicon solar cells is the so-called gravure or rotogravure printing method. Rotogravure is usually applied for the large-scale printing production of packaging and graphic arts products using impressively large printing machines with cylinder widths up to more than 4 m. Gravure printing cylinders consist of a steel core which is coated with a several millimetres thick electroplated copper layer. The printing layout is usually engraved into the comparatively soft copper layer by mechanical or laser engraving methods and subsequently covered by a protective, extremely resistive electroplated chromium layer [97]. The cost-intensive and energy-consuming fabrication of the printing cylinders which is also critical with respect to environmental and safety aspects

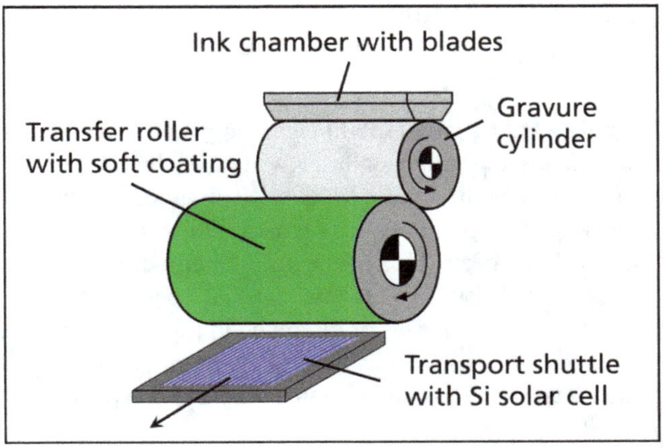

Figure 7.21 *Working principle of an indirect rotogravure printing unit intended for solar cell metallization: The layout is engraved into the gravure cylinder which is inked by an ink chamber with lateral blade knifes. The ink is transferred via the soft-coated transfer roller onto the silicon wafer.*

is a clear drawback of this printing method. Thus, rotogravure processes are only reasonable for extremely large-volume printing applications. Rotogravure printing is on the other hand able to print very fine structures with high, reproducible quality over millions of print runs. The hard and incompressible printing form requires a soft counterpart. On roll-to-roll machines, this is commonly realized by a so-called pressure cylinder which is covered with a soft blanket.

Besides typical large-volume printing applications in the field of graphic arts and package printing, rotogravure is also applied for a variety of technical printing applications ranging from printed electronics [98, 99] to direct printing of semiconductor layers on glass [100]. Focusing on the metallization of silicon solar cells, a directly printed metallization approach using rotogravure is problematic as the hard printing cylinder would destroy the rough and incompressible silicon wafer surface. However, applying a specific variant of rotogravure – the so-called indirect gravure printing or rotary tampon printing process [101] – could be a promising approach for fine-line front-side metallization. This rarely applied printing method uses a rotogravure printing cylinder in combination with a soft-coated transfer cylinder which prints the pattern onto the substrate. Figure 7.21 shows the schematic concept of such an indirect gravure printing unit for solar cell metallization. To the best of the knowledge of the authors, only one study has been carried out so far to investigate this approach [102]. The results of this study revealed the existing challenge to realize an adequate finger height and thus a sufficiently low lateral finger resistance without the necessity of multiple printing steps. However, rotogravure printing can be an interesting option to realize ultra-fine-line front-side contacts of specific solar cell concepts.

7.3 Laser transfer printing

7.3.1 Introduction

The laser-induced forward transfer (LIFT) technology can be categorized as a subset of additive direct laser printing or laser writing technologies. In general, for the LIFT process, a transparent material (donor) is first coated with the layer to be transferred. In the second step, this layer is transferred onto a substrate (target, e.g. a solar cell) using a laser beam [103]. The LIFT method allows for a structured transfer of various materials, such as organic materials, ceramics and metals, which in turn can be present as solids, pastes or dissolved in liquids. A detailed description of the various methods and their applications is given by Piqué and Serra [104]. In the following, the focus will be on metal transfer by LIFT especially for the metallization of solar cells.

One of the first metal LIFT experiments was published by Bohandy *et al.* [105], where a copper film was transferred on silicon in a structured way using an excimer laser. In the following, the basic characteristics of LIFT transfer of metals are summarized.

For laser-induced metal transfer, there is an interaction between the laser and the donor material (whereby the donor material should generally be as transparent as possible), the donor coating (e.g. metal) and a solar cell as a target. For metals, the penetration depth of the laser radiation is relatively small, but the thermal diffusivity is relatively high. Therefore, the ratio of thermal expansion coefficient of the metal, the pulse length, laser fluence and donor layer thickness to each other is essential for the LIFT process of metals. In addition, metals do not exhibit strong wavelength-dependent absorption. This special property makes the LIFT process of metals relatively insensitive to the laser wavelength.

Physically, the LIFT transfer can be categorized into four different processes (a summary is given by Piqué and Serra [104]). The metal can be liquefied by the interaction with the laser (formation of drops) [106, 107], vaporized (metal splashes) [106–108], transferred as solid (LIFT-off) [109] or as a kind of drop by the formation of jets [110]. Usually, these categories of the transfer cannot be separated sharply from each other. Behind each of these categories is a different interaction mechanism between donor, metal and laser, which can be explained using a simple but common model. When the laser hits the metal, a melt metal front spreads through the metal and vaporization occurs at the interface between donor and metal. The resulting vapour pressure causes the peeling of the metal layer from the target (refer, for example, to [104]). As the pulse length increases, the laser-induced heat in the metal has more time to expand. This means that a larger proportion of the metal is melting and evaporating. According to this, with increasing fluence, the portion of vaporized metal increases. Interestingly, the formation of jet-induced droplets is physically not yet fully understood. This phenomenon tends to appear at relatively long pulse lengths and relatively low fluences [110]. A high AR is a characteristic of these jet-transferred metal spots. Often, the droplet mode is preferred as evaporated metal leads to metal splashes at the target, which generates poorly reproducible metal structures and possibly unwanted target damages.

The shape or quality of the metal spots transferred to the target are often evaluated on the basis of their diameter, transferred height and their degree of coverage also called 'optical fill factor'. The optical fill factor is defined as the ratio of the area covered by metal to the total area defined by the laser spot size [111]. This means that the optical fill factor is maximum when the designated area is completely covered with metal. In addition to the influencing factors mentioned above, namely fluence, pulse length and film thickness, the distance between target and donor is also decisive. The edges of the transferred structures become sharper, the optical fill factor increases and the structures become more reproducible with decreasing distance between donor and target [112].

Meanwhile, there is a large number of publications describing LIFT applications for the solar cell metallization using a variety of technical implementations in which, e.g. the donor material is guided over the substrate in a roll-to-roll system [113]. Others are endless loops in which the donor material is directed over four axes or a single axis [114–116]. Also a variety of donor materials (e.g. glass sheets, plastic and kapton foils), donor layers (various metal pastes and inks, e.g. Al, Ag or Cu, and physical vapour deposited (PVD) thin films) and laser parameters were applied which is, e.g. described in the following literature [113, 115–125]. Essentially, three different LIFT methods have become established for the metallization of solar cells. First, the *Laser Transfer Printing* (LTP) technology, which refers to the laser-induced transfer especially of metallic or ceramic inks which is implemented in the four axes system [114, 115]. Second, the *Laser Transfer and Fire* (LTF) technology, the *Laser Transferred Contact* (LTC) technology for the front side and the *All Laser Transferred Contact* (ALTC) for the front-side and back-side metallization, processes in which thin metal films are transferred and used as seed layers for plating processes [111, 119, 124, 126]. Third, the *Pattern Transfer Printing* (PTP) technology, where the donor material is pre-stamped and metal pastes are transferred laser-induced as fingers [117, 121]. The LTF, (A)LTC and PTP approaches are described in more detail below.

7.3.2 The (A)LTC and LTF process

The (A)LTC and LTF laser processes are used to apply seed layers to the solar cell surface and create the contact between Si and metal. Subsequently, an electrochemical copper plating process is used to form contact fingers. This makes the LTF and (A)LTC process an alternative approach to the LCO process. As mentioned above, the LIFT process is generally independent of the laser wavelength used. In contrast to the LCO front-side process, where UV ultra-short-pulse laser sources are used to open the passivation layer, for the (A)LTC and LTF process infrared (IR) laser radiation can be used. IR lasers are cheaper and usually more stable than UV lasers. In addition, the wavelength can be selected in such a way that the damage to the surface next to the metallized surface is as low as possible. A further advantage is the high flexibility of metals, which can be transferred with the help of the LIFT process.

Also, stacks can be transferred. In addition, no high-temperature step is required for contacting, making these processes particularly interesting for temperature-sensitive coatings such as amorphous Si or perovskite. However, a major challenge with the laser processes described here is that molten metal could penetrate through the emitter and shunt the solar cell, which is known as spiking. Therefore, special requirements must be placed on the emitter which is, e.g. shown by Roeder *et al.* [120].

The LTC process was developed by Roeder *et al.* [120, 127] and was applied to metallize the front side of p-type Si-based solar cells. A glass substrate coated with a nickel layer was used as the donor material and a pulsed Nd:YV04 laser with wavelength $\lambda = 532$ nm and pulse duration $\tau = 10$ ns was used for laser transfer. After the nickel and copper plating process, finger widths below 30 μm, ARs larger than 0.25, cell efficiencies of $\eta = 17.4\%$ and fill factor FF = 77.7% were demonstrated. In a next experiment, it was shown that the problem of spiking can be counteracted by using a two-step LTC process [126]. Here, the metal is transferred to the solar cell in a first laser step and the contact is established in a second laser step (refer to Figure 7.21). With that the metal transfer and the contact formation can be optimized separately. For both laser steps, a green ns laser was used with different pulse energies. The laser energy of the first step was higher than that of the second step.

The single laser step LTC process was extended by Wang *et al.* to the so-called 'all-laser-transferred contacts' (ALTC) [124]. Here, the front and back surfaces of a p-Si solar cell was metallized by the laser transfer. Again, glass as the donor material and a green ns laser for the transfer was used. Titanium was laser-induced transferred for the front-side metallization and aluminium for the back side.

Another two-step LIFT process is called 'Laser Transfer and Firing' (LTF) and was developed by Rodofili *et al.* [119] (see Figure 7.22). The LTF process was applied to SHJ solar cells. Here, a commercially available 100 μm thick biaxially oriented polyethylene terephthalate (BoPET) foil (a plastic foil) was used as donor material. This foil is flexible. A sputtered, 100 nm thin NiV (with 7 wt% vanadium) layer as donor coating was used. The laser source for both the first (LIFT) and the

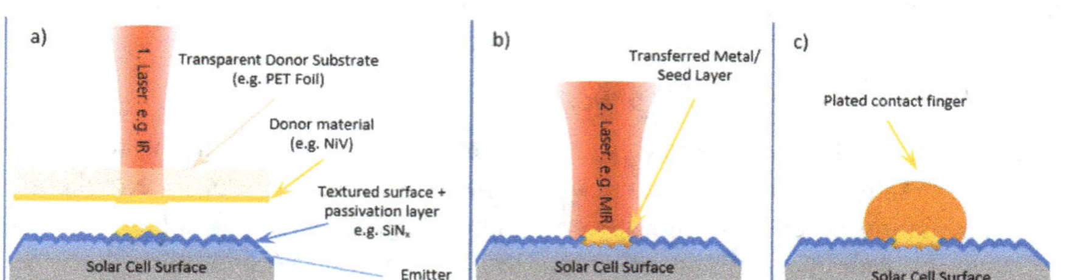

Figure 7.22 Schematic representation of a two-step laser process, which is realized by the LTC or the LTF technologies. The indicated laser wavelengths and the cell design below are exemplary. In the terminology of the LTF process the first laser step (a) is called LIFT and the second step (b)LSH. In the next step (c), the contact finger can be electrochemically grown on the seed layer.

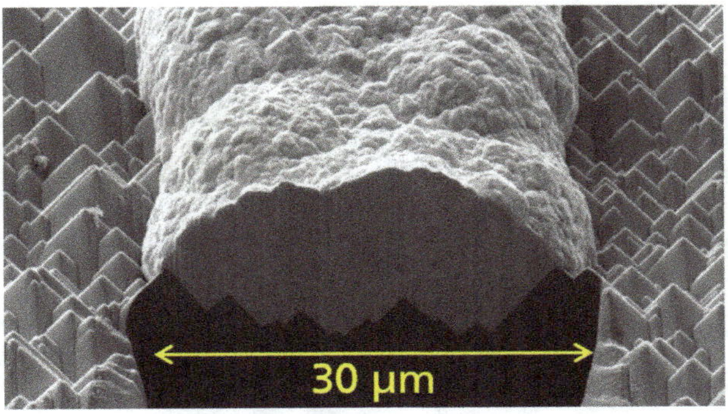

Figure 7.23 *Scanning electron microscope (SEM) image showing the cross section of a plated finger grown on a seed layer applied by the LTF technology. Reprinted, with permission, from [128].*

second ('Laser Selective Heating' (LSH)) laser step was a Q-switched Nd:YVO laser with 1064 nm wavelength and 15 ns pulse length, but different pulse energies for the first and second laser steps [119]. Again, the first pulse was more energetic than the second. In this way, seed layers of less than 10 μm and finger widths after plating of 30 μm could be achieved (see Figure 7.23). The best cell efficiency was 22.2%, which represented an efficiency gain of $0.5\%_{abs}$ compared to the screen-printed reference group.

In a further development, the second laser step in the LTF sequence (LSH laser step) was replaced by a MIR (mid-infrared) laser with a wavelength of 2 μm. In this wavelength range, Si is transparent and therefore the laser-induced damage during the LSH process can be further reduced. This also reduces the challenges of aligning the two laser process steps with each other. Within the frame of a publicly funded project, this LTF technology is implemented on a fully automated role-to-role production line [113].

7.3.3 Pattern transfer printing

The PTP technology has been developed by Utilight Ltd and has proficiencies to deposit narrow finger widths (under 20 μm) with high ARs. Therefore, the main focus of PTP is the front metallization of silicon solar cells. The PTP technology uses transparent polymer foils as the donor material, which consists of pre-embossed trenches. The printing sequence of PTP is illustrated in Figure 7.24. These trenches are filled with the metal paste to be transferred, here silver paste, using two metal blades. In the next step, an infrared laser is used to transfer the paste from the trenches to the solar cell. A detailed description of the PTP printing sequence can be found in Adrian *et al.* [117].

Adequate laser irradiation is a core part of PTP technology. It secures not only the completeness of the transfer process but also affects the finger shape and amount of debris around the deposited finger. The laser irradiation transmits through the

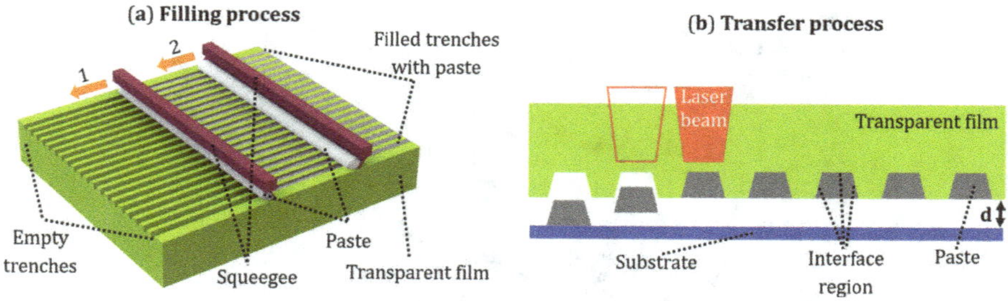

Figure 7.24 *Illustration of PTP printing sequence: (a) Filling process and (b) Transfer process [129]*

transparent film and increases the paste temperature in the interface region between silver paste and trench at which the solvent of the paste evaporates. The optimum laser power (laser threshold) is the minimum laser power required, which generated enough pressure at the interface to overcome the adhesive strength between the paste and the target foil. Therefore, the paste was transferred from the trenches completely onto the substrate without any interruption. When using laser power above this threshold, the generated pressure exceeds not only the adhesive force at the interface but also the cohesive strength within the silver paste, which separates some silver particles from the main paste body during deposition. The separated silver particles (called debris) are scattered around the main paste body (see Figure 7.25). Simultaneously, a high laser power accelerates the drop of the paste towards the solar cell surface and results in broader lines of transferred paste.

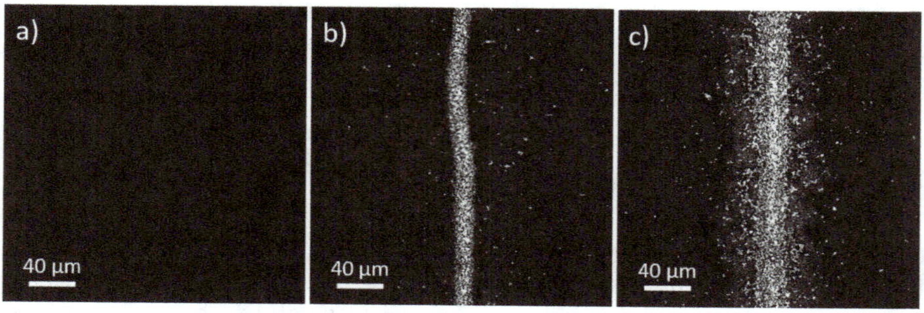

Figure 7.25 *Laser scanning microscopy images of PTP-printed fingers with different laser power; (a) laser power below the threshold: no transfer, the paste was still inside the trenches; (b) laser power same as the threshold: paste was deposited and minor amounts of debris were found around the finger; (c) laser power highly above the threshold: the deposited finger was broader and significant debris amounts were scattered around the finger. © [2020] IEEE. Reprinted, with permission, from Adrian et al. [117].*

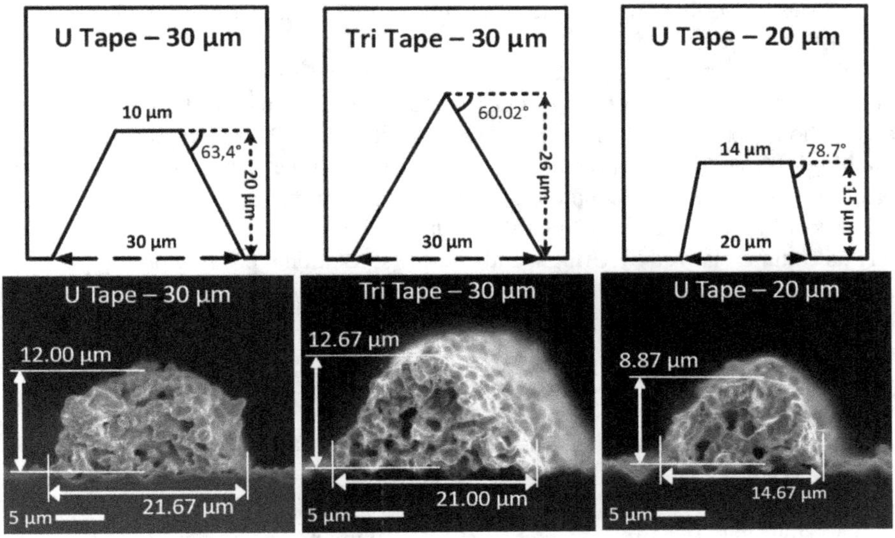

Figure 7.26 Various trench geometries for PTP (top) and scanning electron microscope images of each trench geometry (bottom). © [2020] IEEE. Reprinted, with permission, from Adrian et al. [117]).

Figure 7.26 shows that the deposited finger properties such as the finger shape, paste lay down and cross section are essentially related to the trench geometry and determined by the required laser threshold [117]. Compared with the LIFT process for metal, the trenches of PTP enable fabrication of the higher thick paste and reduce the debris around the transferred paste.

7.4 Chapter summary

When talking about metallization of silicon solar cells, the ITRPV considers mostly screen-printing, potentially stencil-printing on the one side and plating processes on the other. This chapter summarized most relevant alternative metallization approaches including dispensing, three different rotational printing approaches and several laser transfer processes.

The parallel dispensing as well as former co-extrusion print heads, rotational screen printing as well as PTP apply conventional pastes with a high solid content, flexo-printing typically a more diluted ink. The process of contact formation however remains similar to the current industrial standard approach (i.e. drying and subsequent fast firing or curing). Dispensing and PTP both only require minor changes to existing production lines and stand out due to their homogeneous fine lines that allow for efficiency increases and a reduction of Ag consumption. Due to their contact-less printing approaches they allow for wet-in-wet printing and in an initial set-up an existing screen-printing line, they would enhance resulting contact geometries by adding only one new machine or even just insert a dispensing kit in an existing screen printer.

Rotational printing approaches have the potential of a massive boost of production throughput that however goes along with a completely newly developed automation approach that enables cycle times of less than 0.5 s along the whole line. Finally, LTC and LFT approaches could be used to directly penetrate and contact passivated solar cells and promote a seed and plate approach.

According to HighLine Technology, dispensing will tackle field tests in 2021, PTP is commercialized by Utilight Ltd and a first rotational printing prototype from ASYS Automation Systems GmbH is currently investigated at Fraunhofer ISE's PV-TEC in Freiburg. Hence, this chapter shows that the future metallization market might get more diverse regarding technologies, once they proved to push production performance to a higher level than current fine-line screen-printing approaches.

References

[1] Hon K.K.B., Li L., Hutchings L.M. 'Direct writing technology – advances and developments'. *CIRP Annals – Manufacturing Technology*. 2008;**57**(2):601–20.

[2] Chen X.B., Ke H. 'Effects of fluid properties on dispensing processes for electronics packaging'. *IEEE Transactions on Electronics Packaging Manufacturing*. 2006;**29**(2):75–82.

[3] Chen X.B., Li M.G., Cao N. 'Modeling of the fluid volume transferred in contact dispensing processes'. *IEEE Transactions on Electronics Packaging Manufacturing*. 2009;**32**(3):133–7.

[4] Chen X.B., Schoenau G., Zhang W.J. 'Modeling and control of dispensing processes for surface Mount technology'. *IEEE/ASME Transactions on Mechatronics*. 2005;**10**(3):326–34.

[5] Geren N., Ekere N.N. 'Solder paste dispensing in robotic SMD rework'. *Soldering & Surface Mount Technology*. 1994;**6**(1):21–5.

[6] Häussermann T. *Simulation des Underfill-Prozesses bei Flip Chip-Anwendungen*. Dissertation, Universität Stuttgart; 2009.

[7] Othman N. *Entwicklung eines Verfahrens zum präzisen Punkt- und Linienauftrag von hochviskosem Leitkleber mit einem geregelten Mikrodosiersystem*. Dissertation, Universität Stuttgart; 2005.

[8] Razban A., Davies B.L. 'An automated system for dispensing adhesives'. *International Journal of Adhesion and Adhesives*. 1991;**11**(3):187–91.

[9] Wiedenhöfer M. *Prozessgeregeltes Mikrodosieren hochviskoser Klebstoffe*. Dissertation, Universität Stuttgart; 2009.

[10] Frisch L. 'Pem fuel cell stack sealing using silicone elastomers'. *Sealing Technology*. 2001;**2001**(93):7–9.

[11] Nguyen M.N., Grosse M.B. 'Low moisture polymer adhesive for hermetic packages'. *Electronic Components and Technology Conference, 1992. Proceedings., 42nd*; 1992. pp. 239–44.

[12] Vecchio F., Venkatraman V., Shea H., Maeder T., Ryser P. 'Dispensing and hermetic sealing Rb in a miniature reference cell for integrated atomic clocks'. *Procedia Engineering*. 2010;**5**(0):367–70.

[13] Tropmann A., Lass N., Paust N., *et al.* 'Pneumatic dispensing of nano- to picoliter droplets of liquid metal with the StarJet method for rapid prototyping of metal microstructures'. *Microfluidics and Nanofluidics*. 2012;**12**(1-4):75–84.

[14] Patel B.K.C., Morgan H.W., Daniel R.M. 'A simple and efficient method for preparing and dispensing anaerobic media'. *Biotechnology Letters*. 1985;**7**(4):277–8.

[15] Bruneaux J., Therriault D., Heuzey M.-C. 'Micro-extrusion of organic inks for direct-write assembly'. *Journal of Micromechanics and Microengineering*. 2008;**18**(11):115020.

[16] Hanoka J.I., Danielson S.E. 'Method for forming contacts: Mobil solar energy Corporation'.

[17] Specht J., Zengerle K., Pospischil M. 'High aspect ratio front contacts by single step dispensing of metal pastes'. *Proceedings of the 25th European Photovoltaic Solar Energy Conference and Exhibition*; 2010. pp. 1867–70.

[18] Spribille A., Clement F., Erath D., Specht J., Biro D., Preu R. 'Dispensing of etching paste and inkjetting of diffusion barrier for MWT solar cell processing'. *Proceedings of the 25th European Photovoltaic Solar Energy Conference and Exhibition*; 2010. pp. 2654–8.

[19] Chen X., Church K., Haixin Y. 'High speed non-contact printing for solar cell front side metallization'. *Photovoltaic Specialists Conference (PVSC), 2010 35th IEEE*; 2010. pp. 1343–7.

[20] Chen C.-W., Chen X.B., Church K.H. 'High efficiency screen printed low-medium concentrator silicon solar cells with direct printed 50\mum wide fingers'. *Photovoltaic Specialists Conference (PVSC), 2012 38th IEEE*; 2012. pp. 928–31.

[21] Chen X.B., Church K.H., Haixin Y., Cooper I.B., Rohatgi A. 'Improved front side metallization for silicon solar cells by direct printing'. *Photovoltaic Specialists Conference (PVSC), 2011 37th IEEE*; 2011. pp. 3667–71.

[22] Lohmüller E., Thaidigsmann B., Pospischil M., *et al.* '20% efficient passivated large-area metal wrap through solar cells on boron-doped Cz silicon'. *IEEE Electron Device Letters*. 2011;**32**(12):1719–21.

[23] Pospischil M. *A parallel dispensing system for an improved front surface Metallization of silicon solar cells*. Freiburg: Dissertation, Fraunhofer Institute for Solar Energy Systems ISE, Albert-Ludwigs-Universität; 2016.

[24] Pospischil M., Klawitter M., Kuchler M., *et al.* 'Process development for a high-throughput fine line metallization approach based on dispensing technology'. *Energy Procedia*. 2013;**43**:111–16.

[25] Pospischil M., Klawitter M., Kuchler M., *et al.* 'High speed dispensing with Novel 6" Print Head'. *Energy Procedia*. 2016;**98**:61–5.

[26] Pospischil M., Klawitter M., Kuchler M. 'Development of a high-throughput fine line metallization process using CFD-simulation'. *Photovoltaic Specialists Conference (PVSC), 2013 IEEE 39th*; 2013. pp. 2250–3.

[27] Clement F., Lorenz A., Pospischil M. 'High throughput printing for highly efficient cost-effective Si solar cells'. *33rd EU PVSEC*. Amsterdam, The Netherlands; 2017.

[28] Pospischil M. 'PERC meets parallel dispensing: tuning the front side of silicon solar cells by Contact-less printing'. *Constance*. 2017.

[29] Pospischil M., Riebe T., Jimenez A. 'Applications of parallel dispensing in PV metallization'. *AIP Conference Proceedings*; 2019. p. 20005.

[30] Gensowski K., Jimenez A., Freund T., *et al*. 'CIGS mini-modules with dispensed metallization on transparent conductive oxide layer'. *Solar RRL*. 2020;**4**(12):2000475.

[31] Gensowski K., Tepner S., Schweigert S., *et al*. 'Conductive highly filled suspensions for an electrochemical dispensing approach to pattern Full-Area thin metal layers by physical vapour deposition'. *Scientific Reports*. 2020;**10**(1):7409.

[32] Schwirtlich I., von Campe H. '*Verfahren zum Ausbilden einer Struktur*: RWE Schott Solar GmbH, vol. WO2005088730 A3'.

[33] Fork D.K., Hantschel T. Extrusion/dispensing systems and methods. US 2007/0110836A1; 2005.

[34] Beutel M., Lewis A., Prondzinski M., *et al*. 'Fine line metallization by coextrusion technology for next generation solar cells'. *Solar Energy Materials and Solar Cells*. 2014;**131**(0):64–71.

[35] Richter L.P., Fischer G., Sylla L., *et al*. 'Progress in fine-line metallization by co-extrusion printing on cast monosilicon PERC solar cells'. *Solar Energy Materials and Solar Cells*. 2015;**142**:18–23.

[36] Fork D.K., Kalio A. '*Micro-Extrusion Printhead Nozzle with Tapered Cross-Section*: Palo Alto Research Center Incorporated. US8226391B2, published 2008'.

[37] Pospischil M., Kuchler M., Klawitter M., *et al*. 'Dispensing technology on the route to an industrial metallization process'. *Energy Procedia*. 2015;**67**(0):138–46.

[38] Gensowski K., Jimenez A., Tepner S. Dispensing of low temperature silver pastes. *Proc. of the 9th Workshop on Metallization and Interconnection*; 2020.

[39] Pospischil M., Fellmeth T., Brand A., *et al*. 'Optimizing fine line dispensed contact grids'. *Energy Procedia*. 2014;**55**:693–701.

[40] Pospischil M., Specht J., König M., *et al*. 'Paste rheology correlating with dispensed finger geometry'. *IEEE Journal of Photovoltaics*. 2014;**4**(1):498–503.

[41] Shin D.-Y., Yoo S.-S., Song H.-eun., Tak H., Byun D. 'Electrostatic-Force-Assisted dispensing printing to construct high-aspect-ratio of 0.79 electrodes on a textured surface with improved adhesion and contact resistivity'. *Scientific Reports*. 2015;**5**:16704.

[42] Tous L., Beaucarne G., Schubert G., Hoornstra J. *Proceedings of the 8th Workshop on Metallization and Interconnection for Crystalline Silicon Solar Cells*; Konstanz, Germany. [Melville, NY]: AIP Publishing, 13–14 May 2019; 2019.

[43] Kunpai C., Kang M.G., Song H.-eun., Shin D.-Y. 'Fine front side metallisation by stretching the dispensed silver paste filament with graphite nanofibres'. *Solar Energy Materials and Solar Cells*. 2017;**169**(2):167–76.

[44] Mihailetchi V.D., Chu H., Galbiati G., Comparroto C., Halm A., Kopecek R. 'A comparison study of N-type PERT and Ibc cell concepts with screen printed contacts'. *Energy Procedia*. 2015;**77**:534–9.

[45] Astorga E.N., Enriquez M.M., Moreno J.R., Barrado J.R.R. 'Feasibility of dispensing technology to create local aluminum contacts on silicon solar cells'. *2015 10th Spanish Conference on Electron Devices (CDE)*; Aranjuez - Madrid, Spain; 2015. pp. 1–3.

[46] Fellmeth T., Ourinson D., Riebe T., *et al.* 'Al fire through contacts for P-type Bifacial PERC devices'. *Energy procedia*. 2019;**2147**:110002.

[47] Ourinson D., Fellmeth T., Tepner S., *et al.* 'Optimization of Al fire-through contacts for AlO_x–SiN_x rear passivated bifacial p-PERC'. *IEEE Journal of Photovoltaics*. 2020;**10**(6):1523–31.

[48] Erath D., Pospischil M., Keding R., *et al.* 'Comparison of innovative metallization approaches for silicon heterojunction solar cells'. *Energy Procedia*. 2017;**124**:869–74.

[49] Flexographic Technical Association. *Flexography: Principles and Practices 6.0*. 6th edn. USA: Flexographic Technical Association; 2013.

[50] Kipphan H. (ed.). *Handbook of Print Media*. Berlin: Springer; 2001.

[51] Ferd G., Rüesch A.G. Technische Daten Maschinensystem Gallus EM 280. 2015. Available from www.gallus-group.com/docs/default-source/product-documents/gallus-em-280/gallus_em_280_de.pdf?sfvrsn=16.

[52] Krebs F.C., Fyenbo J., Jørgensen M. 'Product integration of compact roll-to-roll processed polymer solar cell modules: methods and manufacture using flexographic printing, slot-die coating and rotary screen printing'. *Journal of Materials Chemistry*. 2010;**20**(41):8994–9001.

[53] Hösel M., Søndergaard R.R., Angmo D., Krebs F.C. 'Comparison of fast roll-to-roll flexographic, Inkjet, flatbed, and rotary screen printing of metal back electrodes for polymer solar cells'. *Advanced Engineering Materials*. 2013;**19**:995–1001.

[54] Zichner R., Baumann R.R. 'Roll-to-roll screen printed radio frequency identification Transponder antennas for vehicle tracking systems'. *Japanese Journal of Applied Physics*. 2013;**52**(5S105):DC24.

[55] Jansson E., Hast J., Petäjä J., Honkala J., Häkkinen J., Huttunen O.H. 'Improving conductivity of rotary screen printed microparticle silver conductors using a roll-to-roll calendering process'. *Journal of Print and Media Technology Research*. 2015;**4**(1):19–26.

[56] Gallus Group. Rotary screen printing – all you need to know about re-using Gallus Screeny screens. 2020. Available from https://www.gallus-group.

com/gallus/whitepaper/2016/03/07/rotativer-siebdruck-das-abc-für-die-wiederverwendung-von-gallus-screeny-sieben.

[57] Gallus Group. *A printing form in 30 minutes: gallus screeny* [online]. 2020. Available from www.gallus-group.com/en/screen-printing/screen-printing/processes/pre-press-process.

[58] Gallus Group. *Printing plate classic line* [online]. 2020. Available from www.gallus-group.com/en/screen-printing/screen-printing/products/classic-line.

[59] Kocher + Beck. *Screen mesh – product information* [online]. 2020. Available from www.kocher-beck.com/produkte-bereiche/printing-technology/tec-screen/siebgewebe/.

[60] Lorenz A., Linse M., Frintrup H. 'Screen printed thick film metallization of silicon solar cells – recent developments and future perspectives'. *Proceedings of the 35th Photovoltaic Solar Energy Conference (EUPVSEC)*; Brussels, Belgium; 2018. pp. 819–24.

[61] Gallus Group. *Parameters in rotary screen printing – the squeegee* [online]. 2016. Available from www.gallus-group.com/en/gallus/whitepaper/2016/03/07/parameters-in-rotary-screen-printing-the-squeegee#:~:text=Squeegee%20rubbers%20with%20three%20different,the%20softer%20the%20squeegee%20rubber.&text=The%20squeegee%20pressure%20should%20always,and%20as%20low%20as%20possible. [Accessed 18 May 2021].

[62] Gallus Group. *Parameters in rotary screen printing – the flow path* [online]. 2020. Available from www.gallus-group.com/en/gallus/whitepaper/2016/03/07/parameters-in-rotary-screen-printing-the-flow-path.

[63] Hahne P. 'Innovative Druck- und Metallisierungsverfahren für die Solarzellentechnologie'. *PhD Thesis, Elektrotechnik, Fernuniversität Hagen, Hagen*. 2000.

[64] Lorenz A., Gredy C., Lehner M. 'Progress with rotational printing for the front side metallization of silicon solar cells'. *Proceedings of the 32nd Photovoltaic Solar Energy Conference (EUPVSEC)*; Munich, Germany; 2016. pp. 413–19.

[65] Lorenz A., Münzer A., Lehner M., *et al.* 'High-throughput front and rear side metallization of silicon solar cells using rotary screen printing'. *Energy Procedia*. 2017;**124**(24):680–90.

[66] Lorenz A., Münzer A., Lehner M., *et al.* 'Towards a high-throughput metallization for silicon solar cells using rotary-printing methods'. *Photovoltaics International*. 2017;**37**(no. 2):48–55.

[67] Lorenz A., Röth J., Zengerle K. 'Project "Rock-Star" – high-speed rotary printing for solar cell metallization: from vision to reality'. *Proceedings of the 37th European Photovoltaic Conference (EUPVSEC), Virtual Conference*; 2020.

[68] Berufsgenossenschaft Druck- und Papierverarbeitung. 'UV technology - a practical guide for all printing processes'. *Wiesbaden*. 2007.

[69] Schube J., Weil M., Fellmeth T., Keding R., Glunz S.W. 'Intense pulsed light meets the metallization of silicon heterojunction solar cells'. *Conference*

Records of the 46th IEEE Photovoltaic Specialists Conference (PVSC); Chicago, IL, USA; 2019. pp. 3231–7.

[70] Lorenz A. *Evaluierung von Rotationsdruckverfahren für die Metallisierung von Silicium-Solarzellen*. Freiburg: Dissertation, Institut für Mikrosystemtechnik und Informatik, Albert-Ludwigs-Universität; 2017.

[71] Tepner S., Ney L., Singler M. 'Screen utility simulation – explaining the evolution and future of screen printed metallization of Si-solar cells'. *Proceedings of the 37th European Photovoltaic Conference (EUPVSEC), Virtual Conference*; 2020. pp. 433–6.

[72] Ney L., Tepner S., Wengenmeyr N. 'Optimization of fine line screen printing using in-depth screen mesh analysis'. *Proceedings of the 8th Workshop on Metallization and Interconnection for Crystalline Silicon Solar Cells, Constance*; Germany; 2019. p. 020006.

[73] Gomez A. *A Historical essay on the development of flexography, Thesis*. Rochester: Rochester Institute of Technology; 2000.

[74] Deutschsprachiger Flexodruck-Fachverband e.V. Flexodruck Geschichte. 2015. Available from www.dfta.de/de/flexodruck/flexodruck-geschichte/geburt-historische-quellen/.

[75] Windmöller & Hölscher. *Technisches Datenblatt Vistaflex C* [online]. 2015. Available from www.wuh-lengerich.de/de/printing/druckmaschinen/flexodruckmaschinen/vistaflex-c/.

[76] Kwak M.K., Shin K.H., Yoon E.Y., Suh K.Y. 'Fabrication of conductive metal lines by plate-to-roll pattern transfer utilizing edge dewetting and flexographic printing'. *Journal of Colloid and Interface Science*. 2010;**343**(1):301–5.

[77] Cole K. *Printability and environmental testing using silver-based conductive flexographic ink printed on a polyamide substrate, Master Thesis*. Rochester: College of Imaging Arts and Sciences, Rochester Institute of Technology; 2007.

[78] Deganello D., Cherry J.A., Gethin D.T., Claypole T.C. 'Patterning of micro-scale conductive networks using reel-to-reel flexographic printing'. *Thin Solid Films*. 2010;**518**(21):6113–16.

[79] Wang Z., Winslow R., Madan D., *et al.* 'Development of MnO_2 cathode inks for flexographically printed rechargeable zinc-based battery'. *Journal of Power Sources*. 2014;**268**:246–54.

[80] Sidén J., Fein M.K., Koptyug A., Nilsson H.-E. 'Printed antennas with variable conductive ink layer thickness'. *IET Microwaves, Antennas & Propagation*. 2007;**1**(2):401.

[81] Spiehl D., Haming M., Sauer H.M., Bonrad K., Dorsam E. 'Engineering of Flexo- and Gravure-printed indium–zinc-oxide semiconductor layers for high-performance thin-film transistors'. *IEEE Transactions on Electron Devices*. 2015;**62**(no. 9):2871–7.

[82] Cushner S., Ni Fan R., Leberzammer E., Shea P., van Zoeren C. 'Laser engravable single-layer flexographic printing element'. *US 5798202 A Us 07/880*. 1998:792.

[83] van Cushner S., Fan R.N., Leberzammer E., Quinn J.A. 'Lasers engravable multilayer flexographic printing element'. *US5804353 A US 07/881,444*. 1998.

[84] Jansen M. 'HD-Flexodruck – mit hohen Farbstärken zum exzellenten Druckbild'. *Flexo + Tiefdruck*. 2018;**6**.

[85] Lorenz A., Gredy C., Senne A., *et al*. 'Flexo-printed busbarless solar cells for multi-wire interconnection'. *Energy Procedia*. 2016;**98**:46–60.

[86] Frey M. *Herstellung und Charakterisierung elektrisch leitfähiger Kontaktstrukturen auf Siliziumsolarzellen mittels Flexodruckverfahren, Master Thesis*. Darmstadt: Technische Universität, Darmstadt; 2010.

[87] Frey M., Clement F., Dilfer S., Erath D., Biro D. 'Front-side metalization by means of flexographic printing'. *Energy Procedia*. 2011;**8**:581–6.

[88] Thibert S., Chaussy D., Beneventi D. 'Silver ink experiments for silicon solar cell metallization by flexographic process'. *Conference Records of the 38th IEEE Photovoltaic Specialists Conference (PVSC)*; Austin, TX, USA; 2012. pp. 2266–70.

[89] Thibert S., Jourdan J., Bechevet B. 'Flexographic process for front side metallization of silicon solar cell'. *Proceedings of the 28th Photovoltaic Solar Energy Conference (EUPVSEC)*; Paris, France; 2013. pp. 1013–16.

[90] Thibert S., Jourdan J., Bechevet B., *et al*. 'Study of the high throughput flexographic process for silicon solar cell metallisation'. *Progress in Photovoltaics: Research and Applications*. 2016;**24**(2):240–52.

[91] Lorenz A., Kalio A., Hofmeister G.T. 'Flexographic printing – high throughput technology for fine line seed layer printing on silicon solar cells'. *Proceeding of the 28th Photovoltaic Solar Energy Conference (EUPVSEC)*; Paris, France; 2013. pp. 1017–23.

[92] Lorenz A., Kalio A., Barnes Hofmeister T., *et al*. 'Developing a high throughput printing technology for silicon solar cell front side metallisation using flexography'. *Journal of Print and Media Technology Research*. 2014;**4**:227–40.

[93] Lorenz A., Kraft A., Gredy C. 'Comprehensive comparison of different fine line printing technologies adressing the seed and plate approach with Ni-Cu-plating'. *Proceedings of the 30th Photovoltaic Solar Energy Conference (EUPVSEC)*; Hamburg, Germany; 2015. pp. 732–6.

[94] Lorenz A., Gredy C., Beyer S., *et al*. 'Flexographic printing – towards an advanced front side metallization approach with high throughput and low silver consumption'. *Solar Energy Materials and Solar Cells*. 2016;**157**:550–7.

[95] Zecher GmbH. *Around the anilox roller* [online]. 2015. Available from www.zecher.com/fileadmin/user_upload/Knowledge/Basics/AniloxBasics/AniloxBasics_DEU.pdf.

[96] Kraft A. *Plated copper front-side metallization on printed seed-layers for silicon solar cells, Dissertation*. Freiburg: Institut für Informatik und Mikrosystemtechnik (Imtek), Albert-Ludwigs-Universität; 2015.

[97] Kapur N., Hewson R., Slegh P.A., Summers J.L., Thompson H.M., Abbott S.J. 'A review of gravure coating systems'. *Convertech & e-Print.* 2011;**1**(4):56–60.

[98] Grau G., Cen J., Kang H., Kitsomboonloha R., Scheideler W.J., Subramanian V. 'Gravure-printed electronics: recent progress in tooling development, understanding of printing physics, and realization of printed devices'. *Flexible and Printed Electronics*. 2016;**1**(2):023002.

[99] de La Fuente Vornbrock A. 'Roll printed electronics: development and scaling of Gravure printing techniques'. *Eletrical Engineering and Computer Sciences, Berkeley, Technical Report No. UCB/EECS-2009-191*. 2009.

[100] Bornemann N., Sauer H.M., Dörsam E. 'Gravure printed ultrathin layers of small-molecule semiconductors on glass'. *Journal of Imaging Science and Technology*. 2011;**55**(4):040201.

[101] Philipp W. 'Rotary tampon printing press'. *US 6619203 B2 DE10119678*. 2003.

[102] Hashimoto K., Ouchi M., Nakamura N., Kobayashi E., Watabe Y. 'Low cost module with heterojunction solar cells applied gravure offset printing and multi-wire technologies'. *Proceeding of the 28th Photovoltaic Solar Energy Conference (EUPVSEC)*; Paris, France; 2013. pp. 1073–6.

[103] Brisbane A.D. 'Pattern deposit by laser'. *Google Patents*.

[104] Piqué A., Serra P. *Laser Printing of Functional Materials: 3d Microfabrication Electronics and Biomedicine*. Hoboken: John Wiley & Sons; 2018.

[105] Bohandy J., Kim B.F., Adrian F.J. 'Metal deposition from a supported metal film using an excimer laser'. *Journal of Applied Physics*. 1986;**60**(4):1538–9.

[106] Banks D.P., Grivas C., Mills J.D., Eason R.W., Zergioti I. 'Nanodroplets deposited in microarrays by femtosecond Ti:sapphire laser-induced forward transfer'. *Applied Physics Letters*. 2006;**89**(19):193107.

[107] Yang L., Wang C.-yue., Ni X.-chang., Wang Z.-jun., Jia W., Chai L. 'Microdroplet deposition of copper film by femtosecond laser-induced forward transfer'. *Applied Physics Letters*. 2006;**89**(16):161110.

[108] Schultze V., Wagner M. 'Laser-induced forward transfer of aluminium'. *Applied Surface Science*. 1991;**52**(4):303–9.

[109] Heise G., Domke M., Konrad J., Sarrach S., Sotrop J., Huber H.P. 'Laser lift-off initiated by direct induced ablation of different metal thin films with ultra-short laser pulses'. *Journal of Physics D: Applied Physics*. 2012;**45**(31):315303.

[110] Willis D.A., Grosu V. 'Microdroplet deposition by laser-induced forward transfer'. *Applied Physics Letters*. 2005;**86**(24):244103.

[111] Franzl Y., Widmaier Y., Kroyer T., Brand A., Kühnhold-Pospischil S., Nekarda J. 'Optical evaluation of laser generated plating seed layers'. *AIP Conference Proceedings*. 2019;**2156**(1):20012.

[112] Sano T., Yamada H., Nakayama T., Miyamoto I. 'Experimental investigation of laser induced forward transfer process of metal thin films'. *Applied Surface Science*. 2002;**186**(1–4):221–6.

[113] Fraunhofer-Institut für Solare Energiesysteme ISE. Projekt »C³PO«: Neues Laser-Kontaktierungsverfahren für Solarzellen entwickelt - Fraunhofer ISE. 2020. Available from www.ise.fraunhofer.de/de/presse-und-medien/presseinformationen/2020/projekt-c3po-neues-laser-kontaktierungsverfahren-fuer-solarzellen-entwickelt.html [Accessed Jul 2021].

[114] Kleine-Jäger F., Kaczun J. Method and printing press for printing a substrate. WO2010069902A1; 2009.

[115] Kleine-Jäger F., Hermes S., Gao X., *et al.* 'BASF's CypoSol metallization inks for newly developed laser transfer printing (LTP) technology'. *Proceedings 2nd Workshop*. 2010.

[116] Sauter T., Hofmann M., Buchner C. Method and device for applying printing substance. German Patent DE102011075025A1; 2011. Available from https://patents.google.com/patent/DE102011075025A1/en?oq=DE102011075025A1.

[117] Adrian A., Rudolph D., Willenbacher N., Lossen J. 'Finger metallization using pattern transfer printing technology for c-Si solar cell'. *IEEE Journal of Photovoltaics*. 2020;**10**(5):1290–8.

[118] Buchner C., Sauter T. Method and apparatus for producing a solar cell. WO2009090098A2; 2008. Available from https://patents.google.com/patent/WO2009090098A2/en?oq=WO2009090098A2.

[119] Rodofili A., Wolke W., Kroely L., *et al.* 'Laser transfer and firing of NIV seed layer for the metallization of silicon heterojunction solar cells by Cu-plating'. *Solar RRL*. 2017;**1**(8):1700085.

[120] Roeder T.C., Hoffmann E., Kohler J.R., Werner J.H. '30 μm wide contacts on silicon cells by laser transfer'. *35th IEEE Photovoltaic Specialists Conference (PVSC)*, 2010: 20 – 25 June 2010; Honolulu, HI, USA; 2010. pp. 3597–9.

[121] Adrian A., Rudolph D., Lossen J., Matusovsky M., Chandrasekaran V. 'Benefits of pattern transfer printing method for finger metallization on silicon solar cells'. *Proceedings 35th European Photovoltaic Solar Energy Conference and Exhibition*; 2018. pp. 434–8.

[122] Yüce C., Okamoto K., Karpowich L., Adrian A., Willenbacher N. 'Non-volatile free silver paste formulation for front-side metallization of silicon solar cells'. *Solar Energy Materials and Solar Cells*. 2019;**200**(12):110040.

[123] Kim H., Kushto G.P., Arnold C.B., Kafafi Z.H., Piqué A. 'Laser processing of nanocrystalline TiO_2 films for dye-sensitized solar cells'. *Applied Physics Letters*. 2004;**85**(3):464–6.

[124] Wang L., Carlson D.E., Gupta M.C. 'Silicon solar cells based on all-laser-transferred contacts'. *Progress in Photovoltaics: Research and Applications*. 2015;**23**(1):61–8.

[125] Kleine-Jeager F., Kaczun J., Lehmann U. Printing machine and method for printing a substrate. Patent number: US8840237B2. 2009. Available from https://patents.google.com/patent/US8840237B2/en?oq=US8840237B2 [Accessed 23 Sep 2014].

[126] Hoffmann E., Röder T.C., Conrad B., Köhler J.R., Werner J.H. Two step process for optimized laser transferred contacts. *In Proceedings 26th EU PV Solar Energy Conference*; 2011. pp. 1640–3.

[127] Röder T.C., Hoffmann E., Konrad B., Köhler J.R. 'Low temperature laser Metallization for silicon solar cells'. *Energy Procedia*. 2011;**8**:552–7.

[128] Rodofili A., Wolke W., Kroely L., *et al.* Laser-transferred NiV-seed for the metallization of silicon heterojunction solar cells by Cu-plating. *33rd European Photovoltaic Solar Energy Conference and Exhibition*; 2017. pp. 402–5.

[129] Adrian A., Rudolph D., Lossen J., Willenbacher N. 'Investigation of thick-film-paste rheology and film material for pattern transfer printing (PTP) technology'. *Coatings*. 2021;**11**(1):108.

Chapter 8

Plated contacts

Jonas Bartsch[1], Andreas Büchler[1], and Sven Kluska[1]

The solar cell metallization approach discussed in this chapter differs from the ones discussed otherwise in this book in its property that the metal contact is composed of different layers, all of which are designed to ideally fulfil their role in the solar cell contact. These roles range from creating the electrical contact with the semiconductor substrate (seed layer, either plated or deposited otherwise), over the function of current conduction (highly conductive plated layer), to the compatibility with the module integration (plated capping layer). The contact system is thus more flexible, better performing and can be adjusted more precisely to the technological needs of the solar cell, but on the other hand the process complexity is typically higher. This chapter focuses on the contacting of silicon in cell concepts where a silicon wafer is coated with a dielectric passivation and antireflection coating. Other cell concepts (Silicon Heterojunction (SHJ) solar cells, tandem solar cells, etc.) may also be metallized with plating and are addressed in Chapter 9.

Metal plating for silicon solar cell contacts has been developed both for making ohmic electrical contact between silicon and metal and for creating well conducting paths to drive the photocurrent to the external circuit. The former approach has even been the first ever attempt to create a metal contact specifically on a silicon solar cell in Bell labs [1], which had already proved to be challenging then. This approach is addressed in Subchapter 8.1. A full overview of the history of plating in silicon photovoltaics (PV) is given by Lennon *et al.* [2]. The approach of just plating a well conducting current transport layer can be combined with any conducting 'seed layer', either plated or deposited otherwise. It is discussed in more detail in the second part of this chapter.

Despite many advantages, plated metallization has not become mainstream in the PV industry so far. Still there are several examples where plating has made its way into production in the past. Also, at present, there is interest in adopting this technology for mass production. This requires concepts and tools, which are discussed in the third part of this chapter, alongside considerations regarding cost and issues related to mass production.

[1]Fraunhofer ISE, Heidenhofstraße 2, Germany

We start with a general introduction into the technology, which will make the specific descriptions in the PV-related subchapters easier to read.

8.1 Fundamental principles and nomenclature

Metal plating is an electrochemical process that is usually denoted as electrolysis. There is extensive literature on electrochemical processes where very detailed descriptions of the process can be found [3–5]. Its fundamental working principle is to immerse a workpiece that is to be coated with a metal into a (typically aqueous) solution of a salt of this metal (called 'electrolyte'). In electrochemistry, such a workpiece that is in contact with an electrolyte is called 'electrode'.

The salt will dissolve and decompose into ions, where one ionic species are metal ions (typically positively charged) and another species are so-called 'counter-ions' that are often of inorganic nature. Typical examples for metal salts are $CuSO_4$ (Cu^{2+} metal ions and SO_4^{2-} counter ions) or $AgNO_3$ (Ag^+ metal ions and NO_3^- counter ions). To deposit a metal layer from such a solution, electrons at the surface of the workpiece discharge the metal ions, which become metallic and form a solid material on the surface.

There are several ways to provide these electrons:

- By providing an external current and voltage (typically denoted as **electroplating**).
- By providing a chemical species that decomposes at the surface and, in that, provides electrons to the surface (typically denoted as **electroless plating**).
- By chemical attack (oxidation, typically dissolution) of the surface material, which provides electrons itself (**displacement deposition**).

Electroless plating and **displacement deposition** are comparatively slow and require more efforts in process control to yield reproducible deposition results. They will be discussed only briefly here for that reason. Historically, electroless plating has been among the first metallization techniques to be demonstrated on solar cells [1] and it has also been commercialized in BP Solar's 'SATURN' cell [6], which featured electroless plating of nickel and copper on both polarities of the solar cell. Especially the process to deposit copper with a layer thickness of several micrometres must have been very time-consuming (>1 hour based on experience with electroless copper plating chemistries). The big advantage of both plating approaches discussed here is their simplicity. They work by simply immersing the solar cell into a suitable plating bath. For electroless plating, the surface to be plated must serve as a so-called 'catalytically active' surface, meaning it should feature the properties suitable for a reducing agent (typically sodium hypophosphite for nickel deposition or formaldehyde or newer replacement chemistries for copper deposition). The decomposition of the reducing agent releases electrons into the surface that are transferred to positions where the metal ions from the solution are discharged and metal layers are deposited. Nickel deposition with such electrolytes works very well on all kinds of differently doped silicon, copper deposition works well on nickel layers. For both types, deposition rates range around 100 nm per minute. Also, seeds of palladium can be deposited on silicon, using tin in different oxidation states as reducing agent.

Displacement depositions also require certain properties from the substrate. The electrolyte is designed in a way that the substrate's surface is dissolved (oxidized, meaning it releases electrons). These electrons can be used to discharge metal onto (or into) the surface. Due to the partial dissolution of the surface, a certain degree of interlocking between metal and substrate is natural in this kind of deposition, which can be beneficial for metal adhesion on silicon [7]. The reaction is self-terminating, as the deposited metal blocks the surface to be oxidized from further contact with the deposition solution. Nickel and palladium are often deposited on silicon (and other substrates) by displacement deposition. Palladium is then often used to serve as a catalytically active surface for subsequent electroless nickel plating. This process is used to deposit metal onto plastics, which is a huge branch of plating industry and thus, this process is very well developed. A detailed overview of the process is given by Mallory and Hajdu [8]. Also, silver plating to cap copper can be done this way, yielding a 100–200 nm thin silver layer that improves solderability and shields copper from corrosion and effects of ageing in the solar module.

As both deposition techniques are slow, they are mostly used to either create a first, thin 'initiation' and adhesion layer on silicon for subsequent electroplating (e.g. for machine types that require a metal contact layer) or they are used in solar cell concepts that require only thin metal layers [9], which are becoming more interesting. The additional efforts associated with electroless plating are so far a too high hurdle to allow industrial implementation.

By far the most effective way to create a metal coating on a workpiece is by **electroplating**, which will be discussed in more detail below. A schematic drawing of a typical setup is shown in Figure 8.1. The workpiece is connected to the negative terminal of the external current/voltage source and denoted 'cathode'; the current passing through this electrode into the electrolyte is called 'cathodic current'. The reaction that takes places at this electrode leads to the reduction of the charge of the positively charged ionic metal species from the solution and is therefore called 'reduction reaction'.

The positive terminal is connected to a counter electrode (anode) that is also immersed in the electrolyte, where a counter reaction (oxidation) that releases electrons into the external circuit takes place.

As the current in electrochemical processes is typically proportional to the electrode surface, it is common to consider current densities rather than absolute currents. For a long time, the rather unusual (compared to other applications) unit 'amperes per square decimeter' (ASD) has been used. It is still common, but the (more graspable for people outside the metal plating community) unit mA cm^{-2} is becoming more and more common; 1 ASD corresponds to 10 mA cm^{-2}.

Electroplating is very simple, there is direct control over the process with the external current/voltage characteristics and it is cheap, well developed and can be operated in a stable fashion, depending on the electrolyte and the operating

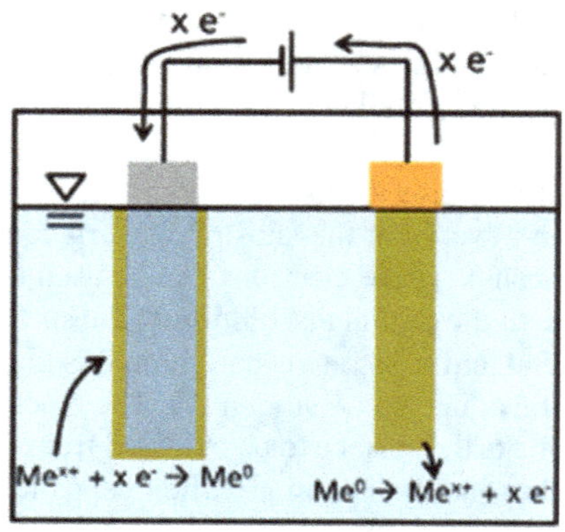

Figure 8.1 *Schematic of the electroplating process with soluble anode. Oxidation reaction of the anode changes for insoluble anode (typically oxidation of solvent)*

conditions, over months or even years. Metal deposited can either be replenished by adding metal salt (typically in combination with an insoluble counter electrode) or by using a soluble anode of the metal to be deposited, which is dissolved in ionic form at that electrode. Mostly, the latter is practically realized by providing a conductive, insoluble container ('anode basket') which is filled with pellets of the metal to be plated.

The deposition rate is one of the highest for metal deposition processes and generally a lot higher as compared to technologies such as Physical vapor deposition (PVD). The deposition rate is governed by Faraday's law, which basically states that a certain amount of charge (96 485 As) corresponds to 1 mol ($6.02214076 \times 10^{23}$) of electrons (Avogadro constant). Thus, the Faraday constant F is 96 485 As*mol^{-1}. Assuming that each electron passing the electrode surface contributes to the deposition of metal, the mass of deposited metal can be precisely calculated:

$$Q^*t = n^*z^*F$$

with Q being the charge passing the electrode, t the time, z the number of electrons needed to fully discharge one metal ion (one for silver and two for copper in the above example) and n the amount of metal in mol.

With n = m/M the mass of metal can be calculated. Assuming a completely homogeneous coverage of the metal to the surface, the current density can by correlated with a growth rate of layer thickness per time considering a metal density, which is given for a few examples below:

Typical current densities range from 1 ASD to 10 ASD (10–100 mA cm^{-2}), but there are also special electrolytes that deposit with current densities of 20 ASD and beyond. The above assumption that each electron contributes to the deposition is often expressed in the so-called 'current efficiency' between 0% and 100% (0% meaning no electron contributing to the desired reaction, e.g. metal deposition, 100% meaning all electrons contributing). Most commercially available electrolytes

Table 8.1 Layer growth rates for different metals by electroplating, depending on the current density

Growth rate	1 ASD 10 mA cm^{-2}	2 ASD 20 mA cm^{-2}	5 ASD 50 mA cm^{-2}	10 ASD 100 mA cm^{-2}	20 ASD 200 mA cm^{-2}	50 ASD 500 mA cm^{-2}
Cu^{2+}	0.22	0.44	1.11	2.22	4.43	11.08
Ag^{+}	0.64	1.28	3.2	6.39	12.79	31.97
Ni^{2+}	0.2	0.41	1.02	2.05	4.1	10.24
Sn^{2+}	0.64	1.28	3.2	6.4	12.8	31.99
Au^{3+}	0.21	0.42	1.06	2.11	4.23	10.57

All values in micrometres per minute.

have current efficiencies of very close to 100% when operated within the specified parameter range, so the values in Table 8.1 are realistic.

Together with tool design and substrate and process conditions, the electrolyte is one of the fundamental influencing factors of the process. It will determine how much mechanical (compressive or tensile) stress is present in the layer, if it grows densely, porously or even dendritic, how uniformly the deposit will be over the surface area, how stably the process runs and so forth. For many years, electrolyte manufacturers have put great effort into understanding the influence of the electrolyte on the deposition characteristics. They are capable of creating electrolytes tailor-made for a customer's application by adding complexing agents, surfactants, levelling agents, inhibitors, etc. These chemical 'additives' may, for instance hinder the deposition to areas with high current density by blocking these surface sites, thus yielding a more uniform deposition by favouring sites that would otherwise be underplated. Even though electrolytes available on the market serve well for current plating applications on solar cells, there might still be a lot of potential in optimized electrolytes in terms of performance, deposition rate, and contact geometry and morphology.

The electrolyte also brings with it one of the fundamental limitations of metal plating in terms of material choice: when dealing with aqueous electrolytes, only metals with an electrochemical standard potential of their reduction reaction within the electrochemical window of water can be electroplated.[a] This is true for the metals in the table above, and several more (e.g. Zn, Co) but for instance, plating of aluminium or plating of titanium is not achievable from aqueous electrolytes. There are organic solvents or ionic liquids that allow such deposition, but these formulations are still quite far from being commercialized and much less mature as compared to aqueous solutions, which will be the focus in this chapter.

[a]Hydrogen evolution as competing reaction, standard potential E_0 of −0.83 V versus standard hydrogen electrode at chemical activity of 1, 25 °C, 1 013 mbar.

There are several huge **advantages** to electroplating. The process is realized at low temperature (typically the electrolyte is operated at room temperature or at temperatures up to 50 °C), the conductivity of the deposited metal is very close to the bulk metal conductivity reported in literature, and the solutions and tools are cheap and can be operated for long durations (up to years, depending on the operation conditions).

One drawback of this method is the necessity to form a low-resistivity electrical contact to each part of the workpiece that is to be coated with metal, to drive electrons into these positions. Electroplating is highly developed for all kinds of metallic surfaces, where it has been intensively used on an industrial scale for many years. Electrical contact and current transport are not so much of an issue on such workpieces.

For solar cells that feature a poorly conducting base material and where very narrow structures often have to be coated, this prerequisite is harder to fulfil. Also creating an electrical contact on a fragile device such as a solar cell is even more challenging, especially if large currents for high current density are desired. The contacting, or strategies to simplify or avoid contacting, will therefore be an important part of the following subchapters. Apart from this, processing conditions and results obtained with these conditions will be discussed. The specific properties of plated contacts and their influence on solar cells will also be covered.

8.2 Direct plating on Si

In this section, different technologies for direct plating are introduced with the understanding that the solar cell surface is electrically passivated before the plating sequence. Assuming an appropriate thickness and density of the passivation layer, it omits leakage currents due to tunnelling and channelling effects. Then, a surface passivation layer is an electrical barrier between plating bath and substrate. Therefore, the passivation layer can be used as mask during the plating process. Local removal of the passivation layer enables local electron exchange and leads to local deposition. The structure of the metal grid for direct plating approaches is determined by local modification or removal of the passivation layer. Unimpaired areas mask the deposition and metal is deposited in the passivation layer openings. The sequence of direct plating process consists of at least two steps: Patterning (definition of contact positions), by local modification or removal of the passivation layer, and the actual plating process. Both steps may be accompanied by pre- and post-treatments, such as removal of surface oxide within the passivation layer openings before metal deposition and a contact formation step after deposition.

In the following, common methods for pattering and deposition are introduced. Boundary conditions for pre- and post-treatments that enable successful sequences are sketched. Contact properties and challenges that are related to process reliability and complexity are discussed.

8.2.1 Patterning methods

The pattern of the plated metal contacts is determined during a structured modification or removal of the solar cell passivation layer that enables local electron exchange of plating bath and solar cell. The patterning process might simultaneously modify the surface near solar cell volume. Different pattering technologies are characterized by the respective precision of the opening width, the positioning of the opening and the modification of the silicon surface.

As the minimum width of direct plated contacts is limited by the size of the contact opening, minimum structure size of the patterning methods is an essential characteristic.

A high position accuracy is of particular importance if the front-side contacts need to be aligned to pre-processes (e.g. for solar cell concepts with implanted or diffused selective emitters).

Given the amount of process variants, the patterning process needs to be understood as a process that defines not only the position of the contact area but may also locally change essential contact properties such as surface roughness, surface oxidation, surface doping density.

Application of costly 'mask and etch'-based approaches and chemical removal of the passivation layer reduces the effect on the surface near the silicon volume and allows for precise control of surface properties. Mask-and-etch approaches are chosen for instance in laboratory attempts for conversion efficiency records.

Laser ablation of the passivation layer is a low-cost, high-throughput mass-scale feasible patterning approach. Laser ablation impedes the surface-near volume. Well-controlled processing enables intentionally improving the surface properties during patterning. Laser-induced passivation layer openings are usually less homogenous than chemically etched passivation layer openings, with respect to surface topography and passivation layer debris, but the laser-induced roughness improves adhesion. Laser ablation might be considered as current state-of-the-art patterning technology for industrial fabrication of directly plated contacts.

8.2.1.1 Wet chemical etching

In order to selectively remove the passivation layer without impeding the solar cell substrate, etches with high selectivity for passivation layer material and (doped) silicon are necessary. Table 8.2 summarizes common etchants for usual passivation layer materials.

Structured removal of the passivation by wet chemical etching is achieved by using etch-resistant masks, structured with a 'negative' pattern of the desired contact structure. In order to prevent unintended modifications of the passivation layer apart from contact openings (which may result in parasitic deposition during plating) the mask needs to feature appropriate density and thickness. Further, underetching which may result from a non-optimal combination of etchant, resin and passivation layer material needs to be either suppressed or included in the design, as it causes a broadening of the passivation layer opening with respect to the mask opening width.

Table 8.2 Common etchants for passivation layer etching in c-Si solar cells

Passivation layer	Etch, process temperature	Etch rate on passivation layer	Etch rate on silicon
SiO_2	10:1 HF, Room temperature	23 nm /min	-
Si_xN_y	10:1 HF, Room temperature	1 nm/min	-
	H_3PO_4, 160 °C	42 nm/min	-
AlO_x	KOH (30%, 80 °C)	>800 nm/min	1 100 nm/min

Etch rates taken from [10, 11].
HF, hydrofluoric acid.

Photolithography is the most accurate way to apply such masks. Up-to-date photolithography enables fabrication of masks with openings widths of down to single digit μm or even below, which can then also result in such small contact openings by applying an optimized etching procedure. Precision, controllability and high resolution for patterning makes the approach inevitable in the context of solar cell research and for attempts to produce highest-efficiency devices. The process is not suitable for mass scale industry due to the low throughput of lithographic process sequences.

Previously, it was suggested to printa resistant mask onto the substrate by ink-jet printing. This process holds the potential to be applied in inline manufacturing. However, both techniques suffer from the need to deposit a large amount of organic resist, which adds to the processing cost as a consumable, and which needs to be stripped from the substrate after the metallization process in an additional step. An alternative that has been explored is printing of etchant pastes, which are printed on only small portions of the solar cell, but are hard to handle in printing processes due to the aggressive nature of the used chemicals towards both the solar cell and the printer materials.

In an optimized wet chemical pattering process, the solar cell surface is not harmed during the structured removal of the passivation layer.

8.2.1.2 Laser ablation

For passivation layer ablation various laser systems are used. Usage of continuous wave (cw) lasers and pulsed systems is reported. Further, different wavelengths, in the UV, green and red range are used. Each configuration leads to a different intensity and distribution of energy, that affects the solar-cell/passivation layer-stack in different ways. The recommended configuration of pulse length and wavelength depends on the intended contact system.

The most prominent approaches are 'minimally invasive' direct ablation of the passivation layer by picosecond-pulsed UV lasers and laser-doped selective emitters, that are driven-in during indirect ablation of the passivation layer by nanosecond-pulsed or continuous wave-lasers.

Laser ablation of the passivation layer always affects the surface-near volume of the solar cell, which might lead to improvement of the contact properties if the modifications are controlled. The following excursion tries to sketch the principles of laser ablation and to derive some guidelines for process variations.

Laser ablation of silicon nitride is generally classified into direct or indirect [12]. In both cases, the laser energy is absorbed in the passivation layer and in the surface near volume, but the share is different. Indirect ablation indicates that most of the laser fluence is transmitted through the dielectric layer. The silicon substrate is heated and blasts off the overlying layer. Indirect ablation is performed by using nanosecond-pulsed lasers or continuous wave-lasers for patterning. Direct ablation relates to passivation layer removal by laser energy directly absorbed in the passivation layer. The high energy density causes evaporation of the layer. This is challenging since the passivation layers for silicon solar cells are optimized for low absorption. Direct ablation is enforced by ultra-short laser pulse times in the picosecond-regime and below (also usage of femtosecond-pulses is reported) and/or by (sub-)UV laser wavelength.

As indirect ablation goes along with melting of the surface near the solar cell, it allows for changes of the contact properties, but it might impact the crystal quality. Hence, it enables divers process variants, but the micro-scale effects on the surface need to be well understood and precisely controlled. Direct ablation usually aims for passivation layer ablation with minimum effect on charge carrier lifetime. If pulsed lasers are used, each laser pulse creates an 'ablation dot'. Line-like passivation layer openings are created by a chain of subsequent ablation dots. The offset of subsequent points is controlled by the velocity of the beam movement on the solar cell surface and the pulse repetition rate. As the beam velocity needs to be in the time scale of the pulse repetition rate, usually galvano scanners are used to guide the laser beam. Depending on the offset and the diameter of the laser ablation dots, subsequent ablation dots in a line-like chain (finger opening), or neighbouring dots in a set of lines (busbar opening) might overlap. Then the surface area is effectively treated by several laser pulses. This requires some more considerations. The energy distribution within the laser beam typically has a Gaussian shape. In the centre of the beam the fluence is above the passivation layer ablation threshold and thus high enough to evaporate the passivation layer. Usually, each ablation laser pulse generates a circular ablation dot in the passivation layer. Line-like contacts are realized by a chain of subsequent dots. Plating allows to create closed metal lines on dots with a certain distance, if the plating process is isotropic. This effect allows for modification of the ratio of metallized and contact area. The ratio of dot diameter and dot distance can be parameterized as pulse overlap. An overlap of 0% means that subsequent ablation dots touch without overlapping. An overlap of 50% means that the middle of a new dot is on the edge of the previous ablation dot. Maximum unmetallized area is achieved by decreasing the overlap (in other words increasing the pulse distance). Homogenous like-line contact openings are created using higher overlaps.

The circular surrounding of each ablation dot is exposed to energy slightly below the threshold for passivation layer ablation. The silicon surface in the circular surrounding of the ablation dots may also be modified. The area modified by the laser is called heat affect zone (HAZ). The crystallinity of the surface near silicon is impaired by laser ablation. Indirect ablation can go along with melting of silicon. Melting and subsequent recrystallization may lead to crystal defects that impede the charge carrier lifetime along the contact opening. Direct ablation is favoured, as the impact on the silicon substrate is expected to be lower [13]. The multiple treatment for laser openings that feature overlap of subsequent pulses increases the modification of the silicon surface, in particular, the oxidation and the melting of the surface structure [14]. The silicon surface is reported to be converted to amorphous silicon (a-Si) within the first few atomic layers [15] during passivation layer ablation using a picosecond-pulse UV laser at the passivation layer threshold. There are no consistent reports whether a-Si is generated throughout the HAZ or if it follows a certain symmetry, but this may depend primarily on the used laser set-up.

It was reported that the effect on charge carrier lifetime (and open circuit voltage of the final solar cell) is more severe for higher pulse overlaps during ablation. It was further shown that this effect is decreased by annealing the solar cell after laser ablation. If laser-structured plated contacts are used in combination with printed metallization (for instance on a Passivated Emitter and Rear cell (PERC) with printed Al back-side), this observation motivates performing laser ablation of the passivation layer before the firing step.

The most important intentional modification during laser ablation is the change of the surface roughness and the surface doping concentration. Roughness affects contact adhesion. Increased doping concentrations shield laser damage and allow for fabrication of ohmic contacts.

There are several attempts to combine passivation layer structuring with the formation of selective emitters. The dopant source can for instance be spun on the passivation layer before laser ablation or the passivation layer may contain the dopant source [16]. For laser doped selective emitters (LDSE), it was highlighted that additional doping shields the laser-induced damage [17]. Different approaches for LDSE, using green or infra-red laser sources, are under discussion [18].

The front side of the monocrystalline silicon solar cells has a random pyramid texture. Melting of silicon causes a modification of the surface texture. The intensity of the modification is related to the wavelength and pulse duration of the laser ablation. Approaches based on indirect ablation such as nanosecond-pulsed or cw-lasers usually induce a melting of the texture pyramids during the impact of the laser which may imply a flattening of the pyramidal shape. For picosecond-pulsed lasers that are operated at the threshold fluence for passivation layer ablation, the laser pulses interact with the substrate in a highly non-linear process. Although this process is referred to as direct and cold ablation [19] the silicon surface is modified by picosecond-pulsed lasers [13]. Interference of incoming and reflected laser light in the instant of the interaction of laser beam and substrate induce a highly periodic exaltation of the laser fluence along the surface. The pattern shows similar shape and distribution as laser-induced periodic surface structures (LIPSS) that are observed for many other applications of ultra-short pulse laser processing [20]. The rough structure is characterized by overhanging silicon rags that create cavities that are filled by metal during plating. Contacts then are 'keyed' in the solar cell surface.

As it enables high adhesion and low effect on charge carrier lifetime, picosecond-pulse UV lasers are currently the favoured laser ablation set-ups beside LDSE approaches. Direct ablation and LDSE are objects of ongoing research [15, 21].

8.2.2 Deposition methods

Direct plating on silicon is challenging due to its semiconducting nature. As introduced above the deposition of metal ions from the electrolyte onto the substrate requires an electron exchange. Plating techniques can be classified according to the mechanism that enforces electron transfer between substrate surface and dissolved ions. The electron transfer can be realized by a pair of red-ox reactions (*electroless plating*), or by displacement reaction or be enforced by the application of an external current (*electrodeposition*).

Initially options that avoid external currents seem more attractive, as the fragile solar cell simply needs to be immersed into an electrolyte solution and deposition starts. The patterning of a grid can be realized by resist masking or by patterning of a dielectric layer, which is insusceptible to all kinds of plating techniques (see section 8.2.1). Indeed, the first approaches to plate onto solar cells have been with electroless plating, often in combination with displacement reactions for seeding. However, these techniques have certain disadvantages, such as higher cost for the electrolyte, shorter electrolyte lifetime and less direct influence on deposition process and higher needed efforts to keep the deposition process stable. Still, these processes were commercialized (and later abandoned) by BP Solar in their Saturn cell production [6]. It is interesting to note that these processes are well established in the microelectronics industry, where the cost per piece is more relaxed as compared to solar PV. Electroless and displacement plating on silicon in general and on solar cells in particular have been studied extensively [7, 22, 23], but have been limited to laboratory scale except for the BP Solar example. The disadvantages of electroless plating processes combined with the needs of the solar industry (high throughput, low machine downtimes, small labour costs) lower the chances of these processes for industrial adoption. Therefore, they will not be discussed in detail here, and the interested reader is referred to the literature sources mentioned above.

To use standard electroplating solutions and processes, the current needs to be distributed in the cell in some way. We distinguish two cases:

a. Metal is present on one face of the solar cell.
b. Both sides are completely unmetallized.

Case b is by far more challenging and will be addressed at the end of this section. Case a has been the relevant case for many years, as monofacial solar cells with full-area rear-side metallization have been dominating the market for a long time. Bifacial solar cells are only recently gaining more interest. In both cases, there are processing routes that consist of creating a first contact on one side of the solar cell by a process other than plating, and in a second step creating the metal contact on the other side of the cell by a plating process (hybrid processing). This allows contacting the already present metal contact to induce the plating current and make use of the cell characteristics to distribute the current. This results in a homogeneous current distribution and keeps the contact away from the electrolyte solution (if the metallized side is kept dry), which is an advantage as it would otherwise suffer from undesired plating. Two cases can be distinguished here:

1. Metal present on the n-type side and plating on the p-type side of the solar cell
2. Metal present on the p-type side and plating on the n-type side of the solar cell.

Historically, case 2 was focused on first, as p-type solar cells with n-type front-side emitters dominated the market for a long time. After first patents with an electronics background [24], the so-called **light-induced plating (LIP)** was studied intensively for solar cell applications. Several PhD theses and journal articles are dedicated to this approach. The working principle is very simple: the solar cell is immersed in the plating electrolyte and illuminated with light that is transmitted by the electrolyte and that induces a photocurrent in the solar cell. As the n-type side of the solar cell is to be plated, the electrons excited by the illumination that reach the n-type surface can serve to induce the deposition process. The required counter process is generally shifted from the p-type side of the solar cell (where metal would be dissolved) to an auxiliary anode, which is connected to the p-type side of the solar cell and which is set to a positive potential with respect to the p-type side of the solar cell. The maximum deposition rate on the n-type side is governed by the photocurrent and thus the illumination intensity. Ideally, the current in the external circuit matches the photocurrent, so there is no net deposition or dissolution of metal on the p-type side of the solar cell. A schematic drawing of LIP is shown in Figure 8.2.

For case 1, either forward bias plating (FBP [25]) or field-induced plating (FIP [26]) was developed. The diode behaviour of the solar cell allows passing current from the n-type to the p-type side in forward bias, which allows closing of the circuit. In this case, the solar cell is processed in the dark and an electrical contact is established to the n-type side. The plating current can then be directly controlled from the external circuit. A schematic drawing of this process is shown in Figure 8.3.

In both cases, the current is distributed quite evenly over the solar cell area. Due to the particularities of plating, a concentration of the plating current density towards the edge of the wafer is known to occur, yielding ~20% increased deposition in these regions unless measures are taken against this behaviour (e.g. shielding). Also, larger structures are generally plated less than small structures, so busbars will be plated with approximately 50%–75% less metal than contact fingers, depending on their width.

In case both sides of the solar cell are unmetallized (case b in the above classification) one way to still realize plating is by depositing an initial layer by electroless deposition (see above), with the discussed limitations of the process. More recently, an alternative pathway by directly establishing an electrical contact to the

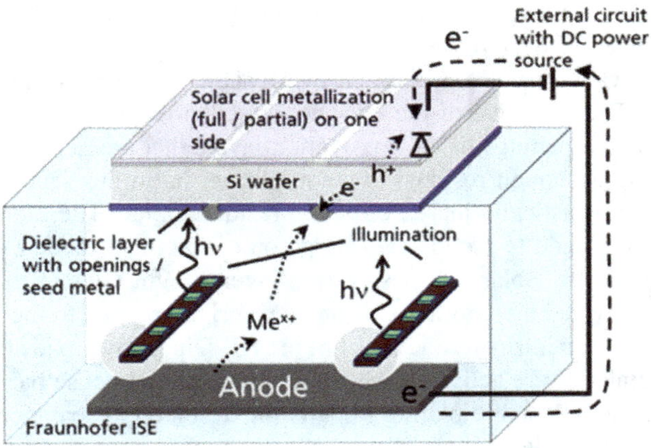

Figure 8.2 *Schematic drawing of LIP of a solar cell in a configuration with single-sided wetting of the cell. The n-type doped side of the solar cell is plated.*

bare silicon has been developed on a laboratory scale [26, 27], where industrial solutions are lacking so far. In any case, the contacting would have to avoid inducing any damage to the solar cell and dielectric layers.

8.2.3 Complete process sequence

The usual plating sequence as applied in mass-scale fabrication includes oxide removal Dip in HF (HF-Dip) before deposition and thermal contact formation step after deposition. While the pattering step might be performed time-wise

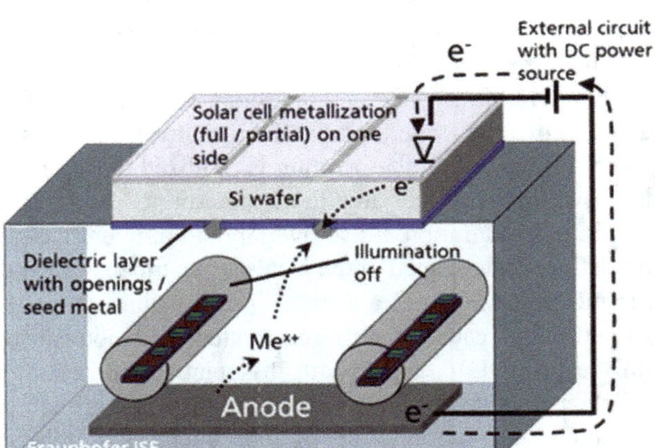

Figure 8.3 *Schematic drawing of FBP of a solar cell in configuration with single-sided wetting of the cell. The p-type doped side of the solar cell is plated.*

independently, the following three steps (oxide removal, deposition and contact formation) must take place shortly after each other.

Oxide removal is usually performed by HF-Dip. Inline baths and bath equipment are used. The objective of the HF-Dip is to remove native oxide in the contact opening (local openings of the passivation layers that are generated during the structuring step). Although residual oxide might not harm the deposition process, contact resistance drastically increases due to residual oxide. Therefore, concentration and etch time need to be adjusted based on contact-resistance measurements, not on plating results. Solar cell passivation layers might contain pinholes. Such pinholes are passivated by natural oxide and a priori not related to the results of the plating process. As an intrinsic side effect of the HF-Dip, native oxide on pinholes is removed and pinholes are activated. Unpassivated pinholes cause parasitic deposition (ghost plating). Parasitic plating impairs the aesthetic appearance of the solar cell and might lower solar cell performance.

Contact formation is achieved by annealing plated contacts at moderate temperatures (200–350 °C). The contact formation step might reduce contact resistance and increase contact adhesion. Formation of nickel silicide at the contact interface has been verified in several microscopic studies. Therefore, the contact formation step is often called silicidation anneal. Unoptimized temperature profiles (too long annealing duration or too high annealing temperatures) lead to shunting of the solar cell device. It was shown that local shunts are caused by deep silicide growth.

Optimum processing conditions of the four processing steps are mutually dependent. The configuration of the HF-Dip needs to be adjusted to the expected oxide thickness in the contact opening. It needs to be considered that the level of oxidation is different on chemically induced openings and on laser-structured contact openings. For chemically induced openings, the thickness of the oxide is a function of the time between passivation layer removal and deposition. In theory, the thickness of the oxide layer can be approximated by the Deal-Grove model [28]. For solar cell surfaces it needs to be considered that the limiting thickness and the growth rate are influenced by silicon surface doping concentration, crystal orientation (mind that (111)-pyramid-surfaces are the actual contact surface) and hydrogen passivation [28, 29]. For laser-structured passivation layer openings it needs to be considered that the surface within the opening is oxidized during the ablation process. The oxide growth during laser ablation is determined by laser settings, such as laser wavelength, pulse duration, pulse overlap and ambient oxygen concentration. Thermal oxidation, as described in the Deal-Grove model [25], occurs for long pulse duration (nanosecond-regime) and light in the visible range. Usage of UV light and picosecond-pulses causes highly accelerated oxidation caused by electron plasmas [30]. In the lower UV range photo dissociation further intensifies oxide growth [31]. A further laser setting, that influences the oxide growth, is the ratio of the laser-pulse diameter and spatial offset of subsequent laser pulses. If the offset is smaller than the pulse diameter, the surface is effectively treated with several laser pulses. While this configuration might ensure more robust passivation layer removal, it also increases the impact of oxidation. Further, the laser-treated silicon surface is roughened and might be amorphous. Both effects influence the growth of native oxide in the passivation layer openings after structuring. There is no theoretical basis for an estimation of the thickness of laser-induced oxide growth on a pyramid-textured silicon surface. In order to avoid harsh etch rates in the HF-Dip, which might increase parasitic plating, it is recommended to suppress native oxide growth by time-wise proximity of patterning and deposition. If laser ablation is used for structuring, ablation configuration needs to be optimized to reduce laser-induced oxidation, in particular by avoiding overlaps of subsequent laser-ablation dots. It was shown that the HF-Dip might be fully skipped, in an 'oxidation'-optimized processing sequence that is characterized by (1) laser-ablation without spatial overlap of subsequent pulses and (2) immediate deposition after laser ablation. The process variant, which is referred to as Easy Plating, prevents parasitic plating, as native passivation on pinholes is not removed before plating.

Post-plating annealing was reported to improve contact resistance and adhesion. Currently, the most common approach is usage of rapid thermal annealing or belt furnaces. Reported plateau temperatures range from 200 °C to 350 °C and plateau durations from 30 s to 15 min. Many authors relate the enhancement of peel-off forces of plated contacts and the reduction of contact resistance, that have been observed to be induced by annealing, to formation of silicide [32, 33]. Formation of silicide is a diffusion-driven process, in which nickel is the moving species [34]. In the mentioned temperature regime two phases are present SiNi and a nickel-rich $SiNi_2$ phase. Silicide formation at the interface of plated layers in the mentioned temperature range is reported to be inhomogeneous. Continuous silicide formation is impeded by interfacial oxide that might also be incorporated during the plating [35], in particular if nickel deposition is not optimized [36]. It was shown, that interfacial oxide layers are a barrier to the nickel silicide formation [37] and local silicide structures are formed at interruptions or pinholes [38] of the oxide layer. This concept also allows for explanation of the typical occurrence of silicide structures in a fried-egg-like configuration of two silicide phases: An inner-core of the nickel-rich phase embedded in a second phase with lower share of nickel. The evolution of this structure might be explained by a necking of the reaction source by the oxide layer [35]. If nanosecond-pulses and continuous wave-lasers are used for structuring, the chemical composition of the surface near silicon volume might change. Incorporation of nitrogen from the passivation layer was reported and it was shown that it impedes silicide formation [39]. In order to enable silicide formation during post-plating annealing, the pre-treatment needs to be well adjusted to the patterning step. Interface oxide needs to be suppressed. For chemically modified surfaces, HF might need to be replaced by buffered HF.

Laser ablation of the passivation layer might go along with local melting and recrystallization of the surface near silicon. Un-optimum recrystallization properties can lead to formation of cracks or other crystal defects. Nickel diffusion is enforced along such defects. As shown by Büchler *et al.* [40], the depth of silicide spikes along defects might be up to ten times deeper than the average formation depth. Formation of deep spikes might be avoided by two-step nickel deposition with an interim anneal. This was proven to work, but the process flow is inconvenient.

An alternative technology to thermally induce silicide formation is application of excimer laser radiation, which was presented by Tous *et al.* [41].

Büchler *et al.* showed that optimized pre-treatment that ensures an oxide-free interface also enables low contact resistance without silicide formation on plated contact structures by laser patterning. In addition to this, it was shown that for interfaces without residual interface oxide, formation of silicide does not further reduce the contact resistance. But, for contacts with interfacial oxide, contact resistance is lowered without formation of silicide. This raised the questions, whether silicidation is the only relevant thermally induced microscopic process during post-plating annealing. Other effects, that still need to be investigated are formation of residual interface oxide [42], phase transformation at the nickel-silicon-oxide interface or any effects related to nickel-induced recrystallization of a-Si [43].

8.2.4 Contact properties

8.2.4.1 Shape

Due to the isotropic layer growth, the contact width $w_{contact}$ and the aspect ratio of plated line contacts are determined by the width of the passivation layer opening $w_{opening}$ and the height h of the plated layer. The final contact width is $w_{contact} = w_{opening} + 2*h$ and the aspect ratio is $R = h / (w_{opening} + 2*h)$. By reducing the width of the contact opening, finger widths of less than 15 μm have been demonstrated [44]. Enforcing anisotropic copper plating can increase the aspect ratio.

8.2.4.2 Contact resistance

Plated nickel or Pd/Ni contacts feature very low contact resistance to silicon with both types of doping and very small surface doping concentrations down to 10^{19} cm^{-3} [30, 31], as shown in Figure 8.4. This has generally been considered one big advantage of plated contacts.

8.2.4.3 Line conductivity

The conductivity of plated metal is typically in the range of literature values for bulk metals, unless unsuitable plating process settings are chosen. Some literature-reported values are summarized in Table 8.3. Both silver and copper have been used as conducting material so far, as these are the metals with highest conductivity. Conductivity has been a critical point when solar cells were still featuring two or three busbars, but has decreased in importance since metallization designs have been adjusted to the needs of screen-printed metallization concepts (multi-wire or multiple busbars). While contact height (plated layer thickness) used to range around 10–15 μm, novel cell designs allow plating of only 5 μm copper, or even below that. On the other hand, solar cells with increasing size and shingling module technology can profit from thick, high conductivity solar cell contacts.

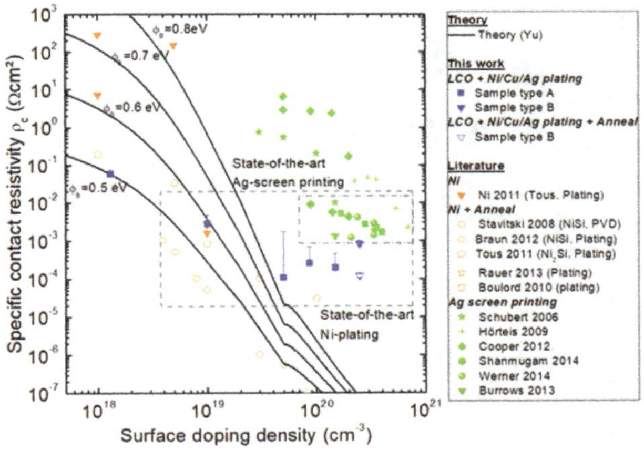

Figure 8.4 *Measured (symbols) and theoretical (lines) contact resistivity of Ni and screen-printed Ag contacts on phosphorus-doped emitters (from Kluska et al. [30])*

8.2.4.4 Solderability/corrosion resistance

The plated contact is typically finished with a capping layer of either tin or silver on top of the copper conducting layer. This layer has a thickness of ~1–2 μm (tin) or 100–200 nm (silver) and promotes solderability and corrosion resistance. There are several reports in literature that show the construction of modules with different capping layers and that also include long-term stability results, which have been found to be compatible with the requirements of PV products [42].

8.2.5 Challenges

8.2.5.1 Adhesion

Figure 8.5 summarizes results of recently published busbar peel force tests. The structuring technology used in the plating procedure determines the colour of the symbol and the symbol itself indicates the angle of the peel force test. This information is necessary, since the peel-off angle influences the magnitude of the measured peel force [48]. The dashed line marks 1 N mm^{-1}, the benchmark according to

Table 8.3 *Literature values for metal conductivity for plated metal on silicon solar cells*

	Copper	Silver
Conductivity (S m^{-1})	1×10^7 [45] 5.78×10^7 (unpublished results @ Fraunhofer ISE)	5.26×10^7 [46]
Bulk value literature	6.086×10^7 [47]	6.301×10^7 [47]

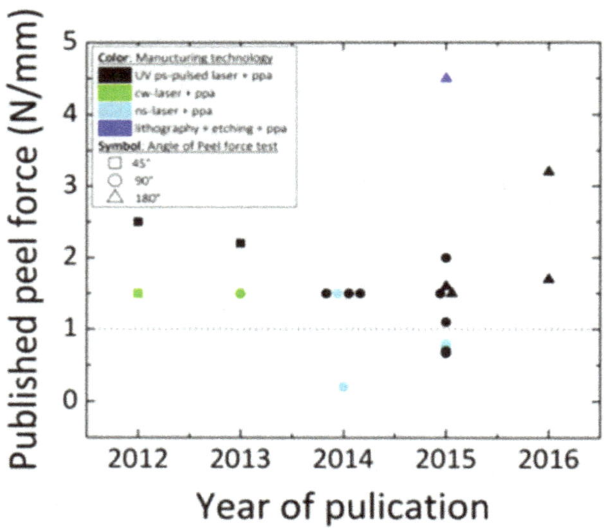

Figure 8.5 *Published peel-off force of plated busbars. Taken from Büchler et al. [32]; full references can be found there.*

European norms. Many examples proved sufficient adhesion of plated contacts. The most successful results in recent years have been obtained by using a picosecond-pulsed UV laser for patterning.

Recent studies on adhesion, can be sorted into two categories: Approaches that tried to reduce the residual stress by optimizing the plating procedure or the post-plating anneal [7,49] and experiments on modifications of the interface, such as silicidation [35] or a beneficial topography [44]. All published positive results on adhesion can be related to one of the following cases: *silicidation of the interface* or to *ensuring suitable surface topography*. Silicidation can increase the adhesion. This requires avoiding an interfacial oxide layer. A suitable surface topography is characterized by a rough surface that features under-cuts or keying-features, such as voids, cracks or laser-induces ripples. Such structures can either be induced during laser ablation or by an additional roughening step.

The only established measurement is the measurement on busbars. A ribbon is soldered to the busbar and the peel force is measured. For plated contacts it needs to be considered that soldering temperatures might affect the nickel-silicon interface and the plated layer stack. Therefore, it is essential to consider the failing interface and break-out mechanisms. Alternative interconnection technologies such as conductive gluing might be more suitable for such tests. Recently, a new method was introduced to quantify the adhesion of fingers: shear test on finger contacts.

8.2.5.1.1 Long-term stability

The long-term stability of plated contacts is always discussed when copper is employed as the main current-conducting layer. Copper can be a very detrimental contamination is silicon, as it is known to diffuse quite fast even at low temperatures and as it forms precipitates in silicon that act as recombination and shunting sites. Typically, long-term stability on cell level has been assessed by heating up solar cells with a plated metallization for long periods of time and by either monitoring the solar cell performance or by carrying out microstructural investigations. Different seed layer materials underneath copper have been tested regarding their ability to withstand copper diffusion, nickel being the most prominent one for solar PV application. Other materials that are effective against copper diffusion are known from microelectronics fabrication (e.g. Ti, Ta and their nitrides).

Another means to study long-term stability is by fabricating modules and subjecting these to accelerated ageing procedures. This has the advantage that apart from copper diffusion, the entire contact stack stability is assessed at the same time. Solar modules of different sizes with fully plated contacts have been subjected to damp heat tests (85 °C, 85% r.h.) and to thermal cycling tests (−40 °C to +80 °C) for several thousand hours and for several hundred cycles, respectively. The modules typically performed at least similar to reference modules metallized with screen-printed silver metallization. Therefore, stability seems to be no major issue in plated solar cell contacts.

In addition, solar modules from BP Solar with fully plated contacts have already been in the field for a long period of time. From one outdoor test field in Toledo, Spain, there is a study [43] comparing module performance after 20 years of operation, where the plated modules perform similarly or superiorly to their reference counterparts. Even though the plating metallization process technology has changed considerably since this time, the mechanisms dominating the ageing process remain the same.

8.2.5.2 Parasitic plating

Plating is based on the mechanism of an electron transfer between the substrate and the surrounding solution. This opens the possibility of influencing the plating position, which is used both here and for the plating approaches discussed in the following by offering a conductive or semiconductive surface next to an isolating surface (i.e. silicon or metal next to the dielectric passivation layer). On the other hand, this mechanism means that any irregularities in the surfaces may result in unwanted deposition behaviour.

Defects in the dielectric layers such as scratches, pinholes, etc. may cause unwanted metal deposition, which will in most cases mean little to no damage to the electrical performance of the solar cell (except for extreme cases, where the current can be affected), but which will influence the optical appearance. As this is not accepted by module makers and end customers, such behaviour needs to be eliminated for plating approaches that are to be introduced into production. Parasitic plating (also called 'Ghostplating' or 'Background plating') has been studied and has been found to be caused by defects (such as pinholes and cracks) and by surface contaminations. Braun *et al.* [Source missing, https://doi.org/10.1016/j.egypro.2011.06.183] give an overview of cases for dielectric layers. A general observation is that parasitic plating is more prominent on exposed features of solar cells, such as the rims (or positions close to the rim), random pyramid tips on monocrystalline solar cells, etc. This is in accordance with electrochemistry fundamentals, as these are sites of increased field density and result in changes in primary and secondary current density. Close spacing of solar cells in production tools can help to overcome this issue. General solution strategies are mainly addressing the quality of the dielectric layer which is deposited on the solar cell. Results from production environments confirm that parasitic plating can be effectively prevented. Some adjustments to the process may be necessary.

8.3 Plating on metal seed layer

8.3.1 Seed layer types

Details of the deposition process on a metallic seed layer depend a lot on the seed layer type and properties. Types of seed layers that have so far been used in plating for PV application comprise:

- Seed layers deposited by PVD processes (sputtering/evaporation)
- Seed layers deposited by electrochemical deposition (see section 8.1)
- Seed layers deposited by printing techniques (e.g. screen printing [50], aerosol jet printing [33], inkjet printing [34], pad printing [35], flexographic printing [36], etc.)
- Seed layers deposited by laser techniques (e.g. laser transfer processes [51], laser micro sintering [37], etc.)

To realize plating on any metal (irrespective of the seeding technology), the topmost layer of the seed layer should consist of a metal that is well susceptible to plating. Typically, these are metals that can be deposited by plating themselves, but also other metals are feasible. In many cases, a pre-conditioning of the seed layer is helpful to facilitate plating and to improve adhesion of the plated layer on the substrate. Literature (e.g. [8]) discusses such **pre-conditioning steps**, but as classical electroplating often takes place in rather unclean environments, the proposed procedures overshoot PV application, where the seed layer and substrate are handled in a clean environment. Layers that can be plated well, and have been used in PV, comprise nickel, copper, silver and palladium. There have also been attempts with iron, aluminium, titanium and other more 'exotic' base materials. Depending on the type of seed layer, the material that forms the base for plating is also the material to contact the semiconductor, or not. Evaporated contacts often feature a layer stack where the first layer in contact with the semiconductor is ideally suited to make a well-adhering and low-resistivity contact to the semiconductor, while the topmost one is ideally suited for subsequent plating.

It is one big advantage of this 'seed and plate' approach that the different functions of solar cell contact, adhesion, contact resistivity, current conduction and connection to the module, do not have to be fulfilled by one material and deposition method simultaneously, but can be tailored along the different metallization steps to be ideally met, combining different materials and techniques.

The seed layer deposition technique does play a role in the preconditioning of the seed layer before plating. Laser techniques, for instance might well result in the formation of an oxide layer on the seed metal. The adhesion between seed metal and plated metal may therefore profit from removal of the oxide in a pre-conditioning step.

Further influencing factors of the seed layer on the subsequent plating process are its **continuity and its conductivity**. A discontinuous seed layer (as for instance can be created by aerosol jet printing) is limited regarding the possibility to be electroplated in a classical way, making an electrical contact at one position with a contacting device. The current flow will be interrupted and only few portions close to the contact will be plated. A continuous and well conducting seed layer, such as a PVD seed layer or a screen-printed seed layer, can be plated well with such a configuration.

Another issue to be considered when matching a seed layer process with a plating process is the **chemical stability** of the seed layer. Especially for screen-printed seed layers, it has been shown that some electrolyte types are capable of chemically attacking the seed layer [50], which may compromise adhesion to the substrate and contact quality. It is likely that the same is true for other printing techniques, where the degree of sensitivity against plating solutions will vary with the paste type, contact porosity, electrolyte selection and processing conditions.

Finally, the seed layer's **capability to act as diffusion barrier** against unwanted metal diffusion into the solar cell has to be considered, especially when copper is amongst the metals to be plated. If the seed layers itself act as a diffusion barrier, copper can be directly plated on top. Otherwise, a metallic barrier layer (typically nickel) has to be deposited beforehand. Dielectric layers such as SiN_x positioned beside the seed layer have been found to act as diffusion barrier layers in these positions [50]. The interplay of diffusion barrier layer, metallization and cell properties is discussed thoroughly in [38].

Table 8.4 gives an overview of frequently used seed layer types and assigns the above discussed attributes to each, giving an overview of the performance and the limitations of each seeding technology, and also gives some more information on the different seeding techniques.

8.3.2 Deposition methods

As discussed above, the classical way to plate is by creating a direct electrical contact between an external contacting unit that will drive the electrons required for metal deposition into the regions to be plated. This is also the distinction between the methods discussed in Subchapter 8.1 and additional options discussed in this chapter. **All the plating techniques discussed above can also be applied on a seed layer deposited before plating, and will generally yield the same characteristics as introduced**. Please refer to Subchapter 8.1 for more details.

Table 8.4 Seed layer technologies for plating

Seeding technique	Adhesion to substrate	Pre-conditioning	Conductivity/ continuity (can be electroplated?)	Chemical stability	Diffusion barrier?	Typical configuration
Thick-film printing (Screen- and Pad-printing)	+ (unless corroded by electrolyte)	+ (no pre-conditioning)	+	-	-	Silver paste printing + firing + Ni + Cu + Ag plating
Thin film printing (Aerosol-, Flexo-, Inkjet-printing)	+ (unless corroded by electrolyte)	+ (no pre-conditioning)	0 (only for multiple passes)	-	--	Silver paste printing + firing + Ni + Cu + Ag plating
PVD seed layer	++ (customizable)	+ (no pre-conditioning)	0 / + (continuous. Patterning prior to plating: poor conductivity)	+	+ (can be customized	PVD of layer stack: contact and adhesion layer (Ti or Al) + diffusion barrier layer (Pd or TiW) + plating start layer (Ag or Cu), plating of Cu and Ag
Laser seed layer	+	0 (may be necessary)	-	+	0 (can be customized, but laser impact must be considered)	Laser transfer of Ni, plating of Cu and Ag
Plated seed layer	+ (depending on patterning approach)	+	0 (continuous, but usually thin)	+	+	Laser ablation of SiNx, plating of Ni, plating of Cu and Ag

In addition the seed layer present on the solar cell enables 'classical' electroplating as an option beyond LIP, FBP and electroless plating. Making the contact to the seed layer remains the most important challenge in this case. One additional challenge is to prevent the contacting unit from suffering to be plated itself, which can (in extreme cases) cause the contacting unit to be firmly connected to the substrate to be plated, where the plated layer itself creates the connection. It is desirable to reduce the amount of contacting device in touch with the electrolyte to as close to zero as possible and to employ materials that are not easily plated (e.g. Ti). In the laboratory, many different ways have been found to achieve this, which will not be discussed in detail here.

One way that may be used both in the lab and in industrial tools is a so-called fountain plater, where the wafer is placed on a so-called cup through which an electrolyte flow is directed onto the wafer side facing downwards. Also, from the bottom, an array of contacts may be placed on the wafer. By aligning these contacts with (relatively wide) seed layer patterns on the wafers side to be plated, electroplating becomes possible. By replacing the metallic supports by plastic ones, LIP and FBP can be realized when contacting the solar cells side facing upwards. Single-cup tools were available and are still used in some labs, and by multiplying the cup in a large machine, industrial-throughput plating was possible, but never made it into production. Direct plating with this strategy has become more challenging with changes in metallization design, as large contact pads such as busbars have continuously been decreased in size, complicating alignment and contacting device design.

Another way to make electrical contact is by holding the solar cell with small metallic clips, either at one edge or at several edges. The former has been realized by the company MECO in a production tool, the latter is typically done in so-called 'rack platers', where the solar cells can be mounted into frames and the frame is entirely immersed into a large plating bath. The company SunPower Corp is known to have been working with such a solution. Both designs will be discussed in more detail in Subchapter 8.4.

In any case, the distribution of the current in standard electroplating is governed by the number and position of the contacting points and by the conductivity of the seed layer. Simulations at Fraunhofer ISE show that a layer thickness (full area) of only 50–100 nm is sufficient to distribute the current evenly over a wafer when contacting at one edge of the wafer [52].

8.3.3 Contact properties

The most important influencing factor for the contact properties in a seed and plate approach is the seed layer. As for direct plating on silicon, the plated layer grows isotropically in all directions. It is desirable to establish narrow contacts with low contact resistivity for each of the techniques introduced above. The contact width will typically increase based on that initial seed layer width by double the height of the plated layer (where the optical width will be lower), according to the formula:

$$W_{contact} = W_{Seedlayer} + 2^{*} h_{Plating}$$

The conductivity of the plated metal will still be that of the bulk material, but the seed layer conductivity also contributes (more or less, depending on the seeding technique) to the resulting overall conductivity of the metal contacts. It is calculated according to:

$$G_{contact} = 1/R_{contact} = 1/R_{Seedlayer} + 1/R_{Plated\ layer}$$

The geometrical properties of the contacts then yield the specific conductivity or resistivity values.

Table 8.5 gives an overview of different seeding techniques and their respective geometrical and electrical properties reported in literature:

8.3.4 Challenges

Apart from parasitic plating, which behaves similarly to that discussed in Subchapter 8.1, the challenges of plating onto seed layers also change with the type of seed layer. One factor that all techniques (except for plated seed layers) have in common is that by combining two different process types, the overall processing sequence and the capabilities needed in a production environment grow.

Many of the seeding techniques tested in labs are not developed to a stage where they can easily be commercialized. There are still challenges associated with these that are not fully known. There are some general trade-offs that can be mentioned here, where each of these can only be optimized to a certain degree in lab environments and where it is likely that significant further progress could be made in volume production.

Adhesion may be a challenge for printed seed layers [50], especially when they are porous. Also, the amount of silver would need to be as low as possible, and a diffusion barrier would need to be plated onto such layers before plating the conducting and capping layers. Porosity in the seed layer is a potential weak spot for diffusion.

Laser-transferred seed layers face the challenge to dose the laser power in a way that does not damage the solar cell, but yields low contact resistivity and high adhesion at the same time.

PVD seeding techniques are very well performing, but suffer from high cost and the need to employ a patterning step on an industrial scale.

A plated starting layer as described in Subcapter 8.1 appears to be the best solution in many respects.

8.4 Production tools for plating and costs of plating processes

8.4.1 Typical tool design, process parameters

State-of-the-art c-Si production lines are equipped with production tools with throughput rates larger than 5 000 wafers/hour and yield rates larger than 95%. The tool designs are usually inline or batch-type machines depending on the solar cell design and the required process properties. There are multiple companies providing plating tools for different cases, e.g. interdigitated back contact (IBC) solar cells,

Table 8.5 Contact properties of different seed layer technologies

Seeding technique	Seed width	Seed height	Contact resistivity	Maturity	Cost	Remarks	Literature
Thick film printing	30 µm	2 µm	0 (moderate doping needed)	++	0 (high material (silver), low machine/process)	1	[35, 53]
Thin film printing	15–20 µm	0.5 µm	0/+ (low doping reported)	0	+	1	[33, 36]
PVD	10–20 µm	0.1–0.2 µm	++ (material choice allows flexible doping)	0 (in PV)	0 (high machine, low material)		[35]
Laser	3–20 µm	0.1–0.3 µm	+	-	+		[39]
Plated	10–20 µm	Typically 1 µm	+	+	+		[23, 32, 49, 54]

Remarks: 1: Printed seed layers may profit from subsequent plating in terms of contact resistivity [35].
PV, Photovoltaics;PVD, Physical vapor deposition.

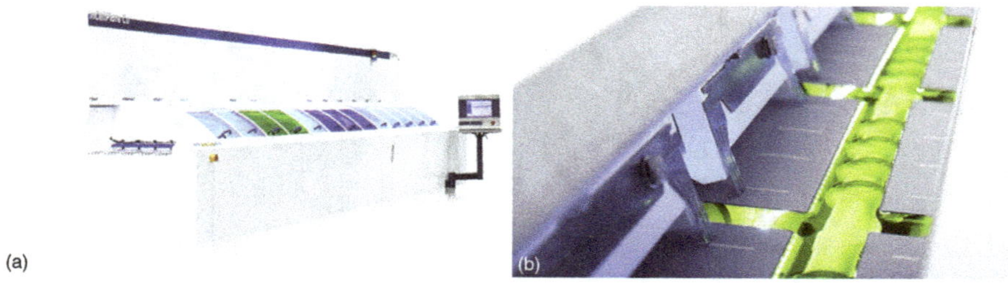

Figure 8.6 *(A) Photograph of an industrial single-side inline plating tool (InCellPlate Cu) for Cu electroplating (light-induced or forward bias plating) of the company Rena Technologies GmbH. (B) The solar cells are only immersed on the front side and the dry rear side is contacted via brushes to provide the plating current.*

bifacial heterojunction solar cells and monofacial (e.g. PERC) or bifacial Tunnel Oxide Passivated Contact (TOPCon) c-Si solar cells fulfilling these requirements.

Each application case has its own set of requirements on contacting design and process specifications, which leads to the fact that there is no established standard in design for state-of-the-art industrial plating tools. The following tool designs have already shown their applicability in industrial production facilities:

- Single-side inline plating (eless, LIP or electroplating)
- Bifacial inline plating (eless, LIP or electroplating)
- Bifacial batch plating (eless, LIP or electroplating).

In the following, examples of each tool design and an example of a typical application case are discussed.

8.4.2 Single-side inline plating

Figure 8.6 a) and b) shows an example of an industrial-type single-side inline plating tool. A typical application case for this kind of tool is direct plating metallization for monofacial (e.g. PERC) or bifacial (TOPCon) c-Si solar cells. One possible process flow for plated contacts on TOPCon solar cells is shown in Figure 8.7. The wafer transport of the laser-patterned solar cell precursor is realized via an inline transport system featuring a single-side immersion of the patterned front side in the plating electrolyte. Depending on the plating process the dry rear side is either electrically contacted (electroplating, FBP) in combination with illumination on the wetted solar cell front side (LIP) or neither contacted nor illuminated for eless plating deposition.

In order to ensure high throughput rates the wafer input can be increased by multiple lanes within the same tool. The benefit of this tool design is homogeneous plating current injection over the whole wafer area. Furthermore, the single-side immersion allows an independent control of the plating process of the front and the rear side. This can be beneficial in the case of PERC solar cells (no etching of

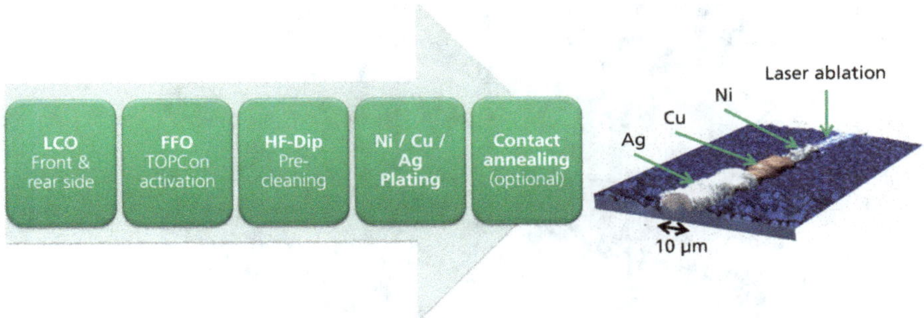

Figure 8.7 Process sequence of laser-defined plated Ni/Cu/Ag contacts for metallization of c-Si solar cells (left) and a composite image of three-dimensional microscope pictures after laser ablation, Ni, Cu and Ag plating (right)

rear-side Al paste) or in the case of TOPCon solar cells (independent deposition mass on the front and the rear side). In addition, the contacting units will not be wetted and thus plated by the electrolyte in this case, which simplifies maintenance.

8.4.3 Bifacial inline plating

Bifacial inline plating sets two main requirements on the plating tools design. First, providing a reliable electrical contact on the front and the rear side of the solar cell precursor to ensure stable injection of the plating current. Second, homogeneous distribution of the plating current along the whole wafer surface. Furthermore, the electrical contact has to be insulated against the plating electrolyte to prevent plating at the contacting unit instead of the solar cell.

The electrical contact can be either realized by contacting a local metal seed layer or a locally masked full area metal seed layer. Full area seed layers have the advantage that this ensures a homogeneous distribution of the plating current along the wafer surface. Especially in the case of metallization for heterojunction solar cells the plating approach of locally masked full area seed layers has been proven to be a reliable alternative to screen printing. A discussion about the process details on seed layer formation and plating metallization of heterojunction solar cells can be found in Chapter 9.

Commercial production tools for bifacial (and optionally also monofacial) inline plating are available, e.g. by the company Meco Equipment Engineers B.V. Figure 8.8 shows an example of an inline bifacial plating tool with vertical wafer transport in an electrolyte bath. This concept was successfully demonstrated for silicon heterojunction solar cell production.

8.4.4 Bifacial batch plating/rack plating/lead frame plating

For bifacial batch plating/rack plating the solar wafers are stacked in a carrier or rack in the electrolyte bath. This plating approach was commercialized by BP

Figure 8.8 Photograph of an industrial bifacial inline plating tool of the company Meco Equipment Engineers B.V. The solar cells are clamped to a transport rail, which vertically transports the solar cells through an electrolyte bath and acts at the same time as the electrical contact to provide the plating current. The solar cells can be metallized in electroplating, light-induced plating (as shown in the image) or forward bias plating mode.

Solar, but later abandoned. In this case the wafers are not moving through the baths and are, in case of electroplating, electrically contacted via the carrier or rack. This plating concept is also applied in other market fields (e.g. decoration/anti-corrosion finishing or printed circuit board fabrication) and offers a large variety of tool suppliers such as Schmid Group, JCU Corp., Technic Inc., Rena Technologies GmbH, Atotech B.V. and more. Although, this tool design is well established in other market fields the adaptation on PV applications was less successful. Batch plating is well established for plating in the production of IBC solar cells. In this case, plating is done into a resist mask which is deposited on top of a full area PVD seed layer. The electrical contacting can be realized by contacting the PVD layer of the busbar structures at the wafer edges.

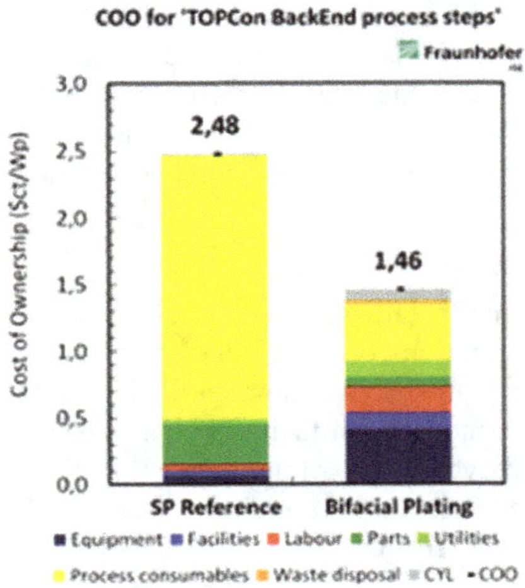

Figure 8.9 *Cost of ownership calculations of the metallization of bifacial TOPCon solar cells assuming either printing of Ag and AgAg paste on the front and the rear side, respectively, or plated Ni/Cu/Ag contacts*

8.5 Challenges

8.5.1 Market penetration/cost calculations

One of the major hurdles for market introduction of plating technology in c-Si PV was the tremendous development of the established Ag screen-printing technology in the last decade. The technological advantage of plated contacts regarding narrow finger thickness, low contact resistivity on lowly doped c-Si surfaces and low line resistivity for front-side metallization of Al-BSF and PERC solar cells was shrinking from year to year. Furthermore, the combination of low silver prices during the years 2013 to 2019 in combination with decreased paste laydowns for screen-printing lowered the cost advantage of plated contacts for monofacial solar cell designs such as PERC. However, the future does not look grim for plating technology. The intrinsic material cost advantage of plated copper contacts can lower dependencies of cell production costs on external factors such as global crisis and price speculations. Volatile silver prices (e.g. triggered due to global crises such as the Covid-19 pandemic) in combination with massive capacity expansions plans towards a Terawatt c-Si PV market increase the risk of volatile manufacturing costs since consumable costs of silver paste make up a significant share of cell production costs. This dependency gets even worse for new bifacial solar cell designs which require a highly conductive (silver) grid not only on one side but on both sides of the solar cell. Printing Ag or AgAl pastes on the front and the rear side further increases the impact of raw material cost on cell production cost. Figure 8.9 shows the distribution of cell metallization costs of bifacial TOPCon solar cells. The calculations assume paste laydowns of 90 mg per wafer side and a silver prize of 564 €/kg. Since technology development in the last decade pushed down tool costs to a minimum, the major share relies on consumables. The cost distribution of plated contacts is mainly driven by equipment costs which show a rather high potential in cost reduction due to scaling effects with increasing market penetration.

8.5.2 Wastewater treatment

The introduction of plating metallization comes along with the need of wastewater treatments due to concentrations of Ni, Cu and Ag in disposed rinse water. Most tool manufacturers offer plating tool equipment including 'zero waste water' solutions using evaporation/condensation facilities. The waste water is evaporated and condensed leading to a recycling rate of up to 99.5% of the diluted waste water and the collection of the metal residue as a highly concentrated solution (0.5%) in the intermediate bulk container (IBC).

8.5.3 Inhomogeneous plating deposition

Depending on the solar cell and plating tool design the deposition rate will not be homogeneous all over the wafer surface. Merhi *et al.* [52] discussed in a comprehensive study the origin and distribution of plating heights for H-patterned solar cell grids plated in a single side or bifacial inline plating tool. He found that for state-of-the-art metal grid designs and plating tool configurations the ratio of minimum to maximum finger height over the grid can be in the range of 50%–75%. The main findings of his work were that increased finger heights appear close to wafer edges, finger height reduction appears close to busbars and when a voltage source is applied only to one side of the grid during electroplating, the entire plated grid heights scale down away from that applied voltage. Although, the work was completely limited to a resistive-based model (no electrochemical effects) the findings demonstrate quite well the behaviour of the experimental work. One of the key findings was also that the appearance of non-homogeneous plating deposition has only a minor influence on the maximum achievable solar cell efficiency compared with a uniform metal finger height.

References

[1] Perlin J. *The Silicon Solar Cell Turns 50 [online]*. 2004. Available from www.nrel.gov/docs/fy04osti/33947.pdf [Accessed 24 Nov 2020].

[2] Lennon A., Yao Y., Wenham S. 'Evolution of metal plating for silicon solar cell metallisation'. *Progress in Photovoltaics: Research and Applications*. 2013;**21**(7):1454–68.

[3] Hamann C.H., Hamnett A., Vielstich W. *Electrochemistry*. 2nd ed. Weinheim: WILEY-VCH; 2007.

[4] Bard A.J., Faulkner L.R. *Electrochemical Methods: Fundamentals and Applications*. 2nd ed. New York: Wiley Textbooks; 2000.

[5] Schlesinger M., Paunovic M. (eds.). *Modern Electroplating*. 5th ed. New York, NY: John Wiley & Sons; 2011.

[6] Eager S., Mason N., Bruton T., Sherborne J., Russell R. 'Environmentally friendly processes in the manufacture of Saturn solar cells'. Conference Record of the Twenty-Ninth Ieee Photovoltaic Specialists Conference; 2002. pp. 62–5.

[7] Rodriguez J., Zhang W., Lim S., Lennon A. 'Improved metal adhesion with galvanic nickel plating to silicon solar cells'. *Solar Energy Materials and Solar Cells*. 2017;**165**:17–26.

[8] Mallory G.O., Hajdu J.B. *Electroless Plating: Fundamentals and Applications*. New York: American Electroplaters and Surface Finishers; 2009.

[9] Cornagliotti E., Tous L., Russell R., *et al*. 'High efficiency bifacial n-PERT cells with co-plated Ni/Ag contacts for multi-wire interconnection'. *Energy Procedia*. 2017;**130**(8):50–4.

[10] Williams K.R., Gupta K., Wasilik M. 'Etch rates for micromachining processing-part II'. *Journal of Microelectromechanical Systems*. 2003;**12**(6):761–78.

[11] Williams K.R., Muller R.S. 'Etch rates for micromachining processing'. *Journal of Microelectromechanical Systems*. 1996;**5**(4):256–69.

[12] Heinrich G., Bähr M., Stolberg K., Wütherich T., Leonhardt M., Lawerenz A. 'Investigation of ablation mechanisms for selective laser ablation of silicon nitride layers'. *Energy Procedia*. 2011;**8**:592–7.

[13] Knorz A., Peters M., Grohe A., Harmel C., Preu R. 'Selective laser ablation of SiN_x layers on textured surfaces for low temperature front side metallizations'. *Progress in Photovoltaics: Research and Applications*. 2009;**17**(2):127–36.

[14] Bonse J., Baudach S., Krüger J., Kautek W., Lenzner M. 'Femtosecond laser ablation of silicon–modification thresholds and morphology'. *Applied Physics A*. 2002;**74**(1):19–25.

[15] Hsiao P.-C., Song N., Wang X., *et al*. '266-nm PS laser ablation for Copper-Plated P-type selective emitter PERC silicon solar cells'. *IEEE Journal of Photovoltaics*. 2018;**8**(4):952–9.

[16] Paviet-Salomon B., Gall S., Monna R., Manuel S., Slaoui A. 'Analysis of laser-doped phosphorus emitters'. *Energy Procedia*. 2011;**8**(7):214–9.

[17] Hameiri Z., Mai L., Puzzer T., Wenham S.R. 'Influence of laser power on the properties of laser doped solar cells'. *Solar Energy Materials and Solar Cells*. 2011;**95**(4):1085–94.

[18] Cornagliotti E., Uruena A., Hallam B., *et al*. 'Large area P-type PERL cells featuring local P+ BSF formed by laser processing of ALD Al2O3 layers'. *Solar Energy Materials and Solar Cells*. 2015;**138**(6):72–9.

[19] von der Linde D., Sokolowski-Tinten K., Bialkowski J. 'Laser–solid interaction in the femtosecond time regime'. *Applied Surface Science*. 1997;**109-110**:1–10.

[20] Bonse J., Kirner S V., Höhm S., *et al*. Applications of laser-induced periodic surface structures (LIPSS). *Laser-based Micro- and Nanoprocessing XI in SPIE Proceedings: SPIE*; 2017. p. 100920N.

[21] Ciesla A., Chen R., Wang S., *et al.* 'High-voltage P-type PERC solar cells with anchored plating and hydrogenation'. *Progress in Photovoltaics: Research and Applications*. 2018;**26**(6):397–401.

[22] Tous L., Russell R., Cornagliotti E., *et al.* '22.4% Bifacial n-PERT cells with Ni/Ag co-plated contacts and Voc ~691 mV'. *Energy Procedia*. 2017;**124**(12):922–9.

[23] Bartsch J. '*Advanced front side Metallization for crystalline silicon solar cells with electrochemical techniques*'. *PhD Thesis*. University of Freiburg, Verlag: Dr. Hut; 2011.

[24] Durkee L.F. '*Method of plating by means of light*'. *US Patent US4144139A*; 1979.

[25] Bartsch J., Kamp M., Hartleb D., *et al.* '21.8% Efficient n-type solar cells with industrially feasible plated metallization'. *Energy Procedia*. 2014;**55**(5):400–9.

[26] Wang X., Allen V., Vais V., *et al.* 'Laser-doped metal-plated bifacial silicon solar cells'. *Solar Energy Materials and Solar Cells*. 2014;**131**:37–45.

[27] Grübel B., Cimiotti G., Arya V., *et al.* 'Plated Ni/Cu/Ag for TOPCon solar cell metallization'. Proceedings of the 36th European Photovoltaic Solar Energy Conference and Exhibition; EU PVSEC; 2019.

[28] Deal B.E., Grove A.S. 'General relationship for the thermal oxidation of silicon'. *Journal of Applied Physics*. 1965;**36**(12):3770–**8**.

[29] Neuwald U., Hessel H.E., Feltz A., Memmert U., Behm R.J. 'Initial stages of native oxide growth on hydrogen passivated Si(111) surfaces studied by scanning tunneling microscopy'. *Applied Physics Letters*. 1992;**60**(11):1307–9.

[30] Kluska S., Bartsch J., Büchler A., *et al.* 'Electrical and mechanical properties of plated Ni/Cu contacts for Si solar cells'. *Energy Procedia*. 2015;**77**:733–43.

[31] Feldmann F. 'Evaluation of topcon technology on large area solar cells'. Proceedings of 33rd European Photovoltaic Solar Energy Conference and Exhibition; 2017. pp. 1–3.

[32] Büchler A., Kluska S., Meyer F., *et al.* 'Interface oxides in femtosecond laser structured plated Ni-Cu-Ag contacts for silicon solar cells'. *Solar Energy Materials and Solar Cells*. 2017;**166**:197–203.

[33] Hörteis M. 'Fine-line printed contacts on crystalline silicon solar cells'. *Dissertation, Fachbereich Physik, Konstanz, Universität, Konstanz*.

[34] Ebong A. 'High efficiency silicon solar cells with ink jetted seed and plated grid on high sheet resistance emitter'. 35th IEEE Photovoltaic Specialists Conference (PVSC), conference proceedings; Piscataway, NJ: IEEE; Honolulu, Hawaii; 2010. pp. 1363–7.

[35] Mette A. *New Concepts for Front Side Metallization of Industrial Silicon Solar Cells*. Freiburg: Faculty of Engineering, Freiburg, Albert-Ludwigs-University; 2007.

[36] Lorenz A. *Evaluierung von Rotationsdruckverfahren für die Metallisierung von Silicium-Solarzellen*. Dissertation, Fraunhofer-Institut für Solare Energiesysteme; Fraunhofer IRB-Verlag; 2017.

[37] Alemán M. 'Laser micro-sintering as a new metallization technique for silicon solar cells'. Lutz-Buob in Preu; 2006. pp. 705–8.

[38] Lennon A., Colwell J., Rodbell K.P. 'Challenges facing copper-plated metallisation for silicon photovoltaics: insights from integrated circuit technology development'. *Progress in Photovoltaics: Research and Applications*. 2019;**27**(1):67–97.

[39] Röder T.C. *Lasermetallisierung für Siliziumsolarzellen*. Dissertation, Fakultät für Informatik, Elektrotechnik und Informationstechnik, Institut für Photovoltaik, Stuttgart, Universität, Stuttgart; 2012.

[40] Büchler A., Kluska S., Brand A., Geisler C., Hopman S., Glatthaar M. 'Micro characterization and imaging of spikes in nickel plated solar cells'. *Energy Procedia*. 2014;**55**:624–32.

[41] Tous L., Lerat J.-F., Emeraud T., *et al*. 'Nickel silicide contacts formed by excimer laser annealing for high efficiency solar cells'. *Progress in Photovoltaics: Research and Applications*. 2013;**21**(3):267–75.

[42] Russell R., Philipsen H., Horzel J., *et al*. 'A simple copper metallisation process for high cell efficiencies and reliable modules'. 27th European Photovoltaic Solar Energy Conference and Exhibition-EUPVSEC; 2012. pp. 538–43.

[43] Alonso-Abella M., Chenlo F., Vela N., *et al*. 'Toledo PV plant 1MWp—10 years of operation'. Proceedings of the 20th European Photovoltaic Solar Energy Conference; 2005. pp. 2454–7.

[44] Büchler A., Kluska S., Bartsch J., *et al*. 'Optimizing adhesion of laser structured plated Ni-Cu contacts with insights from micro characterization'. *Energy Procedia*. 2016;**92**:913–8.

[45] Geissbuhler J., Wolf S.D., Faes A., *et al*. 'Silicon heterojunction solar cells with copper-plated grid : status and comparison with silver Thick-Film techniques'. *IEEE journal of photovoltaics*. 2014;**4**(4):1055–62.

[46] Hörteis M., Bartsch J., Binder S., *et al*. 'Electrical properties of fine line printed and light-induced plated contacts on silicon solar cells'. *Progress in Photovoltaics: Research and Applications*. 2010;**115**(4):n/a–8.

[47] Haynes W.M. *CRC Handbook of Chemistry and Physics*. 93rd ed. AT&T Bell Laboratories, Murray Hill, New Jersey: Taylor & Francis; 2012.

[48] Eitner U., Rendler L.C. 'The mechanical theory behind the peel test'. *Energy Procedia*. 2014;**55**:331–5.

[49] Mondon A. *Nickel Silicide from Plated Nickel for High Adhesion of Fully Plates Silicon Solar Cell Metallization*. Stuttgart: Fraunhofer Verlag; 2015.

[50] Kraft A. *Plated Copper Front Side Metallization on Printed Seed-Layers for Silicon Solar Cells*. Stuttgart: Fraunhofer Verlag; 2015.

[51] Röder T.C., Hoffmann E., Konrad B., Köhler J.R. 'Low temperature laser metallization for silicon solar cells'. *Energy Procedia*. 2011;**8**:552–7.

[52] Merhi H., Fell A., Grubel B., Glatthaar M., Kluska S. 'Inhomogeneity of plated contacts for c-Si solar cells and their impact on solar cell efficiency'. *IEEE Journal of Photovoltaics*. 2020;**10**(5):1455–62.

[53] Kraft M., Würth C., Muhr V., Hirsch T., Resch-Genger U. 'Particle-size-dependent upconversion luminescence of NaYF4: Yb, ER nanoparticles in organic solvents and water at different excitation power densities'. *Nano Research.* 2018;**11**(12):6360–74.
[54] Tous L. 'Nickel/Copper plated contacts as an alternative to silver screen printing for the front side mMetallization of industrial high efficiency silicon solar cells (Elektrochemisch gedeponeerde nikkel/koper contacten ALS alternatief voor zilver zeefdruk voor metallisatie van de voorzijde van hoog-efficiënte industriële silicium zonnecellen'. *PhD Thesis, KU Leuven.* 2014.

Chapter 9
Metallization of specific solar cells

Thorsten Dullweber[1], Armand Bettinelli[2], Agata Lachowicz[3], Antonin Faes[3], Pradeep Padhamnath[4], Ankit Khanna[4], Shubham Duttagupta[4], and Haifeng Chu[5]

9.1 Metallization of industrial PERC and PERC+ solar cells

This subchapter introduces monofacial passivated emitter and rear cells (PERC) and bifacial PERC+ solar cells which are the mainstream solar cell technology in the photovoltaic (PV) industry today. Section 9.1.1 describes the PERC solar cell design as well as major technology development steps in the past decades such as the bifacial PERC+ design. The process technology to manufacture PERC solar cells is covered in Section 9.1.2, whereas the specific aspects of the screen-printed silver (Ag) front and Al rear metal contacts are introduced in detail in Sections 9.1. 3 and 9.1.4, respectively. The busbar design of PERC metal grids for module interconnection is discussed in Section 9.1.5. Finally, Section 9.1.6 provides an outlook on future improvement opportunities of PERC and PERC+ solar cells, in particular regarding the metal contact design and its impact on conversion efficiency.

9.1.1 PERC and PERC+ solar cell design

The PERC solar cell structure was introduced by Blakers *et al.* [1] in the research group of M. A. Green at the University of New South Wales (UNSW), Australia, in 1989. At that time, the goal was not to demonstrate a production-feasible solar cell process but to increase the conversion efficiency to a new world record value of 22.8% applying very expensive laboratory-type processes and materials. As substrate material 0.2 Ω cm p-type float-zone silicon wafers were used with a solar cell size of 4 cm^2. The rear passivation consisted of a thermally grown (~1 000 °C) silicon dioxide layer which was locally opened by photolithography and a masked

[1]Institute for Solar Energy Research Hamelin (ISFH), Germany
[2]National Institute for Solar Energy (INES), France
[3]Centre Suisse d'Electronique et de Microtechnique (CSEM), Switzerland
[4]Solar Energy Research Institute of Singapore (SERIS), Singapore
[5]International Solar Energy Research Center Konstanz e.V., Rudolf-Diesel, Germany

etch-forming point-shaped local rear contact opening. The full-area aluminium (Al) rear metal layer was evaporated followed by a 15-min anneal in forming gas in order to establish good electrical point contacts to the p-type Si wafer, however, without Al-Si alloying. The processing of the front side involved several photolithography and oxide masking steps forming inverted pyramids, and a selective phosphorus emitter. The TiPdAg front contacts have been formed by full-area evaporation followed by photolithographic patterning. Hence, the front contacts cover only 0.5% of the front side and cause only minimal shadowing losses. Titanium provides an excellent electrical contact to the phosphorus emitter, and Ag provides very good lateral conductivity of the narrow front fingers. Hence, the advanced metal contact technology significantly contributed to achieving the excellent conversion efficiency of 22.8% in 1989.

The main device physics of the PERC solar cell is that photogenerated charge carriers are separated by the p–n junction consisting of the phosphorus-doped, n-type, around 0.4 µm shallow emitter and the boron or gallium doped, p-type, around 170 µm thick silicon wafer. The front-side texture in combination with the front and rear surface passivation provides a very good optical design with minimal front surface reflectance losses and high internal reflectance thereby maximizing light absorption in the silicon wafer. The excellent high minority carrier lifetime of today's p-type wafers in combination with a high-quality front and rear surface passivation enables very high effective diffusion lengths of the minority carriers leading to a very good carrier collection across the pn junction and hence, high short circuit currents as well as high open circuit voltages (V_{oc}) of PERC solar cells.

A broad variety of process and technology developments was required in the past decades to develop an industrial PERC manufacturing process sequence applying novel low-cost and high-throughput processes with sufficient quality for high conversion efficiencies. Particular technological breakthroughs were necessary concerning surface passivation, local contact definition and metallization. Whereas in the past two decades many options for industrial PERC processes have been evaluated, e.g. as outlined in the review papers in [2–5], today a mainstream technology has evolved for the production of industrial PERC solar cells as described, e.g. in detail in [4, 5] as well as in section II in this subchapter: as rear passivation around 100 nm thin aluminium oxide (AlO_x)/silicon nitride (SiN_y) layer stacks are deposited by plasma-enhanced chemical vapor deposition (PECVD) with high deposition rates of several nm/s. The rear contact geometry is defined by laser contact opening (LCO) of the rear passivation forming around 30-µm wide dashed-line-shaped or dot-shaped local rear contact openings in the AlO_x/SiN_y rear surface passivation stack. The full-area (PERC) or H-pattern (PERC+) rear aluminium is deposited by screen-printing of the aluminium paste. The local Al-Si contact is alloyed during furnace firing for a few seconds at temperatures around 750 °C–850 °C, where the silicon locally alloys with the aluminium forming a local Si-Al eutectic layer with a surrounding aluminium-doped back surface field (Al-BSF). Front-side processing includes formation of random pyramids by acidic (multi-crystalline p-type Si wafers) or alkaline (mono-crystalline p-type Czochralski (Cz) Si wafers) anisotropic wet chemical etching, phosphorus diffusion of the emitter, SiN_x antireflective coating and front-surface passivation by PECVD. The Ag front contacts are deposited by screen-printing with a typical

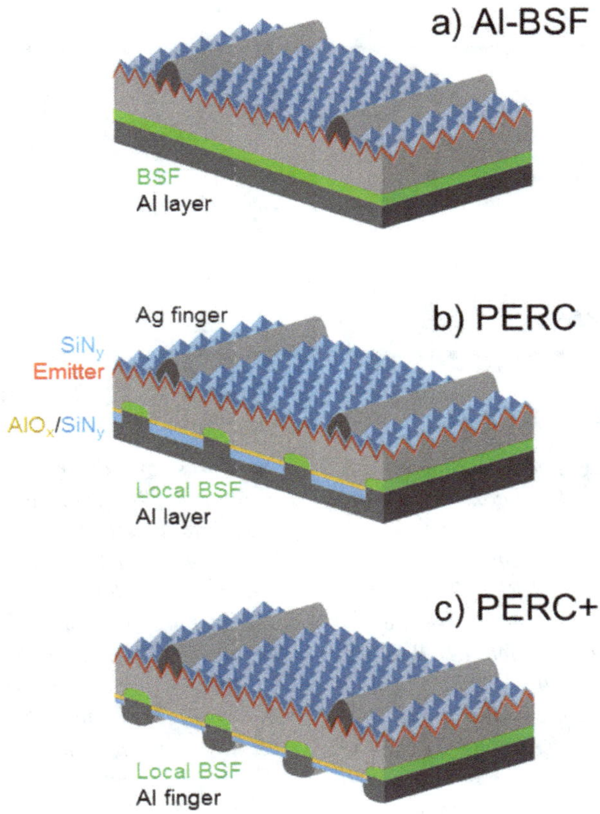

Figure 9.1 *Schematic drawings of (A) an industrial Al-doped back surface field (Al-BSF) solar cell where the screen-printed Al layer alloys with the silicon wafer forming a full-area BSF or Al-BSF; (B) an industrial PERC solar cell with AlO_x/SiN_y rear passivation, local BSF formed by laser contact opening (LCO) and a full-area rear Al layer; (C) a bifacial PERC+ solar cell applying a screen-printed Al finger grid on top of the LCOs on the rear side thereby enabling light absorption from the front and rear sides. In all three solar cell designs, the screen-printed Ag front contact fires through the SiN anti-reflective coating thereby contacting the phosphorus-doped emitter. Figure 9.1 has been reproduced from Dullweber [5] with permission.*

finger width of around 30 µm applying a busbar layout which is compatible with industrial module interconnection technologies.

Historically, most of the PERC manufacturing processes have been developed for and already applied in industrial silicon solar cells with full-area Al rear contact, so-called Al-BSF solar cells. Figure 9.1a shows a schematic cross section of such an Al-BSF solar cell. By introducing wet chemical rear polishing, AlO_x/SiN_y surface

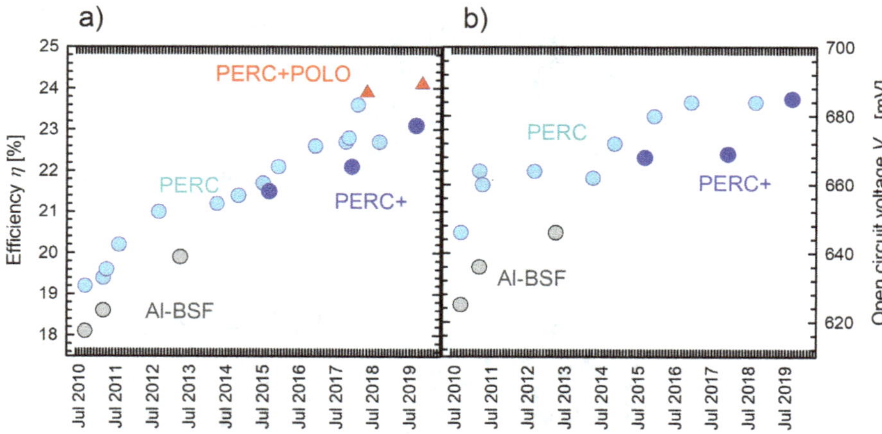

Figure 9.2 *(A) The record efficiency (light blue dots) of industrial PERC solar cells has constantly increased by about 0.5%$_{abs}$./year from 19.2% in 2010 up to 23.6% in 2019. The best bifacial PERC+ cells (blue dots) exhibit similar efficiencies as the monofacial PERC cells. First R&D-type PERC+ POLO cells shown as red triangles achieve efficiencies around 24%. (B) The V_{oc} improved from 625 mV for Al-doped back surface field (Al-BSF) cells (grey dots) in 2010 up to 685 mV for industrial PERC+ cells in 2019 thereby being a main contributor to the increased conversion efficiencies. Unfortunately, the V_{oc} values of the record PERC+ POLO cells have not been published but are expected to be around 700 mV. Figure 9.2 is reproduced from [7] with permission.*

passivation and local LCOs, Al-BSF cells have been upgraded to industrial PERC solar cells as shown schematically in Figure 9.1b. By changing the Al screen-print from full-area to an H-pattern Al finger grid which is printed aligned to the LCOs, bifacial PERC solar cells named PERC+ have been developed [6] as shown schematically in Figure 9.1c. PERC+ solar cells are able to absorb the stray light from the rear side in addition to the direct sunlight from the front side thereby increasing the energy yield of PV power plants. Hence, PERC+ cells are the mainstream solar cell design in the PV industry today.

Figure 9.2 shows the evolution of record conversion efficiencies over the past 10 years obtained with industrial PERC and PERC+ solar cells (large area >230 cm^2 p-type mono-crystalline Cz-grown silicon wafers, screen-printed metal front and rear contacts with industrial busbar layout). In the years 2010 to 2012 the PERC solar cell efficiencies were around 20% thereby increasing the efficiency by 1%$_{abs}$. compared with full-area Al-BSF cells. This increased efficiency in combination with only little additional processing effort and costs was the primary reason for many solar cell manufacturers to build new solar cell production lines applying the PERC design instead of the Al-BSF solar cell design from 2014 onwards. In 2019, PERC

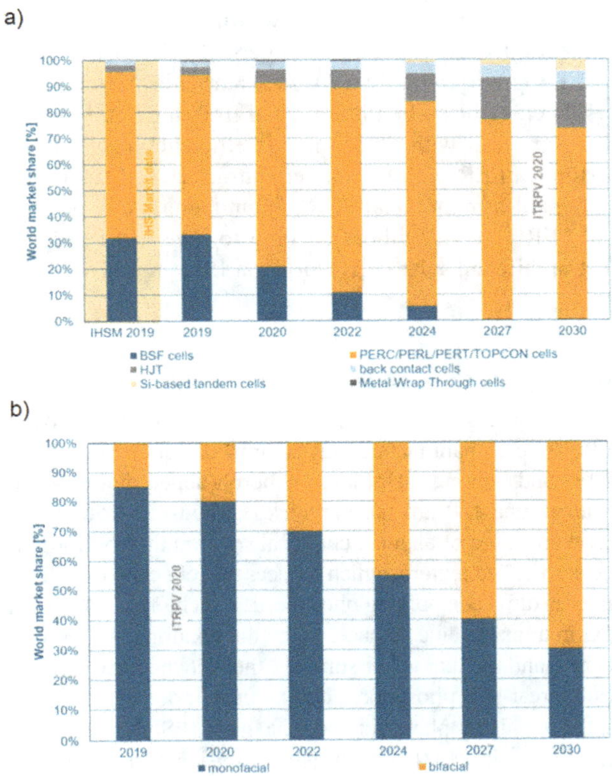

Figure 9.3 *Forecasted worldwide market share of (A) different c-Si solar cell technologies as well as (B) bifacial cells according to the International Technology Roadmap for Photovoltaics (ITRPV) as published in 2020 [8]. PERC cells account for approx. 65% of the world market share of which around 20% are bifacial PERC+ cells. Both market shares are expected to further increase in the next years.*

and PERC+ solar cells accounted for around 65% of the worldwide produced solar cells as shown in Figure 9.3 a which is expected to further increase to 75% within the next years [8]. As can be derived from Figure 9.2, the record efficiencies of industrial PERC cells increased by ~0.5%$_{abs}$ per year as an average value over the past ten years to the present values of around 23%.

An attractive technology option is bifacial silicon solar cells which absorb direct sunlight from the front surface and in addition to the diffuse light from the rear surface of the solar cell. When applied in PV power plants the produced electricity can be increased by up to 20% using bifacial instead of monofacial solar modules [9, 10] depending on the detailed installation conditions, in particular the albedo of the ground and the mounting height. Accordingly, the PV technology roadmap International Technology Roadmap for Photovoltaics (ITRPV) predicts that the market share of bifacial solar cells will increase from the present 20% to 70% within the next 10 years [8]. The Al fingers of PERC+ solar cells with finger widths of around 150 μm account for ~10–20% rear-side shadowing [6]. Hence, the PERC+ conversion efficiency when illuminated from the rear side is a few per cent (absolute) lower compared with the front-side efficiency resulting in around 75% bifaciality [6]. Bifacial PERC+ cells is a very attractive cell technology for bifacial applications and enables monofacial PERC manufacturers to convert their fabs to produce bifacial PERC+ cells without needing to invest in new production tools. Today the industrial bifacial solar cells world market is mainly supplied by PERC+ solar cells.

9.1.2 Production process sequence

Today, the mainstream process flow for industrial production of PERC solar cells is applied by most of the PERC cell manufacturers as summarized in Table 9.1 and described in detail in Dullweber and Schmidt [4]. First, the boron-doped multi- or mono-crystalline silicon wafers with around 180 μm starting thickness and 156×156 mm^2 area are unisotropically etched in an acidic or alkaline chemical solution thereby forming pyramids on the wafer surface, a so-called texture, which reduces the reflection of light. Afterwards, the phosphorus emitter is diffused applying phosphorus oxychloride ($POCl_3$) at temperatures of around 850 °C in a quartz tube furnace. Since the texturing and $POCl_3$ diffusion processes both, the front and the rear wafer surfaces, the wet chemical etching and polishing process removes the rear-side phosphorus doping and smoothens the rear pyramids [11, 12]. Afterwards, the remaining phosphorus silicate glass (PSG) on the wafer front surface is removed in a hydrofluoric acid (HF) etch step and the wafer surfaces are cleaned by following alkaline and acidic chemical steps. Then, the wafer surfaces are coated by surface passivation layer stacks. The rear surface receives a layer stack consisting of 5–20 nm aluminium oxide (AlO_x) and 80–120 nm of SiN_y which are typically deposited by PECVD [13, 14]. AlO_x has been shown to obtain very low surface recombination velocities of minority charge carriers on p-type wafer surfaces due to its built-in negative fixed charges [15, 16]. The SiN_y capping layer is required in order to protect the AlO_x layer against the following Al screen-printing and firing processes, which otherwise would dissolve the AlO_x layer. The phosphorus-doped front surface is passivated by a SiN_y layer, which due to its built-in positive charges [17] is well suited to minimize carrier recombination of n-type wafer surfaces. The front SiN_y layer is typically around 80 nm thick which minimizes the reflection of light from the front surface to below an average of 3% over the whole visible wavelength range due to destructive interference. The rear metal contact pattern is defined by local laser ablation of the AlO_x/SiN_y stack, a process named LCO [18]. Typically, either line-shaped [19] or dashed-line-shaped [20] patterns are used in production. Then, the front and rear metal contacts are deposited by printing Ag- or Al-containing pastes through a mesh screen. The metal pastes consist of an Ag or Al powder with a particle size of a few micrometres (μm) and organic binders and solvents which define the viscosity and printability of the paste. The mesh screen consists of around 15 μm thin metal wires with around 300–400 wires per inch, which are coated by an organic emulsion. At areas where the Ag or Al paste should be printed on the wafer, the emulsion is removed from the mesh

Table 9.1 *Typical production process sequence of industrial PERC and PERC+ solar cells. Whereas the principal process flow of monofacial PERC and bifacial PERC+ solar cells is identical, the individual recipes vary between PERC and PERC+ in order to optimize the front-side efficiency for PERC and the bifaciality for PERC+. Table 9.1 and the related text are reproduced from [5] with permission.*

PERC recipes	PERC/PERC+ process	PERC+ recipes
	Texturing	
	Phosphorus oxychloride ($POCl_3$) diffusion	
3–5 µm polishing removal	Rear chemical polishing	Reduced polishing removal
Thickness ca. 130 nm optimized for passivation and internal reflectance	AlO_x/SiN_y rear passivation	Thickness ca. 80 nm optimized for anti-reflection
	Front SiN_y	
Contact pattern optimized for front-side efficiency	Laser contact opening	Contact pattern optimized for bifaciality, e.g. larger pitch
	Ag rear screen-printing	
Full-area print	Al rear screen-printing	Finger print
	Ag front screen-printing	
	Firing	Reduced set peak temperature

screen allowing the metal paste to penetrate through the screen. First, the rear Ag pads which are later required for module interconnection are screen-printed. Afterwards the Ag paste is dried at 200 °C thereby removing the solvent. The Al rear layer is screen-printed at the rear side and dried. Then, the Ag front contact is screen-printed and dried. The typical Ag front contact layout consists of around 120 very thin Ag conductor lines (fingers) with a finger width of around 30 µm, which are interconnected by up to 12 perpendicular Ag busbars which are required for module interconnection. After drying the Ag paste, the whole PERC process sequence is completed by a rapid thermal process called firing which heats the wafers for a few seconds up to around 800 °C. At this temperature regime the glass frites of the Ag paste etch through the SiN_y front layer allowing the Ag to alloy with the phosphorus-doped silicon forming an electrical contact on the front side. At the same time, the Al paste on the rear side alloys with the silicon wafer at areas where the rear passivation has been removed by LCO thereby forming an electrical contact at the rear side. The Al pastes contain a very small amount of glass frites and hence do not etch the rear passivation layer stack. The resulting industrial PERC solar cell is shown schematically in Figure 9.1b.

The production process sequence described above applies to both, monofacial PERC solar cells as well as bifacial PERC+ solar cells. However, as listed in Table 9.1, several process recipes are optimized differently for PERC versus PERC+ solar cells. Most importantly, the full-area Al layer print of PERC cells is changed to an Al finger print for PERC+ solar cells [6]. The LCO pattern of PERC is optimized for a high front-side efficiency by balancing wafer bulk spreading resistance losses versus recombination losses at the Al rear contacts. For PERC+, an additional optimization criterion is a high rear-side efficiency which favours larger LCO pitches and hence higher spacing of the Al fingers in order to minimize reflection losses of the PERC+ cell when illuminated from the rear side. For the same reason, PERC+ cells apply a thinner rear AlO_x/SiN_y stack around 80 nm [6] which is optimized for anti-reflection and the rear polishing removal may be reduced in order to maintain part of the rear texture, both minimizing reflection losses. For PERC cells, the polishing process and rear passivation thickness are optimized for high front-side efficiencies which typically favour slightly higher polish removal and thicker passivation layers to maximize the internal reflection at the rear. Finally, the set peak temperature of the firing process for PERC+ is typically around 20 °C lower than for PERC since the reduced amount of aluminium paste requires less external heating to obtain the same physical wafer temperature. Figure 9.1c shows a schematic cross section of a PERC+ solar cell.

9.1.3 *Metallization of the Ag front contact*

Since decades, the front-side metallization technology of industrial Al-BSF, PERC and PERC+ cells is screen-printing of the Ag paste followed by a firing step. The main advantage of this approach is that screen-printing is a very fast process enabling a very high throughput of more than 5 000 wafers per hour per screen-printer while at the same time screen-printers are relatively inexpensive, e.g. when compared with evaporation or sputtering tools. Ag has the advantage of very high conductivity, which allows to print very narrow front contact fingers to obtain a small shadowing loss and still maintain a very small grid line resistance. At the same time, Ag pastes are capable of electrically contacting highly phosphorus-doped silicon surfaces by alloying Ag and Si during the firing step. Table 9.2 provides typical parameters of state-of-the-art screen-printed Ag front contacts of today's PERC and PERC+ solar cells.

In terms of process technology, the Ag front metal contact is deposited by printing Ag-containing pastes through a mesh screen. The Ag metal paste consists of Ag powder with a particle size of a few µm and organic binders and solvents which define the viscosity and printability of the paste. Additionally, the Ag pastes incorporate lead-containing glass frits, which during firing dissolve the SiN anti-reflective coating and lower the Ag-Si alloying temperature to around 750 °C. The screen mesh consists of around 13 µm thin metal wires with around 400 wires per inch, which are coated by an organic emulsion. At areas where the Ag paste should be printed on the wafer, the emulsion is removed from the mesh screen by photolithography and selective etching. When the squeegee pushes the paste through the mesh, the high pressure strongly reduces the Ag paste viscosity allowing the Ag paste to flow through the screen to the wafer surface. When the squeegee moves forward, the pressure of the Ag paste reduces and the viscosity increases thereby minimizing the diffluence of the Ag finger. Hence, the thixotropy of the Ag paste enables Ag finger heights h_f of around 20 µm which are close to the Ag finger width w_f of around 30 µm. After screen-printing, the Ag paste is dried at around 200 °C for 30 s, which evaporates the solvent and hardens the Ag paste. The typical Ag front contact layout of industrial PERC and PERC+ solar cells consists of around 120 very thin Ag conductor lines (fingers) with a finger pitch p_f of around 1.3 mm, which are interconnected by 5–12 perpendicular Ag busbars which are required for module interconnection. The finger widths and finger pitch determine the front-side metallized area of the PERC cell which is typically around 3%. In general, the metallized area should be as small as possible to minimize optical shadowing losses while still enabling low series resistance losses of the front grid. Also, since Ag is very expensive, the Ag paste consumption per wafer should be as low as possible.

The PERC and PERC+ metallization process sequence is completed by a rapid thermal process called firing where the wafers are transported on a belt and are heated by infrared irradiation, which triggers several chemical reactions. First, the wafers are heated for around 10 s to temperatures of around 500 °C at which the organic binders are burnt out of the Ag paste. Afterwards, the wafers are heated further to around 800 °C for a few seconds. At this temperature, the Ag particles sinter and form a very dense Ag line conductor with a resistivity around 2 µΩ cm which is close to the specific resistivity of 1.6 µΩ cm of pure Ag. Depending on the exact finger cross-sectional area, screen-printed Ag fingers exhibit a grid line resistance of around 1.0 Ω cm^{-1} which is an important parameter for optimizing the busbar spacing since the current has to be transported through the Ag fingers to the busbars. During the firing step the glass frites of the Ag paste dissolve the SiN_y front layer allowing the Ag to alloy with the phosphorus-doped silicon forming an electrical contact to the solar cells emitter on the front side. The detailed alloying mechanism and contact formation is a very complex topic which is

Table 9.2 Summary of important parameters of the screen-printed Ag front contact and their present typical values in industrial PERC and PERC+ solar cells. The values are taken from [7, 21].

Ag contact parameter	**Industry-typical value**	**Comment**
Finger widths w_f	30 μm ± 5 μm	Depends on screen-opening widths and Ag paste
Finger height h_f	20 μm ± 5 μm	Depends on emulsion height and Ag paste
Finger pitch p_f	1.3 mm ± 0.2 mm	Depends on emitter sheet resistance
Metallized area fraction f_{met}	3% ± 0.5%	Has to be small for little shadowing
Specific resistivity ρ_L	2.0 μΩ cm ± 0.4 μΩ cm	Depends on Ag paste and firing
Grid line resistance R_L	1.0 Ω cm^{-1} ± 0.3 Ω/cm	Depends on finger cross-sectional area
Contact resistance ρ_c	1.5 mΩ cm^2± 0.5 mΩ cm^2	Depends on emitter doping profile and Ag paste
Saturation current density $J_{0,met}$	1 400 fA cm^{-2}± 400 fA cm^{-2}	Depends on emitter doping profile

only briefly described here. More details can be found in Subchapter 6.1 about Ag pastes. Typically, it is assumed that during alloying Ag crystallites grow into the phosphorus-doped Si which establish the electrical contact to the n+ emitter [22]. The Ag crystallites are either in direct electrical contact with bulk Ag or via tunnelling through a very thin glass layer [22]. However, an alternative model is that the electrical contact between the emitter and the Ag bulk is established by nano-Ag colloids dispersed in the interfacial glass layer [23], where the current transport takes place via multi-step tunnelling through the interfacial glass layer. The contact properties of Ag pastes to the n+ Si region depend on various aspects, in particular the phosphorus doping profile [24, 25], SiN thickness, Ag paste composition [26, 27], firing temperature profile and the surface texture [28]. The resulting specific contact resistivity ρ_c is typically in the range of 1.5 mΩ cm^2 and hence much higher compared with, e.g. microelectronic-type alloyed silicided metal-silicon contacts with a ρ_c in the order of μΩ cm^2.

At a metal-silicon contact, both carrier types, electron and holes, can recombine instantly. This is desired for the majority carriers, which in the case of Ag contacts to phosphorus emitters are electrons. However, the minority charge carriers, in this case the holes, should be collected by the pn-junction rather than recombining at the Ag front contact. A parameter quantifying the carrier recombination at the Ag metal contact is its saturation current density contribution $J_{0,met}$ which should be as low as possible in order to obtain high V_{oc}. One known measure is to apply a very deep and high phosphorus doping profile in order to minimize the minority carrier concentration and hence the recombination current at the Ag contact [29, 30]. However, in between the Ag fingers, where the phosphorus emitter is passivated by the SiN, the phosphorus doping should be low in order to suppress Auger and defect recombination [29]. These different emitter optimization requirements are fulfilled by the selective emitter concept [31], where a shallow phosphorus doping below the SiN is applied in combination with strong doping below the Ag contact. The most widely used technology for selective emitter manufacturing is to apply $POCl_3$ diffusion for the phosphorus doping in the passivated area and to locally increase the doping in the metalized area by laser doping [32]. The laser beam melts the PSG and the silicon wafer surface thereby locally diffusing more phosphorus into the silicon wafer surface. A typical value for the saturation current density $J_{0,met}$ of Ag front contacts is in the range of 1 400 fA cm^{-2} as stated in Table 9.2.

The continuous optimization of Ag pastes and screens over the past decades was one of the most important contributors to the steadily increased Al-BSF, PERC and PERC + conversion efficiency as displayed in Figure 9.2. In particular, the shadowing loss of the Ag front grid has been strongly reduced by continuously shrinking the screen-printed Ag finger width from around 100 μm in 2008 to around 25 μm in 2020 as shown in the historic trend chart in Figure 9.4 [21, 33]. The huge progress in screen-printing over the past decade is also nicely displayed in Figure 9.5. Figure 9.5a shows a scanning electron microscope (SEM) image of a 19.4%-efficient PERC solar cell from 2011 with an Ag finger width of around 100 μm and a full-area rear Al layer. In comparison, Figure 9.5b displays a 22.3%-efficient PERC+ solar cell processed in 2019 applying 30 μm narrow front Ag fingers and around 130 μm wide rear Al fingers thereby enabling bifacial light absorption. The last section of

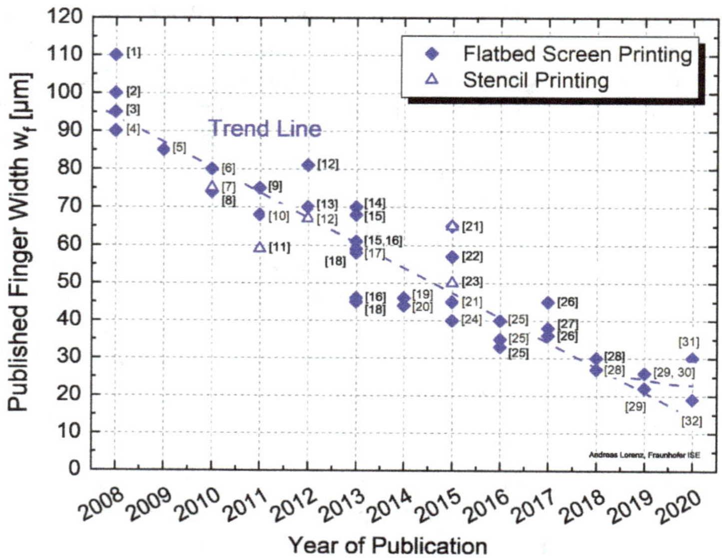

Figure 9.4 *(Source: A. Lorenz, Fraunhofer ISE): From the years 2008 to 2020 the Ag finger widths reduced from 100 µm to around 25 µm by continuously improving the screen-printing and electrical properties of Ag pastes. The reduced finger width minimizes shadowing losses as well as recombination losses of the front metal grid thereby greatly contributing to the increased PERC and PERC+ conversion efficiencies shown in Figure 9.2a. Figure 9.4 is an updated version from [21] to be published in [33] and is reproduced with permission.*

this chapter provides an outlook on future improvement opportunities of the metallization of industrial PERC and PERC+ solar cells.

9.1.4 Metallization of Al rear contact

Similarly as for the Ag front contact, the rear contact technology of Al-BSF, PERC and PERC+ solar cells since decades is based on screen-printing Al paste. The main advantage is, again, that screen-printing is a fast and cheap deposition technology. Also, Al pastes are capable of contacting p-type silicon wafers with low contact resistance and are very inexpensive. As shown in Figure 9.1 a, historically industrial solar cells applied a full area Al rear contact with an Al-doped *P*+BSF and were hence called Al-BSF cells. In this case, the Al layer was screen-printed directly on the full rear surface of the silicon wafer. During furnace firing, the Al layer alloyed with the silicon forming a full-area rear Al-BSF contact. However, the p+/p Al-BSF only marginally supresses carrier recombination at the Al contact with $J_{0,met}$ values around 300 fA cm^{-2} thereby limiting the V_{oc} of Al-BSF cells to around 640 mV and the efficiency to around 20% [5, 7].

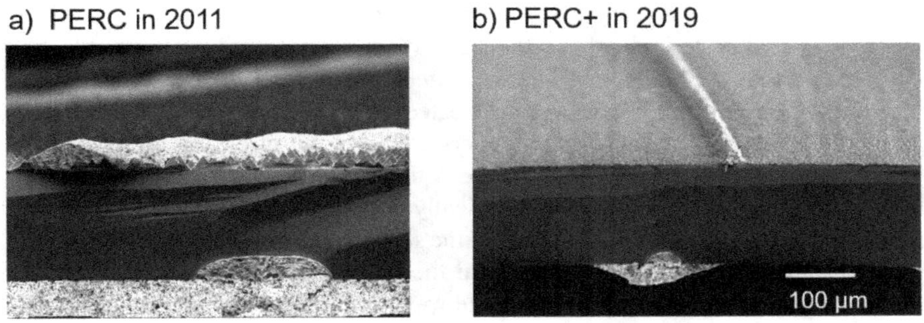

Figure 9.5 *Scanning electron microscope (SEM) images of (A) a 19.4%-efficient ISFH PERC solar cell in 2011 and (B) a 22.3%-efficient ISFH bifacial PERC+ solar cell from 2019. Whereas the Si wafer thickness remained constant at about 180 µm, the Ag finger width has been reduced from around 100 µm in 2010 down to around 35 µm in 2019 as shown in Figure 9.4 thereby minimizing light reflection and front contact recombination. The Al-doped back surface field (Al-BSF) contact area fraction has been reduced from 100% for Al-BSF cells to around 10% for the early PERC cells in 2010 and to around 4% for today's PERC+ solar cells. Whereas the y-axis is the same for both SEM images the x-axis of Figure 9.5a is stretched by a factor 1.4 since the PERC cell has been broken in a 45° angle vs. the metal fingers. In Figure 9.5b, the cell has been broken with a 90° angle perpendicular to the Ag fingers. Figure 9.5 is reproduced from [7] with permission.*

Industrial PERC solar cells passivate the rear surface with AlO_x/SiN_y (see Figure 9.1b) in order to minimize carrier recombination at the rear side of the solar cells. In principal, similar to the Ag front contact, Al pastes with high glass frit content are capable of firing through the AlO_x/SiN_y layer stack in order to form an alloyed Al-Si contact. However, so far the attempts to print a H-pattern with fire-through Al contacts resulted in high $J_{0,met}$ values and thus low V_{oc} values probably because of poor Al-BSF formation at the Al finger edges [34, 35]. Hence, an alternative approach succeeded in the PV industry where the AlO_x/SiN_y layer stack is locally removed by laser ablation forming line-shaped or dot-shaped LCOs. Afterwards, the Al paste is screen-printed on the rear surface full area. In this case, the Al paste contains only a very small amount of glass frits which, during furnace firing, does not dissolve the AlO_x/SiN_y rear passivation layer but only serves to establish a sufficient adhesion of the Al layer to the AlO_x/SiN_y layer. The Al paste consists of a few µm small Al particles, organic binders and solvents which define the viscosity and printability. In addition, today's Al pastes for PERC solar cells contain small amounts of boron and silicon [36] which assist in the formation of a deep and highly doped aluminium (BSF) which is required to obtain low $J_{0,met}$ values. The full-area Al print is less demanding in terms of paste and screen properties and applies a screen mesh with around 15 μm thin wires and 280 wires per inch. The Al paste is dried at around 200 °C for 30–60 s which evaporates the solvents and thus solidifies the Al paste. Afterwards, the Al paste is co-fired with the front Ag paste. First, the wafers are heated for around 10 s to temperatures of around 500 °C at which the organic binders are burnt out of the Ag and Al pastes. Afterwards, the firing temperature increases to around 800 °C for a few seconds. The exact firing temperature profile is typically optimized for the Ag front paste due to the very sensitive reactions described in the section on Ag pastes above. However, the Al contact properties also, in particular the Al-BSF formation, strongly depend on the firing temperature profile. When the wafer temperature during furnace firing exceeds 660 °C, the Al melts, and according to the Al-Si phase diagram [37], silicon from the wafer dissolves in the Al at areas where the rear passivation has been removed by LCO. If the firing temperature and time are sufficiently high, the silicon concentration in the Al melt exceeds the eutectic concentration of 12.6 wt% of silicon in Al. During cool down of the firing process, excess silicon beyond the eutectic concentration epitaxially recrystallizes at the silicon wafer surface, thereby incorporating Al up to its solid solubility limit which is in the range of 5×10^{18} cm^{3}. This local Al-BSF helps to suppress the recombination of charge carriers at the rear contacts due to the p/*P*+built-in electric field. Whereas the basic Al-Si contact alloying mechanism as described above was already well known from full-area Al-BSF solar cells, the locally confined Al-Si alloying of PERC solar cells imposes new challenges. The silicon which dissolves in the Al paste through the LCO openings diffuses into the Al layer several 100 μm beyond the LCO edges. This mechanism reduces the Si concentration at the LCO area which is available for Al-BSF formation leading to thinner or even no Al-BSF in case of PERC cells. Several solutions have been developed to obtain a deep Al-BSF for PERC solar cells such as optimizing the LCO design and adding silicon or boron to the Al paste. A detailed review of the local Al-BSF formation and optimization of PERC cells is published in [4, 5].

Since a few years, bifacial PERC+ solar cells have been developed and are now mass-produced, which apply an Al finger grid on the rear side as shown in Figure 9.1c. The Al fingers are screen-printed and aligned on top of the LCO. In order to print narrow Al fingers, several adjustments were required to the PERC Al screen-printing process. First, the Al paste was modified towards smaller Al particle sizes and increased thixotropy and viscosity in order to reduce the spreading of the Al paste during screen-printing. Second, advanced screens, similar to the Ag screens, were used with thinner wires (ca. 13 μm), higher mesh count (ca. 400 wires per inch) and higher emulsion over mesh thickness of around 12–15 μm. These adjustments continuously reduced the printed Al finger width from the initial 200 μm to the present industry standard of around 140 μm as shown in Figure 9.6 and summarized in Table 9.3. First test prints already demonstrate Al finger widths down to 50 μm approaching the finger widths of today's front Ag fingers. However, in production the alignment accuracy between LCO and Al screen-print is not yet sufficient to apply such narrow Al fingers and hence around 140 μm wide Al fingers are used to account for the alignment tolerances ensuring that all LCOs are covered by Al finger paste in the mass production of PERC+ cells. The Al finger height is

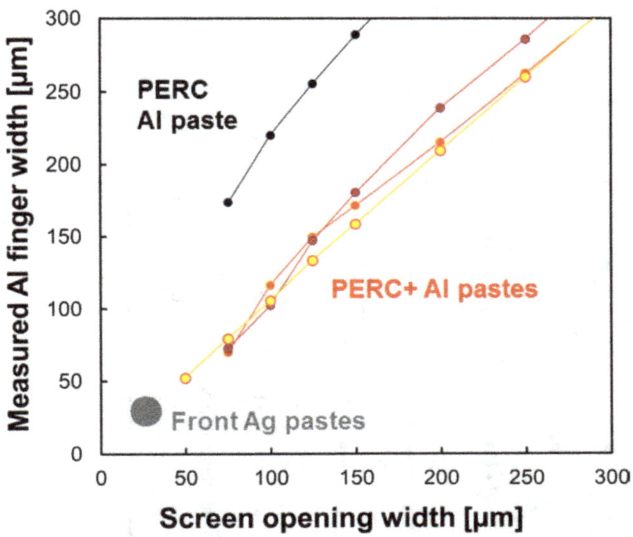

Figure 9.6 *Measured aluminium finger widths in dependence of the screen opening widths. Whereas conventional Al pastes designed for full-area print of PERC solar cells exhibit extreme spreading, recently developed Al pastes for PERC+ finger prints enable Al finger widths close to 50 µm. In comparison, Ag pastes obtain final finger widths around 35 µm with screen opening widths of 25–30 µm. Graph updated from Dullweber [38].*

typically in the range of 20 µm as shown in Table 9.3. The Al finger pitch is chosen as a compromise of achieving a low bulk spreading resistance of the majority carriers favouring narrow finger spacings versus obtaining a high bifaciality and hence little shadowing of the PERC+ rear side which requires a high Al finger pitch. With today's 1 Ω cm p-type Cz wafers and a rear illumination intensity which is around 10% of the front illumination intensity, an Al finger pitch of 1.0 mm is used in the PV industry.

When co-firing the Al fingers with the Ag fingers, the same local Al-Si alloying process occurs at the LCO areas as described above for the PERC cells. However, there is one significant difference. In case of PERC+ the Al fingers confine the Si diffusion from the wafer into the Al paste to the widths of the Al fingers thereby increasing the Si concentration in the Al paste. Hence, during cool-down a deeper Al-BSF recrystallizes for PERC+ cells with Al fingers compared with PERC cells with full-area Al layer [6]. Figure 9.7 shows typical SEM images of local Al contacts of PERC and PERC+ solar cells. Whereas the local Al contacts of PERC cells often show voids and a thin BSF, PERC+ cells exhibit filled contacts and a deep BSF [39]. Figure 9.8 a shows measured Al-BSF depths demonstrating that PERC+ cells obtain up to 8 µm deep Al-BSFs compared with PERC with up to 6 µm. The black line represents a fit applying the model of Mueller *et al.* [40] to the PERC data. The

Table 9.3 Summary of important parameters of the screen-printed Al rear contact and their present typical values in industrial PERC and PERC+ solar cells. The values are taken from [7, 38]

Al contact parameter	Industry-typical value	Comment
Finger widths w_f	140 µm ± 20 µm	Depends on screen-opening widths and Al paste
Finger height h_f	20 µm ± 5 µm	Depends on emulsion height and Al paste
Finger pitch p_f	1.0 mm ± 0.2 mm	Depends on silicon wafer resistivity
Contact area fraction f_{met}	2% ± 1%	Defined by laser contact opening (LCO) geometry
Specific resistivity ρ_L	20 µΩ cm ± 5 µΩ cm	Depends on Al paste
Grid line resistance R_L	1.0 Ω cm^{-1} ± 0.3 Ω/cm^{-1}	Depends on finger cross section area
Contact resistance ρ_c	1.5 mΩ cm^{-2} ± 0.5 mΩ cm^{2}	Depends on Al-doped back surface field (Al-BSF) doping profile
Saturation current density $J_{0,met}$	500 fA cm^{-2} ± 200 fA cm^{-2}	Depends on Al-BSF doping profile

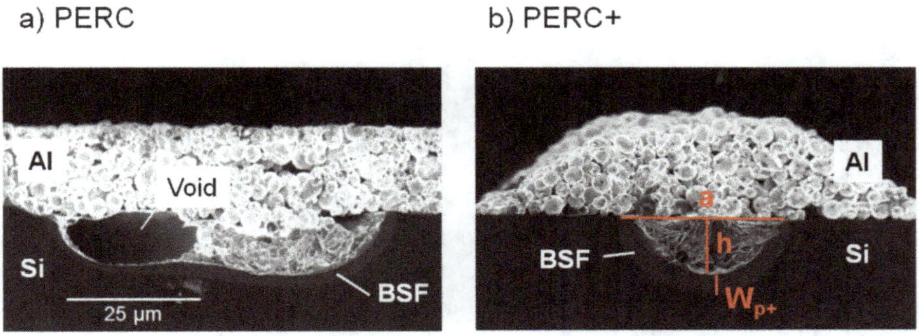

Figure 9.7 Typical scanning electron microscope (SEM) images of locally alloyed aluminium contacts of (A) PERC and (B) PERC+ solar cells. Whereas the local Al contacts of PERC cells often show voids and a thin back surface field (BSF), PERC+ cells exhibit filled contacts and a deep BSF. Al contact geometries used for physical models in Figure 9.8 are shown in image (B) such as the Al contact width a, the contact height h and the Al-doped back surface field (Al-BSF) depth Wp+. Figure 9.7 has been reproduced from Dullweber [5] with permission.

dashed red line shows a fit to the PERC+ data using an extended model that takes the limited Al volume of the Al fingers into account [39]. According to the model, the limited Al volume of the Al fingers leads to a higher silicon concentration in the screen-printed aluminium during furnace firing causing thicker Al-BSFs during the epitaxial re-growth in the cool-down phase. This effect becomes more pronounced for narrow Al contact widths around 50 μm which are industrially preferred in order to increase the throughput of the LCO tool.

Figure 9.8a and b use different symbol styles where filled circles indicate a filled local Al contact, e.g. similar to Figure 9.7b, half-filled circles indicate a partially voided Al contact similar to Figure 9.7a and open circles indicate a fully voided Al contact where the void extends across the whole contact width *a*. According to Figure 9.8b, voids in particular occur for Al contacts with large contact height *h* where the Al-Si eutectic extends more than 20 μm deep into the Si wafer. This finding is explained by an analytical model proposed by Kranz *et al.* that calculates the surface energies of the liquid Al-Si melt, the Si wafer surface and the screen-printed Al particle surface [39]. According to this model, voids form for large contact heights since then a sufficient amount of Al-Si melt is available in order to wet the large surface area of the Al particles rather than the small Si wafer surface area thereby minimizing the total surface energy of the Al-Si system. The PERC+ Al fingers reduce the Al contact height by about 7 μm which is the reason why PERC+ cells do not exhibit voids in contrast to PERC cells. The increased Al-BSF thickness and the reduced amount of voids of PERC+ cells compared with conventional PERC cells result in slightly higher V_{oc} due to reduced rear contact recombination [6].

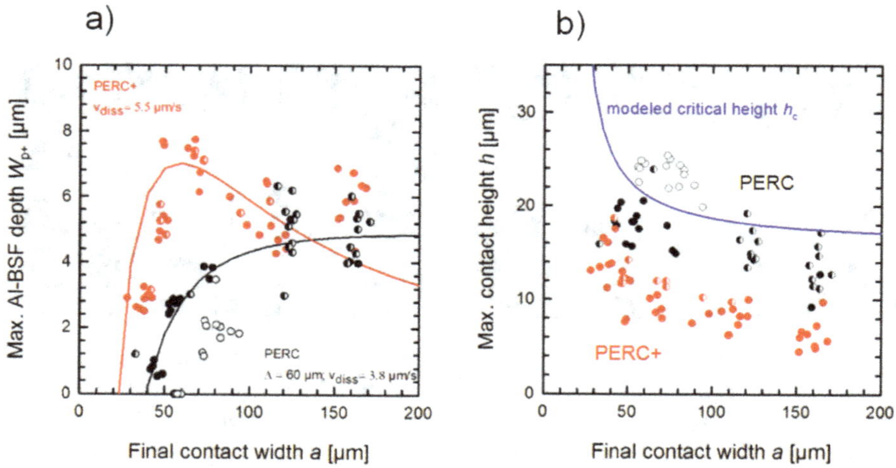

Figure 9.8 *(A) Maximum aluminium-doped back surface field (Al-BSF) depth of PERC (black) and PERC+ (red) solar cells in dependence of the Al contact width a. PERC+ cells obtain a deeper Al-BSF compared with PERC since the Al fingers confine the Si diffusion during firing according to experimental data (dots) and analytical models (lines). (B) Voided Al contacts symbolized by open circles are often found for large contact heights, where the Al alloys deeply into the silicon wafer. This finding is explained by modelling the surface energies of the Si wafer, the Al particles and the liquid Al-Si melt. The Al contact parameters W_{p+}, h and a are defined in Figure 9.7b. Both graphs are reproduced from Kranz et al. [39] with permission.*

Another important parameter of PERC+ solar cells is the busbar design. Since the Al paste after firing exhibits quite a high specific resistivity of around 20 µΩ cm [6], which is about ten times higher compared with Ag pastes, the grid line resistance can become very high limiting the conversion efficiency in case of narrow Al fingers and a wide busbar spacing in, e.g. three or five busbar designs. Figure 9.9 shows the calculated series resistance contribution $R_{s,L}$ of the Al finger grid in dependence of the number of busbars/wires and the Al finger width. In order not to significantly reduce the front-side efficiency when changing from PERC to PERC+, as a rule of thumb the series resistance increase caused by the Al finger grid should remain below 0.05 Ω cm^2. As this is not possible with a three busbar configuration due to the high Al finger length in between the busbars, the five busbar design can be considered as an enabling technology of bifacial PERC+ cells when applying wide Al fingers around 150 µm. When moving to narrow Al fingers below 100 µm width, multi busbar module interconnection technologies with, e.g. 12 wire-shaped busbars per PERC+ cell drastically minimize resistive losses of the Al fingers to below 0.02 Ω cm^2 [38].

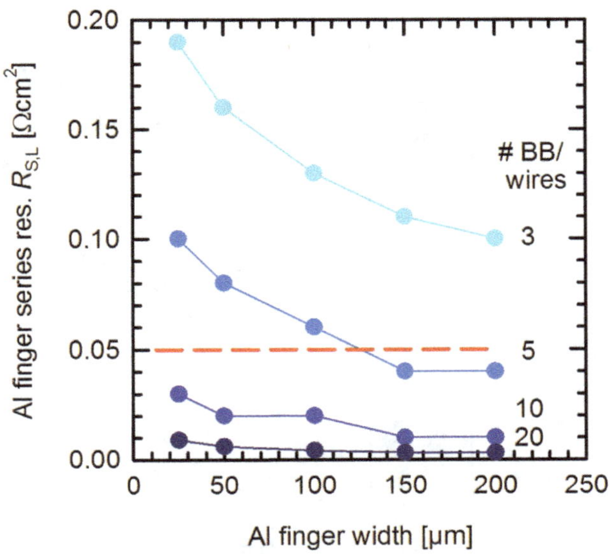

Figure 9.9 *Calculated series resistance contribution Rs,L of the Al finger grid in dependence of the number of busbars/wires and the Al finger width. In order not to reduce the front-side efficiency when changing from PERC to PERC+, the series resistance increase caused by the Al finger grid should remain below 0.05 Ω cm². As this is not possible with a three busbar configuration due to the high Al finger length, the five busbar design can be considered as an enabler of bifacial PERC+ cells. When moving to very narrow Al fingers below 100 μm width, multi-busbar (MBB) technologies with, e.g. 12 wires drastically minimize resistive losses of the Al fingers. Figure reproduced from Dullweber et al. [38]. Copyright (2018) The Japan Society of Applied Physics.*

9.1.5 *Future improvement opportunities*

To predict future technology developments, it is a common approach in semiconductor technology to try to extrapolate technology trends from the past into the future as, e.g. published regularly in the ITRPV Roadmap [8]. In order to continue the quite constant PERC and PERC+ efficiency learning rate in Figure 9.2a of around $0.5\%_{abs}$ per year in the upcoming years, the key will be to further increase the open circuit voltage V_{oc} to beyond 700 mV. Table 9.4 lists the most important technology improvements from typical Al-BSF cells in 2010 to typical monofacial PERC and bifacial PERC+ cells in 2019 enabling the continuous efficiency and V_{oc} increase as shown in Figure 9.2 [8]. Since PERC and PERC+ cells exhibit almost identical saturation current density J_0 values due to their very similar device structure, we show their J_0 values for 2019 in a combined column '2019 PERC/PERC+'. The J_0 values stated in Table 9.4 are either measured for industry-typical PERC and PERC+ cells

Table 9.4 Detailed breakdown of the contributions to reducing the $J_{0,total}$ from 578 fA cm^{-2} in 2010 to 130 fA cm^{-2} in 2019 which enabled a strong increase in V_{oc} as the main driver for the efficiency improvements in Figure 9.2. Furthermore, a projection is provided for the expected J_0 contributions of PERC+ and PERC+ POLO cells in 2022. The individual J_0 values have been either measured at ISFH or taken from the literature, see details in [7]. Table 9.4 is reproduced from Dullweber [7] with permission.

	2010 Al-doped back surface field (Al-BSF)	**Technology improvement**	**2019 PERC/PERC+**	**2022 PERC+**	**2022 PERC+ POLO**
f_{Al} x $J_{0,Al\text{-}BSF}$	100% × 300 fA cm^{-2}	PERC	4% × 400 fA cm^{-2}	1% × 400 fA cm^{-2}	1% × 400 fA cm^{-2}
$J_{0,AlOx/SiN}$	n.a.	PERC	12 fA cm^{-2}	1 fA cm^{-2}	1 fA cm^{-2}
$J_{0,rear}$	300 fA cm^{-2}		28 fA cm^{-2}	5 fA cm^{-2}	5 fA cm^{-2}
$J_{0,E}$	100 fA cm^{-2}	In situ/ex situ oxidation, laser-doped selective emitter (LDSE)	30 fA cm^{-2}	22 fA cm^{-2}	22 fA cm^{-2}
f_{Ag} x $J_{0,Ag}$	7% × 1 400 fA cm^{-2}	Fine line Ag print, 0BB	3% × 1 400 fA cm^{-2}	2% × 1 400 fA cm^{-2}	2% × 35 fA cm^{-2}
$J_{0,front}$	198 fA cm^{-2}		72 fA cm^{-2}	50 fA cm^{-2}	23 fA cm^{-2}
$J_{0,bulk}$	80 fA cm^{-2}	Low O_i, Ga	30 fA/cm^2	8 fA cm^{-2}	8 fA cm^{-2}
$J_{0,total}$	578 fA cm^{-2}		130 fA cm^{-2}	63 fA cm^{-2}	36 fA cm^{-2}
calcul. V_{oc}	636 mV		678 mV	696 mV	711 mV
J_{sc}	37.0 mA cm^{-2}	PERC, fineline Ag print	41.0 mA cm^{-2}	41.0 mA cm^{-2}	41.0 mA cm^{-2}
Bifaciality	0%	PERC+	80%	90%	90%

processed at ISFH or taken from the literature [7]. We add two columns in Table 9.4 extrapolating the short term J_0 reduction and V_{oc} improvement potential for PERC+ and PERC+ POLO cells [7].

Regarding the Al contact properties, in the next few years we expect that improved LCO geometries and Al pastes will further reduce the Al contact area fraction to around 1% without increasing the specific $J_{0,Al\text{-}BSF}$ as already demonstrated [41, 42] with point-shaped LCO contacts in combination with Al pastes with high Si content. This will minimize the area-weighted J_0 contribution of the Al rear contacts to just 4 fA/cm^2 as shown in Table 9.4. At the same time, the AlO_x/SiN_y rear passivation covering almost 100% of the PERC+ rear side contributes a $J_{0,AlOx/SiN}$ of only 12 fA cm^{-2} to the total J_0 [43] due to its excellent passivation properties. We expect that $J_{0,AlOx/SiN}$ will be reduced to 1 fA cm^{-2} in the next few years by continuous process optimization as this value has been measured already at ISFH on test structures. Accordingly, the total rear-side saturation current density $J_{0,rear}$ which is the sum of the area weighted contributions of $J_{0,Al\text{-}BSF}$ and $J_{0,AlOx/SiN}$ reduced from 300 fA cm^{-2} in 2010 to 28 fA cm^{-2} in 2019 and is expected to further reduce to just 5 fA cm^{-2} in 2022.

Another important contribution to increased efficiency and V_{oc} was the reduction of the emitter saturation current density $J_{0,E}$ from around 100 fA cm^{-2} [31, 44] back in 2010 to today's values around 30 fA cm^{-2} [45, 46] by developing advanced emitter doping profiles with reduced phosphorus surface concentrations applying techniques such as in situ [47–50] and ex situ oxidation [29, 46]. The advanced diffusion processes reduce $J_{0,E}$ while maintaining an emitter sheet resistance below 150 Ω/sq. which allows a wide spacing >1 mm of the Ag front contact fingers. As importantly were new developments such as improved Ag pastes and laser-doped selective emitters (LDSEs), which enabled to electrically contact the advanced doping profiles. We expect that this trend will continue and that $J_{0,E}$ values of 22 fA cm^{-2} will be applied in production in the next few years since such low $J_{0,E}$ values have already been published in R&D [45, 46].

One of the most important contributions to increased crystalline silicon solar cell conversion efficiencies in the past decades has been the continuous improvement of the front Ag pastes. Improved Ag paste chemistries enabled much narrower Ag finger widths [29] thereby reducing the optical shadowing loss and metal contact charge carrier recombination loss. Also, improved Ag pastes allowed to contact lightly doped emitters as stated above. The two SEM images in Figure 9.5 demonstrate the huge progress in reducing the front Ag finger widths from ~100 μm in 2010 to around 35 μm in 2019 [21]. Accordingly, the metallization area fraction f_{Ag} of the Ag front fingers reduced from 5% in 2010 to around 2.5% in 2019. Simultaneously, the busbar design also changed during the past nine years. Whereas in 2010 typically three busbars were printed with firing through Ag pastes contacting around 2% of the front surface, today's PERC+ cells often apply non-firing through busbar pastes and/or advanced multi-busbar designs [51] which cover less than 0.5% of the wafer front surface. Hence, the total contacted area fraction f_{Ag} (Ag fingers and busbars) decreased from around 7% in 2010 to around 3%

in 2019 (see Table 9.4). The front Ag contacts exhibit a specific $J_{0,Ag}$ around 1 400 fA cm^{-2} [7]. Assuming a similar $J_{0,Ag}$ of up to 1 400 fA cm^{-2} back in 2010, the area-weighted J_0 contribution of the Ag front contacts reduced from 98 fA cm^{-2} in 2010 to 42 fA cm^{-2} in 2019 as stated in Table 9.4. We expect that it will be possible to further reduce the Ag finger width towards 25 µm in the next years [21] hence further reducing the contacted area fraction f_{Ag} to 2% in 2022. However, it is quite challenging to reduce the high $J_{0,Ag}$ of 1 400 fA cm^{-2}. One solution to suppress Ag contact recombination are selective emitters with very deep doping profile below the Ag contacts [29]. But the industry-typical LDSE forms only a shallow emitter since higher laser intensities required for deep doping melt the silicon pyramids. Other techniques like deep phosphorus diffusion followed by a selective etch back process [31] are likely to be too expensive for industrial production and hence have not yet been adapted in mass production. Another promising approach to reduce metal contact recombination are carrier selective contacts applying polycrystalline silicon on thin interfacial oxide named POLO [52] or TOPCon [53] and their industrial solar cell variants. It has been shown that the carrier recombination at the screen-printed Ag metal contacts can be minimized to 35 fA cm^{-2} by carrier selective poly-Si contacts [54]. When weighted with the contact area fraction of 2%, the recombination at the Ag/poly-Si contact is well below 1 fA cm^{-2}. Hence, in Table 9.4 we introduce a column on the right named 'PERC + POLO 2022' where we suggest to insert a poly-Si finger below the Ag finger of a PERC+ solar cell. The total front side saturation current density $J_{0,front}$ which is the sum of the area weighted contributions of $J_{0,Ag}$ and $J_{0,E}$ reduced from 198 fA cm^{-2} in 2010 to 72 fA cm^{-2} in 2019 thereby being the main V_{oc} limitation of today's PERC and PERC+ solar cells. We expect that advanced emitters and the PERC+ POLO cell design will reduce $J_{0,front}$ to 23 fA cm^{-2} in 2022.

Finally, the saturation current density $J_{0,bulk}$ of the p-type Cz silicon wafer material has also been continuously reduced by improving the wafer quality, reducing the oxygen content and by regeneration procedures minimizing boron oxygen defect recombination, see, e.g. review in Dullweber and Schmidt [4]. We expect that the improvement of the p-type Cz wafer material will continue and enable $J_{0,bulk}$ values as low as 8 fA cm^{-2} as evidenced by today's best R&D Cz wafers applying Ga doping or extremely low oxygen concentrations [55].

Adding up the individual area-weighted saturation current density contributions in Table 9.4 by applying (9.1) results in a total $J_{0,total}$=578 fA cm^{-2} for the Al-BSF cell in 2010.

$$J_{0,total} = f_{Ag} \times J_{0,Ag} + (1 - f_{Ag}) \times J_{0,E} + J_{0,bulk} + f_{Al} \times J_{0,Al} + (1 - f_{Al}) \times J_{0,Al} \quad (9.1)$$

$$V_{oc} = \frac{nkT}{q} \times \ln\left(\frac{J_{sc}}{J_{0,total}}\right) \quad (9.2)$$

When applying a one-diode model according to (9.2) with ideality factor n=1 and a short circuit current density J_{sc} =37 mA cm^{-2}, this $J_{0,total}$ value corresponds to a V_{oc} =636 mV for the Al-BSF cell which fits well to measured IV parameters at that time. For the PERC+ cell in 2019, we obtain a $J_{0,total}$ value of 130 fA cm^{-2} corresponding to a V_{oc} of 678 mV when assuming J_{sc} =41 mA cm^{-2}. This number is quite close to the V_{oc} of around 680 mV of today's typical PERC+ cells. Please note that the largest $J_{0,total}$ contribution of 42 fA cm^{-2} originates from the recombination at the Ag front contacts. Our extrapolation of the PERC+ parameters to 2022 predicts a $J_{0,total}$ value of 63 fA cm^{-2} corresponding to a calculated V_{oc} of 696 mV which is limited, again, mainly by f_{Ag} x $J_{0,Ag}$=28 fA cm^{-2}. Finally, the extrapolation to 2022 for the PERC+ POLO cell results in the smallest $J_{0,total}$ value of 36 fA cm^{-2} corresponding to a calculated V_{oc} of 711 mV. Experimental evidence for this quite high calculated V_{oc} of PERC+ POLO cells is given by measured implied open circuit voltages iV_{oc} up to 712 mV [46] of PERC+ solar cells processed at ISFH without metal contacts which corresponds to a total $J_{0,iVoc}$=35 fA cm^{-2}. According to Table 9.4, iV_{oc} values of PERC+ test structures differ from V_{oc} values of PERC+ POLO cells mainly by the f_{Al} x $J_{0,\ Al\text{-}BSF}$ contribution which is only 4 fA cm^{-2} in the 2022 scenarios. Since $f_{Ag} \times J_{0,\ Ag}$ for PERC+ POLO is <1 fA cm^{-2} and the $f_{Al} \times J_{0,\ Al\text{-}BSF}$ is only 4 fA cm^{-2}, it is expected that the $J_{0,total}$ for PERC+ POLO only increases by 5 fA cm^{-2} from 31 fA cm^{-2} to 36 fA cm^{-2} after metallization. Hence the cell V_{oc} for PERC+ POLO will be close to the iV_{oc} of 712 mV [46] measured for the PERC+ solar cells which corresponds to $J_{0,\ iVoc}$ of 35 fA cm^{-2} (assuming no J_0 contributions from the additional poly-Si process steps). The improvement measures described above have also been recently proposed similarly in a 24% PERC roadmap paper by Trina Solar [56].

9.2 Silicon heterojunction solar cells

9.2.1 Introduction

The silicon heterojunction (SHJ) solar cell structure was patented by Sanyo in the 1990s and commercialized by Sanyo and then Panasonic under the name HIT® from 1997 [57]. The SHJ market share is growing fast and will be around 10% of the global solar market in 2024, which will represent about 20 GWp per year of SHJ module production [58]. SHJ is based on n-type crystallin silicon wafer textured by the KOH process and passivated by intrinsic hydrogenated amorphous silicon (a-Si:H) layers deposited by PECVD. Few nanometres of doped layers are then deposited on the intrinsic a-Si:H passivation. To enable lateral conductivity and good contact resistivity, transparent conductive oxide (TCO) usually based on indium oxide doped by various elements like tin, tungsten or cerium, is deposited by physical vapour deposition (PVD) on the a-Si-doped layer. Finally, the metallization grid is deposited and cured at relatively low temperatures below 220 °C to avoid hydrogen effusion out of the a-Si:H layers or even recrystallization of the amorphous silicon layers [59]. For screen-printing, it obliges to use specific low-temperature pastes (LT pastes) able to be cured around 200 °C or below. The advantages of SHJ cells are high open circuit voltage (V_{oc}), low thermal stresses, low thermal budget processes and good power thermal coefficients. The drawback could be lower current density (J_{sc}) due to higher

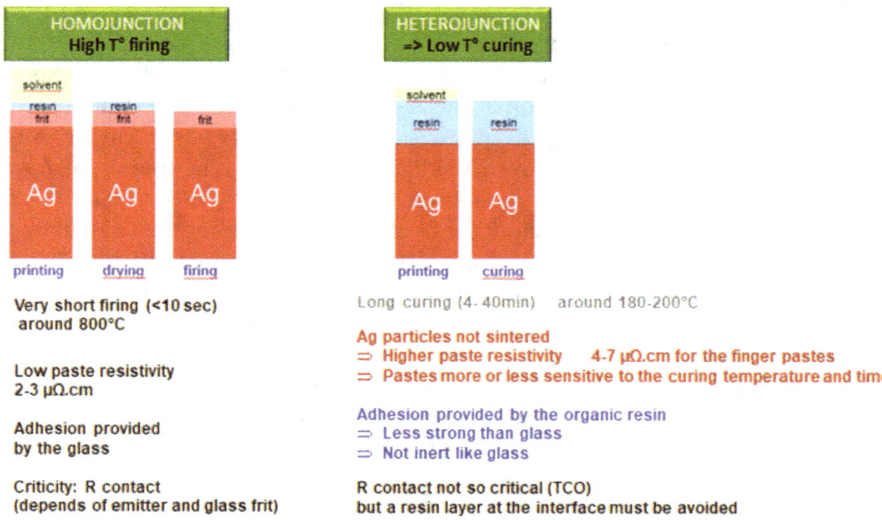

Figure 9.10 Main differences between high-temperature pastes and low-temperature pastes [60]

parasitic losses and constraints on the interconnection as the soldering on non-fired metallization needs a relatively high amount of silver.

9.2.2 Screen-printing of low-temperature pastes

As illustrated in the Figure 9.10 the composition, processing and properties of the high-temperature pastes (HT pastes) and LT pastes are different. For the HT pastes, the resin is dissolved in low concentration; the resin is only a temporary component which burns during the firing, no resin remains on the fired cells. The paste formulation includes a glass frit mixed with the silver particles. During the firing at high temperature (above 700 °C), the glass melts and dissolves the anti-reflective (AR) and passivating layer(s) composed of SiN_x and AlO_x. The glass frit also helps to get a fast sintering of the silver particles (phenomenon of dissolution-reprecipitation of the silver) and it migrates at the interface between the AR and the emitter [61]. The glass frit layer allows a high adhesion of the fired metallization on the cell surface. As the sintered silver is dense, the silver surface is smooth, which makes soldering easier without significant silver dissolution in the solder alloy. Soldered ribbons on HT paste present high peel forces.

In the LT pastes the resin is in a large quantity and remains on the cured cells. The paste formulation does not include any glass frit. The electrical conductivity of the paste is achieved when the resin shrinks during polymerization and pushes the silver particles in contact to each-other. Most of the LT pastes contain large flakes of particles; in this case the electrical conduction is profiting from low resistive losses for the electron transports within the large silver particles. Usually, the LT pastes are not sintered as they include large silver flakes. The printing behaviour of

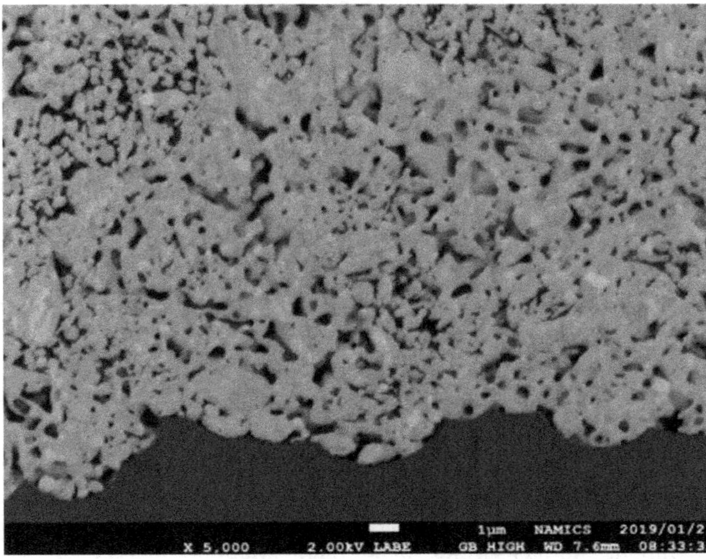

Figure 9.11 Scanning electron microscope (SEM) picture of a cured low temperature (LT) paste (data bar equals 1 µm). Courtesy of Namics Corporation.

pastes having large particles is limited, especially when screens with narrow opening are required. Paste manufacturers are looking for the best mix of small spherical particles and large flakes. In 2010, specific bulk resistivity (ρ) of LT pastes was above 7 µΩ cm. Very long curing time between 20 min and 40 min was mandatory to reach optimal resistivity. Now several suppliers can provide pastes with ρ around 5 µΩ cm and the best products reveal ρ below 4 µΩ cm. In addition, these pastes allow shorter curing times and lower peak temperatures (e.g. interesting resistivities achieved after 4 min at 180 °C). LT pastes usually have a high solid content, often between 91% and 96%.

For the recent formulations, sintering occurs at the particle interfaces. Silver coating or nanoparticles are added to the formulation to densify the paste during the curing. The specific treatment of the silver powders or the addition of nanoparticles can increase the cost of the LT pastes. Figure 9.11 presents the cross-section of a cured LT paste formulated with a mix of small flakes and spherical particles. One can observe some sintering between the grains despite the paste supplier indicating that the paste is formulated without nanoparticles. This paste cured at 200 °C presents an excellent bulk resistivity of 4.2 µΩ cm.

In Figure 9.12, the cross-section of a cured LT paste formulated with a mix of very large flakes and small particles is shown. This type of structure can be observed on cured pastes provided by several paste suppliers and gives resistivities of 5 µΩ cm or below.

One of the specificities of the LT paste is the lack of a glass layer to get the adhesion on the SiN_y layer. The resin present in the cured LT paste enables high

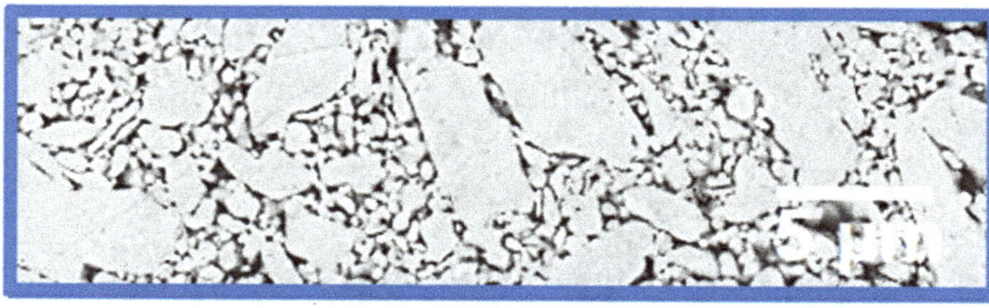

Figure 9.12 Scanning electron microscope (SEM) picture of low temperature (LT) paste for soldering (data bar equals 5 μm) [62]

adhesion on the cell surface. But silver particles must also be in direct contact with the TCO surface in order to get a low contact resistance. The organic chemistry of the LT paste is of high importance and should be well controlled to get low contact resistance. Most of the commercial LT pastes allow acceptable contact resistance for the usual finger widths because the paste is in direct contact with a conductive layer (the TCO) without any layer to be etched (compared with the HT pastes). As the contact resistance is proportional to the surface of contact, it is higher for cell architectures compatible with multi-wire interconnection, like SmartWire Connection Technology (SWCT®) from Meyer Burger, due to the narrower lines [63].

The LT pastes available on the market are mostly based on epoxy or acrylate resins. The storage temperature varies from below −18 °C, to around +5 °C; few types can be stored at room temperature.

9.2.2.1 Screen-printing process

The rheology of the SHJ pastes differs from that of HT paste, first because other resin types and quantity are used to compared with HT paste and second because the silver filler usually includes a mixture of flakes and round particles. Using the same screens with LT paste, the finger deposits are wider and thicker compared with HT paste. The printing speeds are limited to 200 mm/s (600 mm/s usable for the flood), for the pastes dedicated to thick deposits up to 20 μm in thickness with a single print. For thinner deposits, 5 μm to 10 μm, the printing speed can be increased up to 400 mm/s for some pastes [64]. The throughput of the wafer in the metallization line is slower for the LT paste compared with the HT paste.

For the pastes stored at low temperatures, the printability can evolve if the paste stays for a long time at room temperature. During production, it is essential to regularly add fresh products to keep optimal printability.

In 2015, the stainless meshes using wire diameters of 16 μm (325–16 and 360–16) were common for SHJ printing. The mesh opening was large enough to allow the transfer of available commercial pastes using a screen opening of around 40 μm for the five busbar (5BB) cells and down to 30 μm for the busbar-less cells dedicated to SWCT® interconnection. With the development of the continuous mesh, today wire diameters of 14 μm or 13 μm can be used in production; in the near future wire diameters of 12 μm or 11 μm are foreseen. Such small wire meshes can help to get faster printing speeds. Fine wire mesh is mandatory to reduce the screen opening, for example, below 30 μm. Knotless screens are also an interesting option to reduce the screen opening, especially when single printing is considered [65].

9.2.2.2 Curing of LT paste

Because the LT pastes require long curing times, a specific oven must be used. Two main oven technologies are commercially available to get such long curing (typically 10–40 minutes) at high throughput. One solution is to place the cells in a belt oven having a high number of lines (e.g. eight parallel lines). For the second solution, the cells are loaded in metallic cassettes moving through the oven. These two technologies are rather different in terms of curing because in the first solution the cells are exposed directly to the lamps while in the second solution the cells are cured by the hot air coming laterally. In both cases the oven temperature homogeneity is an important criterion, especially for the less reactive LT pastes more sensitive to temperature variations.

Beside these two well-known curing technologies, photonics treatments like IPL (intense pulsed light) are innovative technics proposed for fast curing of the LT pastes. It is reported by Fraunhofer ISE [66] that IPL annealed cells can outperform thermally treated SHJ cells with also the advantage of high throughput and low footprint. Such a curing process can allow better densification due to higher temperature localized in the silver paste but with the risk of damage to the metallization/TCO interface (voids creation), to burn partially the resin or to damage the cell absorber by lowering the open circuit voltage. The photonics curing process should be globally validated by first checking the contact resistance and the paste adhesion and second by checking the module reliability.

The metallization curing step also has an impact on the TCO. Generally, the electrical properties of indium in oxide (ITO) improve as its charge carrier density increases during the curing [67]. This has a drawback on the cell current density as higher parasitic absorption occurs at long wavelengths (>800 nm) due to free carrier absorption [67]. The curing impacts the cell's parameters: first the FF is improved due to three reasons: (1) The finger resistances drop, (2) as do the TCO sheet resistances and (3) the contact resistances at the TCO/silver interfaces. The J_{sc} decreases with curing as mentioned above. The V_{oc} improves until a temperature limit where the hydrogen effusion out from the a-Si:H passivating layer occurs increasing the dangling-bounds density [67].

In order to increase the performance of the SHJ cells, another treatment can be combined with curing at the end of the cell manufacturing process. Applying light soaking or forward bias current modifies the PECVD/PVD layer interfaces and improves the passivation [68]. For example, infra-red (IR) light curing with a combination of the light-soaking and curing process to optimize the performance of the cells.

9.2.3 Metallization of heterojunction cells by electrodeposition of copper

9.2.3.1 Implications of the heterojunction structure for copper metallization

The heterojunction cell structure with layers of amorphous silicon for passivation and a transparent conductive oxide (TCO) on the surface imposes specific requirements for patterning and copper electrodeposition.

Heterojunction cells are intrinsically bifacial and for a cost-competitive process a simple sequence with copper plating simultaneously on both sides is required. The lateral conductivity of the thin conductive oxide is not sufficient for uniform current distribution when contacting at the edges. The TCO sheet resistance is in the same range as for diffused emitters, at 100 Ohm/square and higher. Theoretically light-induced or forward-bias plating could be applied, as used for PERC and TOPCon cells [69], where the contacts are placed at the 'rear' side, i.e. on the cell side opposite to where copper is electrodeposited (see Chapter 8). However, this would require a complex process sequence for bifacial cells, especially for the case of patterning with an organic mask [70]. Moreover, for nickel-copper layers plated directly on TCO, poor adhesion [71] and voids at the TCO-metal interface have been observed [72, 73]. Plating directly on TCO has been applied for busbar-less cells, for interconnection with the SmartWire® technology [70, 74], for which the requirement on adhesion is not as high as for interconnection with soldered ribbons. Methods for ITO pre-treatment to improve adhesion of subsequently plated layers have been described, e.g. ITO reduction by hydrogen plasma [75].

Furthermore, the surface of a heterojunction cell is, contrary to standard diffused cells, conductive and an insulating layer is required to prevent plating on the TCO surface. Whereas laser is the common patterning technique for diffused homojunction cells, to ablate the dielectric anti-reflection coating, it cannot be directly applied on heterojunction cells: the high laser energy would destroy the passivation layers and thus decrease the cell performance.

But the heterojunction structure also offers a huge advantage for copper plating: there is no direct contact between the metal and the silicon and furthermore thin conductive oxides are excellent barriers against copper diffusion [76–78]. Therefore, the reliability concerns related to copper ingress into silicon on homojunction cells, where the electrodeposited metal directly contacts the emitter [79], do not apply for heterojunction cells. Excellent stability of modules having cells with copper-plated metallization has been reported by Silevo (cells with tunnel oxide passivation) [80, 81], Tetrasun [82, 83], Kaneka [84], Sunpreme [78] and by research institutes [85]. The stability of modules with copper metallization currently produced by GS Solar is certified by several institutions [86].

Silevo has used a process with a sputtered seed layer and dry film photoresist for patterning [81, 87]. Tetrasun has also applied at first a sputtered seed layer but patterned it directly to form a 'seed-grid'. A thin dielectric layer, like SiN_y or silicon oxide, has served as mask to prevent plating on the TCO surface [88]. Kaneka has used a process with screen-printed seed-grid and a silicon oxide layer as plating mask [84]. Sunpreme has applied 'typical PCB methods' with photolithography [78] and GS solar uses the concept of 'semiconductor microchip technology' for copper-plated metallization on solar cells [89]. In the PCB and the semiconductor industry, plating on a metal layer, patterned with photolithography, is commonly used.

The industrial processing routes are described in more detail below.

9.2.3.2 Processing and contacting options

The common contacting and processing options are: first, deposition of a seed layer over the entire surface, usually by PVD, and patterning using an organic mask like photoresist or hotmelt inkjet printing [90] and, second, forming a 'seed-grid', i.e. with current conducting material only on the positions of the final metallisation grid lines and deposition of a dielectric layer as plating mask. A seed layer allows to place the contacts independently of the layout and also allows contacting of busbar-less cells. In the case where a 'seed-grid' is formed the contacting for electrodeposition is restricted to the busbars.

A sputtered seed layer offers several advantages: good lateral conductivity, low contact resistance to TCOs and good adhesion [71, 80, 91–93]. A stack of two metals may be used, consisting of a thin adhesion layer and a copper layer on top, which provides the lateral conductivity and is a base for electrodeposition of copper [94].

Photolithography is the standard patterning method used in the PCB and in the semiconductor industry since decades. On solar cells a dry film resist may be applied [81], followed by imaging and development. Well-defined rectangular openings are formed enabling formation of narrow grid lines with a high aspect ratio. By hotmelt inkjet printing fingers of similar dimensions, but slightly irregular shape, are defined. Examples for mask openings and dimensions of copper plated lines are depicted in Figures 9.13 and 9.14. The pattern is realized directly during printing, which means there is only one step in the process for patterning of one cell side. The thickness of both organic masks is about 30 μm and the cost of material for photoresist and hotmelt ink is also in the same range. Since only one process step is required for hotmelt inkjet printing, it may offer an overall cost advantage compared to photolithography comprising resist application, a thermal step, photoimaging and wet chemical development.

Another type of an organic mask is a polymerizable non-photo-sensitive resist. This material is cheaper than a resist containing a photosensitive component. The patterning is realized either by inkjet-printing of an aqueous solution to prevent polymerization [70] or by laser [95].

The second contacting option, formation of a 'seed-grid', is realized either by patterning of a sputtered seed layer ('Tetrasun-process') [88] or through printing a fine line grid by screen or inkjet printing of a metal particle paste [84, 96].

In the Tetrasun-process, first a PVD seed layer is deposited over the entire surface, a grid pattern is formed by inkjet printing of a UV curable ink and the seed layer in between the grid is etched away. In the next step a dielectric layer is deposited by PECVD over the entire surface. Despite the (thin) dielectric layer the UV ink can be removed with a solvent and, at the same time, the remaining seed layer

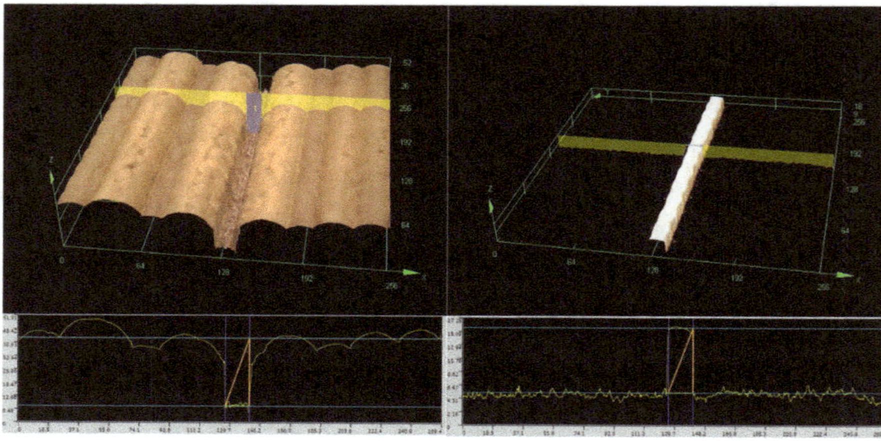

Figure 9.13 *Patterning by hotmelt inkjet printing on a sputtered copper seed layer. On the left: mask opening, width 15 μm, total height 33 μm, with corresponding profile. On the right: copper plated line, width 15 μm, height 10 μm. (Source: CSEM).*

grid is exposed. Copper plating can be carried out simultaneously for both sides with contact at the busbars. The advantage of this process sequence is the low consumption of the organic UV ink for patterning: only the grid area, less than 5% of the cell, is covered by the organic ink. On the other hand, several equipment types are required: PVD, inkjet printer, wet chemical treatment, PECVD for the dielectric

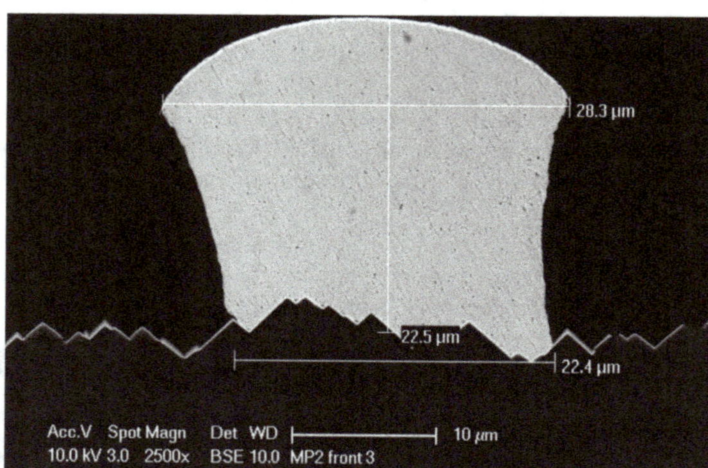

Figure 9.14 *SEM micrograph of a copper line cross section as defined by hotmelt inkjet printing and used for four busbar cells. Line width: 28.3 μm, height 22.5 μm (Source: CSEM).*

layer and a plating line. Consequently, the upfront investment is very high. Despite this complexity the process cost is comparable to the sequence with PVD seed layer and hotmelt inkjet printing (high material cost) and still offers a cost advantage to screen-printing [97].

Contrary to patterning with a thick organic mask, the aspect ratio of lines plated with a dielectric layer as plating mask (thickness ~100 nm) is rather low. The copper is deposited isotropically and the line width increases with plated thickness. Nevertheless, because of internal reflection on the more round finger in the module, the optical finger width is much smaller than the geometrical width. With an optimized finger shape, rough surface and a highly reflecting capping layer the effective shading in the module may be as low as 32% of the geometrical shading [83].

For the formation of a printed 'seed-grid' fine lines are printed either by screen or inkjet printing of a metal particle paste, usually silver paste. The dielectric layer is then deposited over the entire surface. Since the surface of the printed fingers is rough, the dielectric layer on top of the finger is not perfectly dense, the plating current can pass through and copper is deposited on the grid positions [96]. An annealing step after the deposition of the dielectric may enhance the defects in the layer above the printed fingers [84]. This process sequence is quite short: printing, deposition of a dielectric layer, plating. However, the silver paste consumption must be very low in order to maintain a cost advantage in relation to standard screen-printed metallization.

9.2.3.3 Processes in development

The main goal of research work is to further reduce the cost of copper plating, to simplify the process flow and to avoid the necessity of expensive equipment like PVD. The focus is on patterning as it has the highest cost share of the entire process sequence. One approach is to transfer a seed-grid from a foil substrate onto the cell by laser. The process is called laser transfer and firing, 'LTF'. In the first step a dielectric layer, like AlO_x or SiN_y, is deposited on the heterojunction precursor with TCO. Then an NiV seed-grid is transferred from a foil to the cell by laser. In a second laser step the NiV-grid is fired through the dielectric. In this case the energy of the infrared nanosecond pulse laser is absorbed by the metal and does not impact the passivation. For the final metallisation pulse a reverse copper plating process is applied to reduce parasitic plating on the dielectric mask (ghost plating). A promising cell efficiency of 22.2% has been reported [98]. For the LTF process an automated equipment has been developed, with roll-to-roll processing of the foil substrate and capability of transferring different metal layers [99].

Besides on dielectric layers, the LTF process has also been applied for structuring of a sputtered metal seed layer forming a native oxide, like Ti or Al [100]. The thin native oxide layer serves as a plating mask. Mild copper electrolytes and pulse plating are utilized to minimize ghost plating. As alternative patterning methods inkjet printing of a reactive silver ink and inkjet printing of an alkaline etching solution have been tested [101]. Low contact resistivity, good adhesion and promising

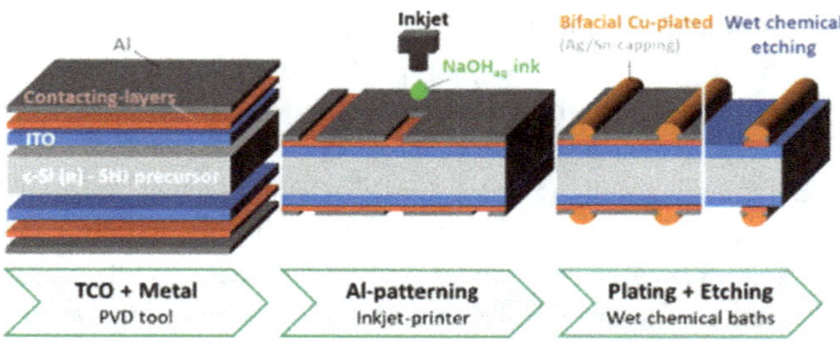

Figure 9.15 Copper plating with native oxide mask (Source: Fraunhofer ISE)

efficiency of 22.7% have been reported with an optimized PVD seed layer stack [92]. The process sequence with inkjet patterning is depicted in Figure 9.15.

The perfect choice would be to plate copper fingers directly on the cell, similarly to dispensing, without any patterning or any additional layer deposition needed. Such a technique is indeed being developed: localized plating of narrow lines directly on the cell surface by micronozzles. It is realized by forming a dynamic confined meniscus between a micronozzle and the cell surface. The micronozzle consists of an electrolyte outlet in the centre and electrolyte suction openings on both sides of the outlet slit. Additionally, the meniscus is confined by air flow (Figure 9.16). The electrolyte flow rate is very high in order to maintain the dynamic meniscus and consequently the solution exchange at the wafer surface is very high. This enables high copper deposition rates, in the range of 1 μm/s. Through ITO reduction before copper electrodeposition the adhesion and contact resistivity have been improved [102]. Copper line plated with DLM is shown in Figure 9.17. The challenges for this technology are the realization of reliable nozzles with micron-size flow channels and their integration into an automatic system to obtain reliable plating of narrow lines at sufficient throughput.

Recently, the application of self-assembling monolayers (SAMs) as plating masks on heterojunction cells has been reported. Since SAMs utilize an extremely small amount of material, the monolayer being only ~2 nm thick, the material reduction compared to a standard organic mask is by a factor of about 1 000. Plating selectivity and patterning width of 20 μm have been achieved on textured industrial size precursors. The patterning technique is to combine with a selective seed

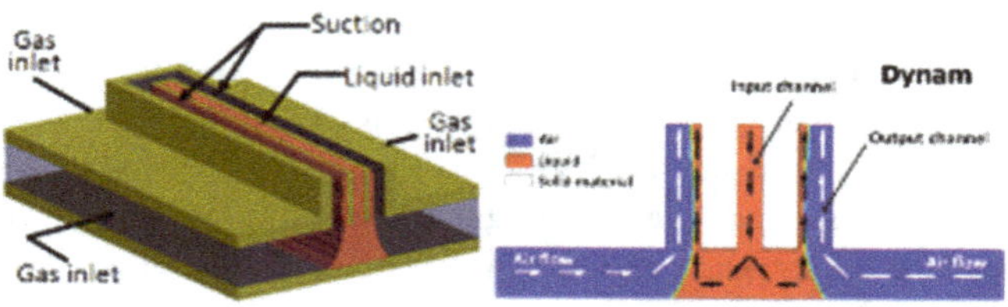

Figure 9.16 Localized plating by dynamic liquid meniscus (Source: RISE Technology)

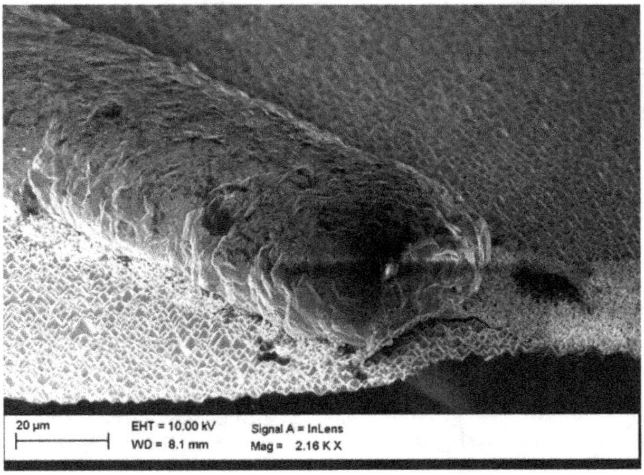

Figure 9.17 *DLM plated copper line on heterojunction cell precursor (Source: RISE Technology)*

layer, formed, e.g. by electrografting of copper in the mask openings for bifacial cells [103].

A simple replacement of the silver paste by copper paste for screen-printing is not possible. The bulk resistivity of commercially available copper pastes is in the range of 10–35 μΩ cm. However, this moderate conductivity is by far sufficient to serve as a seed-grid for copper electrodeposition and can be utilized for a cost-competitive process with a dielectric layer as plating mask. The resistance of the heterojunction structure against copper ingress has been confirmed also for the case of screen-printed copper paste (no impact on photoluminescence signal) and tests of copper electrodeposition on screen-printed copper paste have been started [103].

9.2.3.4 There might not be enough silver for all the solar cells needed in the future

The PV production is expected to grow to terawatt levels in the next decades to enable energy transition to 100% renewables and to meet the climate goals. Annual production of two terawatt is envisaged already for 2030 [104, 105]. For todays annual production of around 100 GW more than 2 000 metric tons of silver are consumed (assuming 100 mg silver consumption and 5 W per cell and 100 GW annual production). In 2019 already 10% of the annual silver supply has been consumed for the PV sector [106]. Even considering a strong reduction in silver consumption per cell (e.g. to 50 mg in 2030) the silver demand for PV will be a considerable part of the entire silver supply. Besides silver, the replacement of other rare materials, like indium, will be crucial for PV production on a terawatt scale. Aluminum zinc oxide is an alternative to the commonly used ITO and contains only abundant elements.

The implementation of plating in production proceeds slowly. The processes and equipment required for copper plating are new to the solar cell manufacturers and the upfront investment is high. The silver paste consumption for bifacial heterojunction cells is high (see Subchapter 9.2.2), particularly high for the standard interconnection with soldered ribbons. Already existing process options for copper plating offer a significant cost advantage to silver screen-printing [92, 94] and the strong increase in silver price in 2020 makes the installation of plating in production even more urgent for heterojunction cell manufacturers relying on ribbon interconnection. High efficiency above 24% has been reported recently for large area heterojunction cells with copper-plated metallization [78, 94], comparable to the efficiency expected for TOPCon solar cells. Besides a simple manufacturing flow, the heterojunction structure offers unique resistance against copper ingress, confirmed by excellent module stability. Could this be an important point for the selection of cell technology for future production on a terawatt scale? First cells with copper plated metallization on aluminum zinc oxide, i.e. silver- and indium-free heterojunction cells, have been demonstrated [107].

9.2.4 *Inkjet printing*

9.2.4.1 Introduction

Inkjet printing is a technology that reproduces digital images on a substrate using ink droplets [108]. Two major technologies have been developed, first the continuous inkjet (CIJ) printing patented in the 1950s and the drop-on-demand (DOD) approach that was commercially developed in the 1970s [108, 109]. Today inkjet printing represents a market of about 100 billion dollars [110]. Inkjet printing uses small holes called nozzles to dispense the ink from a reservoir under pressure. Inkjet printing has been used recently for the fabrication of various solar cell technologies like organic or perovskite [111]. For direct metallization of solar cells, inkjet was already demonstrated in the late 1980s by Teng and West [112]. With the wafer getting thinner, it is clearly advantageous to have a contactless technology like inkjet printing. The second advantage is the optimization of a metallization pattern as the inkjet process is a maskless approach with high versatility in printed design. Also, a high material utilization rate can be achieved in production. Inkjet printing for direct metallization has some potential drawbacks like, first the difficulty to control a constant ink wetting on TCO surface causing line width variation from batch to batch; the sensitivity of the process as the silver load tends to clog the nozzles the low mass deposition per pass; and finally, the relative high cost of inks including nanoparticles or metalorganic decomposition (MOD). These reasons make trends for inkjet printing on solar cells for applications like masking, patterning or seed layer printing more attractive than for direct solar cell metallization [112–114].

9.2.4.2 Inkjet printing for SHJ

As described in previous chapters, copper electroplating on SHJ solar cells needs dielectric masking. By using hotmelt inkjet printing, Lachowicz *et al.* demonstrated certified SHJ solar cell efficiencies above 24.7% on large wafers with copper-plated contacts [90]. Another example of the use of inkjet printing for patterning dielectric layer by local etching has been demonstrated for SHJ application with copper lines as narrow as 20 μm [70].

Silver nanoparticle inkjet printing for direct SHJ metallization was reported in 2014 showing a line width of 32 μm with silver weight down to 9 mg per cell [115]. After cell interconnection using SmartWire (SWCT®), mini-modules showed an active area efficiency of 19.7% comparable to screen-printed reference cells [116]. No module reliability has been reported. A more recent study of SHJ cells metallization compares inkjet printing to screen-printing and flextrail-printing [117]. Inkjet printed cells showed better performances than screen-printed cells with only finger metallization design and measured on PCBTouch® from Pasan. No module performance has been shown for these cells.

9.2.4.3 Optimization of the inkjet printing parameters for SHJ solar cell metallization

The finger width is sensitive to the ink properties (like viscosity and solid loading) and substrate properties (like pyramid size or surface cleanliness). The equipment is also of high importance like the inkjet printer and printing head. Finally, the printing parameter like temperature of the substrate, printing resolution or the number of passes are key for the final line conductivity and width.

The following example describes an optimization procedure used at the CSEM PV-Center to obtain an SHJ module based on inkjet printing metallization. An industrial silver nano-particle ink from Clarian was used in combination with a Pixdro LP50 DOD printer equipped with a Konica-Minolta KM512 head. This head has a drop volume between 3 pico-litres and 14 pico-litres and has two lines of 256 nozzles. The resolution is set by the angle of the head to the moving direction as well as the frequency of the dropping and printing speed. Resolution is given in dots per inch (dpi).

By heating the substrate, the ink's solvent evaporation is faster limiting the drop spearing on the wafer surface as shown in Figure 9.18a. As the distance between the wafer and the printing head is less than 1 mm, a temperature higher than 60 °C heats the inkjet head and induces nozzle clogging during long time processing. The effect of wafer surface finish has an impact only at low temperatures below 60°C as shown by comparing bare textured silicon versus coated with ITO substrate (Figure 9.16). The finger width is also influenced by the printing resolution; with high resolution, the neighbouring drops coalesce and induce larger fingers (Figure 9.18b). Advanced printing strategies reduce the printed line width as it avoids the coalescence of the drops. This reduces the finger width by 40% and is showing in figure impressive aspect ratio of 0.7 after 100 passes (see Figure 9.19). The line resistivity drops with the number of passes and curing temperature as shown in Figures 9.20 and 9.21. To obtain a line resistivity below 10 Ω cm^{-1} a minimum of 15 passes are needed after 20 minutes of curing at 200 °C, where the specific bulk resistivity (ρ) of the paste is about 10 μΩ cm. After 1 hour of curing at 200 °C, ρ is reduced to 6 μΩ cm.

Fabrication of SHJ cells was done on 6-inch cells, with 77 fingers at the front side printed in the first group by inkjet and compared to a second reference group

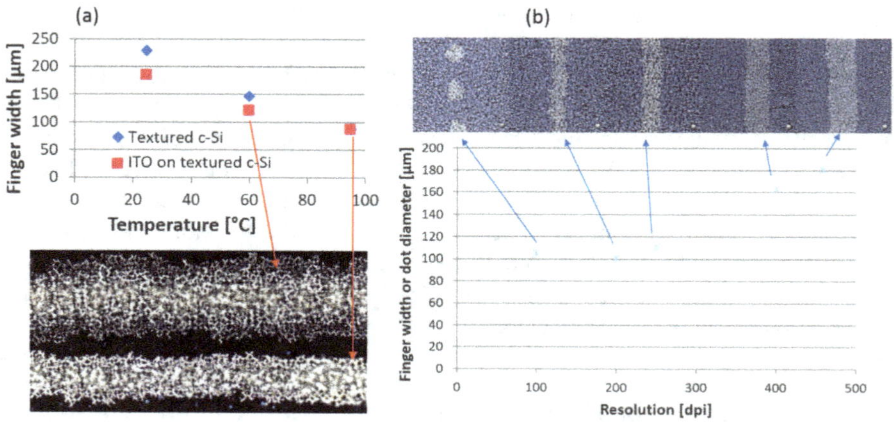

Figure 9.18 Optimization of printing conditions: (A) Substrate temperature optimization for a resolution of 460 dpi and (B) Resolution optimization at 60 °C

done by screen-printing. The backside was similar in all cases and was done by screen-printing on 1 mm square on the full surface. The inkjet was processed with advanced strategies of printing and the curing was set at 200 °C for 30 minutes. The number of passes varied between 5 and 100, which represented printed thicknesses between 2 µm and 28 µm. Each pass comprised 52 000 drops and about 0.7 mg of solid silver deposit. Cell performance versus finger height did not present clear trend as shown in Figure 9.22a. The best cell performance of an inkjet-printed cell was comparable to the screen-printed reference with about 0.15% lower efficiency due to lower current density by about 0.3 mA cm^{-2} (see Table 9.5). After module integration with 18 SmartWire (SWCT® from Meyer Burger) of 0.2 mm in diameter,

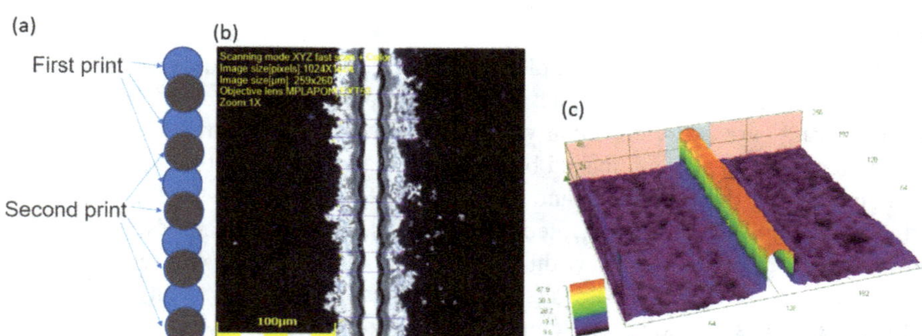

Figure 9.19 Advanced printing strategy (A) Scheme, (B) 2D color microscope image and (C) 3D laser scanning confocal microscope image showing an inkjet printed finger with an aspect ratio of 0.7 after 100 passes

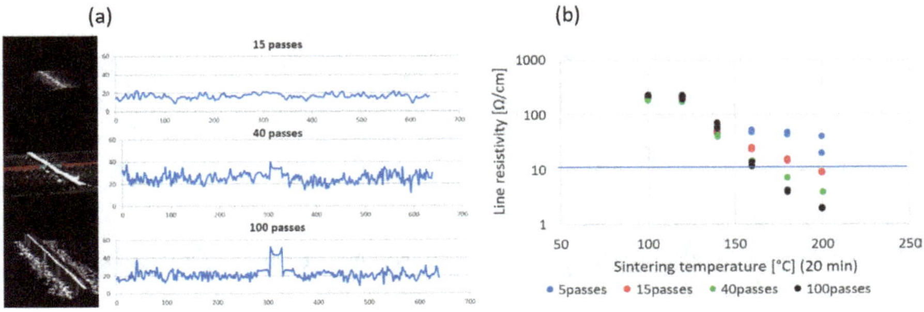

Figure 9.20 Effect of number of passes on (A) Deposited thickness measured by 3D confocal microscope (B) Line resistivity versus sintering temperature

the cell performance of low silver deposit was reduced in module due to more cell to module power losses in the fingers (see Figure 9.22b). Modules with metallization thickness from 5 μm to 20 μm and nearly 30 μm showed similar performance as more conductive thicker lines are wider and so induce more shadowing so one effect compensates the other and the module performance can be comparable at 5 μm and 30 μm finger thickness. All the modules were placed in a climatic chamber for thermocycles (TC between –40 °C to +85 °C following IEC standards). Power degradation after 200 TC showed a clear trend with modules using thinner finger degraded more than the ones with thicker fingers (Figure 9.23a). The degradation for modules using thin fingers was induced by electrical contact losses as shown by the electroluminescence (EL) (Figure 9.23b) after 200 TC. This was also related to a large increase in FF. It must be noted that similar optimization with other silver nano-inks should be done to improve the module reliability for low silver deposit. Moreover, the SmartWire technology might have improved, and this could also have an impact on module reliability.

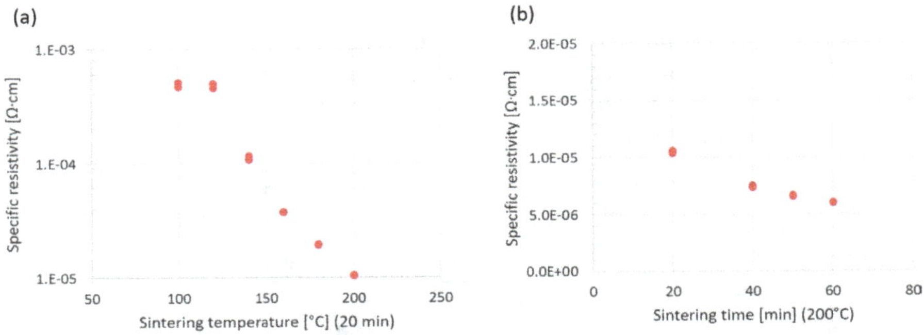

Figure 9.21 Specific bulk resistivity versus sintering time in (A) and sintering temperature in (B) for 40 passes of inkjet printing

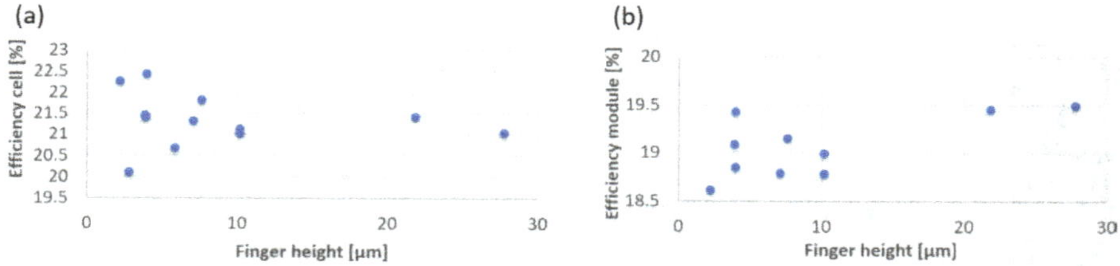

Figure 9.22 *(A) Cells and (B) modules performance versus inkjet printed finger height*

9.2.5 Cell design

The cell design is optimized using metallization performance as described in the previous subchapters, and summaries in Figure 9.24, and by the interconnection technologies that will be discussed in the next subchapter. Figure 9.24a is a summary of the status of SHJ metallization. It shows the finger deposited width versus the line resistivity. The best performing metallizations are found at the bottom left of the graph with a narrow line (small finger width) and low line resistivity. After module integration, the optical width of the finger is reduced, thanks to back-reflection at the glass/air interface [118].

Measurement of the ratio of silver finger optical width in the module with finger width measured on the cell, the ratio is of 58% for screen-printing whereas for plating, inkjet and copper screen-printed fingers the ratio is calculated to be around 90% due to the shape or the lower reflectivity. One can observe that the copper metallization is optimal for long fingers like for busbar cell interconnection the screen-printing is the second-best choice for performance and in particular looks after lamination (Figure 9.24b). Inkjet printing and copper paste still need some more development to be at the same level compared to screen-printing. The choice between plating and silver screen-printing will be done on cost, performance and interconnection possibilities.

Table 9.5 *Current-voltage (IV) curve parameters (Efficiency; Current density at short circuit, J_{sc}; Open circuit voltage, V_{oc}; Fill Factor, FF) of the best cells and modules done with front mettalization by inkjet and by screen-printing*

	Best cell		Best module	
IV param.	**With inkjet**	**With screen-printing**	**With inkjet**	**With screen-printing**
Eff (%)	22.44	22.60	19.49	20.75
J_{sc} (mA cm^2)	38.9	39.3	36.4	38.1
Voltage (V)	736.6	733.6	728.0	728.0
FF (%)	78.26	78.20	73.51	74.30

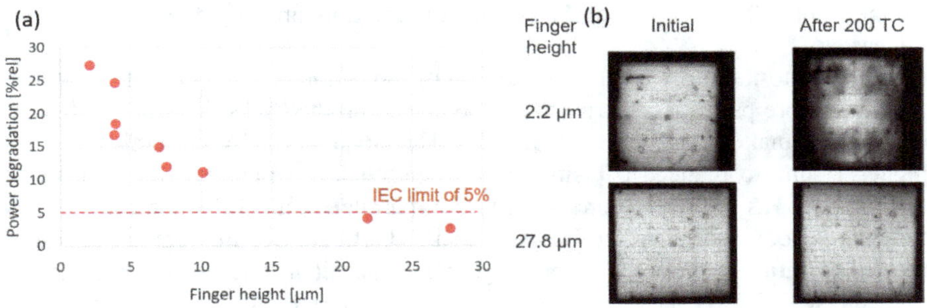

Figure 9.23 (A) Power degradation (positive values are degradation) versus inkjet printed height and (B) electroluminescence (EL) images before and after thermocycles (TC) for two-finger heights

As mentioned earlier, cell interconnection directly influences the cell metallization design. As the paste-specific bulk resistivity is higher for LT pastes, it is generally preferred to use shorter fingers to reduce power losses. Power dissipation losses in the finger, P_f, is inversely proportional to the square of the number of busbars [119–121].

J is the current density, L, the width of the cell, n_f, the fingers number, R_f, the finger line resistance, n_{BB} the busbar number and C a constant. Equation shows the solution for relative high line resistance like for SHJ LT paste, one solution will be to increase the number of busbars (n_{BB}) from 5 BB to 15 BB (with a ratio of 15/5=3), the line resistance (R_f) can be multiplied by 9 (square of the ratio) to reach the same power losses in the fingers.

For the pastes dedicated to cell architectures using a limited quantity of busbars (e.g. 6 busbars or less), the selection of LT pastes formulated for thick deposits is preferred. Spreading is then a risk for the LT paste designed for high aspect ratios (e.g. fingers of 50 µm in width and 15 µm of thickness can be obtained by a single print). Another possibility to achieve thicker fingers is to use double print, especially

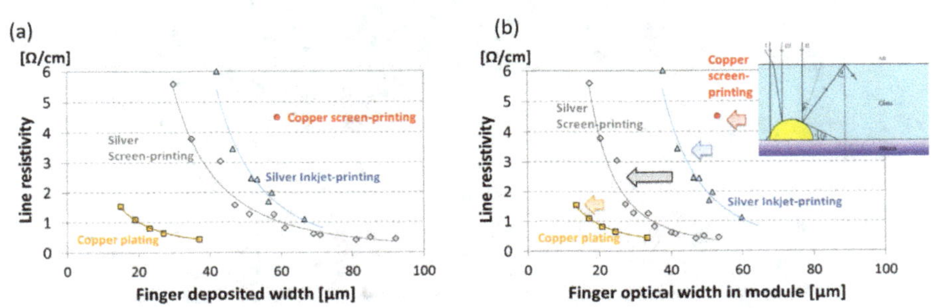

Figure 9.24 (A) Finger deposited width versus the line resistivity and (B) Images before and after thermocycles (TC) for two-finger heights

for the front side. Thickness of 25–30 μm is usually obtained by double printing of LT pastes.

Thanks to the long lifetime of SHJ cells, the p-doped a-Si:H (emitter) which produces more parasitic absorption can be placed on the backside (side not facing the sun), to make 'rear-emitter' SHJ cells. The advantage of a rear-emitter cell is higher J_{sc} due to reduced parasitic absorption and lower metallization density at the font side, as the wafer conductivity is in parallel to the TCO conductivity on the front side. Concerning the backside and in particular for the rear-emitter cell, the metallization is based on a denser grid (smaller pitch) than for the homojunction cells in order to cancel the disadvantage of the more resistive pastes and higher lateral resistance.

The finger resistance problem is amplified for the cells dedicated to shingling. For the 1/6 cell tile format the finger resistance is equivalent to three busbar cells interconnected by conventional ribbons. So even using a high aspect ratio, the double print is mandatory to limit the resistive losses.

In contrast the realization of cells with only fingers (busbar-less cells) dedicated to the wire interconnection from SWCT® allows to strongly reduce the constraint on the finger resistance, as 18 wires are used as for a standard. For such cells, different LT pastes can be used now optimized for printing narrow and thin fingers at high speeds. The reliability can be excellent due to the very high number of connections, especially for deposits around 100 mg per cell (front-side and back-side prints).

Half-cell versus full-cell: Compared to PERC+ cells, SHJ cells have very high passivation and are consequently more affected by the cutting because of the lack of passivation of the sides. Figure 9.25 indicates that the cutting losses decrease as the wafer size increases. For the small tiles like 1/6 dedicated to shingling the simulation integrates the losses on the two sides, but the effective losses after shingling will be smaller due to the fact that one of the two cut sides is overlapped by the next tile.

	Simulated Efficiency (%)			
	M2	M6	M10	M12
Full size	23.07	23.08	23.12	23.14
1/2 cell	22.87	22.98	23.05	23.07
1/3 cell	22.70	22.73	22.85	22.87
1/4 cell	22.53	22.58	22.72	22.75
1/6 cell	22.21	22.27	22.46	22.50

Figure 9.25 Effects of cutting of the silicon heterojunction (SHJ) cell efficiency depending of the wafer size and the number of cuts [122]

Part of the cutting losses can be recovered using a re-passivation process. Obviously, this process must also be done at low temperatures. Shingling on SHJ cells is also more affected by the resistance of the long fingers due to the use of less conductive LT pastes. To reduce these two disadvantages, a research institute is working on a modified shingling technology compatible with larger tiles (1/3 of cell instead of 1/6).

9.2.6 Interconnection

Interconnection of SHJ solar cells is special as the LT paste is of a different composition compared to the HT paste.

For example, the selection of the optimal LT pastes is critical due to antagonist requirements on the paste properties: first, the bulk resistivity needs to be low to reduce the power losses in the fingers; this can be improved by increasing the silver load of the paste. Second, the adhesion of the paste to the TCO surface needs to be improved to support high stresses from soldered interconnection; this can be improved by increasing the resin content of the paste. If a single print is used for fingers and busbars, a trade-off composition of the paste should be found.

During soldering, the melted alloy tends to dissolve some silver from the metallization to reach the thermodynamic equilibrium and is known as silver leaching. To reduce silver leaching, some per cent of silver can be added to the solder alloy, which induces an increase in interconnection cost. As silver flakes of the LT pastes have high surface on volume ratio, the density of LT pastes after curing is lower compared to the sintered silver of the HT pastes. The low density and the high silver surface to volume make LT pastes more sensitive to silver leaching during soldering. The flux interaction with the LT paste could be another reason of lower solderability of this paste type.

Adhesion issues can occur during soldering of ribbons on SHJ cells and even more during soldering of wires in the case of multi-busbar interconnections as the stresses are concentrated in the small pad surfaces. Obtaining high peel forces with soldered ribbons or soldered wires generally requires specific LT pastes dedicated to soldering (pastes designed for higher adhesion having a higher resin to silver content). These pastes must be deposited in rather high thickness, for example, 20 μm or more. The pastes dedicated to the bus or pads are generally not intended to be printed through narrow the screen opening and their resistivities are higher than the pastes for fingers. For the cell front side, a dual printing mode is advantageous for the silver lay-down and to get high thickness for the soldering pads with high adhesion and narrow fingers with low resistivity. Deposits higher than 350 mg are generally mandatory for cells with busbars (as shown in Table 9.6). Strong efforts are seen in China to push the SHJ technology [123]. Multi-busbar soldering on SHJ cells using BiSnAg, PbBiSn or PbSn solder coated wire with peel force higher than 2 N/mm was demonstrated on cells with less than 200 mg of silver deposit [124]. These recent results shown in China clearly demonstrate that SHJ cells and module technology are closing the gap of cost with mainstream PERC but giving about 1% absolute higher efficiency.

Table 9.6 Screen-printed silver paste deposited mass at front and backside for 4, 5, 6 busbars (BBs) for soldering, electrical conductive adhesive (ECA) gluing and wire interconnection grid design

	Soldering			**ECA-gluing**			**Wire Interco**	
	4 BB	**5 BB**	**6 BB**	**4 BB**	**5 BB**	**6 BB**	**reliable**	**light**
Front mg	165	155	145	75	70	45	40	20
Back mg	255	220	190	170	135	100	60	40
Total mg	420	375	335	245	205	145	100	60

The LT paste is less critical in case of gluing of ribbons. For the interconnections using electrically conductive adhesive (ECA), the silver leaching issue is avoided, and the peel force is not linked anymore to the printing thickness. However, the peel force of ribbon glued by ECA on LT paste remains lower than ribbon soldering on HT paste. With most of the ECA, one can observe that the peel force is lower when the ECA is attached on the silver printed areas than when in direct contact with the TCO. So, long bus areas should be avoided. However, because the ECA contact resistance is better on the printed silver than on the TCO, the use of opened pads is the best design to combine a good ECA electrical contact and a high peel force. Another advantage of using open busbars as shown in Figure 9.25 is cost reduction as it significantly reduces the amount of paste required for metallization.

The ribbon gluing is the less critical available technology for the LT pastes because the bonding occurs not only on the metallization but also on the TCO. For this reason, the metallized area can be rather small in the « bus area » where only small pattern can replace the bus used for the soldering technology (see Figure 9.26). With clever designs the weak point is generally not at the cell/metallization interface but on the ribbon side. Around 150 mg of wet LT paste (see Table 9.6) is necessary to print the BB6 cells. Another significant cost is the ECA amount; it can be reduced

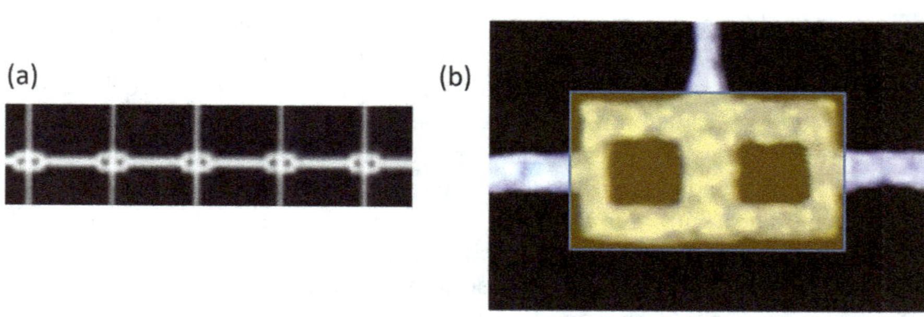

Figure 9.26 Busbar design using open pad patterns dedicated (A) to ribbon gluing and (B) to shingle cell interconnection. Silver busbar and finger is in white and electrical conductive adhesive (ECA) deposition is drawn in yellow.

Table 9.7 Electrical conductive adhesive (ECA) deposit done by screen-printing process for four, five and six busbars (BBs)

	ECA screen-printing process		
	4 BB	5 BB	6 BB
ECA deposit (mg 2 sides) – reliable	80	100	120
ECA deposit (mg 2 sides) – optimized			60

by more than 50% using localized ECA pads. Depending on the ECA selection the optimum ECA deposit must be determined for good module reliability. Reliable modules can be realized using 30 mg of ECA per BB6 half-cells (see Table 9.7).

The external printed lines of these opened pad patterns are also interesting to localize the ECA deposition and avoid its lateral spreading, especially for the connection of very narrow ribbons or for the shingling processes using small overlap values (see Figure 9.26b). The shingling technology is rather new, especially for the SHJ cells using LT pastes. Considering cells to be cut in six tiles, because the finger resistance constraint is then equivalent to 3 BB, rather high deposits are mandatory, i.e. 1/6 cell needs 250 mg of the full cell before cutting. To be reliable, the shingle cell interconnection uses softer ECA with lower elastic modulus compared to solder. The shingling technology has several common points with the gluing technology, especially the same interest to use localized opened pads. Because the interconnection between tiles did not include ribbons, we can take advantage of the very high adhesion of the ECA on the ITO and use very low amount of ECA, significantly lower than for the ribbon gluing. For example, 2 mg of ECA is enough to connect two tiles. So, the interconnection cost can be very low. Good mini-module reliabilities can be achieved with such very low ECA deposits combined with small overlaps of 1 mm.

9.2.7 Module reliability, bifaciality and energy yield

9.2.7.1 Reliability

SHJ module reliability depends on the bill-of-materials of the module. It has been proven by Panasonic that the technology is stable and reliable with more than 20 years testing outside already as the production of HIT module started in 1997. The final product has the best ranking for module reliability [125]. Japanese and European SHJ modules are built using either the ribbon gluing technology or the SmartWire technology (SWCT®). These two interconnection technologies give equivalent module powers (CEA-INES is using the two technologies fabricated modules based on the same cell precursors and report identical power for the two module types). Both technologies can give reliable modules. However, the SWCT technology, because the wires are directly in contact with the narrow fingers, requires more specific materials. On the other hand, thanks to multi-wire interconnection, SWCT can save up to 70% of silver from the ECA and the paste laydown. Recently, the EU Ampere

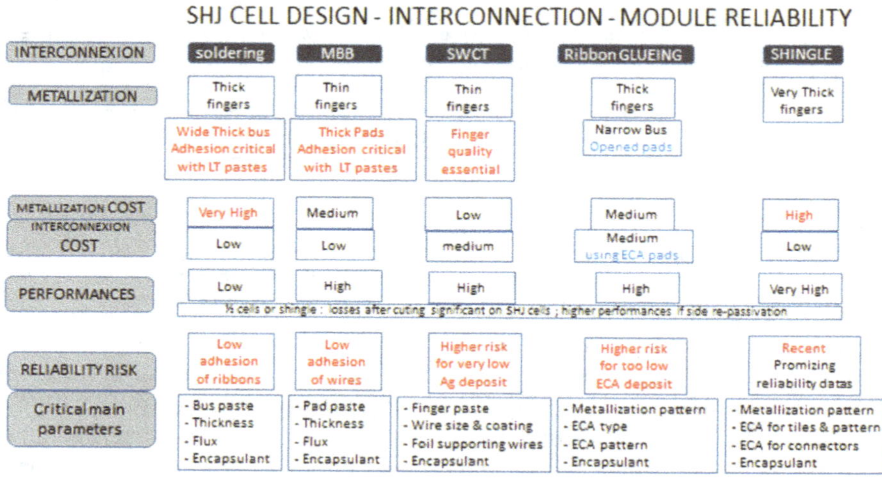

Figure 9.27 *Table giving metallization criteria for various interconnection technologies; courtesy of CEA-INES*

project reported that SHJ cells interconnected by SmartWire (SWCT®) pass more than 1 000 TC from –40 °C to +85 °C and nearly 14 000 hours, which represents one and half years, in damp heat at 85% relative humidity and 85 °C with less than 5% relative power degradation [126]. Chinese SHJ modules are often built using wires soldered on pads. Advantage, disadvantage and parameter criticality are pointed out for each interconnection technology in Figure 9.27.

9.2.7.2 Bifaciality

As the SHJ structure is rather symmetrical with TCO on both sides, it is important to take advantage of its bifacial structure to improve the energy yield of the PV module. Bifacial cells have metallic finger grids deposited on the two sides to contact the TCO. This symmetrical structure avoids stress and cells without camber are generated. Depending on the PECVD and PVD layers, bifaciality of the unmetallized precursors will vary between 90% and 100%. During metallization, the bifacial ratio is generally reduced because of the printing of a denser metallization backside than front side (a dense backside allows to reduce the resistance losses and consequently allows to get a higher cell efficiency). In relation to metallization only, cell and module bifaciality increase with the backside finger pitch, but the cell efficiency can be limited at high bifacial ratio (see Figure 9.28) [127]. Generally, the bifacial coefficient of the SHJ module is between 90% and 94%.

9.2.7.3 Energy yield

As other semiconductor devices, the performance of the c-Si PV cell and module technology varies with temperature. In particular, the open-circuit voltage, related to the bad-gap, drops with temperature increase, whereas the current density increases

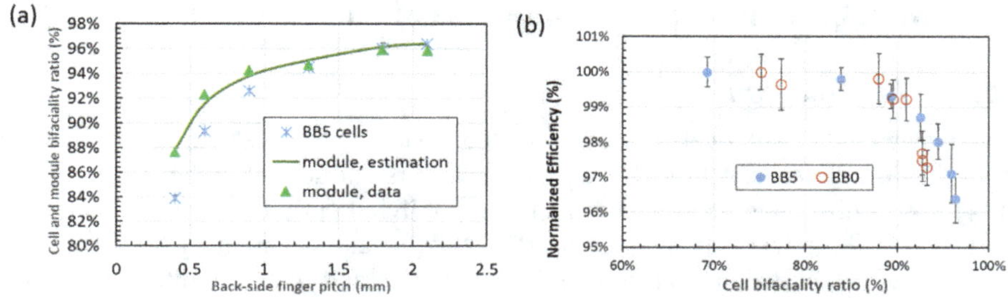

Figure 9.28 *(A) Cell and module bifaciality versus backside finger pitch (front side pitch equals 2.1 mm) and (B) Normalized cell efficiency versus bifaciality ratio [127]*

slightly. The linear variations of the parameter of the IV curve are called temperature coefficients. One advantage is that the thermal coefficient of power (*a*) is better for SHJ cells at –0.3 %/°C compared with diffused c-Si cells –0.45 %/°C [128]. As the high module power occurs at temperatures between 40 °C and 70 °C, the SHJ PV systems present a gain in energy yield compared to standard c-Si technologies. The other advantage is the natural bifaciality of the cell technology. The gain of bifacial PV systems can be from 5% to 35% depending on the PV module mounting and albedo of the ground [129]. The final advantage is the good performance of SHJ modules at low irradiance compared with PERC modules [130].

9.2.8 Summary

SHJ might be the next mainstream PV technology after the PERC only if cell metallization and interconnection are cost-competitive and reliable. Strong efforts are done in China to push down the cost of production of SHJ technology. For example, the multi-busbar interconnection is now possible for SHJ cells on cells with less that 200 mg of silver laydown in total. Similar total laydown is possible using interconnection based on ECA. The SWCT® interconnection enables reduction of silver laydown to ~100 mg in total for SHJ cells, which is only about 30% higher compared to PERC multi-busbar cells. If the price of silver increase or the volume of PV is multiplied by one order, copper plating is also available for SHJ technology. Finally, the inkjet printing technology is today not cost-competitive for direct metallization but could possibly be feasible as seed layer for electroplating.

Acknowledgments

Authors would like to acknowledge for the funding of the AMELIZ project n° SNF-200021L_182101.

9.3 Metallization of poly-Si based passivated contacts solar cells

9.3.1 Introduction

Passivated contact solar cells appear to be the next evolutionary upgrade for industrial silicon solar cells. Although the concept was first introduced in the 1980s [131–134], it was observing its moment in apparent oblivion until it was reintroduced in 2013 [135]. Following the results in 2013, this technology immediately enjoyed comprehensive attention from academic researchers and industry alike.. In 2019, PERC emerged as the market leader for the first time, overtaking Al-BSF technology with ~60% of the world market share of all solar cells manufactured [136]. However, based on reports in 2020, many of the leading PERC manufacturers are already testing passivated contact cells at least in R&D. Jollywood, REC technologies and Trina Solar have already launched modules assembled with cells featuring passivated contact technology in the commercial market.

One of the ways to achieve high efficiency solar cells is to limit the recombination of the charge carriers. There can be various factors or components of recombination in a c-Si solar cell. There is Shockley-Read-Hall (SRH) recombination on the cell's surface and within the bulk. Surface recombination is largely due to the presence of defects, unsatisfied dangling bonds and possible contaminants. Bulk SRH recombination can be attributed to the defects and impurities in the c-Si wafer bulk. Then, there is Auger recombination which occurs in the heavily doped regions. In typical PERC or passivated emitter and rear totally diffused (PERT) cells, there are two such heavily doped regions: i) the emitter and ii) the back or front surface field (which are aptly named depending on their location in a solar cell). Lastly, there is recombination due to the metallization caused by metal-induced gap states [137]. Passivated contact solar cells lead to high conversion efficiencies as they mitigate almost every facet of recombination mentioned above.

The passivated contact technology essentially separates the region of charge carrier generation and their collection by physically isolating contacts from the c-Si bulk. This is achieved by placing an ultra-thin (1–2 nm) layer of a dielectric (e.g. SiO_x) between the charge-generating and charge-collecting regions. The charge-collecting region comprises a heavily doped a-Si or poly-Si layer, onto which the metal contacts are formed. Although different materials can be used in passivating contact architecture based on their passivating properties [138], the combination of a thin silicon oxide layer at the interface and a heavily doped poly-Si layer as the contacting surface has been the most successful combination so far. As an example of a poly-Si passivated contact cell, the cross-sectional schematic of an n-type solar cell with rear passivated contacts is shown in Figure 9.29. Detailed understanding of the underlying principles and functioning of passivated contacts could be found in these excellent reviews [139–143].

As mentioned earlier, the separation of the metal contacts from the c-Si substrate all but eliminates the recombination under the metallized regions. Hence, the metallization techniques which cause least damage to the layer which they are contacting yield the maximum open circuit voltages. Consequently, solar cells fabricated

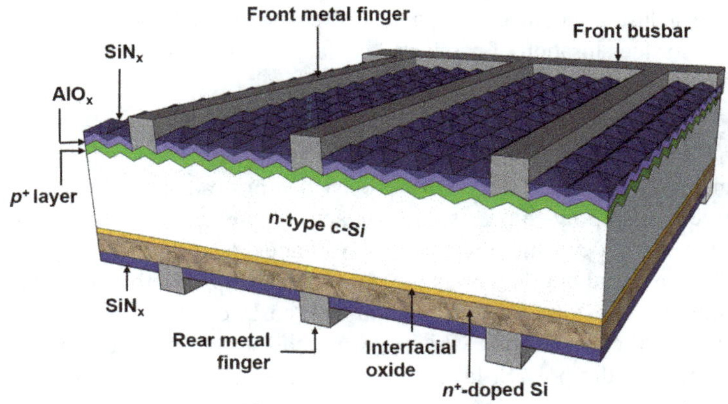

Figure 9.29 *Cross section of a solar cell with poly-Si passivated contact on the rear side and fire-through metallization. Source: SERIS.*

with low-temperature metallization processes, such as PVD, have resulted in the highest efficiencies. Efficiencies of 25.7% (V_{oc}–725 mV) have been achieved for both side contacted solar cells featuring passivated contacts on the rear side [144]. For interdigitated back contact (IBC) solar cells featuring poly-Si passivated contacts for both polarities on the rear side, efficiencies of 26.1% (V_{oc} – 733 mV) have been achieved [145]. Both these record efficiencies were obtained on high-quality float zone (FZ) substrates on small-area solar cells. A low temperature metallization scheme was followed for these solar cells which included patterning by photolithography and metal deposition by thermal evaporation. These results prove the potential of passivating contacts solar cells and set an upper limit to the performance of industrial passivated contact solar cells.

However, for a technology to enter commercialization and mass production by the industry, the processing has to be aligned with prevailing industrial practices. The process needs to be cost-effective, scalable, robust and adaptable to automation. The process should also result in a good throughput (typical expectations from PV manufacturers is a throughput of at least 3 000 wafers/hour). Although, lithography is an industrially accepted process for micro-electronics, it is not suited for the PV industry due to the associated costs and throughput considerations. According to ITRPV, screen-printing is the dominant (>90% world market share) technology employed for commercial solar cells and is likely to remain so for the next ten years [136]. As a result, passivated contact solar cells realized with screen-printed metallization will gain industrial acceptance faster than such solar cells metallized by alternative technologies. Screen-printed metallisation can be applied to passivated contact solar cells either directly on poly-Si capped with a dielectric layer using a high temperature, fire-through (FT) pastes or via a combination of TCO layers and low-temperature non-fire-through (NFT) pastes. While the use of low temperature NFT pastes on TCO layers has been industrialized by the manufacturers of SHJ cells, screen-printing with FT pastes is far more widely used by the commercial Al-BSF and PERC solar cell manufacturers. Therefore, screen-printing of FT metallisation is also the preferred industrial approach for passivated contact cells. Other possible industrial technologies for metallisation of passivated contact cells include plating, inkjet printing or dispensing.

9.3.2 Metallization of poly-Si-based passivated contacts

Various metallization technologies have been demonstrated for the fabrication of c-Si solar cells with poly-Si passivated contacts. However, each technology comes with its own sets of advantages and disadvantages. The applicability of various methods for industrial application is governed by multiple factors: ease of application, technology readiness level, throughput and cost, to name a few. However, the most fundamental consideration which governs the suitability of a specific method is the thickness of the poly-Si layer. The passivated contacts are aptly named so, because the highly doped regions and metal contacts are separated from the absorber region. Hence, it is clear that in order to function as passivated contacts, the interaction of metal should be limited to the poly-Si region. If the metal pierces the passivating interfacial dielectric layer at any location, the contact would no longer remain 'passivated' at least on such locations. Placing metal on top of a poly-Si layer would almost always lead to some degradation in the passivating properties of the layer. Then, the question arises, what is an acceptable trade-off? Which technology with a given set of conditions would provide the best combination of performance and cost? Although the answer to this is not straightforward, some insights could be gained by analyzing the metallization techniques based on the thickness of passivating poly-Si layers.

9.3.2.1 Thin a-Si/poly-Si layers (<100 nm)

The revived passivated contact technology by Fh-ISE, called TOPCon [135, 146], utilized an extremely thin a-Si layer on the rear side of the solar cell. A schematic of the cross section of a cell is shown in Figure 9.30A. The thickness of the a-Si layer in such solar cells ranged from 10 nm to 50 nm. Most developments of similar technology by various research groups such as those in Australia, Europe and the United States employed similar thickness of a-Si layers. These layers were deposited

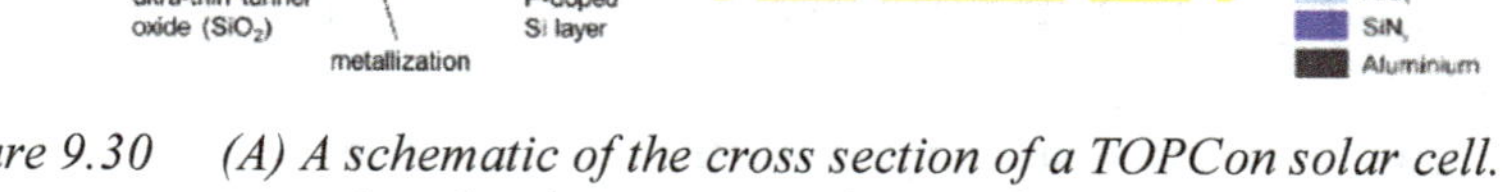

Figure 9.30 (A) A schematic of the cross section of a TOPCon solar cell. Figure reproduced with permission from authors [135]. (B) A schematic of the cross section of a POLO-interdigitated back contact (IBC) solar cell. Figure reproduced with permission from authors [145].

primarily by PECVD and were doped in situ. In some cases, low pressure chemical vapor deposition (LPCVD) was also used to deposit the a-Si [147, 148]. In order to minimize damage to the thin poly-Si layers and passivation of contacts, it was important to prevent the metal from firing through the poly-Si layers. Hence, for such poly-Si layers thermal evaporation of metal contacts became the metallization process of choice. The metal contacts may be supplemented by an electroplating method, to further improve the conductivity. Screen-printing with FT pastes has also been tested on thin poly-Si layers, but it causes high recombination under the metal contacts leading to lower open circuit voltages and poor cell efficiencies [149, 150].

a. Thermal evaporation

Thermal evaporation is a high-vacuum (~10^{-6} Torr) deposition technique that can be used to deposit metal layers with thickness from tens of nanometers to a few microns. This technique involves the use of resistive heating to evaporate metal onto the substrate. Alternatively, an electron beam can also be used to evaporate metal. While metal evaporation is not industrially applied for the metallisation of solar cells, it has often been used as a proof of concept for high-efficiency, small-area cells. On its own, this process results in a full area deposition on the substrate. It is possible to obtain a metallized grid using this process either by using a hard mask or photolithography. The metal can be deposited directly onto the poly-Si surface. The specific contact resistivity for such metallized poly-Si layers (including the tunneling resistance) can range from 0.1 mΩ cm^2 to 10 mΩ cm^2 [151, 152]. Thermal evaporation can also be employed for fabricating IBC cells using laser-patterned dielectric masks [145]. The recombination under evaporated metal contacts is almost negligible and is not expected to contribute significantly to the loss in V_{oc}. Similar metal contacts on doped layers could limit the effective J_0 contribution from the rear side (including the metal) to ~2 fA cm^{-2} while resulting in low contact resistivity of 2–4 mΩ cm^2 [151]. This maintains the advantages of shifting the highly doped regions outside of the absorber. Peak efficiencies of 25.8% (both sides contacted cells) and 26.1% (IBC Polysilicon on Oxide (POLO) cells, see Figure 9.30bB) have been achieved using this thermal evaporation to metallize poly-Si passivated contact cells.

Similar contacts could also be made by first depositing a layer of TCO over the thin poly-Si layer and then depositing metal to contact the TCO [138, 148]. Sometimes the cells are annealed at low temperatures before or after the metal deposition. The annealing temperatures can range from 250 °C to 400 °C. For temperatures greater than 400 °C, the sheet resistance of the TCO usually decreases and the optical properties improve as the TCO becomes more transparent [148]. However, at still higher temperature (600 °C) or by using alternative deposition techniques both conductivity and optical properties of certain TCOs (such as Al:ZnO) might further be improved [153, 154]. The sheet resistance could be maintained by incorporating a capping layer over the TCO, however, the metallization would become complex, involving a method to either open the contacts on the capping layer or etch it away completely, without adversely affecting the performance of the solar cells. These annealing treatments are meant to improve the contact properties and alleviate the damage caused by the sputtered TCO layer, either on the c-Si layer or the poly-Si layer. The poly-Si and the TCO layers could be present either only on the front or both on the front and the rear sides. The TCO layer could help in alleviating the lateral conductivity issues arising due to the high sheet resistances of the thin poly-Si layer. This is especially important for adopting grid metallization to form the passivating contacts. Such cases could arise when the passivating contacts are incorporated on the front side or the cell structure is bifacial, eliminating the full area deposition on the rear.

b. Plating

Plating is another process which could be used to contact thin poly-Si layers. However, plating for poly-Si passivated contact cells would also require an additional patterning step to create openings in a dielectric layer. The patterning step could be done by various methods including laser ablation, inkjet printing, screen-printing or photolithography. Plated metal stacks typically comprise a seed layer, a conductor layer and a capping layer. The seed layer could be a single layer such as an evaporated Ti or plated Ni layer, or a stack such as evaporated Ti-Pd-Ag, Cr-Pd-Ag etc. The plated conductor layer is typically Cu. The capping layer is usually electroplated Ag or Sn to prevent oxidation of the Cu conductor layer. Plating offers the advantage of damage-free contacts with high aspect ratio, and the possibility of replacing Ag metallisation with cheaper Cu. For certain conditions, it could also result in improved fill factors, albeit a lower V_{oc} compared with reference evaporated cells [155–159]. The specific contact resistivity could be as low as 0.1 mΩ cm^2 [156, 159]. The higher FF in some instances as compared to evaporated contacts is attributed to the higher aspect ratio of the plated metal fingers [156]. Plated contacts on passivated contacts can be applied to the entire area creating a monofacial cell [159] or to produce grids for bifacial cells [158]. Plated contacts on phosphorus-doped thin (30–90 nm) poly-Si layers have been demonstrated successfully, where TOPCon cell efficiencies close to 22% were achieved on M2-sized Cz-Si substrates with thickest (90 nm) poly-Si layers [158]. In this study, the thinner poly-Si layers exhibited higher loss in V_{oc} than the thicker layers. This may also have been caused by the laser patterning step used to create the contact openings. Metallization by plating is slightly more complex and time-consuming than screen-printing and hence is likely to result in a lower throughput if adopted in commercial production. The plating process has been discussed in detail in Chapter 8.

c. Screen-printing of LT NFT paste

Screen-printing of NFT pastes followed by low-temperature annealing (typically <250 °C) is used to contact TCO layers for the production of commercial heterojunction solar cells. The process could readily be applied to the poly-Si passivated contact technology without much modifications. Although the contact properties will differ based on the TCO and the type of paste used for metallization, typical specific contact resistivity values can range from 1 mΩ cm^2 to 5 mΩ cm^2 [160]. Heterojunction cells also feature passivated contacts as the carrier collection is spatially decoupled from carrier generation. Besides the structure, one major difference between the two technologies featuring passivated contacts lies in the temperature range of processing. While the passivated contact technology using the combination of iO_x and doped poly-Si layer is stable at temperatures as high as 900 °C, heterojunction cell process temperatures are typically limited to <250 °C to prevent degradation of a-Si thin films. Details about the metallization process of monofacial or bifacial HJT cells are discussed in Section 9.2. Forming contacts by screen-printing of LT pastes involves the use of the TCO layer. Usually such layers are deposited by sputtering which causes damage to the underlying Si layer and is followed by a high temperature annealing step to recover the loss. These processes have been discussed in detail in Sections 5.2 and 5.3. For fabricating solar cells with iO_x and thin poly-Si based passivating contacts on both the front and rear sides, screen-printed metallisation using NFT pastes may be industrially viable. Using the former approach of using thin (~12–15 nm) poly-Si layers on both the front and rear sides, efficiencies up to 22.3% have been demonstrated on large area (M2) Cz-Si substrates [161]. However, if passivated contacts are applied on only the rear side, the use of NFT pastes on the rear and standard FT pastes on the front would lead to a complex process which would appear less attractive for use in commercial production.

9.3.2.2 Thick poly-Si layer (>100 nm)

The motivation of using thin poly-Si layers is to reduce the J_{sc}-limiting parasitic absorption in the poly-Si layer. However, if the poly-Si contacts are employed only on the rear side, the current loss due to parasitic absorption is minimized and takes place only for the near-infrared part of the solar spectrum. Therefore, doped poly-Si layers on the rear side could be thicker than 100 nm. It should be noted that using a thick poly-Si layer on the rear side might lead to some loss in bifaciality of the cell. However, the biggest advantage of using a thick poly-Si layer lies in the ability to use the standard screen-printing process for metallization using FT pastes. The surface dopant concentration for poly-Si layers typically ranges from 1×10^{20} cm^{-3} to 5×10^{20} cm^{-3} which is well within the contacting limits of standard FT pastes. Use of thick n^+ poly-Si layers, low contact resistance and low metal recombination can be achieved simultaneously with standard FT pastes. The specific contact resistivity for a typical metallization with a FT paste on n^+ poly-Si can range from 0.5 mΩ cm^2 to 2 mΩ cm^2 [149, 150, 162–165] while the $J_{0,metal}$ can be as low as 20 fA cm^{-2} for certain conditions [164, 166]. The values of contact resistivity match, and even slightly outperform the corresponding values obtained on the c-Si surface. The values of $J_{0,metal}$ are 1–2 orders of magnitude lower than the counterpart values on c-Si surface [162]. This allows for easy integration of the process by the PV cell manufacturers with minimum modifications to the existing production lines. Based on reports from industry, screen-printed and FT metallization for poly-Si cells are being actively used for commercial fabrication of c-Si solar cells with rear n^+ poly-Si passivated contacts. As explained further in this section, screen-printed metallisation of p^+ poly-Si is currently not as well developed as n^+ poly-Si. Contacts made to *P*+poly-Si with FT Ag-Al pastes exhibit specific contact resistivity ~10 mΩ cm^2 [167] while those with Ag pastes resulted in specific contact resistivity values ranging from 2 mΩ cm² to 10 mΩ cm² [168, 169].

The key considerations for using screen-printed metallization with FT pastes are contact resistance and recombination under the metal contacts. These factors to a certain extent are interrelated and both are dependent on paste formulations. As opposed to evaporation of pure metals, metal pastes include several components other than metal. All FT pastes, include glass frits, among other additives, to etch away dielectric layers. The glass frits melt at high temperature and form a thin layer, which resides in between the bulk metal and the Si substrate. Metal crystallites fire through the glass layer to form the interface where the actual metal-silicon contact takes place. The characteristics of the crystallites hold the key while forming passivated contacts with FT pastes. The main characteristics in this regard are the distribution and size of the crystallites, and the depth up to which they embed into the poly-Si layer. While the distribution and size of the crystallites could be correlated to the contact resistivity and the depth of etch to the recombination current density under the metal contacts, the cumulative effect of both factors could influence both contact resistance and the recombination. For example, higher number of deep etching crystallites would increase recombination under metal contacts. Conversely, if the crystallites etch through the poly-Si layers, they might impact the contact properties severely [149, 163, 170]. In fact, etch depth is the prime concern due to which FT pastes are not currently applied for metallization of thin poly-Si layers.

9.3.2.3 Selection of pastes for FT metallization of passivated contacts

Formation of metal contacts with FT pastes is a complex phenomenon, owing to the composition of the paste. The paste should be fluid enough to pass through the printing screen, but viscous enough not to spread excessively on the cell surface. The pastes should have sufficient viscosity to form fingers with high an aspect ratio. The organic additives in the paste are specially designed to help it retain the physical form while undergoing multiple print cycles, while the glass frits are designed to ensure optimal etching of the ARC layers. All these factors influence the morphology of the contacts, which in turn affect the contact properties. Numerous metal paste manufacturers are actively developing pastes for the metallisation of poly-Si layers. A detailed analysis of the individual components of the metal pastes and their effects is beyond the scope of this chapter. However, in this section, an analysis of the contact properties of commercially available FT pastes on n^+ poly-Si is presented and some underlying reasons for their performance are investigated.

Figure 9.31 shows SEM images of metal crystallites, formed by four different commercial Ag pastes, on 150 nm n^+ poly-Si layers. The metal crystallites are observed here by etching away the bulk of the metallized fingers in hot nitric acid and removing the underlying glass layer by etching in HF. Multiple studies have shown that dense and uniformly distributed crystallites lead to the best values of specific contact resistivity (ρ_c) [162, 163]. The smaller the crystallites formed during the high temperature firing of the cells, the better is their distribution and the lower is supposed to be the contact resistivity. The contact resistivity may not have a strong

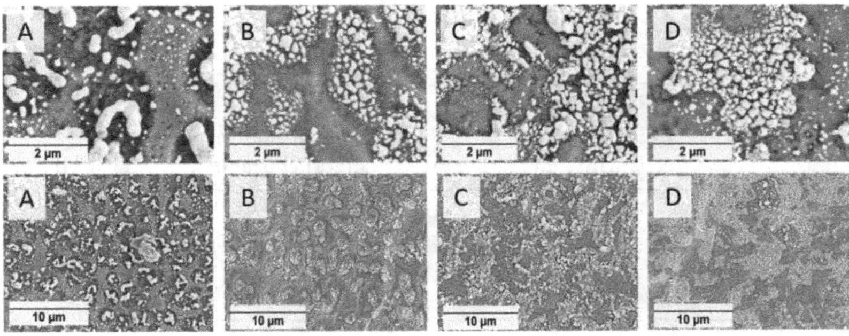

Figure 9.31 Scanning electron microscope (SEM) micrographs of metal cryustallite patterns formed by different commercially avaialble Ag fired-through (FT) pastes (A,B,C,D) with n^+ poly-Si. The images in the top row are magnified images of the ones in the bottom row.

dependence of the thickness of the poly-Si layer as long as the metal crystallites are contained within the poly-Si layer and do not puncture through the iOx layer. Even if they puncture slightly into the c-Si substrate, the dopant tail into the c-Si substrate would prevent any significant rise in the contact resistivity. The increase in resistivity would be clearly visible if the metal crystallites penetrate beyond the poly-Si layers deeper than the region where the dopant concentration falls below 1×10^{19} cm^{-3} [149, 171]. For optimized dopant profiles, resulting in lowest J_0, this range can extend up to 20 nm beyond the poly-Si-iO_x interface. On similar poly-Si layers (similar doping density and thickness), changing the metal paste alone might result in changes in specific contact resistivity values spanning more than two orders of magnitude. For example, the ρ_c for the contacts formed by the pastes shown in the figure above can range from 40 mΩ cm^2 to 50 mΩ cm^2 for paste A to 0.5 mΩ cm^2 for paste D [162, 163].

Another important parameter to pay attention to is the recombination under the metal contacts ($J_{0,metal}$). The $J_{0,metal}$ arises from the complex chemistry of the metal-Si alloyed regions, in addition to the introduction of recombination centres in the forbidden energy gap region. The alloying is not expected to occur significantly in the metallization processes discussed earlier for thinner poly-Si layers, such as thermal evaporation, plating and low temperature polymer pastes or LT NFT pastes. Although the Al paste used in PERC is also NFT (it does not contain glass frits), it does alloy with the c-Si to form the local aluminium and the contact simultaneously. Hence, contacts formed with HT FT pastes on c-Si surface usually lead to high $J_{0,metal}$, which adds to the overall $J_{0,cell}$ and manifests as loss in V_{oc}. The very property of significantly reducing the $J_{0,metal}$ earned the passivated contacts its rightful appellation. To maintain contact passivation it is important that the contacts should be contained well within the poly-Si layer. When properly executed, the $J_{0,metal}$ values can be as low as 20 fA cm^{-2} [166, 172–174]. In such cases the entire contribution of the rear side to the total cell J_0 can be limited to 3–5 fA cm^{-2}. Hence, V_{oc} as high as 715 mV have been achieved in passivated contact solar cells with screen-printed FT

Table 9.8 Summary of contact properties of the four pastes shown in Figure 9.31

Paste	Metallized fraction [%]	Specific contact resistivity (ρ_c) [mΩ cm^2]	Recombination under metal contacts ($J_{0,metal}$) [fA cm^{-2}]
A	25	42.5	20
B	31	12.6	75
C	42	1.9	224
D	51	1.1	30

contacts [173]. Similar to ρ_c, the $J_{0,metal}$ is not expected to have a strong dependence on thickness unless the metal crystallites start approaching the poly-Si/iO_x interface. It has been observed that a change in the poly-Si layer from 100 nm to 50 nm could push the $J_{0,metal}$ higher by three to four times [149, 150]. This also means that more than necessary poly-Si might limit the current (due to parasitic absorption) and hence might limit the efficiencies. While selecting FT metal pastes for manufacturing passivated contact solar cells, attention needs to be given to both ρ_c and $J_{0,metal}$. This is because the pastes which do not contact the poly-Si layers properly, could lead to artificially low $J_{0,metal}$ [162, 163]. To illustrate this point, the characteristics of the four metal pastes are shown in Table 9.8. For example, for the $J_{0,metal}$ for paste A is around 20 fA cm^{-2} which is the lowest of all the four examples. However, that is due to the fact that the paste is unable to form proper contacts with the poly-Si surface. A similar argument holds for paste B. Paste C forms relatively better contacts but leads to high $J_{0,metal}$. This could be due to the aggressive etching of poly-Si layer and some crystallites penetrating the poly-Si-iO_x interface. The fourth paste is an example of an ideal paste exhibiting a combination of both low ρ_c and low $J_{0,metal}$.

9.3.2.4 Other metallization methods

Dispensing of metal pastes or inkjet printing of metal inks could also be applied to the metallization of poly-Si passivated contact cells. A detailed overview of these techniques is given in Chapter 7 of this book. While both these techniques have already been successfully demonstrated on PERC and HJT cells, there are thus far no literature reports demonstrating the application of these techniques on poly-Si passivated contact cells.

9.3.2.5 Summary

To achieve efficiencies higher than 24% in mass production, advanced cell manufacturing concepts, that are robust and cost-effective, have to be adopted. Solar cells with doped poly-Si-based passivated contacts appear to be a viable option and are being considered as the next technological upgrade to PERC. Appropriate metallization of passivated contacts is crucial to successful realization of their potential. Broadly speaking, the selection of a metallization process for poly-Si passivated contact solar cells is dependent on the thickness of the poly-Si layer. For thin layers

(<100 nm), evaporated metal contacts, plating or screen-printing of NFT pastes over TCOs are some of the available options. These technologies have been successfully implemented at the lab-scale to produce highly efficient solar cells. However, all these technologies are currently limited either by the process complexity or cost and hence do not appear to be options for immediate technology transfer to the industry. Screen-printing with FT pastes is the state-of-the-art metallization process currently employed in the industry, and passivated contacts with thick poly-Si (>100 nm) layers appear to be compatible with this process. However, thick poly-Si layers lead to a considerable parasitic absorption and hence limit the cell's J_{sc}. Further research and development of metallization pastes developed specifically for metallization of poly-Si layers could enable contact formation to thin poly-Si layers (50–100 nm) in the near future.

9.3.3 Selected passivated contact cell results from literature with respect to different metallization techniques

In this section we have presented some solar cells results collected from published literature with respect to the different metallization techniques. Given the number of publications and continuous improvements in the results by researchers across the world, we do not intend to claim the completeness of data. These tables are meant to provide the reader with an overview of the general trends in the application of various metallization techniques for fabricating passivated contact solar cells.

Table 9.9 presents data of solar cells incorporating passivated contacts, where the metal contacts were fabricated using the evaporation.

Table 9.10 presents data of passivated contact solar cells where the metal contacts were formed by plating, while Table 9.11 presents data of passivated contact solar cells were screen-printing was used to form contacts to the doped poly-Si layer.

9.4 Interdigitated back contact solar cells

9.4.1 Introduction to IBC cells

The previously discussed both-side contacted solar cells, PERC, PERT, HJT and TOPCon solar cells, share one common loss mechanism: part of the incident light is reflected by the front metal grid, therefore leading to optical shadowing loss. However, this loss can be eliminated by placing all the electrodes to the cell's rear side, and this is the concept of IBC solar cells. A schematic cross section of a typical IBC cell is depicted in Figure 9.32, where the positive and negative electrodes are designed alternately on the cell's rear side. An example of a six-inch IBC cell is shown in Figure 9.33, from which one can see the blue front side and the rear side (with all metal contacts). It was first introduced by Schwartz and Lammert in 1975 [199] and in the following decades numerous studies have been performed to improve the cell's efficiency, to reduce the cost and to bring it to mass production. Today, the world record efficiency of silicon solar cells, 26.7%, has been achieved

Table 9.9 Passivated contact solar cells with evaporated metallization

Sr. no	Cell area [cm^2]	Cell structure	Eff [%]	V_{oc} [mV]	Ff [%]	J_{sc} [mA/cm^2]	Front metallization	Rear metallization	Ref
1	4	Interdigitated back contact (IBC), poly-Si for both polarities	26.1	726.3	84.3	42.6	NA	Evaporated Al	[145]
2	4	Monofacial, SE *P*+Si front, n+poly-Si rear	25.8	724	83.1	42.9	Evaporated Ti/Pd/Ag (lift-off)	Full area evaporated Ag	[175]
3	4	Monofacial, *P*+Si emitter front, n+poly-Si rear	25.6	727.6	83.2	42.2	Evaporated Ti/Pd/Ag (lift-off)	Full area evaporated Ag	[144]
4	4	IBC, poly-Si for both polarities	25.0	722.7	82.6	41.3	NA	Evaporated Al	[176]
5	4	Monofacial, SE *P*+Si front, n+poly-Si rear	24.9	718	83.4	41.5	Evaporated Ti/Pd/Ag seed layer+ Ag electroplating	Full area evaporated Ag	[177]
6	4	Monofacial, SE *P*+Si front, n+poly-Si rear	24.7	704.8	82.6	42.4	Evaporated Cr/Pd/Ag seed layer+ electroplated Ag	Full area evaporated Al	[178]
7	100	Monofacial, *P*+Si emitter front, n+poly-Si rear	24.5	713	83.1	41.3	Photolithography patterning+ evaporated Ag	Full area evaporated Al	[179]
8	4	Monofacial, Selective *P*+FSF, n+poly-Si rear	24.3	714.2	80.8	42.1	Evaporated Ti/Pd/Ag (lift-off)	Full area evaporated Ag	[180]
9	4	IBC, poly-Si for both polarities	24.3	727	80.2	41.7	NA	Evaporated Al	[152]
10	4	IBC, poly-Si for both polarities	23.9	722	78.7	42.1	NA	Evaporated Al	[181]
11	4	Monofacial, SE *P*+Si front, n+poly-Si rear	23.8	711.9	81.1	41.2	Evaporated Ti/Pd/Ag seed layer + Ag electroplating	Full area evaporated Ag	[182]
12	4	IBC, poly-Si for both polarities	23.7	720	79.6	41.3	NA	Evaporated Ti-Pd-Al	[183]

(Continues)

Table 9.9 Continued

Sr. no	Cell area [cm^2]	Cell structure	Eff [%]	V_{oc} [mV]	Ff [%]	J_{sc} [mA/cm^2]	Front metallization	Rear metallization	Ref
13	100	Monofacial, *P*+Si emitter front, n+poly-Si rear	23.4	697	81.2	41.3	Laser contact opening+ Ni/Cu/Ag plating	Full area evaporated Al	[179]
14	4	Monofacial, *P*+Si emitter front, n+poly-Si rear	23.1	694	81	41.1	Evaporated Ti/Pd/Ag (lift-off)	Full area evaporated Ag	[159]
15	4	Monofacial, *P*+Si emitter front, n+poly-Si rear	23.02	698.1	81.1	40.6	Evaporated Ti/Pd/Ag seed layer + Ag electroplating	Full area thermal evaporated Ag	[146]
16	4	Monofacial, *P*+Si emitter front, n+poly-Si rear	23	701	79.9	41.1	Photolithography patterning + evaporated seed layer + plated Ag	Full area evaporated Ag	[184]
17	2	IBC, poly-Si for both polarities	23	701	77.8	42.1	NA	Evaporated Al	[185]
18	100	Monofacial, *P*+Si emitter front, n+poly-Si rear	22.9	694	81	40.7	Ni-Cu-Ag bias plating	Full area evaporated Ag	[186]
19	4	IBC, poly-Si for both polarities	21.7	673	77	41.9	NA	Evaporated Al	[187]
20	4	Monofacial, *P*+Si emitter front, n+poly-Si rear	21.5	693	78.4	39.6	Evaporated Al through shadow mask	Full area evaporated Al	[188, 189]
21	239	Monofacial, *P*+Si emitter front, n+poly-Si rear	21.2	683.4	78.1	39.7	Screen-printed Ag/Al grid	Full area evaporated Ag	[190]
22	9	IBC, poly-Si for both polarities	21.2	692	78.3	39.1	NA	Evaporated Al	[191]
23	9	Monofacial, *P*+Si emitter front, n+poly-Si rear	21	704	73.8	40.4	Sputtered indium thin oxide (ITO) + evaporated Al fingers (patterned by photolithography)	Full area evaporated Ag/Cr/Al	[155]

Table 9.10 Passivated contacts with plated metallization on poly-Si/a-Si

Sr. no.	Cell area [cm^2]	Cell structure	Eff [%]	V_{oc} [mV]	FF [%]	J_{sc} [mA/ cm^2]	Front metallization	Rear metallization	Ref
1	4	Monofacial, *P*+Si emitter front, n+poly-Si rear	22.7	690	83.5	39.4	Evaporated Ti/Pd/Ag (lift-off)	Ni-Cu-Ag plated full area contact	[159]
2	244	Bifacial, *P*+Si emitter front, n+poly-Si rear	22.0	685	80.2	40.0	Screen-printed Ag/Al fired-through (FT) metallization	Ni-Cu-Ag plated grid contact	[158]
3	7.84	Monofacial, n+poly-Si front, *P*+poly-Si rear	19.6	682	75.2	38.2	Ti-Cu plated contact	Full area evaporated Ag/Cr/Al	[156, 157]
4	9	Monofacial, n-a-Si front, *P*+poly-Si rear	19.1	694	74.1	37.1	Ti-Cu plated contact	Full area evaporated Ag	[155]

Table 9.11 Passivated contact solar cells with screen-printed metallization

Sr. no.	Cell area [cm^2]	Cell structure	Eff [%]	V_{oc} [mV]	Ff [%]	J_{sc} [mA/cm^2]	Front metallization	Rear metallization	Ref
1	244	Bifacial, *P*+Si emitter front, n+poly-Si rear	24.6	716.8	84.5	40.6	Screen-printed fired-through (FT) Ag/Al paste	Screen-printed FT Ag paste	[173]
2	244	Bifacial, *P*+Si emitter front, n+poly-Si rear	24.1	714	83.1	40.6	Screen-printed FT Ag/Al paste	Screen-printed FT Ag paste	[173]
3	244	Bifacial, *P*+Si emitter front, n+poly-Si rear	23.05	694	81	41.0	Screen-printed FT Ag/Al paste	Screen-printed FT Ag paste	[172]
4	244	Bifacial, *P*+Si emitter front, n+poly-Si rear	23	701.5	82.2	39.9	Screen-printed FT Ag/Al paste	Screen-printed FT Ag paste	[192]
5	244	Bifacial, *P*+Si emitter front, n+poly-Si rear	23	698	82	40.2	Screen-printed FT Ag/Al paste	Screen-printed FT Ag paste	[164]
6	244	Bifacial, *P*+Si emitter front, n+poly-Si rear	22.8	696	80.9	40.5	Screen-printed FT Ag/Al paste	Screen-printed FT Ag paste	[166]
7	244	Bifacial, *P*+Si emitter front, n+poly-Si rear	22.6	696	80.5	40.3	Screen-printed FT Ag/Al paste	Screen-printed FT Ag paste	[193]
8	239	Bifacial, *P*+Si emitter front, n+poly-Si rear	22.6	702	79.7	40.4	Screen-printed FT Ag/Al paste	Screen-printed FT Ag paste	[174]
9	244	Bifacial, *P*+Si emitter front, n+poly-Si rear	22.5	695	80.5	40.2	Screen-printed FT Ag/Al paste	Screen-printed FT Ag paste	[179]
10	244	Bifacial, *P*+Si emitter front, n+poly-Si rear	22.5	695	80.4	40.3	Screen-printed FT Ag/Al paste	Screen-printed FT Ag paste	[163]
11	244	Bifacial, , n+poly-Si front, *P*+poly-Si rear	22.3	714	81.1	38.5	Sputtered indium thin oxide (ITO), screen-printed non-fire-through (NFT) Ag paste	Sputtered ITO, screen-printed NFT Ag paste	[194]

(Continues)

Table 9.11 Continued

Sr. no.	Cell area [cm^2]	Cell structure	Eff [%]	V_{oc} [mV]	Ff [%]	J_{sc} [mA/cm^2]	Front metallization	Rear metallization	Ref
12	244	Bifacial, n+poly-Si front, *P*+poly-Si rear	22.3	728	80	38.3	Sputtered ITO, screen-printed NFT Ag paste	Sputtered ITO, screen-printed NFT Ag paste	[161]
13	244	Bifacial, *P*+Si emitter front, n+poly-Si rear	22.1	682	81.8	39.6	Screen-printed FT Ag/Al paste	Screen-printed FT Ag paste	[149]
14	244	Bifacial, *P*+Si emitter front, n+poly-Si rear	21.8	691	80	39.4	Screen-printed FT Ag/Al paste	Screen-printed FT Ag paste	[162]
15	244	Bifacial, *P*+Si emitter front, n+poly-Si rear	21.6	685	79	39.9	Screen-printed FT Ag/Al paste	Screen-printed FT Ag paste	[186]
16	244	Bifacial, *P*+Si emitter front, n+poly-Si rear	21.4	672	81.2	39.2	Screen-printed FT Ag/Al paste	Screen-printed FT Ag paste	[195]
17	244	Bifacial, *P*+Si emitter front, n+poly-Si rear	21.3	679	79.9	39.3	Screen-printed FT Ag/Al paste	Screen-printed FT Ag paste	[196]
18	239	Bifacial, SE n+Si front, Rear *P*+poly-Si	20.8	672	79.9	38.7	Screen-printed FT Ag paste	Screen-printed FT Ag/Al paste	[167]
19	239	Bifacial, *P*+Si emitter front, n+poly-Si rear	20.7	675	79.1	38.8	Screen-printed FT Ag/Al paste	Screen-printed FT Ag paste	[197]
20	244	Bifacial, *P*+Si emitter front, *P*+poly-Si under the front metal fingers, n+poly-Si rear	20.4	683	76.5	39.0	Screen-printed FT Ag/Al paste	Screen-printed FT Ag paste	[198]
21	244	Bifacial, *P*+Si emitter front, n+poly-Si rear	20.4	659.7	78.6	39.3	Screen-printed FT Ag paste	Screen-printed FT Ag paste	[169]

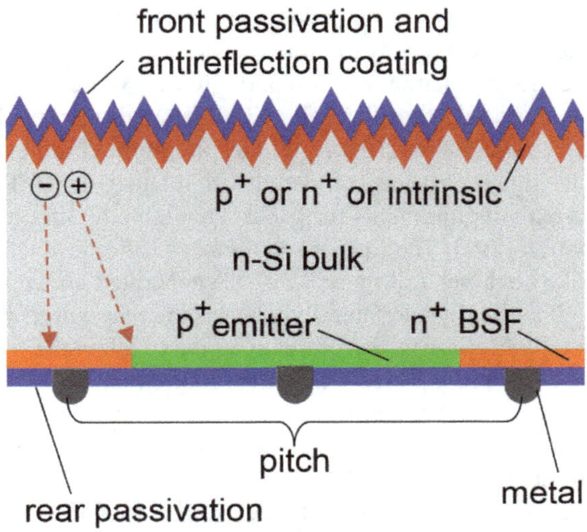

Figure 9.32 *Schematic cross-section of a typical interdigitated back contact (IBC) cell with an n-type Si bulk*

Figure 9.33 *Photograph of a six-inch interdigitated back contact (IBC) cell fabricated at ISC Konstanz. The cell is placed in front of a mirror to show the front and rear sides (copyright at ISC Konstanz).*

by an IBC cell that combines heterojunction solar cell technology [200]. Apart from high efficiency, the IBC cell features a homogeneous front surface which makes it an appealing choice for residential rooftop and building/vehicle integrated applications. These advantages help the IBC cell to hold a small market share, about 2% in 2019 and an estimated 5% in 2030 [201], despite more complex manufacturing processes (and thus higher cost), as compared with mainstream PERC solar cells.

There are two main requirements for IBC cells: high bulk lifetime and low front surface recombination [202]. The former requirement is because when the minority carriers in the bulk (i.e. holes in Figure 9.32) are generated above the BSF region, they must flow both vertically and laterally towards the rear collecting emitters. The wafer with a high bulk lifetime is, therefore, essential to allow the holes to reach the rear emitter before being recombined. The latter requirement is because a substantial percentage of charge carriers are generated near the front surface, a low front surface recombination velocity is needed, whereas the doping of the front surface can be chosen to be p^+, n^+ or intrinsic. The IBC cell structure lifts two restrictions existing in both-side contacted solar cells. The rear side metal grid of IBC cells can be solely optimized to achieve minimum series resistance without the problem of shadowing the front side; the front doping concentration can be solely optimized to achieve minimum recombination without the restriction of maintaining a certain level of contact resistance and lateral conductivity. More detailed discussion of device physics for the IBC can be found in [202, 203].

Over time several different ways have been utilized to form the p^+ and n^+ doped regions for the IBC cell, such as diffusion [204], doped silicate glasses [205], ion implantation [206] and printed dopant paste [207]. The dielectric layer to passivate the IBC cell is the same as other types of cells, where the heavily doped p^+ surface can be effectively passivated by using thermal oxidation, PECVD/ALD AlO_x or PECVD a-Si (amorphous Si), while the n^+ surface can be passivated by thermal oxidation, PECVD SiN_x or PECVD a-Si. The challenge of IBC cell fabrication is the formation of the rear periodically changing doping and metallization pattern. As shown in Figure 9.32, the distance of one n-finger to the next n-finger is called the pitch, which is in the order of hundreds of microns to millimetres. Photolithography offers the best alignment accuracy which is important for the cell with a small pitch (e.g. <600 μm), however, the cost is too high for mass production. Promising results have been achieved with low-cost patterning techniques such as laser ablation, and screen-printing of etching mask or etching paste. For instance, 355-nm and 532-nm picosecond lasers have been demonstrated to be able to directly ablate the SiO_2 layer with only relatively moderate damage to the underlying Si [208, 209], which is an important result as producing local openings in the passivating layer is a requisite for PVD and electroplating.

In terms of metallization, in order to avoid shunting it is crucial to ensure that the p- and n-type electrodes are well aligned to their corresponding doping regions, and that there is no interconnection between the p- and n-type electrodes over the whole cell. Moreover, the design of the busbar layout plays an important role and three basic concepts exist: (1) placing the busbars only at the wafer edges; (2) multi-layer interconnection with continuous fingers and additionally with an insulation layer that separates the p-type fingers with the n-type busbars and vice versa; (3) single-layer interconnection with periodically interrupted fingers. Solar cell results based on these three designs can be found in [210–212] respectively. To replace the conventional busbars of IBC cells, recently research of using wire-based interconnection such as 'SmartWire' has started [213]. Detailed comparison and discussion of the aforementioned metal grid designs can be found in [214, 215].

Typical metallization technologies in the field of photovoltaics, such as PVD, electroplating and screen-printing, have been successfully applied to IBC cells over the years, and the specific process and requirement associated with these three technologies will be discussed in the following sections.

9.4.2 PVD metallization for IBC cells

Aluminium deposition by means of evaporation or sputtering has been widely used to metallize IBC cells from the 1970s to now [216–220]. It can produce a metal grid with sufficiently low series resistance and achieve good ohmic contacts for both p^+ and n^+ Si with low surface concentration. The latter merit has been a key reason that many IBC cells fabricated with the PVD approach outperform those metallized with screen-printed metal pastes. Record high efficiencies of IBC cells fabricated with diffused emitters, or with polysilicon-based passivated contacts, have been reported to reach 25.0% [219] and 26.1% [220], respectively.

The PVD process leads to the deposition of a thin metal layer, e.g. Al, on the entire back surface of the IBC cell as shown in Figure 9.34a, and one needs to further pattern this metal layer to complete the metallization process. Figure 9.33b illustrates a representative rear metal grid design used in many publications [216–221] for the 2 × 2 cm^2 IBC cells, where the p- and n-busbar are located on opposite sides of the cell. Although the metal grid shown here is used for IBC cells with line-shaped diffusions, it can also be applied to IBC cells with dot-shaped diffusions [217, 218].

According to [220, 221] a principle process flow to metallize IBC cells with PVD is schematically shown in Figure 9.35 and summarized as follows:

Step 1: Locally open the rear dielectric layer (e.g. SiO_x or SiN_x) by means of laser ablation; the openings can be dot- or line-shaped.

Step 2: Perform a short dip in HF solution to remove the silicon oxide layer formed after Step 1, or the native oxide that is grown on the Si surface when it is exposed to air.

This step is needed because the presence of any oxide layer between the Si and the subsequent PVD metal can lead to a significant increase of contact resistance [222, 223].

Step 3: Deposit full-area Al layer on the back side of the cell; the thickness of the Al layer typically ranges from 2 μm to 15 μm.

Step 4: Form a thin etch barrier (e.g. SiO_x) on top of the Al layer.

Step 5: Perform laser ablation along the entire borders between the rear emitters and BSFs to remove the etch barrier and a thin layer of the underlying Al. One of the suitable lasers for this process is the 355 nm Nd:YVO4 laser with a pulse length of 20 ns.

Step 6: Locally etch back the Al layer in wet chemical solution (e.g. HCl), while the SiO_x layer acts as an etching mask which is not or barely etched. It is important to ensure the positive and negative electrodes are completely separated over the entire cell's back side.

Step 7: Remove the etch barrier SiO_x in HF solution (optional).

Step 8: Anneal the sample with a low temperature (e.g. 450 °C for 15 minutes [218]) to minimize the contact resistance between Si and Al. This is also an optional step. Some authors include it in their cell process while some do not, and the decision seems to depend on the type of equipment and deposition recipes that are used [224].

Table 9.12 lists the measured specific contact resistivity of PVD Al used for IBC cells in recent years. It shows that with PVD Al metallization one can achieve low specific contact resistivity ranging from 0.01 mΩ cm² to 1 mΩ cm² for both p^+ and n^+ doped surfaces. The low contact resistivity not only reduces the resistive loss associated with the metal contacts but also allows a low contact fraction, resulting in a reduced recombination loss under the metal and hence a higher V_{oc}. For example, the 24.4% cell reported in [226] has a p- and n-type contact fraction of 0.13% and 0.43%, respectively, which corresponds to ~0.04% absolute cell efficiency loss due to contact resistance, and 0.05% absolute efficiency loss due to recombination loss under the metal contacts.

The geometry of the metal grid as depicted in Figure 9.34 has an influence on the solar cell's series resistance and hence the fill factor. One can calculate the series resistance from the metal finger by using the following equations [202, 229]:

$$R_s = \frac{1}{3}\frac{F_p}{WH}\rho L^2 \tag{9.4}$$

$$FF_s = pFF(1 - R_s J_{sc}/V_{oc}) \tag{9.5}$$

where ρ is the finger series resistance, F_p the pitch of the IBC cell, W the finger width, L the finger length, H the finger height, ρ the metal resistivity, FF_s the fill factor when one considers only the finger series resistance while neglecting the series resistances from the other parts of the cell, pFF the ideal fill factor of the cell without any series resistance loss, J_{sc} the short circuit current density and V_{oc} the open-circuit voltage. With these two equations, the influence of the finger height on the finger series resistance and FF_s are calculated and shown in Figure 9.35, in which we compare the series resistance and FF_s for a lab-scale cell of 2 × 2 cm² and a large area cell of 15.6 × 15.6 cm² with various finger heights. The calculation is based on the 2 × 2 cm² IBC cell in [226], where the pitch is 500 μm, and the emitter and BSF finger width are 282 μm and 134 μm, respectively. This IBC cell has the following IV parameters: V_{oc} = 702.5 mV, FF_s = 41.95 mA/cm², FF =82.7% and η=24.4%. The Al resistivity is assumed to be 2.8 × 10^{-6} Ω cm in the calculation, while pFF is assumed to be 85%. One can observe that the 2 × 2 cm² cell needs only 2-μm thick Al fingers to reach an FF_s of 84%, while 64-μm thick Al fingers are needed to reach the same FF_s for the 15.6 × 15.6 cm² cell. Therefore, it is crucial to design a sufficiently high PVD Al layer to avoid the cell being limited by the

Table 9.12 Results of contact resistivity of physical vapour deposition (PVD) Al used for interdigitated back contact (IBC) cells

Junction structure	References	Doping	Surface concentration (cm^{-3})	Sheet resistance (Ω/sq)	Specific contact resistivity ($m\Omega \cdot cm^2$)	Cell efficiency (%)
Crystalline Si homojunction	[218]	n^+	2×10^{19}	50	<0.01	23.0
		p^+	4×10^{18}	90	0.05	
	[225]	n^+	2×10^{20}	25	0.7	22.7
		p^+	1×10^{19}	90	0.05	
	[226]	n^+	Not available	19	0.01	24.4
		p^+	Not available	166	0.015	
Passivated contact (polysilicon)	[227]	n^+	$>1 \times 10^{20}$	118	0.01	24.3
		p^+	2×10^{19}	239	0.4	
	[228]	n^+	Not available	89	0.9	23.0
		p^+	Not available	122	0.3	

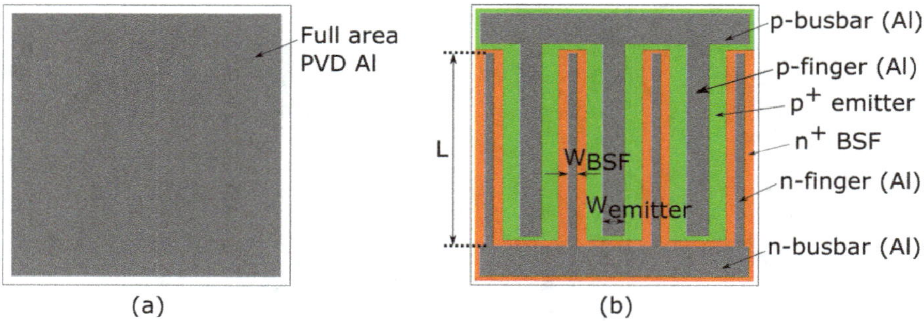

Figure. 9.34 Top view of the back side of an n-type interdigitated back contact (IBC) cell. (A) After the physical vapour deposition (PVD) process that produces a full area Al layer on the back side of the cell; (B) After further structuring into the positive and negative comb-shaped Al electrodes (redrawn according to [217–220]).

metallization series resistance, especially for industrial large area cells. An alternative solution is to replace Al with metal that has lower resistivity such as Cu or Ag.

9.4.3 Electroplating metallization for IBC cells

Compared with PVD metallization, there are much fewer publications about fabricating IBC cells with electroplating [230–233]. Nevertheless, the largest IBC cell company worldwide, SunPower Corporation, has since years produced IBC cells with this technology. Today, the efficiency of the IBC module from them is the

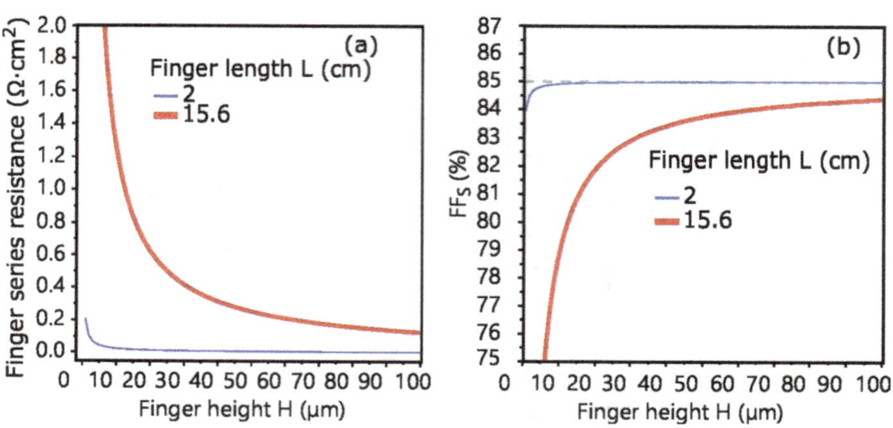

Figure 9.35 Calculated series resistance and fill factor for 2 × 2 cm² and 15.6 × 15.6 cm² interdigitated back contact (IBC) cells with a metal grid design as in Figure 9.34b. For the 15.6 × 15.6 cm² cell much higher fingers are needed to reach an FF_s of 84%.

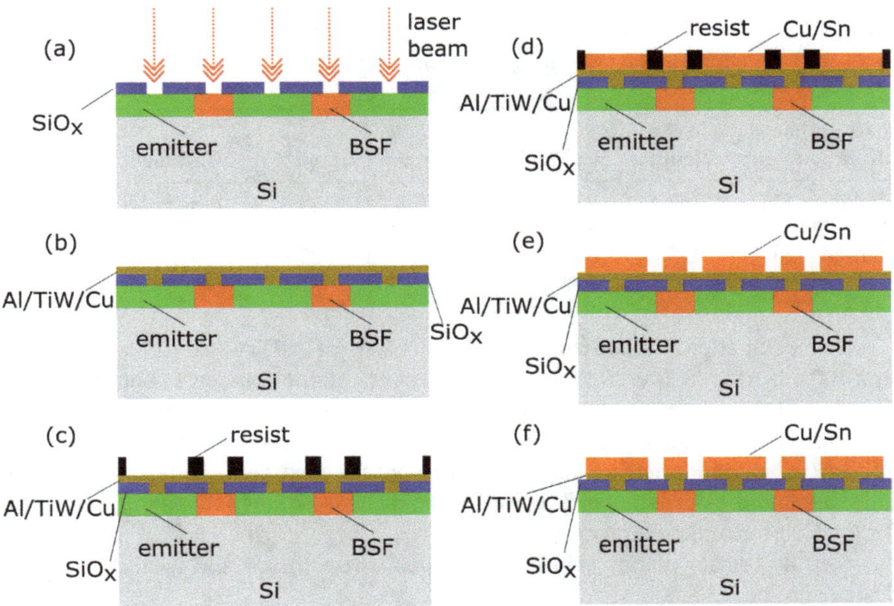

Figure 9.36 *Schematic cross sections of the process flow of electroplating metallization for interdigitated back contact (IBC) cells. (A) Steps 1 and 2: open locally the passivating SiO_x layer for subsequent contact formation, followed by a hydrofluoric acid (HF) dip to remove the oxide layer that presents on the opened Si surface; (B) Step 3: sputter a thin seed layer consisting of Al/TiW/Cu; (C) Step 4: print a resist to define the regions that do not undergo the following plating process; (D) Step 5: perform electroplating of Cu and Sn; (E) Step 6: Strip off the resist; (F) Steps 7 and 8: completely etch away the seed layer that is not covered with plated Cu, followed by forming gas annealing. For simplification only the rear side is shown and the drawings are not to scale (redrawn according to [230–232]).*

highest among all the commercially available modules, reaching a full-area module efficiency of 22.8% [210].

According to [230–232], a 'seed and plate' process flow to metallize IBC cells is schematically shown in Figure 9.36 and summarized as follows. Steps 1, 2 and 8 of this process flow are also used in PVD Al metallization for IBC (see Section 9.4.2).

Step 1: Locally open the rear dielectric layer (e.g. SiO_x or SiN_x) by means of laser ablation or patterned etch resist.

Step 2: Perform a short dip in HF solution if any oxide layer presents on the opened Si surface [222].

Table 9.13 IV parameters of interdigitated back contact (IBC) cells fabricated with Cu electroplating in recent years

First author's affiliation	Area (cm^2)	V_{oc} (mV)	J_{sc} (mA/cm^2)	FF (%)	η (%)	Comment
SunPower [234]	153.5	737	41.33	82.7	25.2	Passivated contact
IMEC [235]	239.1	683	41.15	80.2	22.5	Tube diffusion
Samsung [232]	155.4	679	41.55	81.3	22.9	Ion implantation

Step 3: Sputter a thin metal stack that consists of Al/TiW/Cu [230, 232]. The first 100 nm Al layer is used to form low-resistance ohmic contacts to both p^+ and n^+ Si, and additionally the Al layer contains 1% of Si content to prevent Al spiking in the forming gas annealing step. The second 90 nm TiW layer consists of Ti (10%) and W (90%), and it acts as a barrier to prevent the next deposited Cu layer from diffusing into the bulk silicon. The third 300 nm Cu layer functions as the seed layer for the Cu electroplating process.

Step 4: Print a plating resist on the sites where the p- and n-electrodes are designed to be isolated.

Step 5: Perform electroplating plating of Cu to thicken the metal layer in regions not covered by the plating resist. The Cu thickness is reported to range from 5 μm to 35 μm. Subsequently, a Sn capping layer (about 5–7 μm) can be plated on top of the Cu layer to improve solderability.

Step 6: Strip off the plating resist.

Step 7: Etch back the metal seed layer that is not covered by plated Cu, and after this step the positive and negative electrodes are completely isolated.

Step 8: Perform a low temperature annealing in a forming gas to recover the damage from the previous sputtering process and to reduce the contact resistance.

The specific contact resistivity for the 'seed and plate' metallization is decided by the first layer of the metal stack, i.e. PVD Al. As shown in Table 9.12, low contact resistivity on both p- and n-type Si can be achieved by PVD Al, and therefore it is likely that the IBC cells fabricated with the 'seed and plate' approach as discussed in this section will have similarly low contact resistivity. Table 9.13 lists the *IV* parameters of some of the IBC cells fabricated with Cu electroplating in recent years. All of these cells have a high *FF* above 80%, which suggests that the contact resistance is probably not the dominating loss mechanism for these Cu-plated IBC cells.

9.4.4 Screen-printing metallization for IBC cells

Substantial research of screen-printed IBC cells only started from about 2010 but it has experienced rapid development, partly thanks to the significant improvement of the metal paste technology. Compared with PVD and electroplating metallization for IBC cells, a distinct advantage of screen-printing metallization is that the metal finger can be directly defined by the screen layout, whereas for the other two

Table 9.14 Four typical process flows to metallize interdigitated back contact (IBC) cells with screen-printed metal pastes. After each printing step it is necessary to dry the paste in an oven for a short time, e.g. 20 s at 250 °C. The peak temperature of fast firing usually ranges from about 700 °C to 900 °C.

	Approach 1 [207, 236–238]	**Approach 2 [239, 240]**	**Approach 3: n-type Si bulk [241, 242]**	**Approach 4: p-type Si bulk [243–245]**
Step 1	Print fingers with the firing-through Ag-Al or Ag paste on p^+ Si	Locally open rear dielectric on both p^+ and n^+ Si	Locally open rear dielectric to define emitter regions	Locally open rear dielectric to define BSF regions
Step 2	Print fingers with the firing-through Ag paste on n^+ Si	Print fingers with the non-firing-through Ag paste on both p^+ and n^+ Si	Print fingers with the firing-through Ag paste on n^+ Si	Print fingers with the firing-through Ag paste on n^+ Si
Step 3	Fast firing	Fast firing	Print fingers with the non-firing-through Al paste on the emitter regions	Print fingers with the non-firing-through Al paste on the BSF regions
Step 4			Fast firing	Fast firing

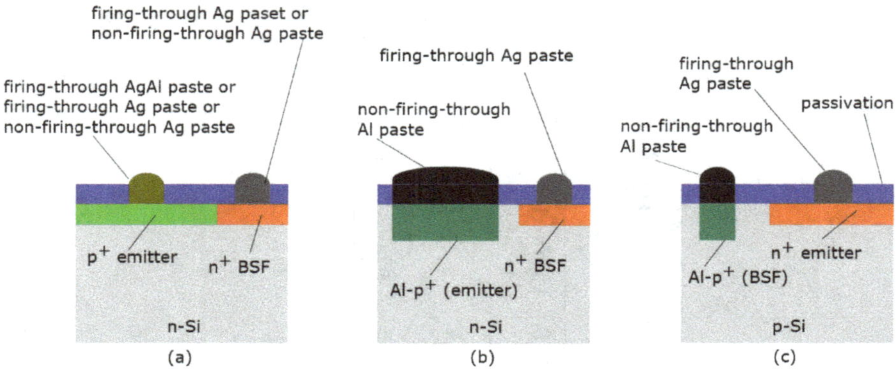

Figure 9.37 *Schematic cross sections of interdigitated back contact (IBC) cells metallized with various metal pastes as listed in Table 9.14. (A) Approach 1 and 2; (B) Approach 3; (C) Approach 4. For simplification only the rear side is shown and the drawings are not to scale.*

metallization technologies a full-area metal layer is firstly deposited and therefore extra processing steps are required to structure this metal layer into isolated positive and negative electrodes. Table 9.14 summarizes four typical approaches to metallize IBC cells with screen-printed metal pastes, and the corresponding schematic cross sections of the cells' rear sides are shown in Figure 9.37. One can observe that screen-printing metallization requires fewer process steps compared with PVD and Cu plating.

As for Approach 1, during the fast firing step the firing-through Ag or Ag-Al paste is able to first etch away the underlying passivating dielectric layer, e.g. SiN_x, and then form an ohmic contact to the heavily doped Si surface. As such, one can save the step to locally open the rear dielectric layer before applying the metal layer. The firing-through Ag-Al paste, which is often used to contact the p^+ front boron emitter of n-PERT solar cells [246], can be printed on the rear p^+ regions of the IBC cell, whereas a conventional firing-through Ag paste used for contacting the phosphorus emitters of p-type cells (e.g. PERC) can be printed on the rear n^+ regions [237]. An alternative way is to use a single firing-through Ag paste to contact both p^+ and n^+ doped regions [236, 238] and thus one printing step can be saved.

Regarding Approach 2, before printing the non-firing-through Ag paste it is required to open locally the rear passivating dielectric either by using laser ablation or printed etching paste, because this type of silver paste cannot penetrate through the passivation dielectric during the firing step. However, excellent specific contact resistivity below 0.5 mΩ cm² can be achieved on both rear p^+ and n^+ regions [239, 240].

For Approaches 1 and 2 the expensive silver paste is used to contact both p^+ and n^+ regions, and hence increases the manufacturing cost of IBC cells. One feasible way to reduce the cost is to replace the Ag paste with Al paste; as such the cost of paste can be reduced significantly, and additionally save one step to create the rear local p⁺ regions. To give an example, for the IBC cell with an n-type Si bulk, one can use the firing-through Ag paste to contact the n⁺ BSF and then print the non-firing-through Al paste to form the p⁺ Al-emitter (see Approach 3). A very similar process can be applied to the IBC cell with a p-type Si bulk where that dielectric opening should be adapted to a smaller one because the Al paste forms the rear p⁺ BSF in this case (see Approach 4). A major loss mechanism of the IBC cells metallized with Approaches 3 and 4 is the comparably high saturation current (J_0) in the Al-emitter or Al-BSF regions, respectively. An upgraded version of this cell concept has been proposed recently, where the n⁺ diffused region is replaced with passivated contacts [244, 245].

There is a laser ablation step to open the dielectric layer for Approaches 2–4. In contrast to PVD metallization, where the interface oxide residue could significantly influence the contact resistance, the interface oxide plays only a minor role when metallized with screen-printed metal pastes, and therefore dedicated etching step of the oxide residue is not necessary. Besides, if the multi-layer metal grid design is chosen (see Section 9.4.1), then after the firing step it is needed to print the insulation layer and subsequently the busbar in order to complete the metallization process.

Compared with the front metal fingers of both-side contacted solar cells, which are designed to be narrow as possible (e.g. <35 μm) to reduce the optical shadowing loss, the requirement of printing narrow fingers is relaxed for the IBC cell. To reduce the metal grid series resistance loss and hence maximize the cell efficiency, one can even intentionally design a very wide finger, e.g. >200 μm, especially when using the non-firing-through Ag paste where the contact width and finger width

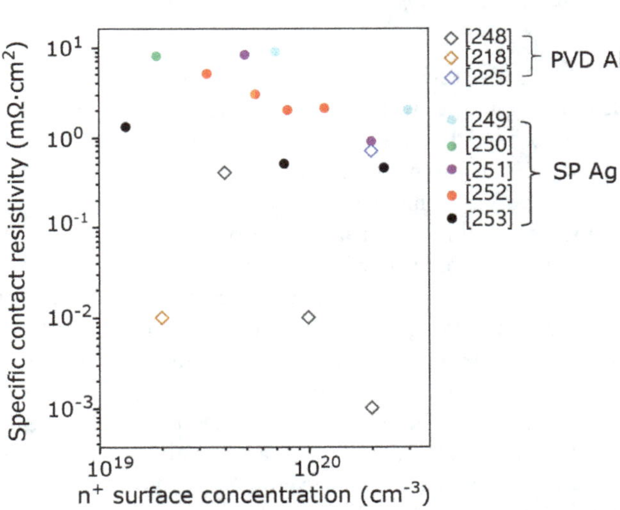

Figure 9.38 *Measured specific contact resistivity as a function of surface doping concentration (n⁺) by using physical vapour deposition (PVD) Al, as well as by using screen-printed and firing-through Ag pastes 218, 225, 248–253 (screen-printing Ag)*

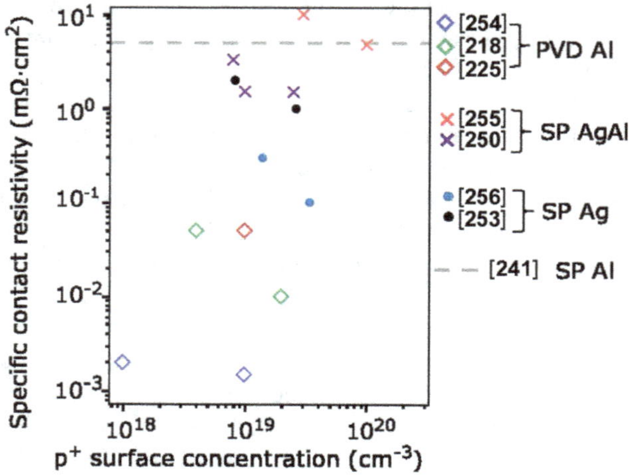

Figure 9.39 Measured specific contact resistivity as a function of surface doping concentration (p^+) by using physical vapour deposition (PVD) Al, screen-printed and firing-through Ag pastes (screen-printing Ag), screen-printed and firing-through Ag-Al pastes (screen-printing Ag-Al), as well as screen-printed and non-firing-through Al pastes (screen-printing Al). The dashed line in this figure (at 5 mΩ cm²) indicates the result achieved with screen-printing Al on a p-type Si wafer with a doping concentration of [218, 225, 243, 250, 253–256] 1.5 × 10^{16} cm^{-3}.

are decoupled [238, 239]. However, the drawbacks of using wide rear fingers are increased paste consumption as well as reduced bifaciality [247].

For any screen-printed IBC cell, it is crucial to find pastes that achieve low contact resistivity for both p^+ and n^+ Si surfaces. Figures 9.38 and 9.39 show the measured contact resistivity with regard to various surface doping concentration for n^+ and p^+ mono-crystalline Si surfaces (no passivated contacts), respectively. In both figures the results obtained with different screen-printed metal pastes and additionally with PVD Al are compared. As expected the contact resistivity generally decreases with increasing surface doping concentration. However for the same category of paste, the achieved contact resistivity at similar surface doping concentration scatters significantly, e.g. over ten times difference among different publications. The cause might be related to the paste composition and the fast firing parameters. One can observe that the contact resistivity achieved with PVD Al is the lowest, followed by screen-printed Ag or Ag-Al paste, whereas the Al paste shows the highest contact resistivity. Nevertheless, many data points achieved with screen-printed metal pastes in Figures 9.37 and 9.38 are below 5 mΩ cm², which are sufficiently low to fabricate a good IBC cell. Until now the best screen-printed IBC cell without using passivated contacts reaches 24.1% [240]. To further improve the cell efficiency, one effective way is to use passivated contacts to replace the diffused

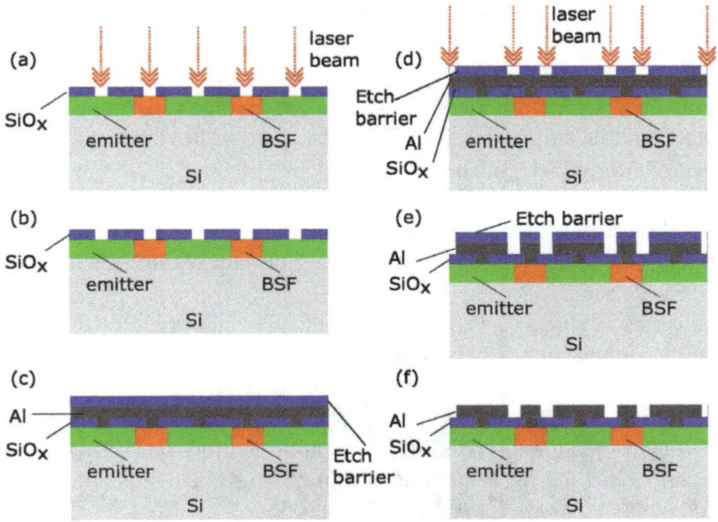

Figure 9.40 *Schematic cross sections of the process flow of physical vapour deposition (PVD) metallization for interdigitated back contact (IBC) cells. (A) Step 1: open locally the passivating SiO_x layer for subsequent contact formation; (B) Step 2: dip in hydrofluoric acid (HF) solution to remove the oxide layer that presents on the opened Si surface; (C) Steps 3 to 4: deposition of an Al layer and an etch barrier; (D) Step 5: locally remove the etch barrier by laser ablation; (E) Step 6: locally etch back the Al layer in wet chemical solution; (F) Steps 7 and 8: remove the etch barrier in HF solution and annealing. For simplification only the rear side is shown and the drawings are not to scale (redrawn according to [220] and [221]).*

Si surfaces, in which a key research topic is contacting the passivated contact layer. Promising results of using screen-printed Ag pastes to contact the heavily doped polycrystalline silicon have been reported [257, 258].

References

[1] Blakers A.W., Wang A., Milne A.M., Zhao J., Green M.A. '22.8% efficient silicon solar cell'. *Applied Physics Letters*. 1989;**55**(13):1363–5.

[2] Metz A., Adler D., Bagus S., *et al*. 'Industrial high performance crystalline silicon solar cells and modules based on rear surface passivation technology'. *Solar Energy Materials and Solar Cells*. 2014;**120**(13):417–25.

[3] Green M.A. 'The passivated emitter and rear cell (PERC): from conception to mass production'. *Solar Energy Materials and Solar Cells*. 2015;**143**(2):190–7.

[4] Dullweber T., Schmidt J. 'Industrial silicon solar cells applying the passivated emitter and rear cell (PERC) concept—a review'. *IEEE Journal of Photovoltaics*. 2016;**6**(5):1366–81.

[5] Dullweber T., Petrova-Koch V. (eds.) 'High efficiency industrial PERC solar cells for monofacial and bifacial applications' in Petrova-Koch V. (ed.). *High-efficient Low-Cost Photovoltaics*. Springer Nature Switzerland; 2020. pp. 65–94.

[6] Dullweber T., Kranz C., Peibst R., *et al*. 'PERC+: industrial PERC solar cells with rear al grid enabling bifaciality and reduced al paste consumption'. *Progress in Photovoltaics: Research and Applications*. 2016;**24**(12):1487–98.

[7] Dullweber T., Stöhr M., Kruse C., *et al*. 'Evolutionary PERC+ solar cell efficiency projection towards 24% evaluating shadow-mask-deposited poly-Si fingers below the Ag front contact as next improvement step'. *Solar Energy Materials and Solar Cells*. 2020;**212**:110586.

[8] International Technology Roadmap for Photovoltaic (ITRPV), 2019 Results [online]. April 2020. Available from http://www.itrpv.net/Reports/Downloads/.

[9] Comparotto C., Noebels M., Popescu L. 'Bifacial n-type solar modules: indoor and outdoor evaluation'. *29th European Photovoltaic Solar Energy Conference and Exhibition*; Amsterdam; 2014. pp. 3248–50.

[10] Janssen G.J.M., Van Aken B.B., Carr A.J., Mewe A.A. 'Outdoor performance of Bifacial modules by measurements and modelling'. *Energy Procedia*. 2015;**77**:364–73.

[11] Kranz C., Wyczanowski S., Baumann U., *et al*. 'Wet chemical polishing for industrial type PERC solar cells'. *Energy Procedia*. 2013;**38**:243–9.

[12] Pysch D., Schmitt C., Latzel B. 'Implementation of an ALD-Al_2O_3 PERC-Technology into a multi- and monocrystalline industrial pilot production'. *29th European PV Solar Energy Conference and Exhibition*. Amsterdam, Netherlands; 2014. pp. 612–16.

[13] Saint-Cast P., Kania D., Hofmann M., Benick J., Rentsch J., Preu R. 'Very low surface recombination velocity on P-type c-Si by high-rate plasma-deposited aluminum oxide'. *Applied Physics Letters*. 2009;**95**(15):151502-1–151502-3.

[14] Sperlich H.-P., Decker D., Saint-Cast P., Erben E., Peters L. 'High productive solar cell passivation on Roth & Rau MAiAs MW-PECVD inline machine—a comparison of Al_2O_3, SiO_2 and SiNx–H process conditions and performance'. *25th European Photovoltaic Solar Energy Conference and Exhibition*; Valencia, Spain; 2010. pp. 1352–7.

[15] Agostinelli G., Delabie A., Vitanov P., *et al*. 'Very low surface recombination velocities on P-type silicon wafers passivated with a dielectric with fixed negative charge'. *Solar Energy Materials and Solar Cells*. 2006;**90**(18-19):3438–43.

[16] Hoex B., Schmidt J., Pohl P., van de Sanden M.C.M., Kessels W.M.M. 'Silicon surface passivation by atomic layer deposited Al_2O_3'. *Journal of Applied Physics*. 2008;**104**(4):044903):1:.

[17] Dauwe S., Mittelstädt L., Metz A., Hezel R. 'Experimental evidence of parasitic shunting in silicon nitride rear surface passivated solar cells'. *Progress in Photovoltaics: Research and Applications*. 2002;**10**(4):271–8.

[18] Preu R., Glunz S.W., Schäfer S., Lüdemann R., Wettling W., Pfleging W. "Laser ablation – a new low-cost approach for passivated rear contact formation in crystalline silicon solar cell technology". *Proceedings of the 16th European Photovoltaic Solar Energy Conference*. Glasgow, United Kingdom; 2000. pp. 1181–14.

[19] Gatz S., Hannebauer H., Hesse R., *et al.* '19.4%-efficient large-area fully screen-printed silicon solar cells'. *Physica Status Solidi (RRL) – Rapid Research Letters*. 2011;**5**(4):147–9.

[20] Tjahjono B., Yang M.J., Wu V., *et al.* 'Optimizing CELCO technology in one year of mass production'. Proceedings of the 28th European Photovoltaic Solar Energy Conference; Paris, France; 2013. pp. 775–9.

[21] Lorenz A., Linse M., Frintrup H. Screen printed thick film metallization of silicon solar cells – recent developments and future perspectives. Proceedings of the 35th European Photovoltaic Solar Energy Conference; Brussels, Belgium; 2018. pp. 819–24.

[22] Schubert G., Huster F., Fath P. 'Physical understanding of printed thick-film front contacts of crystalline Si solar cells—Review of existing models and recent developments'. *Solar Energy Materials and Solar Cells*. 2006;**90**(18-19):3399–406.

[23] Li Z.G., Liang L., Ionkin A.S., *et al.* 'Microstructural comparison of silicon solar cells' front-side Ag contact and the evolution of current conduction mechanisms'. *Journal of Applied Physics*. 2011;**110**:074304.

[24] Shanmugam V., Cunnusamy J., Khanna A., *et al.* 'Electrical and microstructural analysis of contact formation on lightly doped phosphorus emitters using thick-film Ag screen printing pastes'. *IEEE Journal of Photovoltaics*. 2014;**4**(1):168–74.

[25] Cooper I.B., Tate K., Renshaw J.S., *et al.* 'Investigation of the mechanism resulting in low resistance Ag thick-film contact to Si solar cells in the context of emitter doping density and contact firing for current-generation Ag paste'. *IEEE Journal of Photovoltaics*. 2014;**4**(1):134–41.

[26] Hilali M.M., Sridharan S., Khadilkar C., Shaikh A., Rohatgi A., Kim S. 'Effect of glass frit chemistry on the physical and electrical properties of thick-film Ag contacts for silicon solar cells'. *Journal of Electronic Materials*. 2006;**35**(11):2041–7.

[27] Hörteis M., Gutberlet T., Reller A., Glunz S.W. 'High-temperature contact formation on N-type silicon: basic reactions and contact model for seed-layer contacts'. *Advanced Functional Materials*. 2010;**20**(3):476–84.

[28] Khanna A., Basu P.K., Filipovic A., *et al.* 'Influence of random pyramid surface texture on silver screen-printed contact formation for

monocrystalline silicon wafer solar cells'. *Solar Energy Materials and Solar Cells*. 2015;**132**:589–96.

[29] Cuevas A., Basore P.A., Giroult-Matlakowski G., Dubois C. 'Surface recombination velocity of highly doped n-type silicon'. *Journal of Applied Physics*. 1996;**80**(6):3370–5.

[30] Hannebauer H., Sommerfeld M., Müller J., Dullweber T., Brendel R. Analysis of the emitter saturation current density of industrial type silver screen-printed front contacts. *27th European Photovoltaic Solar Energy Conference and Exhibition*; Frankfurt, Germany; 2012. pp. 1360–3.

[31] Hahn G. Status of selective emitter technology. *25th European Photovoltaic Solar Energy Conference*; Valencia, Spain; 2010. pp. 1091–6.

[32] Eisele S.J., Röder T.C., Köhler J.R., Werner J.H. '18.9% efficient full area laser doped silicon solar cell'. *Applied Physics Letters*. 2009;**95**(13):133501.

[33] Tepner S., Lorenz A. 'Printing technology and its impact on the PV industry – a comprehensive review'. *Solar Energy Materials and Solar Cells*. 2020.

[34] Steckemetz S., Metz A., Hezel R. 'Thin Cz-silicon solar cells with rear silicon nitride passivation and screen printed contacts'. *17th European Photovoltaic Solar Energy Conference*. 2001:1902–5.

[35] Cesar I., Romijn I.G., Borsa D.M. 'Benchmark of open rear side solar cell with improved Al-BSF process at ECN'. *23rd European Photovoltaic Solar Energy Conference*. 2008:1770–5.

[36] Rauer M., Woehl R., Rühle K., *et al.* 'Aluminum alloying in local contact areas on dielectrically passivated rear surfaces of silicon solar cells'. *IEEE Electron Device Letters*. 2011;**32**(7):916–8.

[37] Murray J.L., McAlister A.J. 'The Al-Si (aluminum-silicon) system'. *Bulletin of Alloy Phase Diagrams*. 1984;**5**(1):74–84.

[38] Dullweber T., Schulte-Huxel H., Blankemeyer S., *et al.* 'Present status and future perspectives of bifacial PERC+ solar cells and modules'. *Japanese Journal of Applied Physics*. 2018;**57**(08):RA01.

[39] Kranz C., Baumann U., Wolpensinger B., *et al.* 'Void formation in screen-printed local aluminum contacts modeled by surface energy minimization'. *Solar Energy Materials and Solar Cells*. 2016;**158**(1):11–18.

[40] Müller J., Bothe K., Gatz S., Brendel R. 'Modeling the formation of local highly aluminum-doped silicon regions by rapid thermal annealing of screen-printed aluminum'. *Physica Status Solidi (RRL) – Rapid Research Letters*. 2012;**6**(3):111–3.

[41] Peng Z.-W., Buck T., Koduvelikulathu L.J., Mihailetchi V.D., Kopecek R. 'Industrial screen-printed n–PERT-RJ solar cells: efficiencies beyond 22% and open-circuit Voltages Approaching 700 mV'. *IEEE Journal of Photovoltaics*. 2019;**9**(5):1166–74.

[42] Tsuji K., Suzuki S., Morishita N., *et al.* 'Fine line Al printing on narrow point contact opening for front side metallization'. *AIP Conference Proceedings*. 2019;**2147**:040019.

[43] Kranz C., Wyczanowski S., Dorn S., *et al.* 'Impact of the rear surface roughness on industrial-type PERC solar cells'. *27th European Photovoltaic Solar Energy Conference*; 2012. pp. 557–60.

[44] Dullweber T., Hesse R., Bhosle V., Dubé C. 'Ion-implanted PERC solar cells with Al_2O_3/SiNx rear Passivation'. *Energy Procedia*. 2013;**38**:430–5.

[45] Deng W., Ye F., Liu R. 22.61% efficient fully screen printed PERC solar cell. *IEEE 44th Photovoltaic Specialist Conference*; Washington DC, US; 2017.

[46] Jäger P., Baumann U., Dullweber T. 'Impact of the thermal budget of the emitter formation on the pFF of PERC+ solar cells'. *AIP Conference Proceedings*. 2019;**2147**:140005.

[47] Kimmerle A., Woehl R., Wolf A., Biro D. 'Simplified front surface field formation for back contacted silicon solar cells'. *Energy Procedia*. 2013;**38**(8):278–82.

[48] Jäger P., Mertens V., Baumann U., Dullweber T. 'A Detailed Chemical Model for the Diffusion of Phosphorus Into the Silicon Wafer During $POCl_3$ Diffusion'. *IEEE Journal of Photovoltaics*. 2021;**11**(1):50–7.

[49] Piechulla A., Denzer S., Mihailetchi V.D. 'Low pressure diffusions for high quality emitter formation in advanced P- and N-type solar cells'. *Proceedings of the 31st European Photovoltaic Solar Energy Conference*. 2015:420–4.

[50] Dullweber T., Hannebauer H., Dorn S., *et al.* 'Emitter saturation current densities of 22 fA/cm^2 applied to industrial PERC solar cells approaching 22% conversion efficiency'. *Progress in Photovoltaics: Research and Applications*. 2017;**25**(7):509–14.

[51] Braun S., Hahn G., Nissler R., Pönisch C., Habermann D. 'The multi-busbar design: an overview'. *Energy Procedia*. 2013;**43**:86–92.

[52] Brendel R., Dullweber T., Gogolin R. 'Steckenreiter, recent progress and options for future crystalline silicon solar cells, proC'. *Proceedings of the 28th European Photovoltaic Solar Energy Conference*. 2013:676–90.

[53] Richter A., Benick J., Feldmann F., Fell A., Hermle M., Glunz S.W. 'N-type Si solar cells with passivating electron contact: identifying sources for efficiency limitations by wafer thickness and resistivity variation'. *Solar Energy Materials and Solar Cells*. 2017;**173**(6):96–105.

[54] Padhamnath P., Wong J., Nagarajan B., *et al.* 'Metal contact recombination in mono Poly™ solar cells with screen-printed & fire-through contacts'. *Solar Energy Materials and Solar Cells*. 2019;**192**:109–16.

[55] Lim B., Merkle A., Peibst R., Dullweber T., Wang Y., Zhou R. LID-free PERC+ cells with stable efficiencies up to 22.1%. *Proceedings of the 35th European Photovoltaic Solar Energy Conference*; Brussels, Belgium; 2018. pp. 359–65.

[56] Zhang X., Liu W., Chen Y. A roadmap towards 24%-efficiency PERC cells based on screen printing for mass production. *Proceedings of the 37th European Photovoltaic Solar Energy Conference*; 2020.

[57] 20.10.2020. Available from https://eu-solar.panasonic.net/fr/histoire-panasonic-hit-sanyo.htm.

[58] 'ITRPV 2020'.

[59] De Wolf S., Ballif C., Kondo M. 'Kinetics of a-Si:H bulk defect and a-Si:H/c-Si interface-state reduction'. *Physical Review B – Condensed Matter and Materials Physics*. 2012;**85**(11):2–5.

[60] Bettinelli A., Debourdeau M., Diaz J., Pernoud F., *et al.* 'SHJ cell metallization for various interconnection technologies'. *5th Solar Cell Paste and Metallization Forum*. Changzhou; 2019.

[61] Bettinelli A. Cofrittage de l'alumine et du tungstène 1400 °C [PhD Thesis]. France: ISN : 0000 0003 6414 9501, Strasbourg. 1987.

[62] Commault B., Lefillastre P., Bernardis S., Bettinelli A.D.J., Debourdeau M.M., Pernoud F. 'Module integration of SHJ cells by soldering'. *Photovoltaic Solar Energy Conference and Exhibition (PVSEC)*; Marseille, France; 2019.

[63] Faes A., Despeisse M., Levrat J., *et al.* 'Smart Wire solar cell interconnection technology'. *Proceedings of 29th European Photovoltaic Solar Energy Conference and Exhibition (29th EU PVSEC)*; Amsterdam; 2014.

[64] Pingel S., Erath D., Wenzel T., *et al.* 'Low-temperature Ag-paste screening for silicon heterojunction solar cells and modules'. *35th European Photovoltaic Solar Energy Conference and Exhibition*; 2020.

[65] Tepner S., Ney L., Linse M., Lorenz A., Pospischil M., Clement F. 'Studying knotless screen patterns for fine-line screen printing of Si-solar cells'. *IEEE Journal of Photovoltaics*. 2020;**10**(2):319–25.

[66] Schube J., Fellmeth T., Weil M., Nold S., Keding R., Glunz S.W. Intense pulsed light in back end processing of silicon heterojunction solar cells. *Photovoltaic Solar Energy Conference and Exhibition (PVSEC)*; Marseille, France; 2019.

[67] J. eaH. 'Annealing of silicon heterojunction solar cells: interplay of solar cell and indium tin oxide properties'. *IEEE Journal of Photovoltaics*. 2019;**9**(5):1202–7.

[68] Kobayashi E., De Wolf S., Levrat J., *et al.* 'Increasing the efficiency of silicon heterojunction solar cells and modules by light soaking'. *Solar Energy Materials and Solar Cells*. 2017;**173**(6):43–9.

[69] Steinhauser B., Grübel B., Nold S., Arya V., Schmiga C., Kluska S. 'Plating on TOPCon as a way to reduce the fabrication costs of i-TOPCon solar cells'. *Proceeding of the 37th EU PVSEC*; 2020.

[70] Li Z., Hsiao P.-C., Zhang W., *et al.* 'Patterning for plated heterojunction cells'. *Energy Procedia*. 2015;**67**(1):76–83.

[71] Aguilar A., Herasimenka S., Karas J., Jain H., Lee J., Munoz K. 'Development of cu plating for silicon heterojunction solar cells'. *Proceedings of the 43rd Photovoltaic Specialists Conference*; Chicago, USA; 2016.

[72] Geissbühler J., Wolf S.D., Faes A., *et al.* 'Silicon Heterojunction solar cells with copper-plated grid electrodes: status and comparison with silver thick-film techniques'. *IEEE Journal of Photovoltaics*. 2014;**4**:1055–62.

[73] Khanna A., Ritzau K.-U., Kamp M., *et al.* 'Screen-printed masking of transparent conductive oxide layers for copper plating of silicon heterojunction cells'. *Applied Surface Science*. 2015;**349**:880–6.

[74] Papet P., Hermans J., Söderström T., Cucinelli M.A.L., Bätzner D. 'Heterojunction solar cells wtih electroplated Ni/Cu front electrode'. *Proceedings of the 28th EU PVSEC*. 2013.

[75] Herasimenka S., Reginevich M. 'Selective plating of copper on transparent conductive oxide, solar cell structure and manufacturing method'. *Patent US 2016/0359058 A1*. 2016.

[76] Liu C.M., Liu W.L., Chen W.J., Hsieh S.H., Tsai T.K., Yang L.C. 'Ito as a diffusion barrier between Si and Cu'. *Journal of the Electrochemical Society*. 2005;**152**(3):G234–9.

[77] Yu J., Bian J., Duan W., *et al.* 'Tungsten doped indium oxide film: ready for bifacial copper metallization on silicon hetrojunction solar cell'. *Solar Energy Materials and Solar Cells*. 2016.

[78] Ma W., Jackson A., Wang C., Moghadam F., Sinha A. '24% Efficiency hybrid cell technology ingegrating a low-cost, precision-pattern cu metallization used for advanced PCB'. *Proceedings of the 7th World Conference on Photovoltaic Energy Conversion*; 2018.

[79] Lennon A., Colwell J., Rodbell K. 'Challenges facing copper-plated metallisation for silicon phtovoltaics: insights from integrated circuit technology development'. *Progress in Photovolatics*. 2019;**27**:27–97.

[80] Heng J., Fu J., Kong B., Chae Y., Wang W., Xie Z. 'Commrecialization of tunnel oxide junction based cell and module with high efficiency and good reliability'. *Proceedings of the 28th EU PVSEC*. 2013.

[81] Heng J., Fu J., Kong B., Chae Y., Wang W., Xie Z. '>23.1% High efficiency tunnel oxide junction bifacial solar cell with electroplated Cu gridlines'. *Proceedings of the 29th EU PVSEC*; 2014.

[82] Schultz-Wittmann O., De Ceuster D., Turner A., Crafts D., Ong R., Suwito D. 'Fine line copper based metallization for high efficiency crystlalline silicon solar cells'. *Proceedings of the 27th EUPVSEC*; 2012.

[83] Schultz-Wittmann O., Turner A., Eggleston B., De Ceuster D., Suwito D., van Kerschaver E. 'High Volume Manufacturing of High Efficiency Crystalline Silicon Solar Cells with Shielded Metal Contacts'. *Proceedings of the 32nd EU PVSEC*; 2016.

[84] Adachi D., Terashita T., Uto T., Hernández J.L., Yamamoto K. 'Effects of SiOx barrier layer prepared by plasma-enhanced chemical vapor deposition on improvement of long-term reliability and production cost for Cu-plated amorphous Si/crystalline Si heterojunction solar cells'. *Solar Energy Materials and Solar Cells*. 2017;**163**:204–9.

[85] Faes A., Lachowicz A., Bettinelli A., Ribeyron P., Lerat J., Munoz D. 'Metallization and interconnection for high-efficiency bifacial silicon heterojunction solar cells and modules'. *Photovolataics International*. 2018.

[86] GS Solar. gs-solar.com [online]. 2020. Available from http://www.gs-solar.com/en/products-details.php?id=146 [Accessed Oct 2020].

[87] Fu J., Xu Z., Yu C., Heng J. 'Solar cell with metal grid fabricated by electroplating'. *Patent US 2011/A1*. 2010;**0277825**.

[88] Schultz-Wittmann O., Crafts D., DeCeuster D., Turner A. 'Fine line metallization of photovoltaic devices by partial lift-off of optical coating'. *Patent US*. 2011;**8236604**:B2.

[89] GS Solar. gs-solar.com [online]. Available from http://www.gs-solar.com/en/product-Details.php [Accessed Oct 2020].

[90] Hermans J., van Knippenberg R., Kamp E., Brok W., Papet P., Legradic B. 'Inkjet printing for solar cell mass production on the Pixdro JetX platform'. *Proceedings of the 28th EU PVSEC*; 2013.

[91] Ziegler J., Xia L., Zejnelovic R., Borchert D. 'Changes in contact resistance of different metals to magnetron sputtered ITO while annealing'. *Proceedings of the 25th* EU PVSEC: Valencia; 2010.

[92] Hatt T., Bartsch J., Kluska S., Nold S., Glunz S., Glatthaar M. Low-cost Cu-plated metallization on TCOs for SHJ solar cells - Optimization of PVD contacting-layer. *Proceedings of the 47th IEEE PVSC Conference*; 2020.

[93] Ishimura F., Li W., Kobayashi E., Hashimoto K., Sato S., Watabe Y. 'Metal wrap through heterojunction solar cell with plated electrode'. *Proceedings of teh 33rd EU PVSEC*; 2016.

[94] Lachowicz A., Wyss P., Geissbühler J. 'Review on plating processes for silicon heterojunction cells'. *8th Workshop on Metallization on Interconnection, Constance*. 2019.

[95] Allen V. 'A method for forming a photovoltaic cell and a photovoltaic cell formed according to the method'. *Patent WO 2016/A1*. 2015;**000030**.

[96] Ballif C. 'Device comprising electrical contact and its production process'. *Patent EP2489076*. 2010.

[97] Lachowicz A., Descoeudres A., Champliaud J., Faes A., Geissbühler J., Despeisse M. 'Copper plating processes for silicon heterojunction solar cells: an overview'. *Proceedings of the 36th EU PVSEC*; 2019.

[98] Rodofili A., Wolke W., Kroely L., Bivour M., Cimiotti G., Bartsch J. 'Laser-transferred Ni-seed for the metallization of silicon heterojunction solar cells by Cu-plating'. *Proceedings of the 33rd EU PVSEC*; 2017.

[99] Fraunhofer I.S.E. ise.fraunhofer.de [online]. 10.05.2020. Available from https://www.ise.fraunhofer.de/content/dam/ise/en/documents/press-releases/2020/2820_ISE_e_PR_LTF%20Metallization.pdf [Accessed Oct 2020].

[100] Glatthaar M., Rohit R., Rodofili A., Snow Y.J., Nekarda J., Bartsch J. 'Novel plating processes for silicon heterojunction solar cell metallization using a structured seed layer'. *IEEE Journal of Photovoltaics*. 2017;**7**(6):1569–73.

[101] Hatt T., Bartsch J., Franzl Y., Kluska S., Glatthaar M. 'Advances with resist-free copper plating approached for the metallization of silicon heterojunction solar cells'. *AIP Conference Proceedings of the 8th Metallization Workshop*; 2019.

[102] Balucani M., Quaranta S. *A breakthrough in plating for solar cells metallization* [online]. 2018. Available from https://zenodo.org/record/1693191#.X5hRAIhKiUk.

[103] Lachowicz A., Andreatta G., Blondiaux N., Faes A., Badel N., Diaz Leon J. 'Project ameliz: patterning techniques for copper electroplated metallization

on heterojunction cells'. *AIP Conference Proceedings of the 9th Metallization and Interconnection Workshop*; 2020.
[104] Haegel N.M., Atwater H.J., Barnes T. *Terawatt-scale photovoltaics: Transform global energy* [online]. 2019. Available from https://science.sciencemag.org/content/364/6443/836.
[105] Verlinden P. 'Manufacturing of the next generation of high-efficiency solar cells and modules'. *2nd Workshop on Silicon Heterojunction Solar Cells: Science and Industry Technology*. Chengdu; 2019.
[106] The Silver Institute. *World Silver Survey* [online]. 2020. Available from https://www.silverinstitute.org/wp-content/uploads/2020/04/World-Silver-Survey-2020.pdf.
[107] Lachowicz A., Christmann G., Descoeudres A., Nicolay S., Ballif C. 'Silver- and indium-free silicon heterojunction solar cell'. *Proceedings of the 37th EU PVSEC*; 2020.
[108] E. bS.D.H. *Fundamentals of inkjet printing: the science of inkjet and droplets*. John Wiley & Sons; 2016.
[109] ELMQVIST R. 'Measuring instrument of the recording type'. *Patent Patent US*. 1951;**2**(566):443.
[110] Available from https://www.smithers.com/resources/2018/oct/inkjet-printing-builds-towards-$100-billion-market..
[111] Karunakaran S.K., Arumugam G.M., Yang W., *et al*. 'Recent progress in inkjet-printed solar cells'. *Journal of Materials Chemistry A*. 2019;**7**(23):*13873*–902.
[112] Teng K.F., Vest R.W. 'Metallization of solar cells with ink jet printing and silver metallo-organic inks'. *IEEE Transactions on Components, Hybrids, and Manufacturing Technology*. 1988;**11**(3):291–7.
[113] Lennon A.J., Utama R.Y., Lenio M.A.T., Ho-Baillie A.W.Y., Kuepper N.B., Wenham S.R. 'Forming openings to semiconductor layers of silicon solar cells by inkjet printing'. *Solar Energy Materials and Solar Cells*. 2008;**92**(1):1410–15.
[114] Shin D., Micromech J. 'Microeng'. 2010.
[115] Hermans J., Papet P., Pacheco K., *et al*. 'Advanced metallization concepts by inkjet printing'. *Proceedings of the 29th European Photovoltaics Solar Energy Conference*; Amsterdam, The Netherlands; 2014.
[116] Hermans J., Papet P., Pacheco K., Yao Y., W.Brok., Strahm W. 'Inkjet printing of Ag nanoparticle inks for heterojunction solar cell metallization'. *Proceedings of the SNEC PV Power Expo Conference*; Shanghai; 2015.
[117] Schube J., Fellmeth T., Jahn M., Keding R., Glunz S. 'Advanced metallization with low silver consumption for silicon heterojunction solar cells'. *AIP Conference Proceedings*. 2019.
[118] Woehl R., Hörteis M., Glunz S.W. 'Analysis of the optical properties of screen-printed and aerosol-printed and plated fingers of silicon solar cells'. *Advances in Optoelectronics*. 2008.
[119] Papet P. 'Front grid metallization and module interconnections of industrial heterojunction solar cells'. Konstanz: Metallization Workshop; 2013.

[120] Braun S., Micard G., Hahn G. Solar cell improvement by using a multi busbar design as front electrode. *Energy Procedia*; 2013.
[121] Mette A. New Concepts for Front Side Metallization of Industrial Silicon Solar Cells [PhD Thesis]. 2007.
[122] Giglia V. Influence of edge recombinations on the performance of half-, shingled- and full silicon heterojunction solar cells. *Proceedings of 35th European Photovoltaic Solar Energy Conference and Exhibition (36th EU PVSEC)*; Marseille; 2019.
[123] High Efficiency Solar technologies, Heterojunction Cell Technology – Less Production Steps and Higher Output [online]. 2020. Available from http://www.taiyangnews.info/PDF/Agenda_TaiyangNews_Virtual_Conference_on_High_Efficiency_Sola.
[124] HJT2.0 Product Launch by Maxwell [online], on 9th of December 2020. 2020. Available from https://note.youdao.com/ynoteshare1/index.html?id=c24e0b8dd9742d80b1764bbf5eebf6fe&type=note&from=singlemessage#/.
[125] Doyle T., Desharnais R., Erion-Lorico T. '2020 PV module reliability Scorecard, PVEL'. 2020.
[126] Barth V. 'Interconnection technologies for high reliable bifacial heterojunction glass/glass'. *Photovoltaic Modules, 4CO.2.5, Proceedings of 36th European Photovoltaic Solar Energy Conference and Exhibition (36th EU PVSEC)*; 2019.
[127] Danel D. 'Bifaciality optimization of silicon heterojunction solar cells'. *Proceedings of 35th European Photovoltaic Solar Energy Conference and Exhibition (36th EU PVSEC)*; Marseille; 2019.
[128] Abdallah A., Martinez D., Figgis B., El Daif O. 'Performance of silicon Heterojunction photovoltaic modules in Qatar climatic conditions'. *Renewable Energy*. 2016;**97**:860–5.
[129] Shoukry I., Libal J., Kopecek R., Wefringhaus E., Werner J., *et al*. 'Modelling of bifacial gain for stand-alone and in-field installed bifacial PV modules'. *Energy Procedia*. 2016;**92**(1):600–8.
[130] Gerritsen E. 'Bifacial heterojunction pv modules: highest energy yield available and how to measure that'. *PV-Tech Paper*. 2019;**103**:2019.
[131] Green M.A., Blakers A.W. 'Advantages of metal-insulator-semiconductor structures for silicon solar cells'. *Solar Cells*. 1983;**8**(1):3–16.
[132] Green M.A., Blakers A.W., Jiqun Shi., Keller E.M., Wenham S.R. 'High-Efficiency silicon solar cells'. *IEEE Transactions on Electron Devices*. 1984;**31**(5):679–83.
[133] Yablonovitch E., Gmitter T., Swanson R., Kwark Y. 'A 720 mV open circuit voltage SiO_x: c-Si: SiO_x double heterostructure solar cell'. *Applied physics letters*. 1985;**47**(11):1211–13.
[134] Yablonovitch E., Swanson R.M., Eades W.D., Weinberger B.R. 'Electron-hole recombination at the Si-SiO_2 interface'. *Applied Physics Letters*. 01/20/1986;**48**(3):245–7.
[135] Feldmann F., Bivour M., Reichel C., Hermle M., Glunz S.W. 'A passivated rear contact for high-efficiency n-type Si solar cells enabling high Voc's and

FF>82 %'. *28th European Photovoltaic Solar Energy Conference*; 2013. pp. 988–92.

[136] ITRPV. 'The International Technology Roadmap for Photovoltaic Results 2019'. *ITRPV-VDMA*. **11**. Frankfurt,Germany; 2020.

[137] Sajjad M., Yang X., Altermatt P., Singh N., Schwingenschlögl U., De Wolf S. 'Metal-induced gap states in passivating metal/silicon contacts'. *Applied Physics Letters*. 2019;**114**(7):71601.

[138] Feldmann F., Ritzau K.-U., Bivour M., *et al.* 'High and low work function materials for passivated contacts'. *Energy Procedia*. 2015;**77**:263–70.

[139] Glunz S.W., Feldmann F. 'SiO_2 surface passivation layers – a key technology for silicon solar cells'. *Solar Energy Materials and Solar Cells*. 2018;**185**(8):260–9.

[140] Schmidt J., Peibst R., Brendel R. 'Surface passivation of crystalline silicon solar cells: present and future'. *Solar Energy Materials and Solar Cells*. 2018;**187**:39–54.

[141] Cuevas A., Wan Y., Yan D., *et al.* 'Carrier population control and surface passivation in solar cells'. *Solar Energy Materials and Solar Cells*. 2018;**184**:38–47.

[142] Allen T.G., Bullock J., Yang X., Javey A., De Wolf S. 'Passivating contacts for crystalline silicon solar cells'. *Nature energy*. 2019.

[143] Melskens J., van de Loo B.W.H., Macco B., Black L.E., Smit S., Kessels W.M.M. 'Passivating contacts for crystalline silicon solar cells: from concepts and materials to prospects'. *IEEE Journal of Photovoltaics*. 2018;**8**(2):373–88.

[144] Richter A., Benick J., Feldmann F., Fell A., Hermle M., Glunz S.W. 'N-Type Si solar cells with passivating electron contact: identifying sources for efficiency limitations by wafer thickness and resistivity variation'. *Solar Energy Materials and Solar Cells*. 2017;**173**(6):96–105.

[145] Haase F., Hollemann C., Schäfer S., *et al.* 'Laser contact openings for local poly-Si-metal contacts enabling 26.1%-efficient POLO-IBC solar cells'. *Solar Energy Materials and Solar Cells*. 2018;**186**(6):184–93.

[146] Feldmann F., Bivour M., Reichel C., Hermle M., Glunz S.W. 'Passivated rear contacts for high-efficiency N-type Si solar cells providing high interface passivation quality and excellent transport characteristics'. *Solar Energy Materials and Solar Cells*. 2014;**120**:270–4.

[147] Feldmann F., Simon M., Bivour M., Reichel C., Hermle M., Glunz S.W. 'Carrier-selective contacts for Si solar cells'. *Applied Physics Letters*. 2014;**104**(18):181105.

[148] Peibst R. Implementation of n+ and p+ poly junctions on front and rear side of double-side contacted industrial silicon solar cells. *32nd European Photovoltaic Solar Energy Conference and Exhibition*; 2016. pp. 323–7.

[149] Padhamnath P., Khanna A., Nandakumar N., *et al.* 'Development of thin polysilicon layers for application in monoPoly™ cells with screen-printed and fired metallization'. *Solar Energy Materials and Solar Cells*. 2020;**207**(110358):110358.

[150] Çiftpınar H.E., Stodolny M.K., Wu Y., *et al.* 'Study of screen printed metallization for polysilicon based passivating contacts'. *Energy Procedia.* 2017;**124**(1-3):851–61.

[151] Glunz S.W*et al.* 'Passivating and carrier-selective contacts: basic requirements and implementation'. *2017 IEEE 44th Photovoltaic Specialist Conference (PVSC)*; 2017. pp. 2064–9.

[152] Rienäcker M., Bossmeyer M., Merkle A., *et al.* 'Junction resistivity of carrier-selective polysilicon on oxide junctions and its impact on solar cell performance'. *IEEE Journal of Photovoltaics.* 2016;**7**(1):11–18.

[153] Wietler T.F., Min B., Reiter S., *et al.* 'High Temperature Annealing of ZnO:Al on Passivating POLO Junctions: Impact on Transparency, Conductivity, Junction Passivation, and Interface Stability'. *IEEE Journal of Photovoltaics.* 2018;**9**(1):89–96.

[154] Peibst R*et al.* 'Building blocks for industrial screen-printed double-side contacted POLO cells with highly transparent ZnO: Al layers'. *IEEE Journal of Photovoltaics.* 2018;**8**(3):719–25.

[155] Limodio G., Yang G., Ge H., *et al.* 'Front and rear contact Si solar cells combining high and low thermal budget Si passivating contacts'. *Solar Energy Materials and Solar Cells.* 2019;**194**(12):28–35.

[156] Limodio G., De Groot Y., Van Kuler G., *et al.* 'Copper-Plating metallization with alternative seed layers for c-Si solar cells embedding carrier-selective passivating contacts'. *IEEE Journal of Photovoltaics.* 2019;**10**(2):372–82.

[157] Limodio G., Yang G., De Groot Y., *et al.* 'Implantation-based passivating contacts for crystalline silicon front/rear contacted solar cells'. *Progress in Photovoltaics: Research and Applications.* 2020;**28**(5):403–16.

[158] Steinhauser B*et al.* 'Plating on Topcon as a way to reduce the fabrication costs of I-TOPCON solar cells'. *Presented at the 37th European PV Solar Energy Conference and Exhibition.* 2020;**7**:11.

[159] Grübel B. Plated Ni/Cu/Ag for TOPCon solar cell metallization. *36th European Photovoltaic Solar Energy Conference*; 2019. pp. 1–5.

[160] Schube J., Tutsch L., Fellmeth T., *et al.* 'Low-resistivity screen-printed contacts on indium tin oxide layers for silicon solar cells with passivating contacts'. *IEEE Journal of Photovoltaics.* 2018;**8**(5):1208–14.

[161] Larionova Y., Schulte-Huxel H., Min B., *et al.* 'Ultra-thin poly-si layers: passivation quality, utilization of charge carriers generated in the Poly-Si and application on screen-printed double-side contacted polycrystalline Si on oxide cells'. *Solar RRL.* 2020;**4**(10):2000177.

[162] Padhamnath P., Wong J., Nagarajan B., *et al.* 'Metal contact recombination in monoPoly™ solar cells with screen-printed & fire-through contacts'. *Solar Energy Materials and Solar Cells.* 2019;**192**:109–16.

[163] Padhamnath P., Buatis J.K., Khanna A., *et al.* 'Characterization of screen printed and fire-through contacts on LPCVD based passivating contacts in monoPoly™ solar cells'. *Solar Energy.* 2020;**202**:73–9.

[164] Padhamnath P., Khanna A., Balaji N., *et al.* 'Progress in screen-printed metallization of industrial solar cells with SiOx/poly-Si passivating contacts'. *Solar Energy Materials and Solar Cells*. 2020;**218**(2):110751.

[165] Stodolny M.K., Lenes M., Wu Y., *et al.* 'N-Type polysilicon passivating contact for industrial bifacial N-type solar cells'. *Solar Energy Materials and Solar Cells*. 2016;**158**:24–8.

[166] Nandakumar N. 'Approaching 23% with large-area monoPoly cells using screen-printed and fired rear passivating contacts fabricated by inline PECVD'. *Progress in Photovoltaics: Research and Applications*. 2019;**2**:107–12.

[167] Y.-W. O. 'Screen printed large area bifacial N-type back junction silicon solar cells with selective phosphorus front surface field and boron doped poly-Si/SiOx passivated rear emitter'. *Applied Physics Letters*. 2018;**26**:263901.

[168] Mack S., Schube J., Fellmeth T., Feldmann F., Lenes M., Luchies J.-M. 'Metallisation of boron-doped polysilicon layers by screen printed silver pastes'. *Physica Status Solidi (RRL) – Rapid Research Letters*. 2017;**11**(12):1700334.

[169] Mack S., Lenes M., Luchies Jan-Marc., Wolf A. 'P-Type silicon solar cells with passivating rear contact formed by LPCVD p^+ polysilicon and screen printed Ag metallization'. *Physica Status Solidi (RRL) – Rapid Research Letters*. 2019;**13**(7):1900064.

[170] Nandakumar N*et al.* 'Approaching 23% with large-area monoPoly cells using screen-printed and fired rear passivating contacts fabricated by inline PECVD'. *Progress in Photovoltaics: Research and Applications*.

[171] Werner S., Lohmüller E., Maier S., Mourad S., Wolf A. 'Challenges for lowly-doped phosphorus emitters in silicon solar cells with screen-printed silver contacts'. *Energy Procedia*. 2017;**124**:936–46.

[172] Nandakumar N*et al.* 'Investigation of 23% monoPoly screen-printed silicon solar cells with an industrial rear passivated contact'. *2019 IEEE 46th Photovoltaic Specialists Conference (PVSC)*; 2019. pp. 1463–5.

[173] Chen D., Chen Y., Wang Z., *et al.* '24.58% total area efficiency of screen-printed, large area industrial silicon solar cells with the tunnel oxide passivated contacts (i-TOPCon) design'. *Solar Energy Materials and Solar Cells*. 2020;**206**(1):110258.

[174] Huang Y.-Y., Ok Y.-W., Madani K., *et al.* 'Fully screen-printed bifacial large area 22.6% N-type Si solar cell with lightly doped ion-implanted boron emitter and tunnel oxide passivated rear contact'. *Solar Energy Materials and Solar Cells*. 2020;**214**(10):110585.

[175] Richter A. Both sides contacted silicon solar cells: options for approaching 26% efficiency. *36th Eur. PV Sol Energy Conf Exhib*; 2019. pp. 9–13.

[176] Haase F., Kiefer F., Schäfer S., *et al.* 'Interdigitated back contact solar cells with polycrystalline silicon on oxide passivating contacts for both polarities'. *Japanese Journal of Applied Physics*. 2017;**56**(8S2):08MB15.

[177] Hermle M. Approaching efficiencies above 25% with both sides-contacted silicon solar cells. *2015 IEEE 42nd Photovoltaic Specialist Conference (PVSC)*; 2015. pp. 1–3.

[178] Yan D., Phang S.P., Wan Y., Samundsett C., Macdonald D., Cuevas A. 'High efficiency N-type silicon solar cells with passivating contacts based on PECVD silicon films doped by phosphorus diffusion'. *Solar Energy Materials and Solar Cells*. 2019;**193**:80–4.

[179] Feldmann F. Industrial Topcon solar cells realized by a PECVD tube process. *Presented at the 37th European PV Solar Energy Conference and Exhibition*; 2020. pp. 7–11.

[180] Richter A., Benick J., Müller R., *et al.* 'Tunnel oxide passivating electron contacts as full-area rear emitter of high-efficiency P-type silicon solar cells'. *Progress in Photovoltaics: Research and Applications*. 2018;**26**(8):579–86.

[181] Rienäcker M., Merkle A., Römer U., *et al.* 'Recombination behavior of photolithography-free back junction back contact solar cells with carrier-selective polysilicon on oxide junctions for both polarities'. *Energy Procedia*. 2016;**92**(6):412–8.

[182] Rohatgi A., Rounsaville B., Ok Y.-W., *et al.* 'Fabrication and modeling of high-efficiency front junction N-type silicon solar cells with tunnel oxide passivating back contact'. *IEEE Journal of Photovoltaics*. 2017;**7**(5):1236–43.

[183] Reichel C., Müller R., Feldmann F., Richter A., Hermle M., Glunz S.W. 'Influence of the transition region between P- and N-type polycrystalline silicon passivating contacts on the performance of interdigitated back contact silicon solar cells'. *Journal of Applied Physics*. 2017;**122**(18):184502.

[184] Yan D., Cuevas A., Phang S.P., Wan Y., Macdonald D. '23% efficient p-type crystalline silicon solar cells with hole-selective passivating contacts based on physical vapor deposition of doped silicon films'. *Applied Physics Letters*. 2018;**113**(16):061603.

[185] Yang G., Guo P., Procel P., *et al.* 'High-efficiency black IBC c-Si solar cells with poly-Si as carrier-selective passivating contacts'. *Solar Energy Materials and Solar Cells*. 2018;**186**(1):9–13.

[186] Feldmann F. 'Evaluation of topcon technology on large area solar cells'. *33rd European PV Solar Energy Conference and Exhibition*. Amsterdam: The Netherlands; 2017.

[187] Römer U. Ion-implanted poly-Si/c-Si junctions as a back-surface field in back-junction back-contacted solar cells. *Proceedings of the 29th European Photovoltaic Solar Energy Conference*; 2014. pp. 1107–10.

[188] Nemeth W. 'Implementation of tunneling passivated contacts into industrially relevant n-Cz Si solar cells'. *2015 IEEE 42nd Photovoltaic Specialist Conference, PVSC*; 2015.

[189] Nemeth B., Young D.L., Page M.R., *et al.* 'Polycrystalline silicon passivated tunneling contacts for high efficiency silicon solar cells'. *Journal of Materials Research*. 2016;**31**(6):671–81.

[190] Tao Y., Upadhyaya V., Chen C.-W., *et al.* 'Large area tunnel oxide passivated rear contact *n* -type Si solar cells with 21.2% efficiency'. *Progress in Photovoltaics: Research and Applications*. 2016;**24**(6):830–5.

[191] Yang G., Ingenito A., Isabella O., Zeman M. 'Ibc c-Si solar cells based on ion-implanted poly-silicon passivating contacts'. *Solar Energy Materials and Solar Cells*. 2016;**158**(1):84–90.

[192] Chen Y., Chen D., Liu C., *et al.* 'Mass production of industrial tunnel oxide passivated contacts (i-TOPCon) silicon solar cells with average efficiency over 23% and modules over 345 W'. *Progress in Photovoltaics: Research and Applications*. 2019;**27**(0):827–34.

[193] Pradeep P. 'Characterization of fire-through pastes on lpcvd based passivating contacts in monopolytm solar cells'. *36th European Photovoltaic Solar Energy Conference and Exhibition*. **2EO.1.5**. Marseilles, France; 2019. pp. 309–13.

[194] Peibst R. Industrial screen printed double side contacted polo cells. *33rd European Photovoltaic Solar Energy Conference and Exhibition*; Amsterdam; 2017. pp. 4527–30.

[195] Duttagupta S., Nandakumar N., Padhamnath P., Buatis J.K., Stangl R., Aberle A.G. 'monoPoly™ cells: Large-area crystalline silicon solar cells with fire-through screen printed contact to doped polysilicon surfaces'. *Solar Energy Materials and Solar Cells*. 2018;**187**:76–81.

[196] Stodolny M.K., Anker J., Geerligs B.L.J. Material properties of LPCVD processed n-type polysilicon passivating contacts and its application in PERPoly industrial bifacial solar cells. *Paper presented at the 7th International Conference on Silicon Photovoltaics*; Freiburg, Germany: Silicon PV. 2017; 124. 635–42. 2017. Available from http://www.sciencedirect.com/science/article/pii/S1876610217341802.

[197] Stodolny M.K., Lenes M., Wu Y., *et al.* 'N-type polysilicon passivating contact for industrial bifacial N-type solar cells'. *Solar Energy Materials and Solar Cells*. 2016;**158**:24–8.

[198] Stodolny M. Novel schemes of P+ poly sihydrogeneration implemented in industrial 6" bifacial front-and-rear passivating contacts solar cells. *International Conference EU PVSEC for Photovoltaic Research*; Brussels, Belgium; 2018. pp. 1–4.

[199] Schwartz R.J., Lammert M.D. 'Silicon solar cells for high concentration applications'. *Proceedings of the International Electron Devices Meeting; Washington*; 1975. pp. 350–2.

[200] Yamamoto K., Adachi D., Yoshikawa K., Yoshida W., Irie T., Konishi K. 'Record-breaking efficiency back-contact heterojunction crystalline Si solar cell and module'. *Proceedings of 33rd European Photovoltaic Solar Energy Conference and Exhibition*; Amsterdam; 2017. pp. 201–4.

[201] VDMA Photovoltaic Equipment. *International technology roadmap for photovoltaic results* [online]. 2019. Available from https://itrpv.vdma.org/en/ueber-uns [Accessed 23 Sept 2020].

[202] Granek F. High-efficiency back-contact back-junction silicon solar cells [Ph.D. dissertation]. University of Freiburg, Freiburg. 2009.

[203] Reinders A., Verlinden P., van Sark W., Freundlich A., Reinders A. *Photovoltaic Solar Energy: From Fundamentals to Applications*. London: Wiley; 2017. pp. 92–101.

[204] Sinton R.A., Swanson R.M. 'Simplified backside-contact solar cells'. *IEEE Transactions on Electron Devices*. 1990;**37**(2):348–52.

[205] Wehmeier N., Lim B., Merkle A., *et al.* 'PECVD BSG diffusion sources for simplified high-efficiency N-PERT BJ and BJBC solar cells'. *IEEE Journal of Photovoltaics*. 2016;**6**(1):119–25.

[206] Müller R., Schrof J., Reichel C., Benick J., Hermle M. 'Back-junction back-contact n-type silicon solar cell with diffused boron emitter locally blocked by implanted phosphorus'. *Applied Physics Letters*. 2014;**105**(10):103503.

[207] Scardera G., Inns D., Wang G., Dugan S., Dee J., Dang T. 'Screen-printed dopant paste interdigitated back contact solar cells'. *Proceedings of the 42nd IEEE Photovoltaic Specialists Conference*; New Orleans, USA; 2015. pp. 1–4.

[208] Engelhart P., Hermann S., Neubert T., *et al.* 'Laser ablation of SiO_2 for locally contacted Si solar cells with ultra-short pulses'. *Progress in Photovoltaics: Research and Applications*. 2007;**15**(6):521–7.

[209] O'Sullivan B., Debucquoy M., Singh S., Uruena A., Recamán Payo M., Posthuma N. 'Process simplification for high efficiency, small area interdigitated back contact silicon solar cells'. *Proceedings of the European Photovoltaic Solar Energy Conference and Exhibition*; Paris France; 2013. pp. 956–60.

[210] SunPower Corporation. Available from https://us.sunpower.com/products/solar-panels [Accessed 10 Sep 2020].

[211] Kopecek R., Libal J., Lossen J., Mihailetchi V.D., Chu H., Peter C. 'ZEBRA technology: low cost bifacial IBC solar cells in mass production with efficiency exceeding 23.5%'. *Proceedings of the 2020 IEEE 47th Photovoltaic Specialists Conference (PVSC); Virtual meeting to be published*; 2020.

[212] Schinke C., Kiefer F., Offer M., *et al.* 'Contacting interdigitated back-contact solar cells with four busbars for precise current-voltage measurements under standard testing conditions'. *IEEE Journal of Photovoltaics*. 2012;**2**(3):247–55.

[213] Faes A., Paviet-Salomon B., Tomasi A., Lachenal D., Badel N., Christmann G. 'Multi-wireinterconnection of back-contactedsilicon heterojunction solar cells'. *Presented in the 7th Workshop on Metallization & Interconnection for Crystalline Silicon Solar Cells*; Konstanz, Germany; 2017.

[214] Hendrichs M., Padilla M., Walter J., Clement F., Rech B. 'Screen-printed metallization concepts for large-area back-contact back-junction silicon solar cells'. *IEEE Journal of Photovoltaics*. 2016;**6**(1):374–83.

[215] Huyeng J.D., Spribille A., Rendler L.C., Ebert C., Eitner U., Keding R. 'Influence of interconnection concepts for IBC solar cell performance by simulation'. *AIP Conference Proceedings*. 1999;**2018**(1):20011.

[216] Lammert M.D., Schwartz R.J. 'The interdigitated back contact solar cell: a silicon solar cell for use in concentrated sunlight'. *IEEE Transactions on Electron Devices*. 1977;**24**(4):337–42.

[217] Swanson R.M. 'Point-contact solar cells: modeling and experiment'. *Solar Cells*. 1986;**17**(1):85–118.

[218] Reichel C., Granek F., Hermle M., Glunz S.W. "Back-contacted back-junction n-type silicon solarcells featuring an insulating thin film for decoupling charge carrier collection and metallizationgeometry". *Progress in Photovoltaics: Research and Applications*. 2013;**21**(5):1063–76.

[219] Kho T.C., Fong K.C., Stocks M., *et al.* 'Excellent ONO passivation on phosphorus and boron diffusion demonstrating a 25% efficient IBC solar cell'. *Progress in Photovoltaics: Research and Applications*. 2020;**28**(10):1034–44.

[220] Hollemann C., Haase F., Schäfer S., Krügener J., Brendel R., Peibst R. '26.1%-efficient POLO-IBC cells: Quantification of electrical and optical loss mechanisms'. *Progress in Photovoltaics: Research and Applications*. 2019;**27**(11):950–8.

[221] Teppe A., Engelhart P., Müller J. '*Methods for the Contact Separation of Electrically-Conducting Layers on the Back Contact of Solar Cells and Corresponding Solar Cells*. Patent No. WO/2006/042698'.

[222] Tao Y., Payne A., Upadhyaya V.D., Rohatgi A. '20.7% efficient ion-implanted large area n-type front junctio*n* silicon solar cells with rear point contacts formed by laser opening and physical vapor deposition'. *Progress in Photovoltaics: Research and Applications*. 2014;**22**(10):1030–9.

[223] Du Z., Zhang C., Li F., Zhou R., Hong M. 'Impact of laser-induced oxidation on silicon wafer solar cells' performance'. *IEEE Journal of Photovoltaics*. 2016;**6**(3):617–23.

[224] Heinemeyer F., Mader C., Münster D., Dullweber T., Brendel R. 'Inline high-rate thermal evaporation of aluminium for novel industrial solar cell metallization'. *Proceedings of the 25th European Photovoltaic Solar Energy Conference and Exhibition; Valencia*; Spain; 2010. pp. 2066–8.

[225] Chen Y., Yang Y., Zhang X., *et al.* 'Forming aluminum electrodes by screen printing and electron-beam evaporation for high performance interdigitated back contact solar cells'. *Solar Energy Materials and Solar Cells*. 2015;**143**:205–11.

[226] Franklin E., Fong K., McIntosh K., *et al.* 'Design, fabrication and characterisation of a 24.4% efficient interdigitated back contact solar cell'. *Progress in Photovoltaics: Research and Applications*. 2016;**24**(4):411–27.

[227] Rienacker M., Bossmeyer M., Merkle A., *et al.* 'Junction resistivity of carrier-selective polysilicon on oxide junctions and its impact on solar cell performance'. *IEEE Journal of Photovoltaics*. 2017;**7**(1):11–18.

[228] Yang G., Guo P., Procel P., *et al.* 'High-efficiency black Ibc c-Si solar cells with poly-Si as carrier-selective passivating contacts'. *Solar Energy Materials and Solar Cells*. 2018;**186**(1):9–13.

[229] Green M.A. *Silicon Solar Cells: Advanced Principles and Practice*. Sydney: Centre for Photovoltaic Devices and Systems; 1995.

[230] Mulligan W.P., Cudzinovic M.J., Pass T., Smith D., Swanson R.M. 'Metal contact structure for solar cell and method of manufacture. Patent US7388147B2'. *Patent US7388147B2*. 2008.

[231] Singh S., O'Sullivan B., Kyuzo M., Jambaldinni S., Tous L., Russell R. 'Application of seed and plate metallization to 15.6cm × 15.6cm IBC cells'. *Photovoltaics International*. 2015;**29**:62–70.

[232] Kim Y.-S., Mo C., Lee D.Y., *et al.* 'Gapless point back surface field for the counter doping of large-area interdigitated back contact solar cells using a blanket shadow mask implantation process'. *Progress in Photovoltaics: Research and Applications*. 2017;**25**(12):989–95.

[233] Song N., Romer U., Gentle A., *et al.* 'Metallization method for interdigitated back-contact silicon solar cells employing an Insulating resin layer and a Ti/Ag/Cu metal stack'. *IEEE Journal of Photovoltaics*. 2018;**8**(4):916–22.

[234] Smith D.D., Reich G., Baldrias M., Reich M., Boitnott N., Bunea G. 'Silicon solar cells with total area efficiency above 25%'. *Proceedings of the 43rd IEEE Photovoltaic Specialists Conference; Portland*; USA; 2016. pp.3351–5.

[235] O'Sullivan B.J., Jambaldinni S., Singh S., Kyuzo M., Debucquoy M., Szlufcik J. 'Loss analysis and design optimization of Large-Area high-efficiency Back-Contacted silicon solar cells'. *IEEE Journal of Photovoltaics*. 2016;**6**(4):810–16.

[236] Guillevin N., Mewe A., Spinelli P., Burgers A., Janssen G., van de Loo B. Mercury: industrial IBC cell with front floating emitter for 20.9% and higher efficiency. *Proceedings of the Proceedings of the 31st European Photovoltaic Solar Energy Conference and Exhibition*; Hamburg Germany; 2015. pp. 976–82.

[237] Lu G., Wang J., Qian Z., Shen W. 'Development of back-junction back-contact silicon solar cells based on industrial processes'. *Progress in Photovoltaics: Research and Applications*. 2017;**25**(6):441–51.

[238] Galbiati G., Chu H., Mihailetchi V.D., Libal J., Kopecek R. 'Latest results in screen-printed IBC-ZEBRA solar cells'. *Proceedings of the 7th World Conference on Photovoltaic Energy Conversion (WCPEC-7)*; Waikoloa USA; 2018. pp. 1540–3.

[239] Zhang X., Yang Y., Liu W., Zhang K., Chen Y., Li Z. 'Development of high efficiency interdigitated back contact silicon solar cells and modules with industrial processing technologies'. *Proceedings of the 6th World Conference on Photovoltaic Energy Conversion*; Kyoto, Japan; 2014.

[240] Xu G., Yang Y., Zhang X., Chen S., Liu W., Chen Y. 'Large-area (6 inch) screen-printed IBC solar cells with efficiency above 24.1% without passivated contacts'. *Proceedings of the 33rd European Photovoltaic Solar Energy Conference and Exhibition*; Amsterdam, Netherlands; 2017. pp. 428–30.

[241] Woehl R., Rüdiger M., Biro D., Wilde J. 'All-screen-printed back-contact back-junction silicon solar cells with aluminum-alloyed emitter and demonstration of interconnection of point-shaped metalized contacts'. *Progress in Photovoltaics: Research and Applications*. 2015;**23**(2):226–37.

[242] Wohrle N., Kimmerle A., Greulich J., Wolf A. 'All-diffused back-contact back-junction solar cell with aluminum-alloyed emitter—experiment and simulation'. *IEEE Journal of Photovoltaics*. 2016;**6**(3):641–8.

[243] Peng Y.-W., Chen C.-H., Li L.-Y *et al*. 'Design, fabrication, and characterization of n-Si IBC solar cells using PERC technology'. *IEEE Journal of Photovoltaics*. 2020;**10**(2):383–9.

[244] Bende E.E., Guillevin N., Burgers A.R. Wu Y. P-IBC: 'Combining PERC and TOPCON'. *Presented in the 8th Workshop on Metallization & Interconnection for Crystalline Silicon Solar Cells*; Konstanz, Germany; 2019.

[245] Haase F., Hollemann C., Schafer S., Krugener J., Brendel R., Peibst R. 'Transferring the record p-type si POLO-IBC cell technology towards an industrial level'. *Proceedings of the IEEE 46th Photovoltaic Specialists Conference (PVSC)*; Chicago, USA; 2019. pp. 2200–6.

[246] Rodriguez J., Wang E.-C., Chen N., *et al*. 'Towards 22% efficient screen-printed bifacial n-type silicon solar cells'. *Solar Energy Materials and Solar Cells*. 2018;**187**:91–6.

[247] Libal L., Kopecek R. (eds.). *Bifacial Photovoltaics: Technology, Applications and Economics*. London: Institution of Engineering and Technology; 2018.

[248] Yu A.. 'Electron tunneling and contact resistance of metal-silicon contact barriers'. *Solid-State Electronics*. 1970;**13**(2):239–47.

[249] Dullweber T., Hannebauer H., Dorn S., *et al*. 'Emitter saturation current densities of 22 fA/cm^2 applied to industrial PERC solar cells approaching 22% conversion efficiency'. *Progress in Photovoltaics: Research and Applications*. 2017;**25**(7):509–14.

[250] Werner S., Lohmüller E., Wolf A., Clement F. 'Extending the limits of screen-printed metallization of phosphorus- and boron-doped surfaces'. *Solar Energy Materials and Solar Cells*. 2016;**158**:37–42.

[251] Li H., Hallam B., Wenham S., Abbott M. 'Oxidation drive-in to improve industrial emitter performance by $POCl_3$ diffusion'. *IEEE Journal of Photovoltaics*. 2017;**7**(1):144–52.

[252] Werner S., Lohmüller., Maier S., Mourad S., Wolf A. 'Challenges for lowly-doped phosphorus emitters in silicon solar cells with screen-printed silver contacts'. *Energy Procedia*. 2017;**124**:936–46.

[253] Chu H., Preis P., Lossen J., *et al*. 'Impact of the presence of busbars during the fast firing process on contact resistances'. *IEEE Journal of Photovoltaics*. 2018;**8**(4):923–9.

[254] Carver G.P., Kopanski J.J., Novotny D.B., Forman R.A. 'Specific contact resistivity of metal-semiconductor contacts – a new, accurate method linked to spreading resistance'. *IEEE Transactions on Electron Devices*. 1988;**35**(4):489–97.

[255] Fritz S., Riegel S., Gloger S., *et al*. 'Influence of emitter properties on contact formation to p+ silicon'. *Energy Procedia*. 2013;**38**:720–4.

[256] Fritz S., Engelhardt J., Ebert S., Hahn G. 'Contacting boron emitters on N – type silicon solar cells with aluminium-free silver screen-printing pastes'. *Physica Status Solidi (RRL) - Rapid Research Letters*. 2016;**10**(4):305–9.

[257] Mewe A., Stodolny M., Anker J., Lenes M., Pagès X., Wu Y. 'Full wafer size IBC cell with polysilicon passivating contacts'. *AIP Conference Proceedings*. 1999;**2018**(1):40014.

[258] Xu G., Deng M., Chen S., Chen D., Hu Y., Zhang X. '25% cell efficiency with integration of passivating contact technology and interdigitated back contact structure on 6" wafers'. *Proceedings of the IEEE 46th Photovoltaic Specialists Conference (PVSC)*; Chicago, USA; 2019. pp. 1452–5.

Chapter 10

Module interconnection technologies

Henning Schulte-Huxel[1], Achim Kraft[2], Torsten Roessler[2], and Angela De Rose[2]

To manufacture PV modules for solar power generation, the solar cells have to be interconnected in series to solar cell strings to enable the desired voltage output. The encapsulation serves as protection of the fragile solar cells and the interconnection against environmental impacts. In PV industry, the solar cell interconnection is realized by soldering, with a market share of around 95% as stated by the ITRPV [1]. Besides soldering presented in Section 10.1, solar cells can also be interconnected by the use of electrically conductive adhesives (ECAs). This alternative interconnection technology is discussed in detail in Section 10.2. Additionally, welding is also investigated as an alternative interconnection technology for solar cells and described in Section 10.3.

10.1 Solder interconnection of solar cells

Common solar cell strings are manufactured by the connection of the front side electrode of one solar cell with the rear electrode of the next solar cell, using copper ribbons with a nearly rectangular cross section or copper wires with a circular cross section. Figure 10.1 shows a sketch of the series interconnection of three solar cells with five interconnectors.

In the following subsections, the materials and the automated industrial process for solar cell interconnection by soldering are described in detail.

10.1.1 Materials for solder interconnection

10.1.1.1 Solar cell metallization

On the front side of PERC solar cells, the electrical contacts are realized as sintered metal grid made of silver. For a more detailed description of solar cell metallization, compare Chapters 6 and 7. The grid consists of a large number of fine contact

[1]Institute for Solar Energy Research in Hamelin (ISFH), Hamelin, Germany
[2]Fraunhofer Institute for Solar Energy Systems ISE, Freiburg, Germany

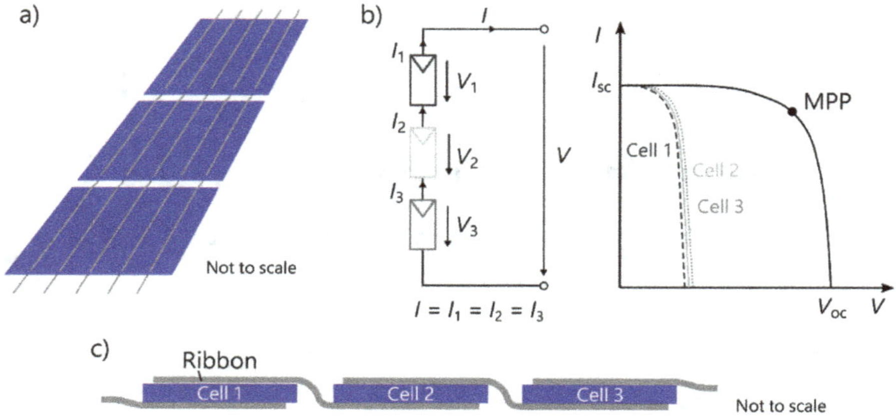

Figure 10.1 *Tilted top view sketch (a) and side view (b) of the series interconnection of three solar cells*

fingers (approx. 50 µm width), to collect the current from the cell and, depending on the used interconnection technology, a number of busbars (usually 5, approx. 600 µm width), which serve as solderable electrode in perpendicular orientation to the fingers. These silver busbars function as joint partner on cell side. The busbar shape shows multiple designs depending on the cell manufacturer. Usually, a number of wider parts, working as solder pads, are connected by line structures. For the wire interconnection, a row of small pads is arranged on a redundancy line.

10.1.1.2 Interconnectors

For the interconnection of solar cells, copper interconnectors coated with solder are used and work as second joint partner. Figure 10.2a shows the cross section of a solder coated copper interconnector with the dimensions of 0.90 × 0.22 mm, as it is

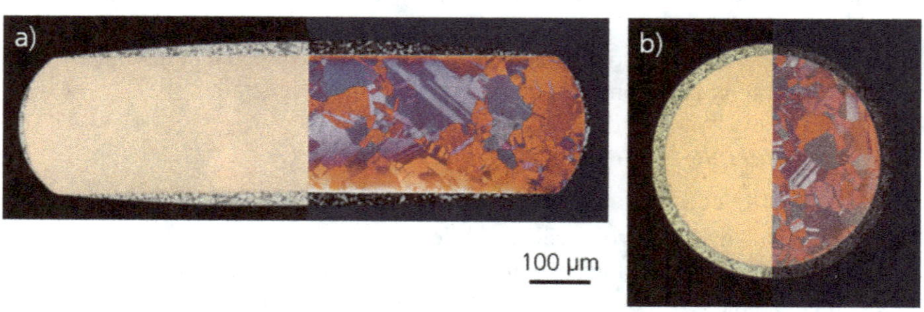

Figure 10.2 *Optical microscope image of the metallographic cross section of (a) a solder-coated copper ribbon and (b) a solder-coated copper wire. The left part of the images shows the cross section as polished, in the right part, the grain structure in the copper core is made visible*

commonly used for the interconnection of solar cells with five busbars. Figure 10.2b shows the cross section of a solder-coated copper wire with a diameter of 0.4 mm. In both images, the left part shows the bare copper core as polished; on the right, the structure of the copper grains is made visible by etching.

The solder coating of the interconnectors is applied by a hot dip tinning process. The thickness of the solder coating can generally be adjusted by the driving speed of the ribbon/wire through the liquid solder bath. The higher the feed speed, the thicker the coating and vice versa, because the liquid solder does not have the time to run back into the solder bath before its temperature decreases below the solidification temperature. In industrial production, the hot dip tinning process is operated by driving speeds of approx. 150 m/min for high throughput. The thickness of the coating is adjusted by the arrangement and setting of a compressed air jet at the exit of the liquid solder bath. Typical solder thicknesses for ribbon interconnectors are in the range of 10–25 and 5–15 μm for wires. The thickness of the solder on wires is usually smaller because of the different flow characteristic of the liquid solder on the circular cross section. For higher solder thicknesses, the production of an accurate concentricity of the resulting solder layer is challenging.

An important characteristic of copper interconnectors is the yield strength $R_{p0.2}$. It ranges from approx. 50 to 150 mPa. Interconnectors with a higher yield strength are stiffer. Interconnectors with a lower yield strength are softer and therefore show the ability to perform elastic deformation to a higher extent. In the last years, interconnector suppliers effectively reduced the yield strength of the interconnectors by adapting their production processes. The yield strength of the interconnectors correlates with the resulting copper grain size distribution [2]. This correlation is known as Hall–Petch relation [3, 4]. Interconnectors with small grains show higher yield strength because the higher number of grain boundaries inhibit dislocation slipping, which is necessary for plastic deformation to occur [5]. Figure 10.3 shows the average grain size of three different copper wires (cross-section view) and the associated yield strength [2]. The grain size distribution of these wires is plotted in Figure 10.4.

However, the yield strength of copper interconnectors is not only influenced by its grain size distribution. Grain orientation, impurities and dislocations are additional influence factors [6, 7].

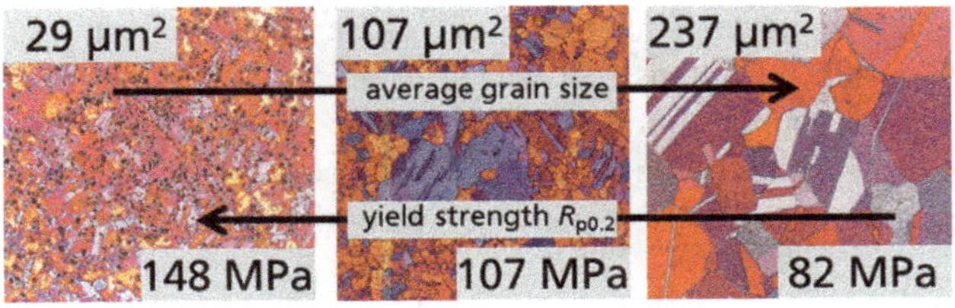

Figure 10.3 Optical microscope images of metallographic cross sections after etching of the copper core of three wires with different average grain size and yield strength [2]

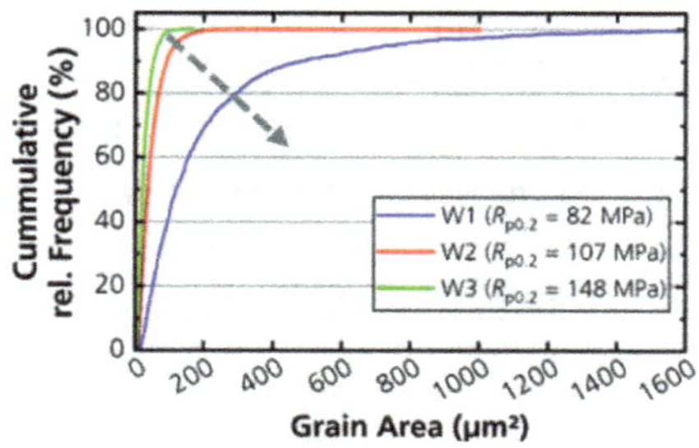

Figure 10.4 *Grain size distribution of the three copper wires shown in Figure 10.3*

The mechanical properties of the copper interconnectors and especially the yield strength will have an influence on the thermomechanical stress in the soldered interconnection and may consequently affect the stability of the solar cell interconnection especially under thermal cycles in climate chamber tests.

10.1.1.3 Solder

The characteristic material of a solder joint is the used solder alloy. The key attribute of the solder is its melting point, which needs to be lower than the melting points of the joint partners. In case of solar cell interconnections, the joint partners are copper (on the interconnector side) and silver (on the cell side). Due to the following boundary conditions for industrial interconnections of solar cells, the amount of solders in operation is reduced to a few alloys.

Requirements for solder alloys in industrial solar cell interconnection:

- suitable for very short soldering times <2 s
- cheap and highly available materials
- formation of diffusion zones by interaction with copper and silver surfaces
- long-term stability of mechanical and electrical properties under thermal cycles and moisture
- stable against chemical corrosion.

Due to a valid exclusion for the photovoltaic industry from the 'restrictions of the use of certain hazardous substances in electric and electronic equipment' (RoHS) [8] in the European Union, the use of lead containing solders is still allowed in the production of photovoltaic modules. This, in combination with extremely high cost pressure, causes tin–lead solders to be nearly the only solder type currently in operation.

Tin-based solders already have been used over decades in electronics to a very high extend and were thereupon adapted for solar cell interconnection. The reason is the ability of the metallurgical binary system copper–tin to form intermetallic compounds (IMC) at low temperatures with strong metallic bonding. In SnPb solders,

Table 10.1 Different tin-based solder alloys and their melting temperatures

Solder alloy(%wt)	Melting point/range (°C)	Eutectic	References
Sn99.9	232	Yes	[9–11]
Sn96.5Ag3.5	221	Yes	[9–11]
Sn96.5Ag3Cu0.5	217–220	No	[11, 12]
Sn60Pb40	183–190	No	[9, 11]
Sn62Pb38	183	Yes	[9–12]
Sn62Pb36Ag2	179	Yes	[9, 11]
Sn60Bi40	139–170	No	[11]
Sn42Bi57Ag1	138–140	No	[11]
Sn43Bi57	139	Yes	[9–11]

The solder most frequently used in PV-industry for solar cell interconnection is highlighted in green.

only tin reacts with the copper surface, whereas lead does not form any IMC. Lead is used in the solder for the following reasons:

- to lower the melting point (compare Table 10.1)
- to reach a higher ductility
- to support spreading behaviour of the solder
- to achieve a shiny appearance and
- eutectic SnPb solders do not show Sn whisker growth.

Two key advantages are the main reason why the use of eutectic solder alloys is common for solder interconnection:

1. For most of the alloys, the melting point of the eutectic alloy is lower than the melting point of its components.
2. The high throughput requirements in PV industry make very short soldering times indispensable. This can be realized by the use of solder alloys with a small melting range or even a single melting point.

This means that the entire solder melts when the temperature exceeds the eutectic temperature and the surfaces of both joint partners are joint simultaneously when the heat decreases below the eutectic temperature.

Figure 10.5 shows the lead–tin phase diagram with the eutectic point (black, empty circle) at the crossing point of the liquidus line (orange) and the solidus line (green) at 183 °C and a weight percentage share of 61.9% tin.

In the past, tin–lead solder alloys containing a small percentage of silver were widely used for solar cell interconnection in the PV industry, e.g., Sn62Pb36Ag2. Due to the high cost pressure and the associated requirement to reduce material costs, the silver-containing solders have totally disappeared from the market. Currently an eutectic tin–lead solder, which is usually labelled as Sn60Pb40, consisting of ~62% tin and ~38% lead, is widely used.

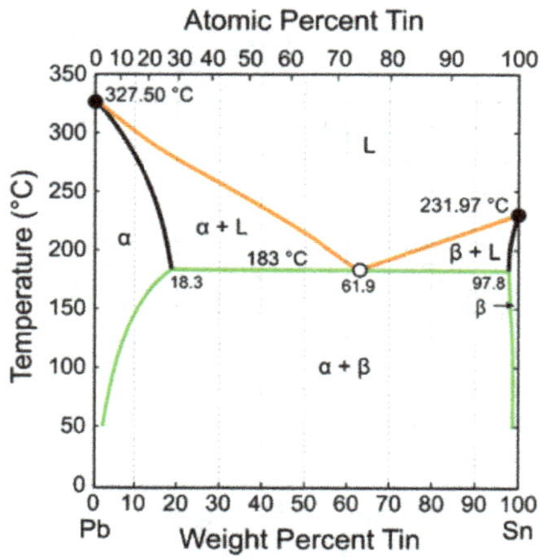

Figure 10.5 *Phase diagram of lead and tin with eutectic point (white circle) at the interception of the liquidus line (orange) and the solidus line (green) at 183 °C and a weight percentage share of 61.9% tin (redrawn Fraunhofer ISE)*

To realize lead-free interconnections or to enable low-temperature soldering for temperature sensitive solar cells (e.g. heterojunction [13, 14] or tandem solar cells), bismuth–tin alloys are also in operation. A study of the microstructure of lead-free solders used for solar cell interconnection can be found in [15]. Table 10.1 shows a selection of tin-based solder alloys and their melting temperatures. The melting temperature increases with the Sn content.

Table 10.1 shows the most relevant solders that are used for solar cell interconnection. In general, also alloys with variant Sn or Bi share or other metals that feature melting points in a suitable range can be used for the interconnection of silicon solar cells. For example, solders with indium, cadmium or gallium [9, 10], like the eutectic Sn48In52, with a melting point of 117 °C [9].

10.1.1.4 Flux

Flux describes a non-metallic material that is used in solder processes to remove the existing oxide layers on the surfaces of the join partners and to avoid their regeneration during the process. This leads to an improved wettability of the surfaces by the liquid solder. The three general types of fluxes are resin based, organic and inorganic.

For solar cell interconnection, so-called ‘no clean’ fluxes are used. This describes fluxes that do not show residues on the cell surface, which would require an additional cleaning process after interconnection. For this reason, the fluxes have low solid contents of ~2% and are based on solvents, e.g., isopropanol and acids, e.g., the organic adipic acid. Depending on the composition, every flux has its characteristic effective temperature range of operation. The temperature range starts

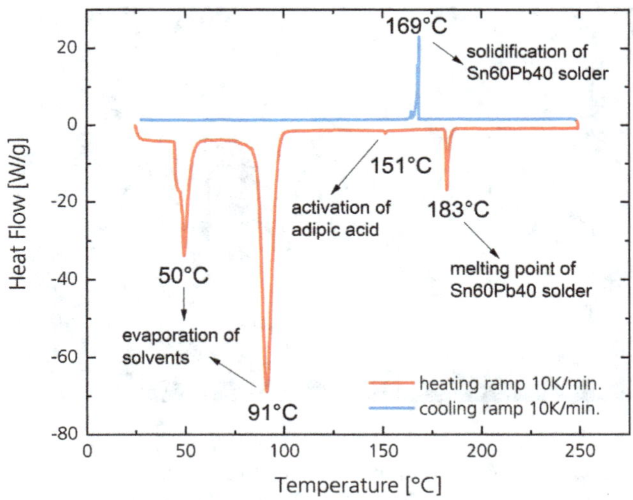

Figure 10.6 *DSC diagram of the supplied and absorbed heat quantity of flux and a copper interconnector coated with Sn60Pb40 solder at a heating ramp from room temperature to 250 °C and back to room temperature with a heating and cooling rate of 10 K/min (exotherm reactions up)*

with the melting temperature of the acid followed by a period of reactions and is completed with the saturation where the effect tones down. The adjustment of the effective temperature range of the used flux and the solder process, especially the melting temperature of the solder, is important. To achieve good interconnection results, the activation temperature of the flux has to be ~50 K lower than the melting temperature of the solder. Figure 10.6 shows the measurement result of differential scanning calorimetry (DSC) of flux based on solvents and adipic acid and a copper interconnector coated with Sn60Pb40 solder. The materials were subjected to a heating ramp from room temperature to 250 °C and a cooling ramp back to room temperature with 10 K/min. The diagram shows the heat flow over the temperature, which is negative for endothermic reactions, e.g. evaporation or melting of materials (red part in Figure 10.6), and positive for exothermic reactions, e.g. solidification (blue part in Figure 10.6). The two negative peaks at 50 and 91 °C represent the evaporation of the solvents from the flux. The small peak at 151 °C marks the activation of the adipic acid, and the peak at 183 °C shows the melting of the solder. The solidification of the solder was observed at 169 °C during the cooling ramp. The difference between melting and solidification temperature (here 14 °C) is called undercooling effect.

To improve the alloying of the metals, activation agents are added to the flux, which generate a higher concentration of mobile ions in the liquid flux film at the surface of the joint partners.

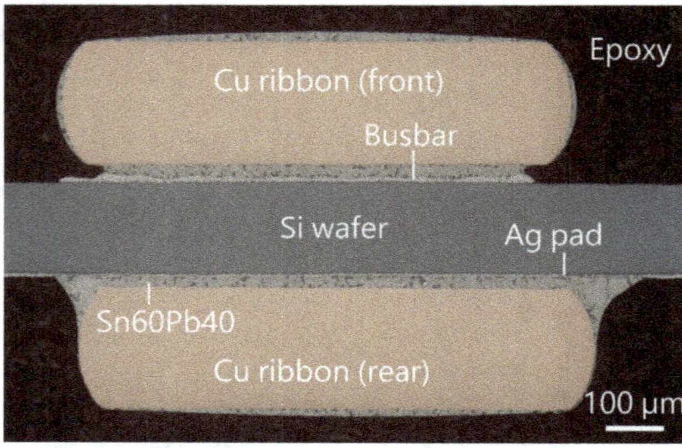

Figure 10.7 *Optical microscope image of a metallographic cross section of copper ribbons soldered on the silver metallization of a monofacial PERC solar cell on front and rear side (Fraunhofer ISE)*

10.1.1.5 Solder joints and intermetallic compounds

A solder joint between two join partners is in general indicated by the presence of two joint interfaces and diffusion zones at both interfaces.

Figure 10.7 shows the optical microscope image of a metallographic cross section of copper ribbons soldered on the front and rear side of a monofacial PERC solar cell. The joint interfaces are the interface between copper and solder and the interface between solder and the silver metallization on the solar cell. At both interfaces, intermetallic phase growth can be observed.

The appearance and growth kinetics of intermetallic phases for different solder alloys that can be observed in solar cell interconnection are described in detail in [16]. The formation of homogeneous IMC layers at the joint interfaces after the soldering process represents an indication for an accurately formed metallurgical bond [17, 18].

At the copper–solder interface, a Cu_6Sn_5 layer already develops following the hot dip tinning process during the ribbon manufacturing. During the soldering process, this intermetallic phase is remolten and formed again after cooling. Usually, a ~0.5-µm-thick Cu_6Sn_5 compound layer of scallop type morphology can be observed [16]. The thickness of the intermetallic phase depends on the dissolution rate of Cu in the molten solder, which is 2 µm/min at 200 °C for the given materials [10]. Additionally, the binary phase of Cu_3Sn is most likely to form between Cu and Cu_6Sn_5. The IMC formation begins to proceed when the local equilibrium solubility of Cu is exceeded.

At the tin–silver interface, a 0.2–1.2-µm-thick Ag_3Sn phase is formed during the soldering process [16]. The dissolution rate of Ag in the molten solder is 36 µm/min at 200 °C [10], which is significantly higher than the dissolution rate of Cu in SnPb solder.

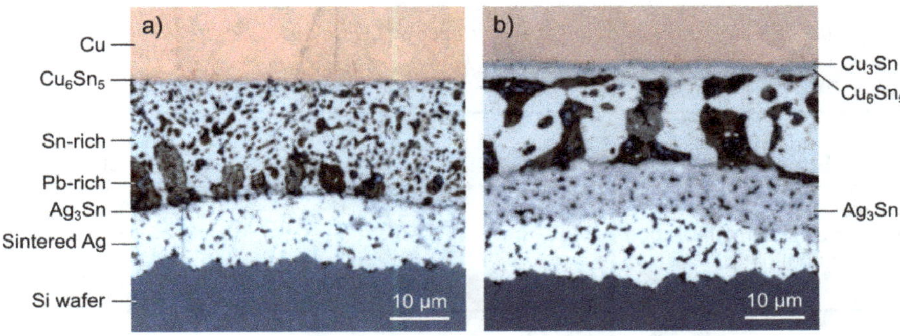

Figure 10.8 *Optical microscope image of a metallographic cross section of a Sn60Pb40 solder-coated copper ribbon soldered to the sintered front side silver busbar of a silicon solar cell. Solder joint directly after the soldering process (a) and after isothermal ageing for 500 h at 130 °C in nitrogen atmosphere [16].*

During soldering and also after cooling, the microstructure of the solder joint changes its appearance depending on time and temperature. Figure 10.8 shows a cross-sectional microscope image of a copper ribbon soldered to a silver busbar of a silicon solar cell using Sn60Pb40 solder. Directly after the soldering process (Figure 10.8a), the Pb-rich (black) and the Sn-rich (grey) parts within the solder layer are distributed nearly homogenously with some coarse Pb-rich structures at the busbar interface. The Ag_3Sn phase at the interface to the silver busbar and the Cu_6Sn_5 layer at the solder–copper interface are visible. After isothermal ageing of 500 h at 130 °C in nitrogen atmosphere (Figure 10.8b), the Ag_3Sn phase has significantly increased to a thickness of around 10 µm. At the solder–copper interface, the Cu_6Sn_5 phase also shows an increased thickness up to around 1.5 µm and in addition a Cu_3Sn phase with some Kirkendall voids [19] that develop under the consumption of the Cu_6Sn_5 phase [20]. Within the solder layer, a consequent grain coarsening is visible. The Sn-rich and the Pb-rich structures tend to separate from each other reducing the phase surface area. The driving force for this process is the reduction of the Gibb's free energy.

10.1.1.6 Thermomechanical properties of the interconnection materials

The described materials used to establish the solar cell interconnection have different coefficients of thermal expansion (CTE). This induces thermomechanical stress in the interconnection during heating and cooling. Table 10.2 shows the CTE values of the involved materials.

The CTE of copper is 6.5 times higher than the CTE of silicon. This leads to a significant difference in expansion and contraction during temperature changes. Directly after the solidification of the solder following the soldering process, the

Table 10.2 Coefficients of thermal expansion (CTE) for the materials involved in the common interconnection process for silicon solar cells

Material	CTE (10^6/K)	References
Silicon	2.6	[21]
Silver front side metallization	10.4	[22]
Solder (Sn60Pb40)	25.2	[23]
Copper	17	[24]

material compound is joint and the different contraction of the materials during the subsequent cooling down to room temperature leads to thermomechanical stress.

Figure 10.9 shows the photo of a solar cell with three copper ribbons soldered on the silver busbars on the cell's front side. Due to the difference in CTE of copper and silicon, the solar cell shows a severe bow of ~2–3 cm. This is a visible indicator for the thermomechanical stress induced during the interconnection process and can be used as a measure of its magnitude [25].

Because of the symmetric arrangement of the interconnectors on front and rear sides of the solar cell, the tension is balanced and no deformation occurs for common solar cell strings. Nevertheless, the thermomechanical stress does exist and can be analysed, e.g., by FEM simulations. Rendler *et al.* found the tensile stress to be up to 145 MPa in ribbon interconnectors and 183 MPa in wires [26]. This exceeds the typical yield strength of the interconnectors and indicates that the copper material is subjected to plastic deformation, most probably resulting in cold hardening after the soldering process, regardless of the interconnection technology.

A direct contacting of metal fingers of the front side metallization with the interconnectors would have the advantages of eliminating the silver busbar to save silver and to reduce the recombination losses on cell level. The described thermomechanical stress makes it impossible, regardless of the mechanical stability of the interconnection because the small contact area between wire and contact finger does not

Figure 10.9 Image of a three busbar solar cell with soldered interconnector ribbons only to the cell's front side (Fraunhofer ISE)

resist typical shear stresses after the soldering process. Even for wire interconnection, certain solder pad sizes are required and their positions on the solar cell have to be considered [27].

Possible methods to reduce the stress in the interconnection are:

- the use of soft interconnectors with a low yield strength [28]
- the introduction of stress relief structures into the interconnectors, as it is suggested for back contact solar cells by Walter *et al.* [29] and presented to realize the direct interconnection of contact fingers of PERC solar cells by Rendler *et al.* [30]
- the use of low soldering temperatures (compare Table 10.1)
- the use of slow cooling rates after the soldering process [31].

10.1.2 Soldering process

To establish solder interconnections for silicon solar cells with an industrial soldering process, some general conditions and process details are required:

- positioning of the joint partners → precise alignment of solar cell busbar and interconnector
- intimate contact between the joint partners → nearly eliminated clearance between interconnector and busbar surface during the soldering process
- sufficient wettability of the surfaces of the joint partners by the liquid solder
- effectual coordination of soldering time and soldering temperature on the basis of the requirements of the used materials (flux, solder and solar cell metallization).

In PV module production, automated stringer machines usually perform the solar cell interconnection. The throughput ranges from ~1 000 to 5 500 solar cells per hour. This corresponds to a fast interconnection process with soldering times below 2 s per cell. In machines with a very high throughput, a parallel interconnection of two or more cells is realized.

The industrial interconnection process is composed of the following process steps: (a) separation of the solar cells from a cell supply stack, (b) check for cell breakage, (c) application of flux (to the cell or to the ribbon), (d) drying of the applied flux, (e) pull-out of the interconnector from the spool, (f) cutting of the interconnector to the desired length, (g) stretching of the interconnector, (h) imprint of the crimp to the ribbon, (i) placing the interconnector onto the conveyor belt, (j) alignment of the cell on the belt on top, (k) positioning of the second interconnector on the cell, (l) holding down the interconnector during the soldering process, (m) pre-heating of the solar cell and the interconnector, (n) heating the solder on the interconnector above the melting temperature, (o) post-heating of the string after the soldering process to regulate the cooling ramp, (p) string pick-off from the conveyor belt.

Depending on the machine concept, several of these processes can be performed in parallel. Usually, the solar cell supply for the stringer machines is realized by manual or automated feed of cell stacks in boxes. The cell at the top of the stack is picked up by a vacuum gripper and moved to the breakage control camera under backlighting. To ensure that only one cell is picked from the stack, separation is mostly assisted by an airstream. The camera control rejects cells with broken edges and recognizes the position of the metallization layout. To ensure sufficient wetting of the busbar by liquid solder in the soldering process, the busbars are sprayed with flux on both sides of the solar cell. In some machine concepts, the interconnectors are alternatively driven through a flux bath. However, a sufficient amount of flux needs to be distributed homogenously and remains on the busbar or the interconnector surface to be available for activation during the soldering process. Mostly, the fluxed cells are heated on a hotplate or the fluxed interconnectors by hot air to evaporate the solvents and force the remaining solid components to stick on the surface in crystalline form.

The interconnectors are provided on spools and have to be unrolled by clamp and pull mechanisms. After cutting to the desired length, the interconnectors are stretched by usually around 0.5% of their initial length to straighten them and make them easy to handle. A crimp is imprinted by a stamp. The crimp avoids crack formation at the cell edge in the interconnector transition from the front side of one cell to the rear side of the next cell. The interconnector handling in the stringer machine (e.g. the transportation by deflection rollers) and the stretching affects their yield strength, particularly for interconnectors with low initial $R_{p0.2}$ values. Depending on the implementation of the interconnector handling in the stringer, the resulting yield strength of the interconnector on the cell can be significantly higher than the yield strength labelled on the spool by the manufacturer.

The interconnector and the solar cell are united on a conveyor belt by a pick and place process to realize the serial interconnection. The precision of this process is essential for high interconnection quality and becomes more challenging for higher numbers of smaller busbars and interconnectors. After the positioning of cell and interconnector, a holding down device keeps the parts in place during the soldering process. The assembly is transported over several fixed hotplates to pre-heat the solar cell and the interconnectors in a stepwise process up to ~150 °C. The pre-heating and post-heating steps enable a low-temperature difference for the final soldering step and a fast and precise process. The remaining temperature difference between pre-heating and the melting point of the used solder is closed by the heat source in the soldering unit. Several soldering technologies have been used for solar cell interconnection in industrial stringer machines:

- contact soldering
- induction soldering
- hot air soldering
- laser soldering
- infrared soldering.

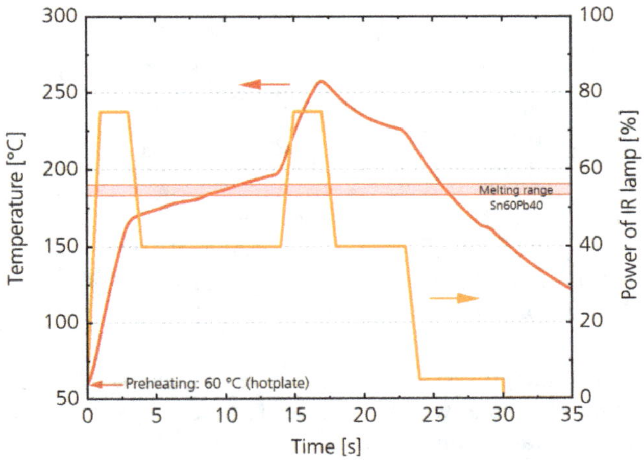

Figure 10.10 *Diagram of the power setting at the infrared lamps (orange) and the resulting temperature curve on the solar cell (red) over time (Fraunhofer ISE)*

Today, most of the stringer machines use infrared soldering. This means that the IR light is coupled into the solar cell and the cell is heated. Heat transfer from the cell to the interconnector indirectly heats the solder. The solar cell temperature is controlled by continuous temperature measurements on the cell surface using pyrometers. For a correct measurement, the positioning of the pyrometer needs to consider reflected infrared heating rays. Additionally, the emission value has to be adapted to the measured cell surface. The measured temperature is used as control signal for the infrared lamps. Figure 10.10 shows the power setting of the infrared lamps (orange curve) and the resulting temperature curve on the solar cell (red curve) for an infrared soldering process on a hotplate at a laboratory set up with a pre-heating of 60 °C. The soldering duration is ~15 s in this example. For industrial interconnection, solder durations <2 s are used. Within this time, the solder is liquid and wets the solar cell metallization. To ensure a reliable interconnection, the set soldering temperature has to be ~50 K higher than the melting temperature of the used solder. The melting range of the used solder alloy (Sn60Pb40) is marked in red. The soldering process can be divided in the pre-heating phase, the soldering phase and the cooling phase. In the pre-heating phase, the solar cell is heated up to a set-temperature of ~150–200 °C. In the soldering phase, the temperature is increased to reach the desired soldering set-temperature, which is 250 °C in the example in Figure 10.10 and ensures the exceeding of the liquidus temperature of the solder. In the cooling phase, the infrared lamps are switched of and the solar cell cools down, whereas the temperature falls below the solidification temperature.

In general, the soldering process using tin–lead solders is comparatively easy to establish due to a relatively wide process window. The convenient flow characteristics of lead excuse small inaccuracies in the time-temperature setting. For lead-free

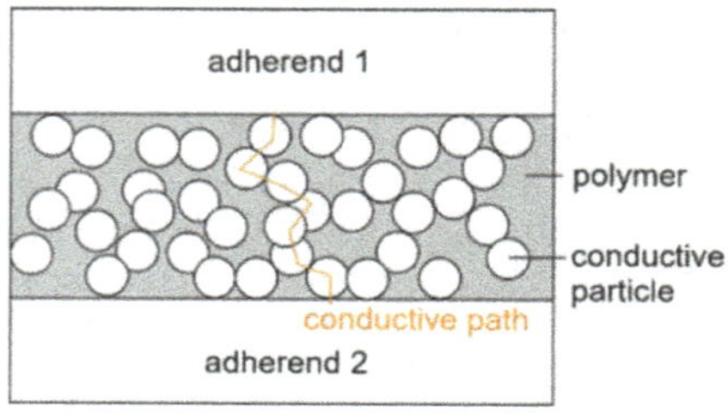

Figure 10.11 *Schematic drawing of the operation principle of ECA interconnections (not to scale)*

tin–bismuth solders, the process settings are more challenging. The reduced ability of liquid bismuth to flow and wet the surfaces requires higher demands on precise positioning and temperature–time settings.

10.2 Electrically conductive adhesives

10.2.1 Introduction

ECAs are polymer-based pastes or tapes that provide electrical contact and mechanical bonding at the same time. Mechanical adhesion is achieved by the polymer and electrical conduction by the added electrically conductive particles called fillers. A schematic drawing of the theory of operation is shown in Figure 10.11. The first use of ECAs in electronics was reported in the 1950s [32]. Since then they have been applied successfully in hybrid, die-attach and display assembly [33]. ECAs are well-established materials in small passive chip attachment in automative electronics, for RFID tags or LED manufacturing [34]. Furthermore, they were candidates as solder replacement for hazardous tin–lead solders in electronics but other advantages, such as low processing temperatures compared to soldering processes, no-flux, no-clean and simple processing, made them attractive for flip-chip bonding and surface mount technology as well [35].

ECAs were first introduced in 2001 as an interconnection technology for solar cells [36, 37]. They have been proposed as an alternative to conventional ribbon soldering to remove toxic tin–lead and to reduce thermomechanical stress in the interconnection joint. The cross-sectional view of an interconnection scheme is shown in Figure 10.12. The ECA (yellow) is applied on both sides of the solar cell metallization (light grey). The ribbons (red), made of a copper core with a thin silver coating

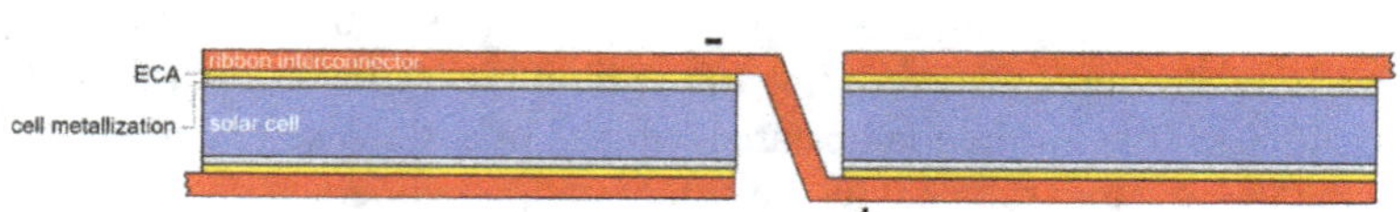

Figure 10.12 *Solar cell series interconnection by a copper ribbon using an ECA instead of soldering (not to scale)*

to serially connect the cells (blue) to strings, are placed on the ECA. The stack is cured with moderate pressure using a heat source.

Using ECAs instead of conventional solder allows to decrease the process temperature from around 240–300 °C to a range of 120–180 °C. This reduces the thermomechanical stress that results from the thermal expansion mismatch of silicon and copper ribbon. Since the ECA is a polymeric material it is softer and therefore more compliant than a solder bond and can compensate mechanical stress to some extent. Thus, ECAs are a potential solution for processing thin and fragile solar cells [38, 39]. However, due to the additional costs of ECA, which are mainly driven by their silver content, they could not replace soldering for conventional solar cell interconnection in the past.

Nevertheless, ECAs play an important role in applications where soldering is difficult or even impossible. The materials have been studied and used for the development of back-contact solar cells and PV modules, particularly the metal-wrap-through concept [40]. Additionally, they can be applied to metallization, which does not easily adhere to the solar cell surface. For instance, low-temperature metallization pastes for silicon heterojunction (SHJ) solar cells often do not stick well to the wafer.

Today's most dominant areas for ECA application in the solar industry are the interconnection of SHJ solar cells. The other application is the shingle interconnection. Both approaches will be described in the following sections.

10.2.2 Applications of ECAs for solar cell interconnection

10.2.2.1 Interconnection of silicon heterojunction solar cells

The benefits of ECAs, such as low-temperature processing, mechanical flexibility and process simplicity, are especially attractive in applications where the temperatures of the soldering processes cannot be used without severe cell degradation or where the stiff solder bond cause damages by temperature changes. They have been proposed for the interconnection of SHJ solar cells [41, 42], because the involved amorphous silicon layers can degrade at temperatures >200 °C (conventional lead-based soldering is operated at approximately at 240 °C). The cells are thinner (<180 μm) and often more fragile. Additionally, low-temperature metallization pastes show lower adhesion to the cell surface than conventional fired metallization of PERC solar cells. This is due to the missing glass phase formed in a conventional high-temperature sintering process. Although there are commercially available low-temperature metallization pastes, which allow also to be soldered [43], they can suffer from other disadvantages, such as reduced conductivity inconvenient printability or higher costs. A prominent alternative for the interconnection of SHJ cells is the Smart Wire Connection Technology [44]. It was recently announced to be solely used by Meyer Burger in the future [45]. Thus, ECA technology could now gain in attractiveness for other SHJ cell manufacturers.

The company 'teamtechnik' commercialized a fully automated ECA ribbon stringer for the interconnection of SHJ solar cells [46]. Several companies use this technology to manufacture PV modules based on SHJ cells [47, 48]. Hevel LLC claims a module power of more than 390 Wp with 22.8% efficient SHJ solar cells featuring a five-busbar ECA technology used to realize a module configuration with 72 cells. However, they also admit that costs of interconnection are high and a conventional process would be preferable [49].

The basic concept of the interconnection of SHJ cells is shown in Figure 10.12. Recent results about aspects regarding industrialization of this technology have been reported in our earlier publications [50, 51]. The core features of this technology are that the ECA is screen printed on the cell metallization. The metallization grid comprises of two framing fingers (glue fingers) with a distance of 0.7 mm perpendicular to the contact fingers at the busbar positions. They limit the bleeding of resin and offer electrical redundancy along the interconnection. The missing busbars save up to 100 mg of silver paste per cell depending on the number and original height of the busbars. The ribbon is coated with an approx. 1-μm-thin layer of silver and has a grooved surface structure, which increases the internal reflection in the module and the short circuit current by around 2% [52]. Accelerated ageing tests (600 thermal cycles, TC, from 85 to −40 °C and 1 000 h of 85 °C/85% RH 'damp heat', DH) as well as 1-year outdoor monitoring showed promising results [50].

Key challenges for ECAs used for ribbon interconnection of SHJ solar cells are:

- adequate peel strength
- fast curing speed
- reduced silver content for low material costs.

The latter can also be achieved if the silver coating around the ribbons could be removed and the printing pattern was optimized. Of course, the optimizations must meet the necessary IEC certification standards IEC 61215 and IEC 61730.

As tandem solar cell technologies based on SHJ bottom cells (e.g. Perovskite-Si-SHJ [53]) will enter the market in the coming years, the ECA ribbon technology will potentially be used to integrate these even more temperature-sensitive devices into modules.

10.2.2.2 Shingle interconnection

ECAs gained much attention during the market introduction of PV modules with shingled solar cells. Industrial manufacturers for shingled solar modules are, e.g., Sunpower and Solaria [54, 55]. The solar cells are cutted by laser systems into rectangular pieces having the length of the full wafer and a width of 22–31 mm. The pieces are serially connected by slightly overlapping at their long sides and an intermittent layer of ECA as a bonding material between the shingles (see Figure 10.13a,b). This concept maximizes the active area in the PV module, does not require interconnector ribbons and, hence, results in more efficient PV panels with valuable aesthetics [56, 57]. The shingled solar cell interconnection was originally invented several decades ago [58, 59]. It has been used for space applications, solar cars and for thin-film PV interconnections [60].

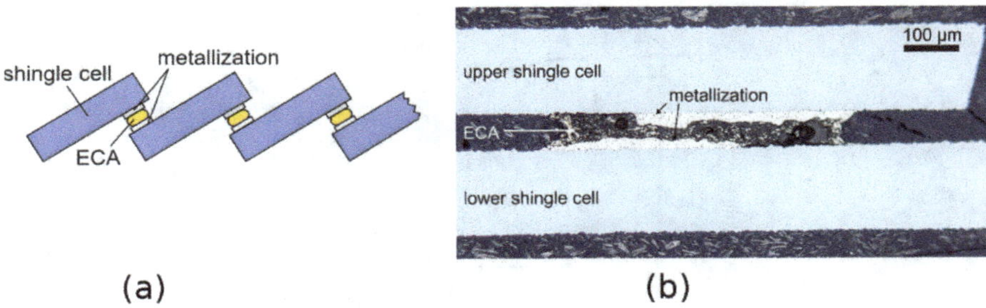

Figure 10.13 (a) Schematic drawing of the shingle interconnection concept and (b) metallographic cross section of shingled solar cell joint

The function of ECA between the shingles is to provide electrical connection and the necessary mechanical flexibility for the entire string. As silicon is very brittle, the ECA allows bending of the shingle strings without cell breakage when mechanically loaded in the PV module. Furthermore, ECAs can absorb shear deformations during thermal cycling of the PV module without bond rupture.

The challenges of shingled solar cells are the minimization of edge recombination losses as well as an integrated optimization of cell metallization considering the ECA bond design, its mechanical and electrical properties and the module encapsulation used [61–64]. A further optimization of the active area in the PV module is offered by the matrix shingle concept in that the shingles are positioned similar to a masonry wall [65, 66].

In addition, shingle interconnect ECAs need to be optimized with regard to fine-printing or dispensing capability to allow high precision automated stringing with low material consumption, high throughput and low reject rate [63]. Automated shingle stringer equipment is developed by, e.g., Applied Materials Inc. [67] or M10 Industries AG.

10.2.3 ECA material systems

10.2.3.1 Operation principle and conduction types

Pure epoxy, acrylate or silicone materials have a resistivity of 10^{15}–10^{12} Ω cm and are practically insulators. When mixed with electrically conductive particles, the so-called fillers, above a certain volume fraction, the resistivity of the composites drops several orders of magnitude. The necessary volume fraction of fillers to achieve conductivity of the composite is called the percolation threshold. This is where the first conductive path between the fillers is formed [68]. The percolation threshold depends on many factors, such as metal type, particle size [69], size distribution [70], shape and alignment of particles [32] and the adhesive thickness [71]. For a silver-filled epoxy, the threshold is in the range of 25–30% vol (75–85% wt) [72]. ECAs with such high filler content are designed to conduct current in all directions and are called isotropically conductive adhesives (ICAs). Sometimes, the term isotropically conductive paste is also used. Figure 10.14. illustrates the relationship

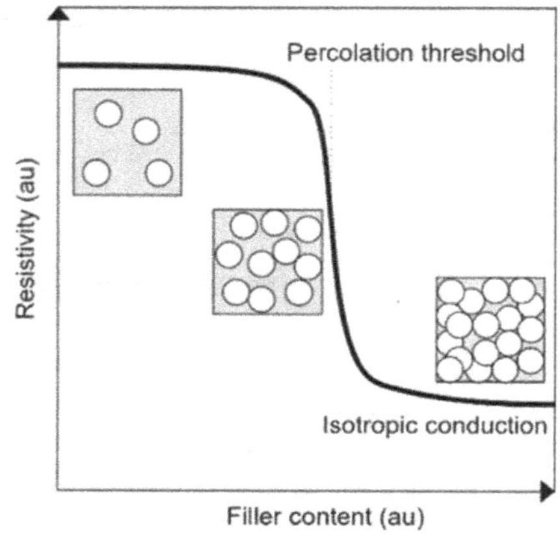

Figure 10.14 Relationship of filler content and resistivity of a polymer filled with conductive particles. (Adapted from Morris and Wang [68]).

between filler content and resistivity. ICA is presently the most common category of adhesives used in PV module manufacturing.

A prerequisite for the development of mechanical adhesion and electrical conduction is the curing process, which is initiated and driven mainly by heat. The uncured, pasty ECA composed of monomers or prepolymers is transformed to a thermoset or elastomer by crosslinking reactions (see Section 10.2.4.1). Moreover, the filler particles are forced into closer contact and form a three-dimensional electrical network. Other than low-temperature metallization pastes for SHJ solar cells, ECAs are reactive materials, and usually no substantial solvent evaporation is involved in the curing process.

Conduction of the resulting material is influenced by two processes occurring in parallel to the crosslinking reaction. The first is the thermal decomposition of lubricants around the metal fillers [73]. These lubricants are required to maintain a uniform dispersion of fillers in the paste. However, they form a thin insulating barrier around the particles that need to be removed during the curing process. Second, the volume shrinkage of the polymer during crosslinking further reduces the distance between the particles and lowers the final resistivity [74].

There are ECAs that work below the percolation threshold. They only conduct current in *z*-direction and not laterally. These materials are called anisotropic conductive adhesives (ACA). Since they often come as films or tapes, they are also called anisotropic conductive films (ACFs) or simply conductive films (CFs). The filler content is in the range of only 5–10 vol% [75]. The anisotropic conduction mechanism is achieved by mechanically deforming and locking the particles between upper and lower substrate using heat and pressure. It was shown that CFs can be used to interconnect solar cells in PV modules [76–78]. Due to the reduced filler content, ACAs can potentially be more cost-efficient than ICAs. However, the additional requirements on filler particle characteristics (uniformity of size and distribution within the polymer) as well as processing conditions (i.e. time, pressure) could outweigh the practical cost benefits.

10.2.3.2 Base polymers for ECAs

Besides the categorization according to the conduction mechanism, an alternative way is to consider the group of polymers that the ECA is made of. ECAs exist on the market based on epoxies, acrylates, silicones, polyurethanes, polyimides and hybrids of those materials. Among those, the first two are most significant for solar cell interconnection nowadays. Epoxies and acrylates have in common that their rheology and mechanical properties can be adjusted relatively easy according to the specific application. Both polymer types make ECAs that cure in seconds with a stiffness ranging from very soft and rubbery to rather rigid and strong. Moreover, both types do neither require a considerable amount of solvents nor form unwanted by-products during curing.

Epoxies can form sufficient bonding strength and are resistant against heat and moisture as well as chemical attacks [79, 80]. The curing reaction mechanism is called polyaddition, because two chemically different monomer systems, the resin and the hardener, are combined to result in the solidified polymer. Hundreds of different resin and hardener combinations are possible. Hence, in principle, a vast amount of resulting materials can be obtained. The curing can be tailored to take place within seconds or minutes and is usually slower than for acrylates.

Acrylates cure rapidly once the reaction is initiated because of the underlying free radical chain polymerization. The main strength of acrylates is said to be the ability to bond unprepared metals fast [81]. Processing seems to be more sensitive to moisture and temperature variation [64]. Acrylic ECAs are often proprietary formulations that contain a variety of ingredients to control curing, handling, shelf life, physical properties and adhesion.

Silicone-based ECAs are in between organic and inorganic compositions as the carbon backbone is replaced by silicon–oxygen–silicon bonds (siloxanes). The main advantages are temperature stability to more than 200 °C and mechanical flexibility to a temperature below −50 °C. Furthermore, they are resistant to moisture and chemical attacks [82]. The main weakness of this material type is the long curing time of several minutes at elevated temperature that makes it unsuitable for many high-throughput stringing applications.

10.2.3.3 Fillers

The mostly used filler material is silver. It offers highest conductivity at moderate costs, is easily malleable and its oxide is conductive, which makes these ECAs relatively resistant to ageing [83]. The particles have the form of flakes or granules in a size of 1–15 μm. An attractive alternative to that is the use of silver-coated copper [79]. However, due to higher manufacturing costs these fillers may not be cheaper than pure silver. The use of pure copper as a filler material requires a thorough removal of the oxide before mixing into an ECA and additional organic-based coatings to reduce corrosion [72, 84].

The use of alternative filler materials, such as carbon nano tubes to replace silver, is reported by Zemen *et al.* with promising first results for PV cell interconnection [85]. Lastly, nickel might also be a relevant candidate as filler, but the limitations are its hardness and limited malleability. However, products with nickel are also available that possess lower conductivity [35].

10.2.4 Mechanical and electrical properties of cured ECAs

10.2.4.1 Dynamic mechanical behaviour

The mechanical properties are dominated by the polymeric nature of the materials. Polymers are organic or semi-organic macromolecules and consist of entangled molecular chains with mostly repetitive structures. The molecular chains are made of carbon, oxygen, hydrogen, nitrogen, silicon, chlorine, fluorine and sulphur [86].

Polymers can be categorized in three groups: thermoplastics, thermosets and elastomers. Thermoplastics only contain physical bonds and no chemical crosslinks between the molecular chains. They are solid at room temperature and can repeatedly be molten at elevated temperatures and solidified by cooling. Currently, thermoplastic ECAs are not used for solar cell interconnection.

Thermosetting polymers contain a high density of crosslinks in addition to the physical bonds, which makes them hard and brittle at room temperature. They cannot be reshaped or molten but softened at elevated temperature due to the reduced physical bonding strength that acts simultaneously to the chemical bonding mechanism of the crosslinks. The crosslinks prevent the material from purely flowing. Many commercially available ECAs belong to the group of thermosets.

Elastomers have a low density of crosslinks. They are elastic at room temperature but, like thermosets, cannot be molten by heat. Thanks to their ability to easily deform under mechanical load elastomers make an important group of ECAs for solar cell interconnection.

Polymers have three different aggregate states:

- energy-elastic or glassy state
- entropy-elastic or rubbery state
- viscous state or flow state.

A simplified view on the mechanical elasticity of polymers with respect to the operating temperature is shown in Figure 10.15. Polymers in the energy-elastic state are hard, brittle and do not allow significant elastic deformation. Moving, sliding and turning of the molecular chains are practically impossible. Thermosets at room temperature are in this state. In the entropy-elastic state, polymers are elastically deformable to some extent. Polymer chains can move, slide and turn. The thermal expansion coefficient is approximately one magnitude higher than in the energy-elastic state. Elastomers operate in this state. The transition between energy elastic and entropy elastic is called glass transition region. This is where the steepest change of mechanical properties takes place. The glass transition temperature is the

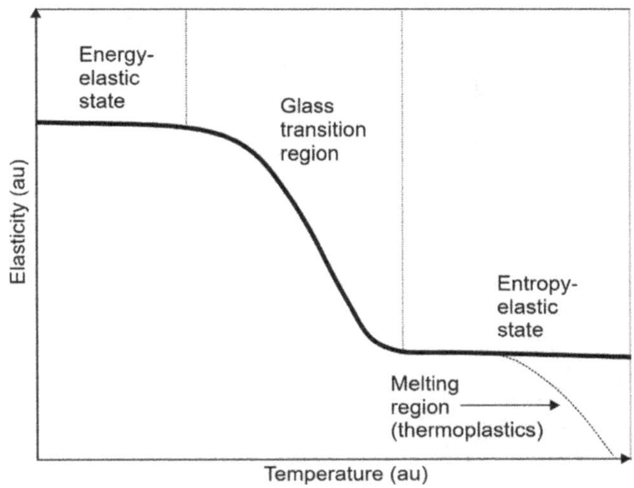

Figure 10.15 *Diagram of the qualitative change of elasticity of polymers with respect to their aggregate state*

characteristic temperature within this region because it determines its flexibility at operating temperature, thermal expansion and electrical properties.

The mechanical properties of ECAs are often studied with frequency and temperature dependent dynamic-mechanical analysis (DMA) rather than quasi-static testing. Figure 10.16 shows DMA curves of ECAs (A–D) with different temperature dependent mechanical behaviour. The cuboid samples were fully cured and measured in tensile mode at 1-Hz-excitation frequency. The storage modulus is a simplified estimate for the tensile modulus in a quasi-static tensile test. For reference, PET, as a typical material for plastic bottles, has a tensile modulus of around 3 000 MPa at room temperature. Low-density polyethylene is a common material for disposable plastic bags and has a tensile modulus of around 200 MPa.

ECA A in Figure 10.16 is a typical thermoset and has a constant storage modulus of around 2 500 MPa over the considered temperature range of −25 to 80 °C. The glass transition region is above 80 °C (not shown here). The material is stiff and relatively brittle in many situations during operation of the PV module. ECA B is also rather stiff (~2 000 MPa) up to room temperature, but the glass transition sets in at approx. 40 °C. At the higher end of a module's operating temperature material B is soft and compliant (~200 MPa). The glass transition region of ECA C is around 25 °C. The material is soft (200 MPa) at 40 °C but rigid at 0 °C (~3 000 MPa). Eventually, material D is an elastomer with mechanical flexibility in the entire considered temperature range. The storage modulus appears to have a plateau at room temperature. The material may even have two glass transition regions (indicative for a co-polymer), one at around −20 °C and the other at 50 °C.

Each ECA shows different mechanical behaviour, which depends on the operating temperature, on the frequency of the mechanical load, the duration and its

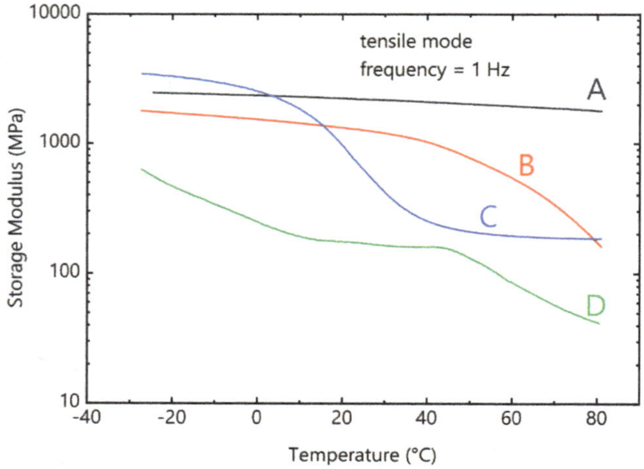

Figure 10.16 *Diagram of the storage modulus – as an estimate of the tensile modulus – of four ECAs based on different polymers measured with DMA in tensile mode at a frequency of 1 Hz. The labels A–D indicate different ECAs with individual temperature-dependent mechanical behaviour.*

intensity. To obtain an accurate understanding requires case-by-case analysis, sophisticated dynamic thermal-mechanical analysis of the materials and viscoelastic modelling [87–89]. Future research in this field needs to address the relationship of mechanical ECA properties and long-term stability of the PV module.

10.2.4.2 Adhesion of ECA bonds

Mechanical strength of an ECA joint is determined by cohesion of its ECA bulk and adhesion of the glue to the bonding surfaces. While cohesion is attained through covalent bonds of the backbone, crosslinks and Van der Waals forces acting between the intertwined molecules, the adhesion phenomenon is another complex interaction of physical, chemical and thermodynamical effects. There is no single explanation for the different modes of adhesion so that commonly different superimposing concepts are adduced to understand the adhesion phenomenon. Baldan summarizes six models for adhesion: the mechanical interlocking model, the electrostatic theory, the weak boundary layer theory, the adsorption theory, the diffusion theory and the chemical bonding theory [90]. The adsorption theory, stating that adhesion is a result of attracting primary and secondary molecular forces, is most widely applicable [91].

In electronics, the adhesion strength of ECAs is often analysed using lap-shear or die-shear tests. The range for lap-shear strength ranges from 3 to 10 MPa [92] and sometimes a lap-shear strength of 20 MPa is reported [83]. Bosco and Springer measured the debond fracture energy of an ECA with a single cantilever beam test. While in a dry environment it is stable at around 350 J/m^2, the fracture energy declines to 100 J/m^2 or less in humid environments [93].

For PV modules, interconnects are commonly evaluated with peel tests. Interconnector ribbons are bonded to the solar cell metallization and are then peeled off in 90° or 180° angle at 50 mm/min. In many cases, an inferior peel strength for ECA interconnections of below 1 N/mm is measured (force normalized to bond width) [94–96]. Although this result often does not correlate to the evaluated stability of the so produced PV modules in thermal cycling tests (see [97]) the reduced adhesion compared to solder joints can be a challenge during automated handling of cell strings in industrial production and, generally, for the acceptance of ECA technology in the PV industry. In the last years, improving the peel strength has been addressed by the ECA manufactures so that nowadays ECAs achieve the required peel strength of 1 N/mm at least partially [50, 98, 99].

Besides fully curing the ECA, the peel strength of an ECA bond depends on the applied ECA amount and bonding pattern [100]. Reducing the amount usually also reduces the peel strength but does not necessarily influence the electrical properties. This highlights the potential to reduce the material consumption. Moreover, the impact of mechanical and electrical bond failure on PV module performance depends on the location of failure within the overall bond. For ribbon-interconnected solar cells, a bond failure at the edges is more critical than a failure in the centre [101]. Obtaining an adhesive with sufficient peel strength for automated handling and long-term stability requires testing of many different products, optimizing the ECA bonding patterns and finding a trade-off with the overall material consumption.

10.2.4.3 Electrical properties

The electrical characteristics of ECA bonds are a superposition of the ECA's bulk or volume resistivity and the contact resistivity at the interfaces to the substrates. The measurement of the volume resistivity is rather straight forward by using a four-point probe setup on a cured film of ECA [102].

Highly silver or silver-coated copper filled materials have a resistivity of 10^{-4}–10^{-3} Ω cm or even slightly lower [50]. For reference, the resistivity of Sn60Pb40 solder is 1.5×10^{-5} Ω cm. The temperature coefficient of the volume resistivity of a silver-filled epoxy-based ECA was shown to be in the range of 2×10^{-3} to 3×10^{-3} K^{-1}, which is close to the temperature coefficient of elementary silver [103]. This might be different if the silver content is reduced.

There is no linear relationship between filler content and resistivity. As was shown in Figure 10.14, the polymer is insulating without fillers, and once the filler content exceeds the percolation threshold the resistivity after full curing reaches a stabilized value. It cannot be altered much by adding more filler particles.

However, the curing temperature itself influences the resulting resistivity of the ECA. Curing at elevated temperature leads to more pervaded curing, polymer shrinkage, internal stress of polymer chains, effective lubricant decomposition, closer particle contact and therefore better conductivity [103]. Usually, the drawback is a more rigid bond line and higher thermomechanical stress due to the temperature difference. The curing temperature as a method to control the resulting resistivity of an ECA was used to determine the impact on PV module performance. It was shown that a moderate resistivity of around 10^{-3} Ω cm is sufficiently low for PV modules [101]. High conductivity and high filler content can be sacrificed for material and cost savings. There seems to be already a tendency of ECA developers to introduce materials with lowered filler content that have semi-anisotropic electrical properties. They are significantly less conductive laterally, but still a proper electrical bond in *z*-direction on solar cells can be fabricated [100].

The contact resistivity subsumes any interfacial oxides or imperfections that impede the current flow from the bulk ECA into the substrate. The situation in a real ECA bond is quite complex because different substrate materials may be involved (for instance cell metallization and ribbon surface) and a contribution of the bulk resistivity is difficult to exclude from practical measurement setups. Several methods for the measurement of the contact resistivity have been proposed in the past. The most commonly used approach is the transfer length method (TLM) ([104] as cited in [105]). Furthermore, the contact end resistance method (CER) can be applied [106]. A rather simple 'stacked Greek cross' test structure was also used [79]. It is beyond the scope of this chapter to discuss the details of these measurement approaches. A comparison of TLM and CER for one adhesive can be found in [107]. The different measurement approaches led to varying and inconsistent contact resistivity values for ECAs. Table 10.3 aims to give an overview of the range of reported values.

Mostly, the contact resistivity is very low and stable on noble metals, such as silver, gold, palladium and platinum. Moreover, the contact resistivity on these surfaces can be as low as for soldered contacts [112]. On the contrary, bonds on non-noble surfaces, such as copper, tin, tin–lead or aluminium, are affected by increased contact resistivity [113, 114]. One reason is the natural presence of dielectric oxide layers on these metals.

Several 'acceptance criteria' for the contact resistivity have been proposed in the past to guide the selection of suitable material combinations. Recommendations are below 0.1 or 0.5 m Ω cm^2 [98, 101]. Of course, this threshold depends on the total interconnected area in the architecture of the final bond. If the contact area is rather large, a higher contact resistivity may be allowable. Furthermore, Zwolinski defines a criterion in terms of stability of the contact resistance: it is required that the junction resistance may not increase by 20% after 500 h 85%/85 r.h. exposure [92].

10.2.4.4 Moisture ingress and mechanical/electrical ageing

Water is regarded as the most problematic substance for adhesive joints. Once moisture enters the joint it can lead to the following changes [115]:

- reversible plasticization of the polymer matrix
- irreversible hydrolysis, cracking and crazing
- displacing the adhesive at the interface or hydrating the metal surfaces
- reversible swelling.

Table 10.3 Overview of reported contact resistivity values for ECA interconnection of silicon solar cells from literature

Adhesive	Adherend 1	Adherend 2	Measurement method	Contact resistivity (m Ω cm^2)	References
Not specified	Ag-coated ribbon	Low-temperature Ag busbar	TLM	0.1	[42]
Ag-filled acrylate	Ag-coated ribbon	Sintered Ag busbar	TLM	0.09	[39]
Ag-based ECA	Ag-coated ribbon	Sintered Ag busbar	TLM	Ag: 0.29–0.55 CNT: 3.9	[98]
Ag-filled silicone	Ag-coated copper OSP-treated copper	Not specified	Not specified	Ag: <0.1 Cu: <0.5	[108]
Two ECAs	Ribbon (coating not specified)	Sintered Ag busbar	TLM	0.07–0.6	[109]
Four ECAs	Evaporated Ag	Sn(Pb)Ag	TLM	0.1–1.5	[110]
Ag-filled acrylate Ag/Cu-filled epoxy Solder-filled epoxy	Sn-coated ribbon	Ag grid fingers	TLM	Ag-filled acrylate: 0.18 Ag-filled epoxy: 0.11 Solder-filled epoxy: 0.12	[111]
Ag-filled acrylate	Ag-coated ribbon Sn-coated ribbon Cu-coated ribbon	Ag-coated ribbon Sn-coated ribbon Cu-coated ribbon	Stacked Greek cross	Ag: 0.01 Sn: 0.11 Cu: 5.14	[101]
Two different ECAs	Ag-coated ribbon Cu-coated ribbon	Ag-coated ribbon Cu-coated ribbon	Stacked Greek cross	Ag: 0.0003 Cu: 0.0006–0.03	[50]
Three different ECA chemistries	Ag-coated ribbon	Not specified	TLM	0.10–0.81	[64]
One ECA	Sintered Ag paste	none	TLM TLM-CER	0.12–0.36 3.04–6.83	[107]

Plasticization depresses the glass transition temperature of the material and reduces its modulus and strength [116]. The adhesive is displaced if water diffuses along the interface between adhesive and substrate. There it ruptures Van der Waals forces and may form a weak boundary layer due to hydrolysis of the materials. Furthermore, swelling may lead to a loss of cohesion and deterioration of the electrical resistivity as the distance between the conductive particles in the bulk of the material could increase. Dry heat ageing alone may lead to post-curing crosslinking, which can cause an increase of the glass transition temperature of the adhesives along with the stiffness.

To better understand the effect of heat and moisture on the bulk electrical characteristics, commercially available ECAs were analysed in terms of the change in resistivity during a 1 000-h DH exposure test for PV modules. Three out of four unencapsulated ECAs did not show an increase in resistivity despite the harshness of the test conditions. One ECA showed significantly increased resistivity and also clear signs of embrittlement after exposure [50]. Pitta Bauermann *et al.* found indications that water uptake of bulk ECA in an acetic acid solution (which emulates the decomposition of ethylene–vinyl acetate foil in the PV module) could relate to a significant performance drop of PV modules with this ECA under DH conditions [117]. Hence, bulk resistivity degradation due to moisture uptake and chemical decomposition can be a relevant failure mode for some ECAs.

Besides bulk degradation, many of the frequently observed failure modes in ECA bonds, such as oxide formation, corrosion, cracking and delamination, are associated with the presence of non-noble surface metals [118]. Contact resistance instability caused by oxidation of surfaces may be one reason. However, galvanic corrosion is considered as another major failure mechanism when two dissimilar metals with different electrochemical potential, an electrolyte solution such as ions in water, oxygen and elevated temperatures, are present [83, 119].

Non-noble surface metals do not only impair the stability under hot and humid conditions they also cause a more likely failure in thermal cycling tests [120–122]. Schneider *et al.* studied an ECA with respect to its stability on differently coated ribbons for PV module integration [78, 99]. When the coating was silver, this ECA showed no additional power loss and good stability in TC and DH tests comparable to soldering. While copper seemed to be an interesting alternative with similar performance, tin-coated ribbons were prone to degradation.

In our past research, we have also shown stable module performance of one ECA in combination with copper, but a second ECA led to noticeable degradation in TC tests (600 cycles) and DH exposure (1 000 h) [50]. In an earlier study, we successfully used tin-coated ribbons in combination with an Ag/Cu-filled epoxy. We demonstrated a power degradation of maximum 3% in TC 200 and DH 1 000 tests, which was still acceptable. However, particularly in TC, the degradation was higher than the modules with silver-coated ribbons (only maximum 1% degradation) [111]. Hence, it is very difficult to generally predict the performance of ECAs in combination with non-noble metal surfaces without experimental testing. Generally, degradation can be reduced when the ingress of moisture in the PV module is circumvented by a polyolefin encapsulant and a moisture barrier within the backsheet [99].

Eventually, the outgassing behaviour and chemical interaction of ECAs during lamination of PV modules and subsequent ageing were studied by Oreski *et al.* [123]. They did not find detrimental interactions of ECAs with the module encapsulation foil. Moreover, the phenomenon of silver migration is mostly not considered a problem in silver-filled ECA interconnections as summarized by Jagt [118].

Stable module performance in extended climatic chamber tests with ECA-interconnected solar cells in various applications has been demonstrated many times when the surface metal was silver [39, 42, 49, 98, 124, 125]. A careful choice of the encapsulant materials is sometimes required to obtain good long-term stability [50, 64, 99, 126].

10.2.5 Processing aspects of ECA

Conductive adhesives are typically shipped as one-component products in jars or syringes frozen with dry ice. They need to be stored between −20 and −40 °C. The 'dry-ice-logistics' requires relatively short delivery times. The advantage is that sensitive mixing is not required by the PV module manufacturer.

Once the ECA is thawed, the pot life is around 24–48 h but can also be lower for some products. The materials are mainly not intended to be refrozen. Particularly, the viscosity is affected by prolonged exposure to room temperature. For instance, an increase of viscosity from around 30 000 to 60 000 mPa s (shear rate = 10/s, plate–plate geometry) was measured for one ECA [51]. Besides that, particle segregation may take place during room temperature exposure. The ECA manufacturers usually provide detailed handling instructions for their materials.

ECAs are applied by dispensing, screen or stencil printing, ink-jet printing and jetting [63, 79]. After the application process, resin bleeding on the solar cell surface is a known effect [36]. Currently, the amount of applied ECA for a five-busbar solar cell is between 50 and 100 mg per cell [100]. For a shingle joint, the industry aims for well below 3 mg per shingle [64]. These numbers depend, of course, on the density of the material and the printing pattern.

Full curing of ECAs is an important prerequisite for the formation of proper adhesion and electrical connection. From our experience, poor bonding strength is often a result of incomplete curing. The curing can be conveniently analysed by DSC and simulated as well [127].

Practically, curing of the materials can be done in a furnace, on hot plates and with additional infrared lamps. It is reported that microwaves can be used to cure thermosetting materials, and also photo-curable ECAs are possible [128, 129]. While in SHJ interconnection and shingling applications, a fast curing speed of below 30 s at below 180 °C is required, and ECAs for back-contact module fabrication are designed to cure along with the lamination process. They are optimized for curing in several minutes at approx. 150 °C.

ICAs do not require a high contact pressure during curing, but some pressure is helpful to keep the adherends in place and to assist the wetting of the surfaces. In contrast, ACAs or ACFs require a relatively high bonding pressure (2 MPa) and temperature (180 °C for 10 s) during the main curing step [76].

10.2.5.1 Comparison of ECA ribbon with tin–lead solder ribbon interconnection

Although it is not easy to directly compare conventional tin–lead soldering with ECA interconnection technology for PV modules, because either one will be used for different applications, Table 10.4. aims to help the reader to understand the main differences between the two approaches.

10.2.6 Summary

ECAs are polymers filled with conductive particles. The filler particle content must exceed the so-called percolation threshold to achieve isotropic conductivity of the composite. Due to their low processing temperature and softness, ECAs are especially suited for the interconnection of temperature sensitive solar cells like SHJ solar cells. Moreover, the mechanical flexibility makes ECAs the material of choice to interconnect shingle solar cells. ECAs are lead free and can be formulated from various base polymers like epoxies or acrylates and further chemically adjusted to meet different requirements for viscosity, temperature dependent elasticity, pot life, curing speed and electrical resistivity. The main drawback is additional costs due to their significant content of silver, which limits the use of ECAs to applications where soldering is impossible. ECAs offer an easy and reliable interconnection method for innovative new cell and module technologies.

This chapter explained the operation principle, general characteristics and main material groups of ECAs. The polymeric nature of the materials determines their mechanical response to heat as well as mechanical load. Viscoelastic theory is required to understand and model their mechanical behaviour. The electrical performance of the entire interconnection depends on the volume resistivity of the ECA bulk material but, sometimes even more important, the contact resistivity to the attached surfaces. While noble metals like silver are technically the best choice as a bonding surface, non-noble metals like copper or tin can cause significantly higher contact resistivity. Some ECAs can be adjusted to form a proper bond to a certain non-noble metal, but generally this is more difficult to achieve.

It has been shown many times that ECAs can achieve a similarly good interconnection for solar cells in terms of bonding strength, electrical performance and long-term stability as soldered joints. It is important to understand the main material changes caused by ageing and leading to possible joint degradation. Degradation is often associated with the ingress of moisture into the ECA bond. Possible degradation modes are plasticization, water diffusion and delamination at the interface of the ECA to the substrate, hydrolysis of the ECA itself or the substrates, bulk swelling, thermomechanically induced crack formation, corrosion and electrochemical

Table 10.4 *Comparison of the main material and process properties for ECA and solder interconnection*

	ECA	Solder	References
Process temperature (°C)	120–180	220–300	[39, 130]
Process time (s)	5–30 ('snap cure' ECAs)	1–2	[39, 50]
Material costs (EU/kg)	500–2 000	~10	[131, 132]
Bulk resistivity (Ω cm)	10^{-4}–10^{-3} (isotropic ECAs)	1.5×10^{-5}	[50, 95, 133]
Contact resistivity (TLM) (m Ω cm²)	On Ag surfaces: 0.07–0.8 Non-noble surfaces: variable	0.02–0.08	Table 10.3., [109]
Young's modulus (GPa)	Elastomers: 0.01–0.5 Thermosets: 5–10	31–39	[134, 135]
Shear strength (MPa)	3–10	20–24	[92, 134]
Peel strength (N/mm)	0.2–1.5	1–7	[50, 97, 136, 137]
Requirements to cell metallization	Front: no busbars or 'glue fingers' Rear: conventional Ag-pads	Full silver busbars or pad metallization on front and rear	[78]
Requirements to module lamination	Conventional	Conventional	
Reported indoor reliability testing stability	TC 800, DH 2 500–3 000	TC 800 and more, DH 2 000–3 000	[98, 125, 138, 139]
Reported field experience	~1 year	>20 years	[50, 140]
Availability of equipment	Equipment similar to conventional soldering, similar throughput or slightly less	Mainstream equipment with high throughput (2 100 cells per hour on single line)	[46, 141]
Logistic impact	String not easily reworkable Transportation and storage of frozen ECA	String is reworkable	
Main application	SHJ solar cell interconnection Shingle interconnection Back-contact interconnection	Conventional PERC interconnection	
Potential future application	Tandem solar cell interconnection	SHJ solar cell interconnection	

corrosion. The ECA selection itself and sometimes even the selection of the encapsulants need to be considered more carefully than for conventional solar modules.

10.3 Welding-based solar cell interconnection

10.3.1 Introduction

Welding is a process for joining materials by heating them above their melting temperature or by enabling a coalescence by the application of pressure with or without heat. In contrast to the two previous described interconnection techniques in Sections 10.1 and 10.2, no additional materials, e.g., solder alloy having a lower melting point as the joining partners or ECAs, are required.

The melting points of common metals used for metallization of solar cells such as silver and aluminium are 962 °C and 660 °C, respectively. Since these melting temperatures are surpassed, e.g., during laser welding, the temperatures occurring during the welding process are significantly higher than of typical soldering processes used for interconnection of solar cells with <230 °C, see Section 10.1. Therefore, the heat must be applied highly localized in order to avoid damaging the solar cells [142]. Melting part of the cell metallization for joining by localized heat results in different process parameters and requirements compared to soldering processes. Soldering is a well-proven process widely applied for interconnection of solar cells, but still some technical aspects have to be addressed [142]:

- Soldering or contacting with ECA typically requires surfaces that are free of a chemically stable oxide for a reliable connection [143]. These are in general cost intensive noble metals, like silver, or a stack of different metals [144], which complicates the process. Aluminium forms a chemically stable oxide and is not solderable with the common processes, but can be welded by different technologies [142, 143]. Additionally, for the soldering process unwanted oxides from the surface are removed by flux agents. These chemicals have the risk to interact with the lamination foil leading to bubble formation in the laminate [145].
- Various metals are involved in the soldered interconnections, like tin, silver and lead in the solder, copper in the ribbon and silver as well as aluminium in the pastes on the cells. Due to their different oxidation potentials [146], this enables electrochemical processes, like galvanic corrosion, which have a negative impact on the modules' long-term stability [146, 147]. Additionally, the large variety of metals requires several process steps during recycling [148, 149].
- During the relatively slow soldering process typically the whole solar cell and ribbons are heated above 180 °C. Due to the difference of the linear thermal expansion coefficient of silicon (2.6×10^{-6} K^{-1}) [150] and copper (16.8×10^{-6} K^{-1}) [151], the soldering process induces mechanical stress after cooling down to room temperature, which might induce cracks into the solar cell [152–156]. By heating only the joining partners locally, mechanical stress can be significantly reduced [142].

Therefore, there is substantial research effort to apply and develop different alternative joining processes.

Since the 1970s, welding was used and investigated as an alternative to soldering for interconnection of Si solar cells [157–161]. Back then, parallel gap resistance welding, thermal compression bonding or ultrasonic bonding/welding was in focus of research. Welding in PV was pioneered by AEG Telefunken in 1968. Mainly due to the fact that solder joints have a high breakage risk in PV arrays for space application [157], they ceased soldering for their array production in 1971 [161] and used parallel gap resistance welding. The technique is still used for the interconnection of solar cells for space application. AEG Telefunken also applied welding for terrestrial application of PV modules since the 1970s. The used solar cells featured Ti/Pd/Ag metallization stacks and were series connected with silver-based interconnects [162]. In the 1980s, AEG modules based on aluminium-metallized solar cells and aluminium interconnects were interconnected by ultrasonic welding. The direct contacting of aluminium by ultrasonic bonding drew again attention in the recent years [143] due to cost and environmental advantages by reducing or avoiding the usage of silver-, copper- and lead-containing solders.

For the same reasons, different laser welding processes have been investigated over the last two decades. The Laser Zentrum Hannover worked together with the Institute for Solar Energy Research in Hamelin (ISFH) on a laser welding process for an interconnection scheme similar to the standard soldering process of both side contacted solar cells. They joined 150-μm-thick Cu interconnect ribbons to the screen-printed Ag metallization or Al PVD metallization using ms laser pulses [163, 164]. Since the thickness ratio of the interconnect and the cell metallization is challenging with respect to a damage-free contacting of solar cells, several groups worked on thin (~10 μm) foil-based interconnections. Ehrhardt *et al.* [165] contacted GaAs solar cells with a thin silver foil. In their process, the Ag foil is ablated using an ns laser until it is thinned down so that the thermal diffusion length l_T is larger than the remaining foil thickness and a weld spot can be formed. Al-foil for interconnection of rear-contacted solar cells was investigated by different groups [166, 167]. It was proven to enable damage-free contacting of aluminium PVD-metallized solar cells [168]. Laser welding of Al-foils can be also used to contact screen-printed Ag [167, 169] or Al cell metallization [170].

Present research on welding technologies for interconnect of silicon solar cells focuses on laser welding, since it allows short process times and there is no need of high mechanical pressure for clamping. For completeness, parallel gap welding and ultrasonic welding are briefly discussed in the following.

10.3.2 Parallel gap welding

Parallel gap welding is a resistance welding process where the weld electrodes are placed close to each other and are in contact with the joining partner on one surface, as shown in Figure 10.17a. Between these electrodes, a voltage is applied of typically about 0.5–0.75 V resulting in a current flow of several tens of amperes [171–174]. Thereby, the interface between the joining partners is heated and the

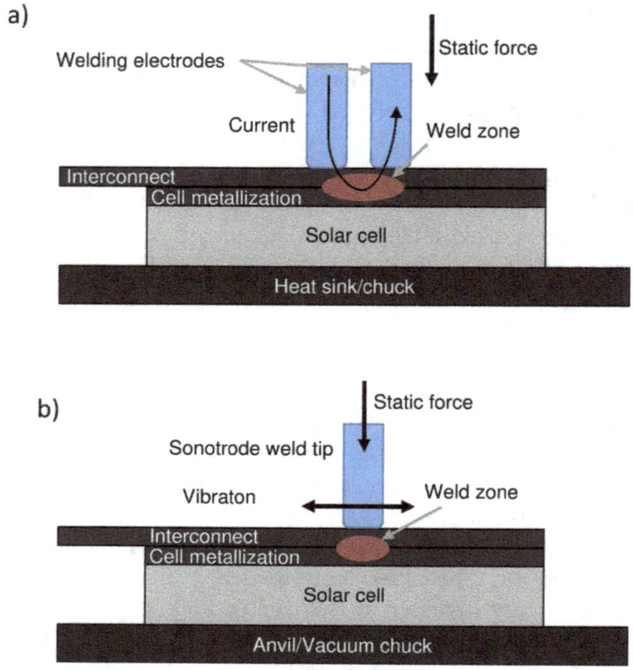

Figure 10.17 *Schematics of (a) parallel gap welding and (b) ultrasonic welding process*

weld is formed. Depending on the process parameters, the metal at the interface between the joining partners melts and forms a fusion weld nugget. If the temperatures stay below the melting point, e.g., a temperature of about 80% of the melting temperature, the weld is formed by brazing or the formation of a solid-state weld [174]. It is critical to ensure an intimate contact between the joining surfaces by pressing the partners together. This requires high forces in the order of some N/mm² [171–173] resulting in the risk of cell breakage.

As the heat is generated due to the resistances at the interface between the joining partners, the surface roughness is a crucial aspect for parallel gap welding. Thus, the typical surface morphology of industrial silicon solar cells may lead only to partial or even insufficient welding [175] and mirror-like surfaces are preferred.

For space application, commonly silver or gold plated surfaces of Kovar, Invar or molybdenum interconnects [176] are welded to the silver- or gold-coated rear side of III–V multi-junction solar cells. Such interconnections are tested differently compared to terrestrial applications. The weld joints are tested in accelerated ageing of several 10 000 thermal cycles (~−110 °C to +100 °C) or even larger thermal ranges and no significant decrease of the mechanical properties is observed [158, 159, 176]. In 45° peel tests, typically 3.5 N is reached per interconnector [171, 177, 178] or more [158].

Due to the risk of cell breakage, the application of cost intensive materials, the need of a compact cell metallization, i.e., evaporated or plated contacts, and the high requirements on the surface morphology inhibit the application for terrestrial c-Si modules nowadays.

10.3.3 Ultrasonic welding

During ultrasonic welding, a high frequency movement is generated at the interface plane between the joining partners. Commonly, frequencies of some tenth of kHz are introduced by a sonotrode in the upper joining partner, while the lower one is fixed on an anvil, see Figure 10.17b. The relative motion of some micrometres results in the breakage of thin oxide layers on the surface of the joining partners, a plastic deformation of the surface and shearing of surface asperities. This results in solid-state welding between the surfaces [179]. During the bonding process, the surfaces are heated, but not molten. Since the heat is generated at the interface between the joining partners and relative low temperatures are used, the risk of thermal damage is reduced. For the same reason only compact cell metallization like evaporated metal layers or silver screen printing can be contacted. However, the required application of pressure during welding may cause mechanical damages [143].

In the 1980s, ultrasonic welding was successfully used for commercial modules. For example, AEG produced modules with solar cells metallized by aluminium and aluminium interconnects.[a] These modules are working properly up to now in the year 2020. Figure 10.18 shows a photography of an example of ultrasonic weld spots on a test module.

Since ultrasonic welding enables direct contact of Al layers, in 2012 the process was adapted to modern solar cells in order to decrease the consumption of silver for cell fabrication. Heimann *et al.* [143] demonstrated the bonding of Al interconnects to screen-printed silver metallization and evaporated Al layers of both side contacted solar cells. Peel forces of 2 N for 2-mm-wide ribbons on Ag metallization of the front side of their solar cells are reported, which coincides with the required 1 N/mm in the standards (DIN EN 50461). On the cells, rear side forces >5 N on Al PVD layers (2 μm thick) are measured. By optimization of the interconnection process, the rate of cells featuring micro cracks is reduced to <2% for fabrication of full-sized modules. The modules are tested under artificial ageing according to ICE 61215. After damp heat testing of 1 000 h, the modules show a power loss of <2% and for a combination of 50 thermal cycles and 10 humidity freeze cycles of <3%. Only during ageing in 200 temperature cycles, the power of the modules is reduced by 6% relative and, thus, more than the limit of 5% according to the IEC standards. However, the main loss is in current and the fill factor reduces only by 2.9%. Details on the contact resistance are not given. However, as stable metal–metal contacts are

[a]Information based on modules installed in front of ISFH. The nameplates indicate them as modules fabricated by AEG in 1988. The analysis of the interconnects is done by energy dispersive X-ray spectroscopy (EDX) in-house.

Figure 10.18 *Photography of a solar cell with ultrasonic welded interconnect. The interconnect is welded to the cell's front side busbar at three locations*

formed it is expected that they are in the order of other interconnection technologies like soldering and laser welding and not limiting the module efficiency.

10.3.4 Laser welding

Laser welding is the welding technique with the largest gain in market share over all industries due to the ability of high weld speeds, an easy automation, and its flexibility [180]. It can be classified in two types based depending on the process dynamics and laser intensity [142, 181]:

1. Conduction-limited welding: the material of the joining partners locally melts by moderate laser intensities without significant evaporation. The heat transport is dominated by conduction and convection within and around the melt pool.
2. Deep-penetration or keyhole welding: intensive continuous wave laser light results in local evaporation and forms a deep vapour cavity (keyhole) surrounded by a melt pool. Therefore, keyhole welding can form deep weld seams and reach large aspect ratios. Multiple reflections and high absorption within the keyhole result in an efficient welding process.

For solar cell interconnection, metal structures with thicknesses of some 100 μm or less are joined. Therefore, conduction-limited welding is used. Due to the opaque metal layers, the heat is applied on the outer surface of one joining partner instead to the interface as in the case of parallel gap or ultrasonic welding. Laser welding is a well-proven process in industry. However, welding of such thin metal layers is in focus of research and still challenging as material distortion, crack formation and material drop-out are critical [182–185]. Especially when joining thin metal layers on thermally sensitive substrates as in the case of solar cells or polymer layers [165, 186, 187].

Figure 10.19 shows different approaches to interconnect solar cell by laser welding. All approaches have to face the same challenges:

- Finding the trade-off between providing sufficient energy to form a weld spot between both joining partners, i.e., melting the interface between the interconnect and the cell metallization, and limiting the melt front within the cell metallization layer in order not to damage the passivation layers between the cell metallization and the Si bulk material.
- For reliable and stable weld spots, the joining partners need to be in good contact, and sufficient material needs to be provided to fill any gap between the joining partners caused by surface roughness.

The approach in Figure 10.19a is close to the conventional soldering tabbing-stringing approach, where a solder-coated Cu ribbon is laser welded to the busbar of a screen-printed solar cell. The main driver for investing this approach is to reduce the process time for cell interconnection [163, 188]. Conventional soldering processes require several hundred ms to allow the heat to diffuse through the materials, whereas welding can be done in some ms or less.

The approach of welding cell interconnection ribbon to solar cells was presented by Moalem *et al.* [163, 164]. They show that weld seams formed by laser with a duration between 5 and 10 ms lead to good weld results for solder-coated Cu cell interconnection ribbons (d_{int} = 150 μm) and evaporated cell metallization (d_{cell} = 15 μm Al + 3 μm Ag). They use circular weld seams and weld speeds between 430 and 760 mm/s, and pulse peak powers between 1.1 and 1.5 kW at a wavelength of 1064 nm. No clamping was used to avoid mechanical stress. Due to the relatively thick ribbon, sufficient material is molten to fill any gaps between the joining partners. Weld spot by individual laser pulses did not allow a reliable welding. The weld seams reach peel forces of 2–10 N under a peeling angle of 90°, but the width of the ribbon is not stated. No crystal damage after peel-off is reported for the above-mentioned parameters. However, analysis about the laser-induced damage on solar cells' surface passivation is not shown to the best of our knowledge. Electrical contact resistances of 1–20 μ Ω cm^2 are achieved. However, after 107 thermal cycles the resistances increase by a factor of 5 [164]. Details about the behaviour after the standard test of 200 cycles or other test sequences are not given. Promising results were also shown for 15 μm Al metallization. Circular weld seams formed by 15 ms laser pulses and pulse peak powers of 0.65–0.725 kW at 270 mm/s allow peel forces of up to 4 N. The contact resistance increases from 5 to 25 μ Ω cm^2 after 107 thermal cycles. However, on screen-printed pastes the laser welds did not fulfil the required peel force of 2 N, since the porous paste collapses during welding.

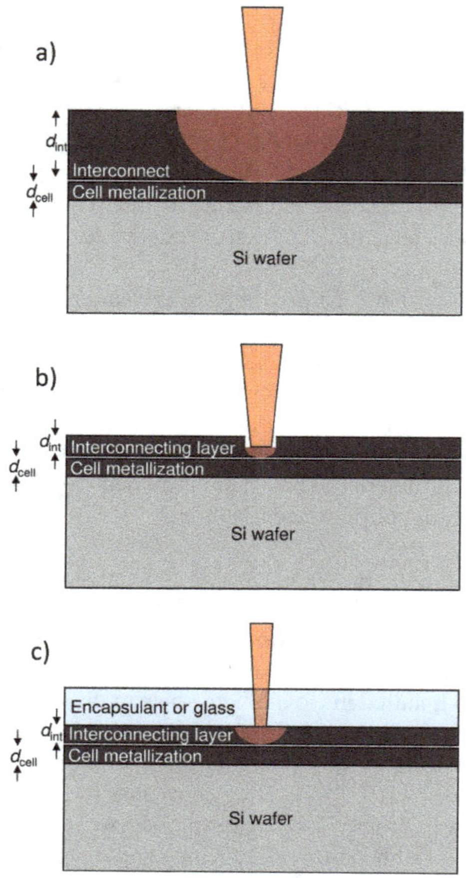

Figure 10.19 *Schematics of different laser welding approaches for c-Si solar cells: (a) a process joining copper interconnection ribbons to the cell metallization of both side contacted solar cells close to the conventional soldering tabbing-stringing approach, (b) a laser welding process based on the evaporation recoil pressure and the ablation of the interconnecting metal foil and (c) a laser welding process driving the molten material towards the interface between the joining partners by the application of a transparent substrate.* d_{int} *and* d_{cell} *are the thicknesses of the interconnects and cell metallization, respectivly*

Rofin-Bassel improved the welding process by using a bi-wavelength process [188]. Instead of soldering a Cu ribbon to the cell metallization, it is welded by two overlapped laser beams. One of the lasers emits light at 532 nm and the second at 1 064 nm. During the bi-wavelength welding, the green laser (532 nm) with a low efficiency initiates the melting. This decreases the reflectivity of Cu or solder for the IR light (1 064 nm), such that the more efficient IR laser can introduce the high power required for welding into the Cu ribbon.

For evaporated metal contacts, different foil-based approaches have been developed. Ehrhardt *et al.* [165] weld a silver foil (d_{int} = 8 μm) to the busbar of GaAs solar cells . They use a KrF laser with 25 ns pulse duration at 248 nm. The Ag foil is ablated by multiple laser pulses (in the order of 100), as indicated in Figure 10.19b until it is thinned down to about 3 μm such that the thermal diffusion length is larger than the foil thickness. Thereby, a weld spot can be formed. The process benefits from the recoil pressure of the ablated material in order to press the joining partners together. The mechanical properties for the Ag–Ag joints were tested in a 0° peel test [186] instead of the commonly used 90° or 180°. The measured shear tensions for this test are 0.62 and 1.1 N/mm² depending on the material combination. Joining Cu and Ag to Al, Ag and Cu is shown using a laser fluency of about 3.5 J/cm² [186]. The only reported detail on the electrical properties is the efficiency of the GaAs solar cells. It reduces from 26.4% before interconnection to 25.4% after joining [165].

Reliable and damage-free contacts between thin foils and the cell metallization have been realized by the process depicted in Figure 10.19c. The process called aluminium-based mechanical and electrical laser interconnection (AMELI) [189] is to the best of our knowledge the only laser welding process for interconnection of solar cells, whose investigation and application is published in detail. The close contact between the joining partners, e.g., aluminium foil and evaporated aluminium contacts of the solar cell, is realized by using a substrate that is transparent with respect to the laser irradiation and by generating a vacuum in the gap between the joining partners. Thereby, the joining partners are pressed homogeneously to each other. In contrast to the above-mentioned epitaxially grown GaAs solar cells, the surface of industrial silicon solar cell is not polished to the same degree and a surface roughness of several micrometre is present. As shown in Figure 10.20, this gap can be closed by directing the liquid aluminium of the foil towards the second joining partner, i.e., the cell metallization, and avoiding the evaporation in an open half space. During the laser welding, the chemically stable native oxide of the aluminium breaks up. Since the cell metallization in general is rather thin, d_{cell} is in the order of 10 μm, the thermal impact on the underlying Si surface and its passivation must be minimized. This can be achieved by using short laser pulses due to their short thermal diffusion length l_T. The process is performed with a laser having a pulse duration between 20 ns and 1.2 μs resulting in a thermal diffusion length between 3 and 22 μm [49].

A schematic of the laser welding process for cell interconnection of rear contacted solar cells is shown in Figure 10.21. An Al layer serving as an interconnect in the final module is attached to a substrate and placed in contact to the cell metallization. The substrate needs to be transparent with respect to the laser light. Depending on the application, it can be a glass substrate or the lamination foil, e.g., ethylene–vinyl acetate, silicone or others, needed for the encapsulation of the solar cells. Additional to its critical role for the welding, the substrate enables an easier handling and positioning of the interconnects. After placing the substrate and the solar cells on a vacuum chuck, the edges are sealed using an adhesion foil. By

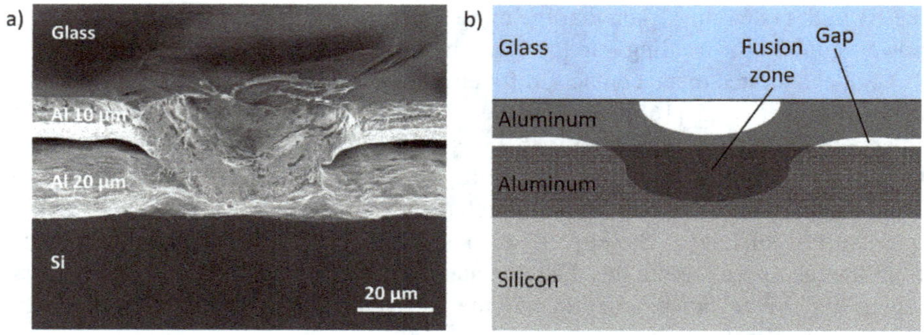

Figure 10.20 *(a) Scanning electron microscope (SEM) image of a cross section of a laser weld spot. The first joining partner is a 10-µm-thick Al layer evaporated on a glass substrate, and the second joining partner is a 20-µm-thick evaporated Al layer on a Si wafer. The weld spot is formed by using eight pulses with a pulse duration of 20 ns at 40 J/cm2. (b) Schematics of the SEM image in (a). (Adapted from [142])*

evacuating the space between the two Al joining partners they are pressed together by the external air pressure. The laser beam is focused on the Al layer on the substrate and melts and fuses the Al layers, as shown in Figure 10.20. This requires a close contact between the two joining partners, since the molten material should fuse the two Al layers and form a mechanically stable and electrically conductive interconnection. If the gap is too wide, molten aluminium may splash into the gap between the joining partners without enabling a contact.

In the case of an interconnecting layer thickness d_{int} of 10 µm, the laser welding process allows to contact solar cells with a cell metallization thickness d_{cell} down to 1 µm without inducing any laser damage and forming mechanical stable weld spot [142, 190]. In case of a lamination foil as substrate, the mechanical properties of the welded interconnection are enhanced by the encapsulant filling the gap between the

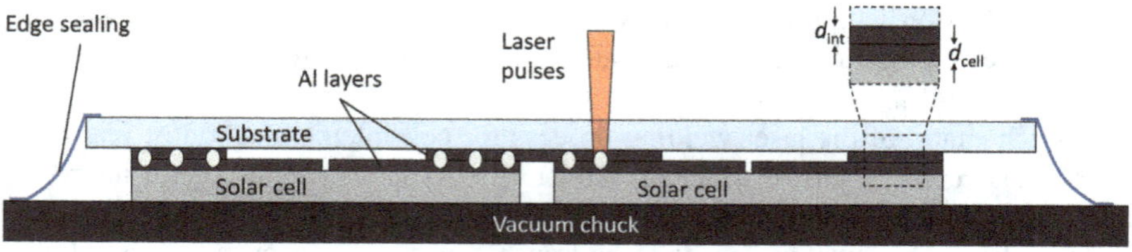

Figure 10.21 *Schematic of the AMELI process used for interconnection of solar cells. In the shown example, two rear-contacted solar cells are interconnected. The magnification indicates the thickness of interconnects and cell metallization* d_{int} *and* d_{cell}*, respectively. (Adapted from [142])*

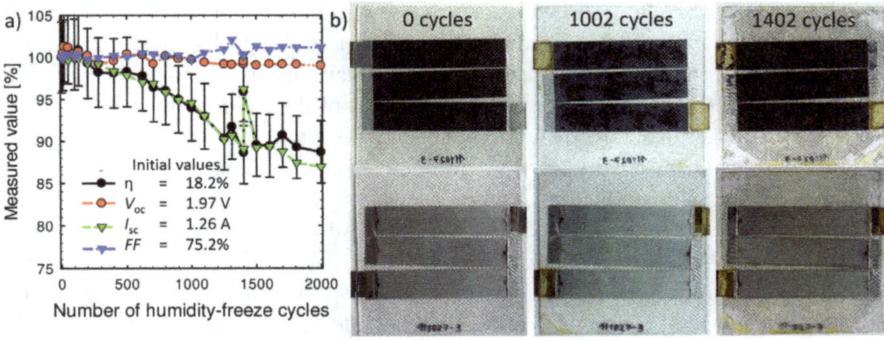

Figure 10.22 *(a) Relative I–V parameters of a module stressed under humidity-freeze cycles. The values are relatively compared to initial values in dependence on the number of cycles. At 1 402 cycles, the transparent front sheet is fully delaminated. The module is measured with and without the front sheet, which is kept in front of the module afterwards for consistency. The latter measurements show an increase in current of 7%. (b) Photographs of the front (upper row) and rear side (lower row) of the same module after 0, 1 002 and 1 402 humidity-freeze cycles. The photographs at 1 402 cycles are without the front sheet142. (Adapted from [142]).*

bonding partners during the final lamination of the module. The encapsulant entering the gap supports the weld spot by its adhesive properties [168], reaching perpendicular tear-off stresses for a spot distance of 250 μm of 380 ± 80 kPa. Please note, that a classical peel-off test is not possible with the Al-foil since the foils ruptures before the weld spots fail.

The mechanically strong weld spots combined with the usage of only one type of metal avoiding electrochemical processes allow to build long-term stable solar modules [191]. Figure 10.22 shows that the interconnection is stable under 2 000 humidity-freeze cycles [1]. Only the module current decreases due to the degradation of the encapsulant and the delamination of the transparent front sheet. After fully delamination of the front sheet, the module is measured with and without front sheet. Without the front sheet and its optical impact, the power loss after 1 400 cycles is only 4.2%. However, the fill factor, which would be influenced by a contact degradation, is stable up to the 2 000 cycles tested, which by far exceed classical test procedures of some tenth of cycles or the ten humidity-freeze cycles required in the IEC 61215 standard. The fill factor increase relative to the initial value is caused by the reduced module current idue optical effects. Consequently, the resistive losses at the maximum power point decrease leading to a rise in fill factor.

Another advantage of laser spot welding is the high spatial precision of the contact points. As shown in Figure 10.20a the diameter of the weld spot is about 60 μm, depending on the specific optical components used. It allows to contact small structures, such as narrow fingers [166] or point contacts [192], for busbar-less cell interconnection. Thereby, the losses associated with the busbars of rear contacted solar cells [166, 193] can be avoided. The busbar-less design has also the advantage that the interconnecting foil covers the full rear side of the back-contact cell with local contacts, supporting the current transport of the cell metallization. A thin layer of encapsulant isolates the non-contacted areas and is penetrated during laser welding. This concept, also known as multilevel metallization, reduces the series resistance in the cell metallization significantly [194] and allows to reduce the metallization thickness of back contacted solar cells. Additionally, the resistive losses due to the lateral transport of majority charge carries in the silicon , which are generated above the emitter busbar, and the current transport in the busbar are omitted. In sum, an increase in fill factor from cell to module of 5% is demonstrated [166]. Since the laser welding does not damage the passivation layer underneath the metallization, no change in the open circuit voltage is observed. Thereby, the cell to module losses can be minimized and only result from the decrease of the current due to the change of the optical environment.

The contacting by the AMELI laser welding process is not limited to evaporated metal layers. It can also be applied to silver screen-printed metallization [169] and aluminium screen-printed metallization [170]. In order to simplify the process, the welding can be also done after laminating the module within the laminate, when using a transparent module rear side [169]. The laser contacts cannot only be used to interconnect solar cells, but also to form the metal-semiconductor contacts simultaneously. By such a laser processing laser-fired and bonding contacts (LFBC) on the passivated rear side of the solar cells can be formed, similar to the laser-fired contacts (LFC) [195] to substitute the Al-screen printing and firing of Al pastes. In contrast to the LFC process, for formation of LFCBs the Al-foil is embedded between two polymeric layers, one sealing the gap between the cells rear side and the Al-foil for enhanced long-term stability and the second on the outer side of the Al-foil similarly to the transparent substrate as shown in Figure 10.21. This directs the molten aluminium towards the cell's rear side and enables to form less combinative contacts with 1000^{+1000}_{-500} cm/s in comparison to LFCs [169]. The contact resistance to *p*-type 0.5 Ω cm silicon is determined to be 3.52 ± 0.74 m Ω cm².

Recently, Fisher *et al.* presented also an in-laminate welding process [167] similar to the above-mentioned AMELI process. They showed the ability of performing laser welding within laminated modules to contact Al-foils to screen-printed Ag busbars of rear-contacted solar cells [167]. To ensure long-term stability, the area between the foil and the cell's rear side was sealed with an additional layer of encapsulant. An embossed Al-foil enables the contact though this encapsulant layer. This improved the long-term stability of the interconnection, but a damage-free interconnection is not shown [167].

In summary, various laser welding processes have been investigated and developed over the last two decades without any permanent breakthrough in terrestrial application so far. Challenges, which have to be faced, are either the mechanical pressure or the high temperatures required for welding. Promising processes use thin foils for cell interconnection which are suitable for interconnection of rear contacted solar cells. With the need to replace silver and a gain in market share for rear contact solar cells [1] laser welding might enter the market as announced by first companies [196].

References

[1] ITRPV. 'International Technology Roadmap for Photovoltaic (ITRPV)'. *2019 Results*. 11th ed; 2020.

[2] Walter J., Stegmaier J., Kraft A., Eitner U. 'Analysis of grain-size distribution and yield strength of interconnector ribbons and wires at different stretching conditions using color etching'. Proceedings of the 35th European Photovoltaic Solar Energy Conference and Exhibition (EU PVSEC); Brussels, Belgium; 2018. pp. 58–66.

[3] Hall E.O. 'The deformation and ageing of mild steel: III discussion of results'. *Proceedings of the Physical Society. Section B*. 1951;**64**(9):747–53.

[4] Petch J. 'The cleavage strength of polycrystals'. *Journal of the Iron and Steel Institute*. 1953;**174**:25–8.

[5] Schiotz J. Simulations of nanocrystalline metals at the atomic scale. What can we do? What can we trust? 22nd Risoe International Conference on Materials Science; Roskilde, Denmark; 2001.

[6] Rösler J., Harders H., Bäker M. *Mechanisches Verhalten der Werkstoffe*. 4th ed.. Wiesbaden: Springer Fachmedien Wiesbaden; 2012.

[7] Petukhov B.V. 'A theory of the effect of impurities on the yield stress of silicon crystals'. *Semiconductors*. 2004;**38**(4):369–75.

[8] RoHS-Richtlinie 2011/65/EU. 'Beschränkung Der Verwendung bestimmter gefährlicher Stoffe in Elektro- und Elektronikgeräten'. 2011.

[9] Klein Wassink R.J. *Soldering in Electronics*. 2nd ed. Saulgau/Württ: Leuze; 1991.

[10] Humpston G., Jacobson D.M. *Principles of Soldering*. Materials Park, H: ASM International; 2004.

[11] Baker H. *Alloy Phase Diagrams: ASM Handbook*. 3rd ed. Materials Park, Ohio: ASM International; 2016.

[12] Tu K.-N. *Solder Joint Technology: Materials, Properties, and Reliability*. New York: Springer; 2007.

[13] De Rose A., Erath D., Geipel T., Kraft A., Eitner U. 'Low-temperature soldering for the interconnection of silicon heterojunction solar cells'. Proceedings of the 33rd European Photovoltaic Solar Energy Conference and Exhibition (EU PVSEC); Amsterdam, Netherlands; 2017. pp. 710–4.

[14] De Rose A., Geipel T., Eberlein D., Kraft A., Nowottnick M. 'Interconnection of silicon heterojunction solar cells by infrared soldering – solder joint analysis and temperature study'. Proceedings of the 36th European Photovoltaic Solar Energy Conference and Exhibition (EU PVSEC); Marseille, France; 2019. pp. 229–34.

[15] Geipel T., Eberlein D., Kraft A. 'Lead-free solders for ribbon interconnection of crystalline silicon PERC solar cells with infrared soldering'. *AIP Conference Proceedings*. **2156**. Konstanz; 2019. p. 20015.

[16] Geipel T., Moeller M., Walter J., Kraft A., Eitner U. 'Intermetallic compounds in solar cell interconnections: microstructure and growth kinetics'. *Solar Energy Materials and Solar Cells*. 2017;**159**(2):370–88.

[17] Laurila T., Vuorinen V., Kivilahti J.K. 'Interfacial reactions between lead-free solders and common base materials'. *Materials Science and Engineering: R: Reports*. 2005;**49**(1-2):1–60.

[18] Evans J.W. *A Guide to Lead-free Solders*. London: Springer; 2007.

[19] Zeng G., Xue S., Zhang L., Gao L., Dai W., Luo J. 'A review on the interfacial intermetallic compounds between Sn–Ag–Cu based solders and substrates'. *Journal of Materials Science: Materials in Electronics*. 2010;**21**(5):421–40.

[20] So A.C.K., Chan Y.C., Lai J.K.L. 'Aging studies of Cu-Sn intermetallic compounds in annealed surface Mount solder joints'. *IEEE Transactions on Components, Packaging, and Manufacturing Technology: Part B*. 1997;**20**(2):161–6.

[21] Lyon K.G., Salinger G.L., Swenson C.A., White G.K. 'Linear thermal expansion measurements on silicon from 6 to 340 K'. *Journal of Applied Physics*. 1977;**48**(3):865–8.

[22] Kohn C. Influence of the metallization process on the strength of silicon solar cells. in Proceedings of the 24th European Photovoltaic Solar Energy Conference; Hamburg, Germany, 2009; 2009. pp. 1419–23.

[23] Kraemer F., Wiese S. 'Assessment of long term reliability of photovoltaic glass–glass modules vs. glass-back sheet modules subjected to temperature cycles by FE-analysis'. *Microelectronics Reliability*. 2015;**55**(5):716–21.

[24] Wiese S., Meier R., Kraemer F. 'Mechanical behaviour and fatigue of copper ribbons used as solar cell interconnectors'. Proceedings of the 11th IEEE International Conference on Thermal, Mechanical and Multi-Physics Simulation and Experiments in Microelectronics and Microsystems; Bordeaux, France; 2010. pp. 1–5.

[25] Zemen Y., Prewitz T., Geipel T., Pingel S., Berghold J. 'The impact of yield strength of the interconnector on the internal stress of the solar cell within a module'. Proceedings of the 25th European Photovoltaic Solar Energy Conference and Exhibition/ 5th World Conference on Energy Conversion; Valencia, Spain; 2010. pp. 4073–8.

[26] Rendler L.C., Kraft A., Ebert C., Wiese S., Eitner U. 'Investigation of thermomechanical stress in solar cells with multi busbar interconnection by finite element modeling'. Proceedings of the 32nd European Photovoltaic Solar Energy Conference and Exhibition; Munich, Germany; 2016. pp. 94–8.

[27] Rendler L.C., Romer P., Beinert A.J., *et al*. 'Thermomechanical stress in solar cells: contact pad modeling and reliability analysis'. *Solar Energy Materials and Solar Cells*. 2019;**196**(2):167–77.

[28] Meier R., PanderM., KlengelR., *et al.* 'Reduction of soldering induced stresses in solar cells by microstructural optimization of copper-ribbons'. *Proceedings of SPIE 8112, Reliability of Photovoltaic Cells, Modules, Components, and Systems IV*; San Diego, California; 20112011. p. 811206.

[29] Walter J., Rendler L.C., Halm A., Mihailetchi V., Kraft A., Eitner U. 'Ribbon interconnection of 6" BC-BJ solar cells'. *Energy Procedia*. 2017;**124**(6):504–14.

[30] Rendler L.C., Walter J., Kraft A., Ebert C., Wiese S., Eitner U. 'Ultra-Soft wires for direct soldering on finger grids of solar cells'. *Energy Procedia*. 2017;**124**:478–83.

[31] Yao D., Shang J.K. 'Effect of cooling rate on interfacial fatigue-crack growth in Sn-Pb solder joints'. *IEEE Transactions on Components, Packaging and Manufacturing Technology*. 1996;**19**(1):154–65.

[32] Nicolics J., Mündlein M. 'Electrically conductive adhesives' in Suhir E., Lee Y., Wong C.P. (eds.). *Micro- and Opto-Electronic Materials and Structures: Physics, Mechanics, Design, Reliability, Packaging*. USA: Springer; 2007. pp. B571–610.

[33] Morris J.E. 'Isotropic conductive adhesives: Future trends, possibilities and risks'. 5th International Conference on Polymers and Adhesives in Microelectronics and Photonics, Polytronic; Wrocław, Poland; 2005. pp. 233–4.

[34] Battermann A. 'Joining technology that's fit for the future'. *Adhesion – Adhesives and Sealants*. 2012;**2012**:40–3.

[35] Li L., Morris J.E. 'An introduction to electrically conductive adhesives'. *International Journal of Microelectronics and Electronic Packaging*. 1998;**1**:159–75.

[36] Beier B., Mau S., Metz A., Hezel R. 'Electrical conductive adhesives: a novel reliable method for interconnecting crystalline silicon solar cells'. Proceedings of the 17th European Photovoltaic Solar Energy Conference and Exhibition; Munich, Germany; 2001. pp. 812–5.

[37] Frisson L., Lieten K., Roberts S., Bruton T. 'Conductive adhesives as an interconnection technique for very thin solar cells'. Proceedings of the 17th European Photovoltaic Solar Energy Conference and Exhibition; Munich, Germany; 2001. pp. 414–6.

[38] Lalaguna B., Sánchez-Friera P., Mäckel H., Sánchez D., Alonso J. 'Evaluation of stress on cells during different interconnection processes'. Proceedings of the 23rd European Photovoltaic Solar Energy Conference; Valencia, Spain; 2008. pp. 2705–8.

[39] Bennett I.J. 'Low-stress interconnection of solar cells'. Proceedings of the 22nd European Photovoltaic Solar Energy Conference and Exhibition; Milano, Italy; 2007. pp. 2674–8.

[40] de Jong P.C. 'Single-step laminated full-size PV modules made with back-contacted MC-Si cells and conductive adhesives'. Proceedings of the 19th European Photovoltaic Solar Energy Conference; Paris, France; 2004. pp. 2145–8.

[41] Scherff M.L.D., Schwertheim S., Ma Yue M.T., Fahrner W.R. '10 x 10 cm² HIT solar cells contacted with lead-free electrical conductive adhesives to solar cell interconnectors'. Proceedings of the 4th IEEE World Conference on Photovoltaic Energy Conversion; Waikoloa, Hawaii, USA; 2006. pp. 1384–7.

[42] Späth M. 'Hetero-junction module technology'. Proceedings of the 26th European Photovoltaic Solar Energy Conference and Exhibition; Hamburg, Germany; 2011. pp. 3121–4.

[43] De Rose A., Geipel T., Eberlein D., Kraft A., Nowottnick M. 'Interconnection of silicon heterojunction solar cells by infrared soldering: Solder joint analysis and temperature study'. Proceedings of the 36th European Photovoltaic Solar Energy Conference and Exhibition; Marseille, France; 2019. pp. 229–34.

[44] Soederstroem T., Papet P., Ufheil J. 'Smart wire connection technology'. Proceedings of the 28th European Photovoltaic Solar Energy Conference and Exhibition; Paris, France; 2013. pp. 495–9.

[45] Meyer Burger Technology Ltd., *Press release*. [online]. Available from www.meyerburger.com/en/newsroom/news/meyer-burger-technology-ltd-aims-to-transform-itself-from-a-supplier-of-production-equipment-to-a-technologically-leading-manufacturer-of-solar-cells-and-modules-capital-increase-targeting-gross-proceeds-of-chf-165-million-planned/# [Accessed 07.12.2020].

[46] teamtechnik Maschinen und Anlagen GmbH. *Innovative adhesive technology for industrial serial production: Stringer TT1600ECA*. [online]. Available from www.teamtechnik.com/en/solar/stringer-tt/stringer-tt1600-eca/ [Accessed 2 Dec 2020].

[47] teamtechnik Maschinen und Anlagen GmbH. *Press release: Lower costs, more efficiency: Russian module manufacturer and* PV project developer Hevel LLC switches to adhesive technology from teamtechnik for its HJT modules [online]. 2018. Available from www.teamtechnik.com/en/news/lower-costs-more-efficiency-russian-module-manufacturer-and-pv-project-developer-hevel-llc-switches-to-adhesive-technology-from-teamtechnik-for-its-hjt-modules/ [Accessed 15 May 2021].

[48] teamtechnik Maschinen und Anlagen GmbH. Press release: *Major contract for teamtechnik, Italian PV manufacturer orders stringers for production of high-efficiency modules with HJT cells [online]*. Available from www.teamtechnik.com/en/news/major-contract-for-teamtechnik-italian-pv-manufacturer-orders-stringers-for-pro-duction-of-high-efficiency-modules-with-hjt-cells/ [Accessed 12 Apr 2018].

[49] Andronikov D. Silicon heteronjunction cells and modules in mass production and utility scale operation. 3rd SHJ workshop; virtual workshop, 19 - 21 October 2020; 2020.

[50] Geipel T. 'Industrialization of the ribbon interconnection of silicon heterojunction solar cells with electrically conductive adhesives'. Proceedings of the 36th European Photovoltaic Solar Energy Conference and Exhibition; Marseille, France; 2019. pp. 916–25.

[51] De Rose A., Geipel T., Erath D., Kraft A., Eitner U. 'Challenges for the interconnection of crystalline silicon heterojunction solar cells'. *Photovoltaics International*. 2018;**40**:78–86.

[52] Schneider J., Turek M., Dyrba M., Baumann I., Koll B., Booz T. 'Combined effect of light harvesting strings, anti-reflective coating, thin glass, and high ultraviolet transmission encapsulant to reduce optical losses in solar modules'. *Progress in Photovoltaics: Research and Applications*. 2014;**22**(7):830–7.

[53] Al-Ashouri A., Köhnen E., Li B., *et al*. 'Monolithic perovskite/silicon tandem solar cell with >29% efficiency by enhanced hole extraction'. *Science*. 2020;**370**(6522):1300–9.

[54] Sunpower Corporation. Sunpower performance solar panels [online]. Available from sunpower.maxeon.com/int/solar-panel-products/sunpower-performance-solar-panels [Accessed 6 Dec 2020].

[55] Solaria Corporation. Solaria Technology [online]. Available from www.solaria.com/ [Accessed 6 Dec 2020].

[56] Mittag M. 'Cell-to-module (CTM) analysis for photovoltaic modules with shingled solar cells'. Proceedings of the 44th IEEE Photovoltaic Specialists Conference; Washington DC, USA; 2017. pp. 1531–6.

[57] Rabanal-Arabach J. 'Cell-to-module conversion loss simulation for shingled-cell concept'. Proceedings of the 33rd European Photovoltaic Solar Energy Conference and Exhibition; Amsterdam; 2017. pp. 178–82.

[58] Zhao J., Wang A., Abbaspour-Sani E., Yun F., Green M.A. 'Improved efficiency silicon solar cell module'. *IEEE Electron Device Letters*. 1997;**18**(2):48–50.

[59] Dickson D.C. (Jr.). *Photovoltaic semiconductor apparartus or the like*. US 2938938 A; 1960.

[60] Klasen N., Mondon A., Kraft A., Eitner U. 'Shingled cell interconnection: a new generation of bifacial PV-modules'. 7th Workshop on Metallization and Interconnection for Crystalline Silicon Solar Cells; Constance, Germany; 2017.

[61] Baliozian P. 'PERC-based shingled solar cells and modules at Fraunhofer ISE'. *Photovoltaics International*. 2019;**43**:63–79.

[62] Rudolph D. 'Cell design optimization for shingled modules'. Proceedings of the 33rd European Photovoltaic Solar Energy Conference and Exhibition; Amsterdam; 2017. pp. 880–3.

[63] Henckens A. *'Application requirements for electrically conductive adhesives as interconnect materials for shingled solar cells'*. 8th Workshop on Metallization and Interconnection for Crystalline Silicon Solar Cells; Constance, Germany; 2019.

[64] Estruga M., Theunissen L., Ardizzone A., Willems B., Henckens A. 'Electrically conductive adhesives for cell interconnection in shingled module technology: the impact of material properties on mini-module performance'. Proceedings of the 35th European Photovoltaic Solar Energy Conference and Exhibition; Brussels, Belgium; 2018.

[65] Schmidt W., Rasch K.-D. 'New interconnection technology for enhanced module efficiency'. *IEEE Transactions on Electron Devices*. 1990;**37**(2):355–7.

[66] Mondon A., Klasen N., Mittag M., Heinrich M., Wirth H. 'Comparison of layouts for shingled bifacial PV modules in terms of power output, cell-to-module ratio and bifaciality'. Proceedings of the 35th European Photovoltaic Solar Energy Conference and Exhibition; Brussels, Belgium; 2018. pp. 1006–10.

[67] Tonini D. 'Shingling technology for cell interconnection: technological aspects and process integration'. Proceedings of the 33rd European Photovoltaic Solar Energy Conference and Exhibition; Amsterdam, Netherlands; 2017. pp. 38–41.

[68] Morris J.E., Wang L. 'Isotropic conductive adhesive interconnect technology in electronics packaging applications' in Mittal K., Ahsan T. (eds.). *Adhesion in Microelectronics*. Hoboken, NJ, USA: John Wiley & Sons, Inc; 2014. pp. 173–210.

[69] Wu H., Wu X., Ge M., Zhang G., Wang Y., Jiang J. 'Effect analysis of filler sizes on percolation threshold of isotropical conductive adhesives'. *Composites Science and Technology*. 2007;**67**(6):1116–20.

[70] Lyons A.M. 'Electrically conductive adhesives: effect of particle composition and size distribution'. *Polymer Engineering and Science*. 1991;**31**(6):445–50.

[71] Sancaktar E., Bai L. 'Electrically conductive epoxy adhesives'. *Polymers*. 2011;**3**(1):427–66.

[72] Rabilloud G. *High-Performance Polymers – 1 Conductive Adhesives: Chemistry and Applications*. Paris: Éditions Technip; 1997.

[73] Lu D.D., Tong Q.K., Wong C.P. 'A study of lubricants on silver flakes for microelectronics conductive adhesives'. *IEEE Transactions on Components and Packaging Technologies*. 1999;**22**:365–71.

[74] Lu D.D., Wong C.P. 'Effects of shrinkage on conductivity of isotropic conductive adhesives'. *International Journal of Adhesion and Adhesives*. 2000;**20**:189–93.

[75] Liu J., Lai Z., Kristiansen H., Khoo C. 'Overview of conductive adhesive joining technology in electronics packaging applications'. Proceedings of 3rd International Conference on Adhesive Joining and Coating Technology in Electronics Manufacturing; Binghamton, NY USA; 1998. pp. 1–18.

[76] Kopecek R. 'Aluminium rear emitter large area n-type Cz-Si solar cells for industrial application'. Proceedings of the 25th European Photovoltaic Solar Energy Conference and Exhibition/ 5th World Conference on Energy Conversion; Valencia, Spain; 2010. pp. 2381–6.

[77] Zemen Y., Alam R., Pingel S. 'Degradation mechanism and reliability of modules built with new interconnection approach'. Proceedings of the 28th European Photovolatic Solar Energy Conference; Paris, France; 2013. pp. 3114–18.

[78] Schneider A., Harney R., Aulehla S., Lemp E., Koch S. 'Progress in interconnection of busbar-less solar cells by means of conductive gluing'. *Energy Procedia*. 2013;**38**:387–94.

[79] Li Y., Lu D.D., Wong C.P. *Electrical Conductive Adhesives with Nanotechnologies*. New York, Dordrecht, Heidelberg, London: Springer; 2009.

[80] Habenicht G., Ahner C. *Applied Adhesive Bonding: A Practical Guide for Flawless Results*. Weinheim, Chichester: Wiley-VCH; 2009.

[81] Damico D.J. 'Reactive acrylic adhesives' in Pizzi A., Mittal K. (eds.). *Handbook of Adhesive Technology*. 2nd ed. New York: CRC Press; 2003. pp. 747–61.

[82] Lower L.D., Klosowski J.M. 'Silicone adhesives and sealants' in Pizzi A., Mittal K. (eds.). *Handbook of Adhesive Technology*. 2nd ed. New York: CRC Press; 2003. pp. 813–21.

[83] Li Y., Wong C.P. 'Recent advances of conductive adhesives as a lead-free alternative in electronic packaging: materials, processing, reliability and applications'. *Materials Science and Engineering: R: Reports*. 2006;**51**(1-3):1–35.

[84] Yim M.J., Li Y., Moon K.S., Wong C.P. 'Oxidation prevention and electrical property enhancement of copper-filled isotropically conductive adhesives'. *Journal of Electronic Materials*. 2007;**36**(10):1341–7.

[85] Zemen Y., Schulz S.C., Trommler H., Buschhorn S.T., Bauhofer W., Schulte K. 'Comparison of new conductive adhesives based on silver and carbon nanotubes for solar cells interconnection'. *Solar Energy Materials and Solar Cells*. 2013;**109**:155–9.

[86] Ehrenstein G.W. *Polymeric Materials: Structure - Properties - Applications*. Munich: Hanser; 2001.

[87] Klasen N., Romer P., Beinert A., Kraft A. 'FEM simulation of deformations in strings of shingled solar cells subjected to mechanical reliability testing'. 8th Workshop on Metallization & Interconnection for Crystalline Silicon Solar Cells; Constance, Germany; 2019. p. 20016.

[88] Springer M., Bosco N. 'Linear viscoelastic characterization of electrically conductive adhesives used as interconnect in photovoltaic modules'. *Progress in Photovoltaics: Research and Applications*. 2020;**28**(7):659–81.

[89] Pander M., Schulze S.-H., Ebert M. 'Mechanical modeling of eletrically conductive adhesives for photovoltaic applications'. Proceedings of the 29th European Photovoltaic Solar Energy Conference and Exhibition; Amsterdam, Netherlands; 2014. pp. 3399–405.

[90] Baldan A. 'Adhesion phenomena in bonded joints'. *International Journal of Adhesion and Adhesives*. 2012;**38**(2):95–116.

[91] Kinloch A.J. *Adhesion and Adhesives: Science and Technology*. 1st ed.. Netherlands: Springer; 1987.

[92] Zwolinski M., Hickman J., Rubin H., *et al.* 'Electrically conductive adhesives for surface mount solder replacement'. *IEEE Transactions on Components, Packaging, and Manufacturing Technology: Part C*. 1996;**19**(4):241–50.

[93] Bosco N., Springer M. 'Investigation of failure modes, mechanisms and driving forces for electrically conductive adhesives as interconnects in PV modules'. Proceedings of the 37th European Photovoltaic Solar Energy Conference and Exhibition; online; 2020. pp. 845–7.
[94] Eitner U., Geipel T., Holtschke S.-N., Tranitz M. 'Characterization of electrically conductive adhesives'. *Energy Procedia*. 2012;**27**:676–9.
[95] Hinz I., Schulze S.-H., Barche F., Schak M., Wenger F. 'Characterization of electrical conductive adhesives (ECA) for the photovoltaic-industry'. Proceedings of the 27th European Photovoltaic Specialist Conference and Exhibition; Frankfurt, Germany; 2012. pp. 3474–8.
[96] Schneider A. 'Conductive gluing as interconnection technique towards solar cells without front busbars and rear pads'. Proceedings of the 27th European Photovoltaic Specialist Conference and Exhibition; Frankfurt, Germany; 2012. pp. 335–9.
[97] Hoffmann S., Geipel T., Meinert M., Kraft A. 'Analysis of peel and shear forces after temperature cycle tests for electrical conductive adhesives'. Proceedings of the 33rd European Photovoltaic Solar Energy Conference and Exhibition; Amsterdam, Netherlands; 2017. pp. 183–6.
[98] Zemen Y. 'Innovative and gentle interconnection technique for high efficiency c-Si solar cells and cost-of-ownership-analysis (COO)'. Proceedings of the 26th European Photovoltaic Solar Energy Conference and Exhibition; Hamburg, Germany; 2011. pp. 3125–32.
[99] Schneider A., Harney R., Aulehla S., *et al.* 'Comprehensive study of material dependency for silver based conductive glues'. *Energy Procedia.* 2014;**55**:509–18.
[100] Kaiser C., Nikitina V., Geipel T., Kraft A. R'eduction of ECA amount for the ribbon interconnection of heterojunction solar cells'. Proceedings of the 37th European Photovoltaic Solar Energy Conference and Exhibition; online; 2020. pp. 1086–90.
[101] Geipel T., Meinert M., Kraft A., Eitner U. 'Optimization of electrically conductive adhesive bonds in photovoltaic modules'. *IEEE Journal of Photovoltaics*. 2018;**8**(4):1074–81.
[102] Test methods standards: microcircuits. MIL-STD-883H. **Department of Denfense USA**; 2010.
[103] Inoue M., Muta H., Yamanaka S., Suganuma K. 'Temperature dependence of electrical resistivity of isotropic conductive adhesive composed of an epoxy-based binder'. International Symposium on High Density Design Packaging and Microsystem Integration; Shanghai, China; 2007. pp. 1–4.
[104] Shockley W.B. *Research and investigation of inverse epitaxial UHF power transistor: Technical documentary report no. AL TDR 64-207*. Ohio: AF Avionics Laboratory; 1964.
[105] Schroder D.K. 'Contact resistance and Schottky barriers'. *Semiconductor Material and Device Characterization.* IEEE Press, Wiley-Interscience; 2006. pp. 127–84.

[106] Chern J.G.J., Oldham W.G. 'Determining specific contact resistivity from contact end resistance measurements'. *IEEE Electron Device Letters.* 1984;**5**:178–80.

[107] Devoto M.I., Timofte T., Halm A., Tune D. 'Measuring the contact resistivity of ECA-based joints'. Proceedings of the 37th European Photovoltaic Solar Energy Conference and Exhibition; online; 2020. pp. 1001–8.

[108] Beaucarne G., Broek K., Chislea B., *et al.* 'Study of compatibility of silicone-based electrically conductive adhesives and conductive Backsheets for MWT modules'. *Energy Procedia.* 2014;**55**:444–50.

[109] Sánchez-Friera P., Ropero F., Lalaguna B., Caballero L.J., Alonso J. 'Power losses in crystalline silicon PV modules due to cell interconnection'. Proceedings of the 23rd European Photovoltaic Solar Energy Conference and Exhibition; Valencia, Spain; 2008. pp. 2701–4.

[110] Schwertheim S., Scherff M.L.D., Mueller T., Fahrner W.R., Neitzert H.C. 'Lead-free electrical conductive adhesives for solar cell interconnectors'. Proceedings of the 33rd IEEE Photovoltaic Specialists Conference; San Diego, CA, USA; 2008. pp. 1–6.

[111] Geipel T., Huq M.Z., Eitner U. 'Reliable interconnection of the front side grid fingers using silver-reduced conductive adhesives'. *Energy Procedia.* 2014;**55**:336–41.

[112] Geipel T., Kumm J., Moeller M., *et al.* 'Low-temperature interconnection of PVD-aluminium metallization'. *Energy Procedia.* 2016;**98**(5):125–35.

[113] Ivanov V., Boehme B., Wolter K.-J. 'Resistance of conductive adhesive joints on non-noble surface finishes'. 35th International Spring Seminar on Electronics Technology (ISSE); Bad Aussee; 2012. pp. 254–9.

[114] Wong C.P., Lu D.D., Meyers L., Vona S.A., Jr., Tong Q.K. F'undamental study of electrically conductive adhesives (ECAs)'. Proceedings of the 1st IEEE International Symposium on Polymeric Electronics Packaging; Norrkoping, Sweden; 1997. pp. 80–5.

[115] Kinloch A.J. (ed.). *Durability of Structural Adhesives.* Netherlands: Springer; 1983.

[116] Xu S., Dillard D.A. 'Environmental aging effects on thermal and mechanical properties of electrically conductive adhesives'. *The Journal of Adhesion.* 2003;**79**(7):699–723.

[117] Bauermann L.P., Fokuhl E., Stecklum S., *et al.*. ' 'Qualification of conductive adhesives for photovoltaic application - accelerated ageing tests''. *Energy Procedia.* 2017;**124**:554–9.

[118] Jagt J.C. 'Reliability of electrically conductive adhesive joints for surface mount applications: a summary of the state of the art'. *IEEE Transactions on Components, Packaging, and Manufacturing Technology: Part A.* 1998;**21**(2):215–25.

[119] Lu D.D., Tong Q.K., Wong C.P. 'Mechanisms underlying the unstable contact resistance of conductive adhesives'. *IEEE Transactions on Electronics Packaging Manufacturing.* 1999;**22**:228–32.

[120] Mo Z., Lai Z., Li S., Liu J., Johan L. 'Thermal fatigue cracking of surface mount conductive adhesive joints'. *Soldering & Surface Mount Technology*. 2004;**16**(1):48–52.

[121] Périchaud M.G., Delétage J.Y., Frémont H., Danto Y., Faure C. 'Reliability evaluation of adhesive bonded smt components in industrial applications'. *Microelectronics Reliability*. 2000;**40**(7):1227–34.

[122] Frisk L., Lahokallio S., Mostofizadeh M., Kiilunen J., Saarinen K. 'Reliability study of isotropic electrically conductive adhesives under thermal cycling testing'. 63rd IEEE Electronic Components and Technology Conference (ECTC); Las Vegas, NV, USA; 2013. pp. 1794–9.

[123] Oreski G. 'Reliability of electically conductive adhesives'. *Photovoltaics International*. 2019;**42**:35–48.

[124] Eerenstein W. 'Climate chamber test result of MWT back contact modules'. Proceedings of the 25th European Photovoltaic Solar Energy Conference and Exhibition/ 5th World Conference on Energy Conversion; Valencia, Spain; 2010. pp. 3854–7.

[125] Ebner R. 'Increased power output of crystalline silicon solar modules by application of new module concepts'. Proceedings of the 29th European Photovoltaic Specialist Conference and Exhibition; Amsterdam, Netherlands; 2014. pp. 171–6.

[126] Ullmann I., Rudolph D., Rabanal Arabach J., Schneider A. 'Investigation on the quality of adhesive joints of shingled solar cells by accelerated lifetime testing'. Proceedings of the 35th European Photovoltaic Solar Energy Conference and Exhibition; Brussels, Belgium; 2018. pp. 1210–13.

[127] Geipel T., Eitner U. 'Cure kinetics of electrically conductive adhesives for solar cell interconnection'. *IEEE Journal of Photovoltaics*. 2013;**3**(4):1208–14.

[128] Adamietz R. 'A Microwave curing system for microelectronics assembly'. 4th Electronic System-Integration Technology Conference; Amsterdam, Netherlands; 2012. pp. 1–5.

[129] Cheng W.-T., Chih Y.-W., Yeh W.-T. 'In situ fabrication of photocurable conductive adhesives with silver nano-particles in the absence of capping agent'. *International Journal of Adhesion and Adhesives*. 2007;**27**(3):236–43.

[130] Wirth H. 'Crystalline silicon PV module technology' in Willeke G.P., Weber R.E. (eds.). *Semiconductors and Semimetals Volume 89: Advances in Photovoltaics Part 2*. Amsterdam: Academic Press; 2013. pp. 135–97.

[131] Eerenstein W., Bennett I.J., Späth M., Dekker N.J., Rosca V. 'Back contact module technology'. SolarCon/CPTIC China; Shanghai; 2012.

[132] Francis M. 'Latest developments in conductive adhesives for back contact applications'. Engineered Conductive Materials (ECM), 5th MWT Workshop; Amsterdam, Netherlands; 2013.

[133] MatWeb. *matweb LLC, material property data, blacksburg, virginia* [online]. 2020. Available from www.matweb.com [Accessed 30 Nov 2020].

[134] Klein Wassink R.J. *Soldering in Electronics*. 2nd ed. Saulgau/Württ: Leuze; 1991.

[135] Geipel T., Rendler L.C., Stompe M., Eitner U., Rissing L. 'Reduction of thermomechanical stress using electrically conductive adhesives'. *Energy Procedia*. 2015;**77**:346–55.

[136] Eitner U., Rendler L.C. 'Peel testing of ribbons on solar cells at different angles: Consistent comparison by using adhesive fracture energies'. Proceedings of the 29th European Photovoltaic Solar Energy Conference and Exhibition; Amsterdam, Netherlands; 2014. pp. 3406–8.

[137] Geipel T., Eberlein D., Kraft A. "Lead-free solders for ribbon interconnection of crystalline silicon PERC solar cells with infrared soldering". *AIP Conference Proceedings*. **2156**. Konstanz; 2019. p. 20015.

[138] Pingel S., Korth H., Winkler M., Weiss P., Geipel T. 'Investigation of damp heat degradation mechanisms and correlation to an accelerated test procedure (HAST)'. Proceedings of the 27th European Photovoltaic Specialist Conference and Exhibition; Frankfurt; 2012. pp. 3137–41.

[139] Hacke P. 'Test-to-Failure of crystalline silicon modules'. Proceedings of the 35th IEEE Photovoltaic Specialists Conference; Honolulu; 2010. pp. 244–50.

[140] Jordan D.C., Kurtz S.R. 'Photovoltaic degradation rates-an analytical review'. *Progress in Photovoltaics: Research and Applications*. 2013;**21**(1):12–29.

[141] Ecoprogetti SRL. ETS2100: *Stringer Machine [online]*. Available from https://ecoprogetti.com/product/ets2100/ [Accessed 12 Feb 2020].

[142] Schulte-Huxel H. *'Laser microspot welding for interconnection of back-contacted silicon solar cells'. Dissertation*. Hannover: Leibniz Universität Hannover; 2015.

[143] Heimann M., Klaerner P., Luechinger C., *et al*. 'Ultrasonic bonding of aluminum ribbons to interconnect high-efficiency crystalline-silicon solar cells'. *Energy Procedia*. 2012;**27**:670–5.

[144] Jung V., Köntges M. '{Al/Ni:V/Ag} metal stacks as rear-side metallization for crystalline silicon solar cells'. *Progress in Photovoltaics: Research and Applications*. 2012;**21**(5):876–83.

[145] Köntges M., Kurtz S., Packard C., Jahn U., Berger K.A., Kato K. *Performance and reliability of photovoltaic systems: subtask 3.2: review of failures of photovoltaic modules*. International Energy Agency (IEA); 2014.

[146] Kim J.-H., Park J., Kim D., Park N. 'Study on mitigation method of Solder corrosion for crystalline silicon photovoltaic modules'. *International Journal of Photoenergy*. 2014;**2014**(7):1–9.

[147] Duerr I., Peike C., Hoffmann S., Koehl M., Weiß K.-A. 'X-ray study on damp-heat induced cell degradation (DHID)'. Proceedings of the 28th European Photovoltaic Solar Energy Conference and Exhibition; Paris, France; 2013. pp. 3300–2.

[148] Palitzsch W., Loser U. 'Economic PV waste recycling solutions – Results from R&D and practice'. Proceedings of the 29th European Photovoltaic Solar Energy Conference and Exhibition; Amsterdam, The Netherland; 2014. pp. 628–31.

[149] Klugmann-Radziemska E., Ostrowski P., Drabczyk K., Panek P., Szkodo M. 'Experimental validation of crystalline silicon solar cells recycling by thermal and chemical methods'. *Solar Energy Materials and Solar Cells*. 2010;**94**(12):2275–82.

[150] Roberts R.B. 'Thermal expansion reference data: silicon 300-850 K'. *Journal of Physics D: Applied Physics*. 1981;**14**(10):L163–6.

[151] Mayr M. *Technische Mechanik: Statik, Kinematik, Kinetik, Schwingungen, Festigkeitslehre*. 2nd ed.. München: Hanser; 1999.

[152] Gabor A.M., Ralli M., Montminy S., Alegria L., Bordonaro C., Woods J. 'Soldering induced damage to thin Si solar cells and detection of cracked cells in modules'. Proceedings of the 21st European Photovoltaic Solar Energy Conference and Exhibition; Dresden, Germany; 2006. pp. 2042–7.

[153] Wendt J., Träger M., Mette M., Pfennig A., Jäckel B. 'The link between mechanical stress induced by soldering and micro damages in silicon solar cells'. Proceedings of the 24th European Photovoltaic Solar Energy Conference and Exhibition; Hamburg, Germany; 2009. pp. 3420–2.

[154] Sander M., Henke B., Schwarz H., Dietrich S., Schweizer S., Ebert M. 'Characterization of PV modules by combining results of mechanical and electrical analysis methods'. Proceedings of SPIE 7773, Reliability of Photovoltaic Cells, Modules, Components, and Systems III; San Diego, CA, USA; 2010. p. 777308.

[155] Dietrich S., Pander M., Sander M., Zeller U., Ebert M., Dhere N.G. 'Stress analysis of encapsulated solar cells by means of superposition of thermal and mechanical stresses'. Proceedings of SPIE 8825, Reliability of Photovoltaic Cells, Modules, Components, and Systems VI; San Diego, CA, USA; 2013. pp. 882505–10.

[156] Budiman A.S., Illya G., Handara V., *et al.* 'Enabling thin silicon technologies for next generation c-Si solar pv renewable energy systems using synchrotron X-ray microdiffraction as stress and crack mechanism probe'. *Solar Energy Materials and Solar Cells*. 2014;**130**:303–8.

[157] Curtin D.J., Billerbeck W.J. 'Development of advanced intereonnectors for solar cells'. *COMSAT Technical Review*. 1974;**4**(no. 1):53–68.

[158] Buhs R. 'Layout and Technology of the CTS Solar Array Blanket'. Proceedings of the 10th IEEE Photovoltaic Specialists Conference; Palo Alto, CA, USA; 1973. pp. 296–306.

[159] Reinhartz K.K., Capart J.J. 'Status of Welded Solar Cell Module Technology at ESRO'. Proceedings of the 9th IEEE Photovoltaic Specialists Conference; Silver Springs, MD, USA; 1970. pp. 287–92.

[160] Mesch H.G. 'Parallel Gap Welding of Silver Plated Solar Cells'. Proceedings of the 10th IEEE Photovoltaic Specialists Conference; Palo Alto, CA, USA; 1973. pp. 275–80.

[161] Rauschenbach H.S. *Solar Cell Array Design Handbook: The Principles and Technology of Photovoltaic Energy Conversion*. New York: Van Nostrand Reinhold; 1980.

[162] Buhs R. 'Silicon Solar Cell Arrays Technical Status, Terrestrial Applications, and Future Developments for Low Cost Production'. Proceedings of the 13th IEEE Photovoltaic Specialists Conference; Washington, DC, USA; 1978. pp. 262–7.

[163] Moalem A., Schoonderbeek A., Kling R., Stute U., Denkena B. 'Laser micro welding in silicon solar module manufacturing'. Proceedings of the 8th International Symposium on Laser Precision Microfabrication; Vienna, Austria; 2007.

[164] Moalem A., Schoonderbeek A., Kling R., Ostendorf A. 'Advancing laser joining in solar module manufacturing'. Proceedings of the 17th international Photovoltaics Science and Engineering; Fukuoka, Japan; 2007. pp. 438–9.

[165] Ehrhardt M., Lorenz P., Scheit C., Wehrmann A., Zimmer K., Gerlach L. 'Low-damage micro-interconnection with nanosecond laser pulses for high-efficient GaAs solar cells'. Presented 11th International Conference on Laser Ablation; Playa del Cramen, Mexico; 2011. p. PMO79.

[166] Schulte-Huxel H., Blankemeyer S., Merkle A., Steckenreiter V., Kajari-Schröder S., Brendel R. 'Interconnection of busbar-free back contacted solar cells by laser welding'. *Progress in Photovoltaics: Research and Applications*. 2015;**23**(8):1057–65.

[167] Fisher K., Meng X., Hartweg B., Mony S., Bertoni M., Holman Z. 'Novel foil interconnects for back-contact silicon solar cells'. Proceedings of the 46th IEEE Photovoltaic Specialists Conference; Chicago, IL, USA; 2019. pp. 126–30.

[168] Schulte-Huxel H., Blankemeyer S., Bock R., Merkle A., Kajari-Schröder S., Brendel R. 'Al-Foil on Encapsulant for the interconnection of Al-metalized silicon solar cells'. *IEEE Journal of Photovoltaics*. 2013;**3**(1):77–82.

[169] Schulte-Huxel H., Petermann J.-H., Blankemeyer S., Steckenreiter V., Kajari-Schroeder S., Brendel R. 'Simultaneous contacting and interconnection of Passivated emitter and rear solar cells'. *Energy Procedia*. 2016;**92**(6):515–22.

[170] Schulte-Huxel H., Blankemeyer S., Steckenreiter V., Kajari-Schroeder S., Brendel R. 'Laser-welded interconnection of screen-printed Si solar cells'. *Energy Procedia*. 2014;**55**(1):356–60.

[171] Wu Z., Yu H., Su B., Liu S., Lin J., Cui X. 'Develop solderability and welding heat source of the Kovar/Ag interconnector based on finite element simulation'. Proceedings of the 21st International Conference on Electronic Packaging Technology (ICEPT); Guangzhou, China; 2020. pp. 1–3.

[172] Zhan X., Zhang Q., Zhu Z., Wei Y. 'Numerical simulation of resistance welding of solar cell using a thermal-electrical-mechanical coupled model'. *Journal of Mechanical Science and Technology*. 2018;**32**(1):269–76.

[173] Fendrock J.J., Hong L.M. 'Parallel-gap welding to very-thin metallization for high temperature microelectronic interconnects'. *IEEE Transactions on Components, Hybrids, and Manufacturing Technology*. 1990;**13**(2):376–82.

[174] Moore T.J., Watson G.K., Baraona C.R. 'Microstructural analysis of solar cell welds'. *Space Photovoltaic Research and Technology: High Efficiency, Radiation Damage, and Blanket Technology*. Washington, DC, USA: National Aeronautics and Space Administration (NASA); 1982. pp. 237–49.

[175] Brennan P.C., Jamieson R.A., Garcia A., Malachesky P.A., Simburger E.J. 'Evaluation of welded interconnects on thermally cycled silicon solar cells'. Proceedings of 1994 IEEE 1st World Conference on Photovoltaic Energy Conversion - WCPEC (A Joint Conference of PVSC, PVSEC and PSEC): IEEE; Waikoloa, HI, USA; 1994. pp. 2026–9.

[176] La Roche G.J., Oxynos-Lauschke C., Wehner K.-H. 'Gold interconnectors for solar generators in low earth orbits'. Proceedings of 1994 IEEE 1st World Conference on Photovoltaic Energy Conversion - WCPEC (A Joint Conference of PVSC, PVSEC and PSEC): IEEE; 1994. pp. 2196–9.

[177] Cavicchi B.T., Lillington D.R., Garlick G.F.J., Glenn G.S., Tobin S.P. 'GaAs on Ge cell and panel technology for advanced space flight application'. Conference Record of the Twentieth IEEE Photovoltaic Specialists Conference: IEEE, Vol 2; 1988. pp. 918–23.

[178] Chu C.-L. Weldability of GaAs solar cells on either GaAs or Ge substrates. Conference Record of the Twentieth IEEE Photovoltaic Specialists Conference: IEEE, Vol 2; Sep. 1996 – Sep. 1988. pp. 968–73.

[179] Ruge J. *Handbuch der Schweißtechnik: Band II: Verfahren und Fertigung*. Berlin, Heidelberg: Springer Berlin Heidelberg; 1980.

[180] Feldmann K., Schöppner V., Spur G. (eds.) *Handbuch Fügen, Handhaben, Montieren*. 2nd ed. München: Hanser; 2014.

[181] Bäuerle D. *Laser Processing and Chemistry*. Berlin, Heidelberg: Springer; 2011.

[182] P'ng D., Molian P. 'Q-switch Nd:YAG laser welding of AISI 304 stainless steel foils'. *Materials Science and Engineering: A*. 2008;**486**(1-2):680–5.

[183] Patschger A., Bliedtner J., Bergmann J.P. 'Approaches to increase process efficiency in laser micro welding'. *Physics Procedia*. 2013;**41**:592–602.

[184] Ventrella V.A., Berretta J.R., de Rossi W. 'Pulsed Nd:YAG laser welding of Ni-alloy hastelloy C-276 foils'. *Physics Procedia*. 2012;**39**:569–76.

[185] Hirano T., Demura M., Kishida K., Minamida K., Xu Y. 'Laser spot welding of cold-rolled boron-free Ni_3Al foils'. *Metallurgical and Materials Transactions A*. 2007;**38**(5):1041–7.

[186] Ehrhard M., Zimmer K. 'Laser micro joining of thin metal films on flexible substrates for mechanical and electrical connections'. Proceedings SPIE 7921, Laser-based Micro- and Nanopackaging and Assembly V; San Francisco, CA, USA; 2011. p. 792109.

[187] Ismail M.I.S., Okamoto Y., Okada A., Uno Y., Ueoka K. 'Direct micro-joining of flexible printed circuit and metal electrode by pulsed Nd:YAG laser'. *International Journal of Precision Engineering and Manufacturing*. 2012;**13**(3):321–9.

[188] Hendel R., Lehner B., Rüttimann C. 'Faster and reliable joining of solar cells: Bi-Wavelength laser welding for photovoltaic module integration'. *Laser Technik Journal*. 2012;**9**(1):20–3.

[189] Schulte-Huxel H., Bock R., Blankemeyer S., Merkle A., Brendel R. 'Aluminum-based mechanical and electrical laser interconnection process for module integration of silicon solar cells'. *IEEE Journal of Photovoltaics*. 2012;**2**(1):16–21.

[190] Schulte-Huxel H., Blankemeyer S., Kajari-Schröder S., Brendel R. 'Laser microwelding of thin Al layers for interconnection of crystalline Si solar cells: analysis of process limits for n and μs lasers'. *Journal of Photonics for Energy*. 2014;**4**(1):041597–13.

[191] Schulte-Huxel H., Blankemeyer S., Bock R., Merkle A., Kajari-Schröder S., Brendel R. 'Aging behaviour of laser welded Al-interconnections in crystalline silicon modules'. *Solar Energy Materials and Solar Cells*. 2012;**106**(1):22–6.

[192] Schulte-Huxel H., Römer U., Blankemeyer S., *et al*. 'Two-Level metallization and module integration of point-contacted solar cells'. *Energy Procedia*. 2014;**55**(5):361–8.

[193] Hermle M., Granek F., Schultz-Wittmann O., Glunz S.W. 'Shading effects in back-junction back-contacted silicon solar cells'. Proceedings of the 33rd IEEE Photovoltaic Specialist Conference; San Diego, CA, USA; 2008. pp. 412–5.

[194] Verlinden P., Swanson R.M., Sinton R.A., Kane D.E. 'Multilevel metallization for large area point-contact solar cells'. Proceedings of the 20th IEEE Photovoltaic Specialist Conference; Las Vegas, NV, USA; 1988. pp. 532–7.

[195] Schneiderlöchner E., Preu R., Lüdemann R., Glunz S.W. 'Laser-fired rear contacts for crystalline silicon solar cells'. *Progress in Photovoltaics: Research and Applications*. 2002;**10**(1):29–34.

[196] Sunflexsolar. *A new way to interconnect back-contact solar cells*. 2020. Available from http://sunflexsolar.net/ [Accessed 2 Nov 2020].

Chapter 11
Module design

Yang Li[1], Pei-Chieh Hsiao[1], and Alison Lennon[1]

11.1 Module design

Photovoltaic (PV) module design is the process of determining the electrical connection configuration of solar cells to deliver a desired current and voltage with the supporting structures of the module being configured for maximal light absorption, environmental protection and mechanical strength [1]. It directly impacts the performance, lifetime and levelized cost of PV-generated electricity. Over recent years, PV module design has assumed a growing focus, as the challenges for modules are becoming as critical as they are for cell design. This is because modules are composite optical and electrical structures which require cross-disciplinary research in physics, chemistry, materials science and engineering.

Crystalline Si PV modules were pioneered by Bell Laboratories for the application of powering remote telecommunication applications in 1955 [2]. However, practical uses were limited to space applications over the following 20 years due to their higher value proposition. Then, triggered by the oil crisis in the 1970s, interest returned to the manufacture of modules for terrestrial applications with companies such as Sharp and Philips becoming engaged. The next ten years saw the evolution of the modern era of modules with inventions including the glass cover, the lamination process with ethylene vinyl acetate (EVA) and the use of quasi-square crystalline wafers. Standards were also established to test the durability of manufactured modules. Module design then stabilized whilst focus centred on improving cell efficiency and the cost-effectiveness of the manufacturing process for volume production. However, as solar cell technology has matured and reduced massively in cost during the last decade, a new phase of innovations has occurred in module manufacturing resulting in advances such as half-cut cells for high power modules, multi-busbar (MBB) interconnection and bifacial modules.

Modules comprise a number of interconnected solar cells encapsulated between protecting materials, such as EVA, silicone, thermoset polyolefin elastomer (POE) and polyvinyl butyral (PVB), and front/back covers comprising glass and backsheet (see Figure 11.1). The multiple layers protect the solar cells and their interconnecting

[1]University of New South Wales (UNSW), Kensington, Australia

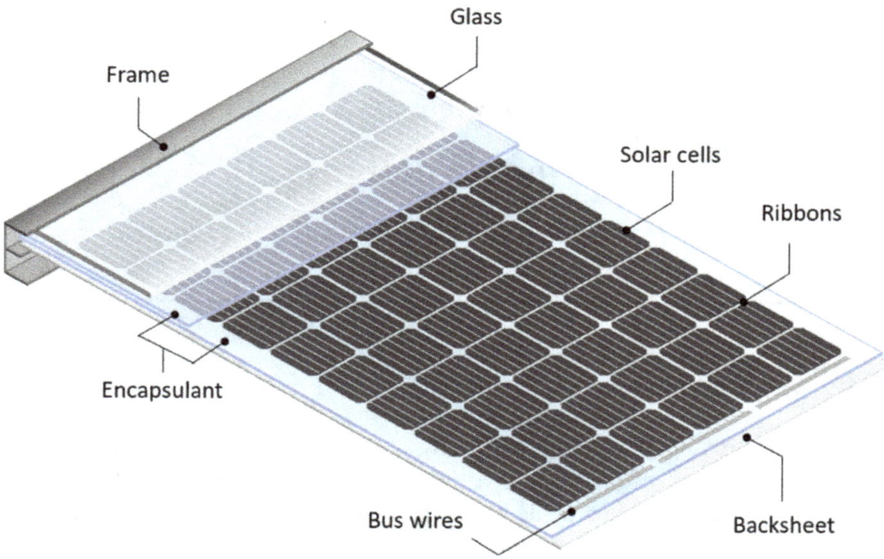

Figure 11.1 *Schematic of a solar module with solar cells laminated in encapsulant, backsheet and front glass*

wires from the harsh environmental conditions in the field. They provide mechanical support for wind load, hail and other impacts, and protection for metal contacts against corrosion, which can be initiated by moisture ingress and corrosive degradation products of the module materials. Metal frames are used to seal the edge and provide extra mechanical support and facilitate connections to racks.

The design of a module needs to accommodate the requirements of purchasers, including typical electrical loads and installation scenarios (e.g. manual or automated installation). From these requirements, the specifications for power, size and weight, can be determined. Initial concept designs can be evaluated using modelling and simulation methods and then, after an analysis of manufacturability and cost, prototypes can be fabricated for performance measurement and stability testing.

The effectiveness of transferring the power and efficiency of solar cells into those of a module can be evaluated using a cell-to-module (CTM) ratio [3, 4]. Losses/gains incurred in this transfer include reflection at various interfaces (e.g. air-glass, glass-encapsulant and encapsulant-cells), parasitic absorption by the front glass and encapsulant material, coupling gain of scattered light and electrical losses from interconnection and mismatch. Figure 11.2 illustrates these losses graphically as a waterfall diagram. CTM analysis is a systemic framework to evaluate the design of a module and identify opportunities for optimization. However, the methodology becomes challenging to implement when solar cells are difficult to characterize, for example when busbar-less cells are used.

Mechanical designs are also critical to ensure the long-term performance of a module. It is necessary to understand the properties of encapsulants,

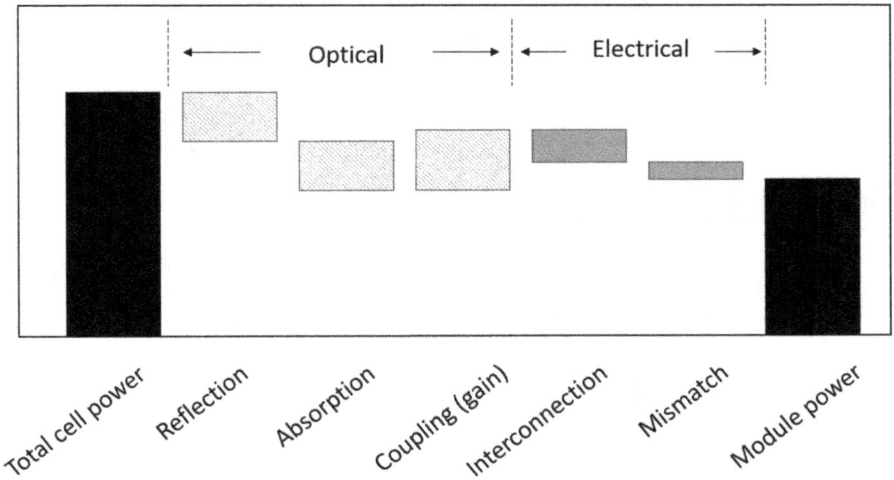

Figure 11.2 *Example of a waterfall diagram for the analysis of a solar module*

thermomechanical stress in module-making processes, supporting structures and possible failure modes. Based on a physical understanding of these processes, the module design, material choices and fabrication processes can be optimised for longer operating lifetimes. The detailed design processes and considerations for electrical, optical and mechanical designs will be discussed in this chapter.

11.2 Electrical design

The electrical parameters of modules are routinely measured under the standard test condition (STC), which is defined as nominal cell temperature of 25 °C, irradiance of 1 000 W m^{-2} at spectral distribution of air mass 1.5. Every PV module data sheet provided by the manufacturer includes the ratings of: maximum power (P_{max}), rated voltage and current at P_{max} (V_{mp} and I_{mp}), short-circuit current (I_{sc}), open-circuit voltage (V_{oc}) and module efficiency (η_m). In addition, these ratings should be provided with a nominal operating cell temperature (NOCT), which is defined as the equilibrium cell temperature corresponding to an open-circuit module operating in a reference environment of 800 W m^{-2} irradiance, 20 °C ambient air temperature with a 1 m/s wind speed and an open back side mounting. These parameters need to be measured at STC and NOCT for cross-referencing and standardisation. However, it is not always straightforward to use these parameters to reflect the actual module performance in the field, as performance depends on factors additional to electrical parameters measured at STC and NOCT.

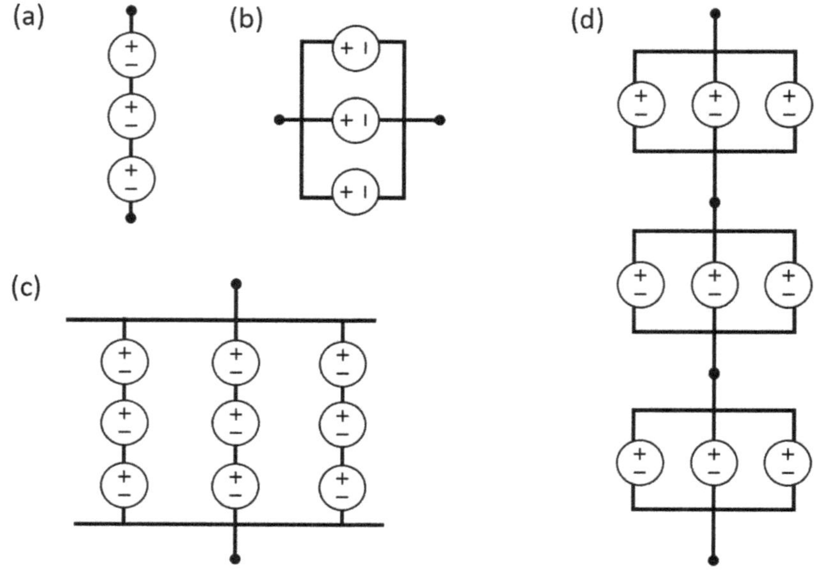

Figure 11.3 *Solar cells can be series connected in a string (A) in parallel as a group (B), grouping strings together (C) or stringing groups (D)*

11.2.1 Module circuit design

Series and parallel interconnection

Individual solar cells can be connected in series, parallel or hybrid arrangements as required for different applications, as illustrated in Figure 11.3. When cells are stringed in series by connecting the positive terminal of one cell to the negative of the next, the voltage is the sum of all cells in the string. On the other hand, cells can be connected in parallel where all the positive terminals and negative terminals are connected. In this way, the module current is the sum of the solar cell current in the group and equivalent to fabricating a cell with a larger surface area. In hybrid arrangements, cells in parallel groups can then be placed in series with parallel groups.

For a configuration with M strings of N identical cells per string, the module short circuit and maximum power point current ($I_{sc,module}$ and $I_{mp,module}$) and open circuit and maximum power point voltage ($V_{oc,module}$ and $V_{mp,module}$) are:

$$I_{sc,module} = I_{sc,cell} \times M \tag{11.1}$$

$$I_{mp,module} = I_{mp,cell} \times M \tag{11.2}$$

$$V_{oc,module} = V_{oc,cell} \times N \tag{11.3}$$

$$V_{mp,module} = V_{mp,cell} \times N \tag{11.4}$$

Historically, solar modules were designed for stand-alone applications which required a voltage higher than 15 V to charge 12 V batteries. As $V_{mp,cell}$ was ~0.5V at normal operating temperatures, most early PV modules were designed to have 36 cells in series. As grid-connected residential, commercial and utility scale systems have evolved to become more dominant, modules with 60 and 72 cells in series are most commonly available in the market. Modules can be connected in a combination of series and parallel circuits to achieve practically any desired final voltage and current.

Electrical power loss calculation

The electrical power losses for cell interconnection into modules for the typical H-pattern cell designs with front and backside contacts can be expressed as:

$$\Delta P = \Delta P_{ribbon} + \Delta P_{cellgap} + \Delta P_{buswire} \tag{11.5}$$

where ΔP_{ribbon} , $\Delta P_{cellgap}$ and $\Delta P_{buswire}$ are power losses in the ribbon on the front and backside of solar cells, cell gap and bus wires, respectively. By increasing the packing density of the solar cells in a module, both $\Delta P_{cellgap}$ and $\Delta P_{buswire}$ can be reduced. However, the ohmic loss in ribbons, ΔP_{ribbon}, is usually the largest electrical loss [5, 6]. It can be expressed in detail as:

$$\Delta P_{ribbon} = n\sum_{k=1}^{n_{bb}} \frac{1}{3} \times I_k^2 \times R_{bb+ribbon} = n\sum_{k=1}^{n_{bb}} \frac{1}{3} \times I_k^2 \times \frac{\bar{\rho}_{bb+ribbon}}{A} \times l \tag{11.6}$$

where n is the number of cells in the module, n_{bb} is number of busbars, I_k is the current collected on the kth busbar, $R_{bb+ribbon}$ and $\rho_{bb+ribbon}$ are the electrical resistance and effective resistivity (Ω m) of the parallel connection between the busbar and the ribbon, and l and A are the length and cross-sectional area of the ribbon. From (11.6), it is evident that power loss can be effectively reduced in modules by reducing the current collected by each ribbon. This can be achieved by increasing the number of busbars and using half-cut cells, two trends that have become apparent in module manufacturing since 2019 [7].

For a module with 72 five-busbar full cells in series and a ribbon length of l, the electrical power loss on ribbons is:

$$\Delta P_{ribbon,\ 5bb,full} = 72 \times \frac{5}{3} \times \left(\frac{I_{mp}}{5}\right)^2 \times \frac{\bar{\rho}_{bb+ribbon}}{A} \times l \tag{11.7}$$

When the number of busbars is increased to nine, the loss becomes:

$$\Delta P_{ribbon,9bb,full} = 72 \times \frac{9}{3} \times \left(\frac{I'_{mp}}{9}\right)^2 \times \frac{\bar{\rho}_{bb+rib}}{A} \times l \approx 0.56\Delta P_{ribbon,5bb,full} \tag{11.8}$$

where I'_{mp}, the rated current for nine-busbar full cells, may differ from I_{mp} due to different shading. Furthermore, if half-cut cells are used in addition to nine busbars, the loss can be further reduced to:

$$\Delta P_{ribbon,9bb,half} = 144 \times \frac{9}{3} \times \left(\frac{I'_{mp}/2}{9}\right)^2 \times \frac{\bar{\rho}_{bb+rib}}{A} \times \frac{l}{2} \approx 0.14\Delta P_{ribbon,5bb,full} \tag{11.9}$$

However, it should be noted that narrower ribbons are typically used when the number of busbars is increased. This increases the ribbon's effective resistance. Additionally, when cells are cut in half, power losses are incurred from the increased number of cell gaps and passivation losses from the cut cell edges.

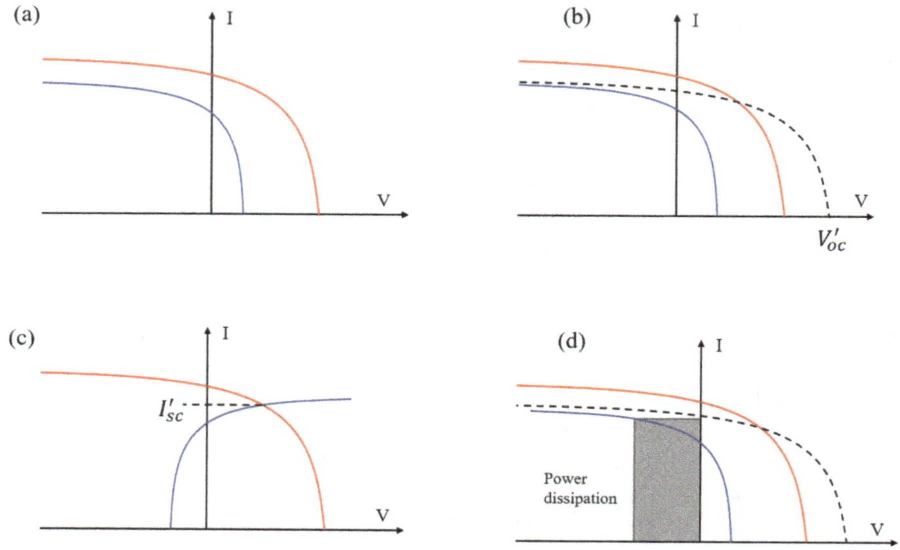

Figure 11.4 *Example* I-V *curves for: (A) Four similar non-shaded series-connected solar cells (red) and a single shaded (low* I*) cell (blue); (B)* V_{oc} *of series connected string (dashed line); (C)* I_{sc} *for the string; and (D) power dissipated in the shaded cell when the string is short-circuited (adapted from Roche et al. [8])*

Current mismatch

If the cells in a module are not matched in power, then the output power of the module can be less than the sum of the maximum power of individual cells. This difference is mismatch loss and can also contribute to the electrical losses of modules. Additionally, current mismatches can impact long-term stability and operational safety of modules.

A module can suffer mismatch due to manufacturing processes. Even though cells are routinely sorted into groups at the end of the production line based on their current-voltage (*I-V*) performance, some differences can still occur depending on the tolerance of the bin setting. Mismatch can also occur due to operational factors. For example, thermal gradients can arise when cells at the edge of the module receive greater air flow and experience more effective heat exchange or receive different irradiance from ground reflections. Cells in a module can also degrade at different rates. Finally, mismatches can arise through shading from leaves, bird droppings and uneven soiling [8].

As most modules comprise series-connected cells, series mismatches are the most common type of mismatch loss. Figure 11.4 illustrates the *I-V* curves that can result when a single poor performing cell (e.g. caused by shading) is series connected with four non-shaded cells. The *I* and *V* of the shaded cell significantly limits the power of the string. The string V_{oc} is obtained by adding the voltages of the

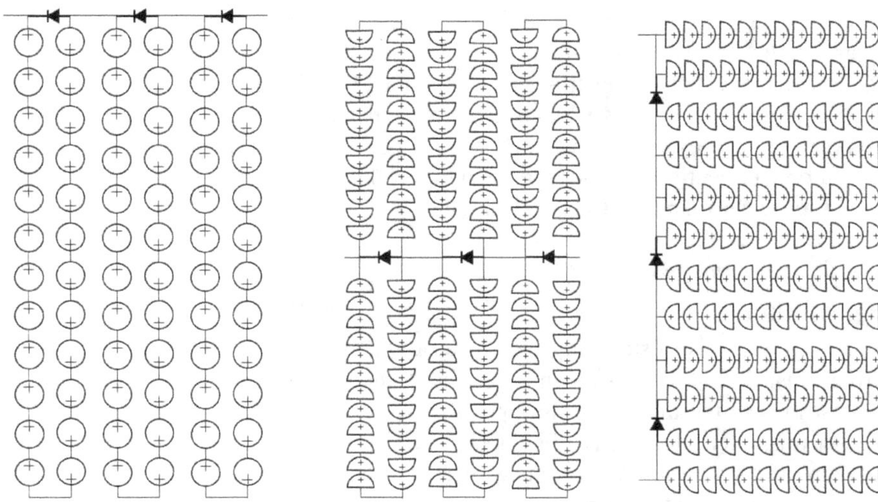

(a) (b) (c)

Figure 11.5 Photovoltaic (PV) module layout for: (A) A standard 72 full cell module with three bypass diodes; (B) A configuration with three series blocks, each block consisting of two parallel strings of 24 half cells, with three bypass diodes; and (C) A configuration with a series of six blocks, each block consisting of two parallel strings of 12 half cells, with three bypass diodes

shaded and non-shaded cells at the I flowing through the string. The I_{sc} of the string can be calculated by reflecting the I-V curve of the shaded cell in the I axis and identifying the point of intersection of the two I-V curves. As the current flowing through all the cells in a string must be the same, the total I from a mismatched string cannot exceed that of the poor, or shaded cell.

When all of the cells are operating in forward bias with minimal mismatches, the I_{mp} of the total string will operate close to the average of all the cells [9, 10]. The mismatch loss is proportional to the square of relative standard deviation of I_{mp} of the cells [9]. However, when operating at low voltages, the poorly performing or shaded cell can be operating under reverse bias with the excess power generated from the good cells being dissipated in the poor cells. When operating under short circuit, the entire module power can be dissipated in this way by the shading of a single cell as shown in Figure 11.4 d. This can result in localized heating, which can severely damage the shaded cell, melt solders, crack encapsulant materials and cause module failures. This phenomenon is known as hot-spot heating.

There are two techniques widely used for mitigating mismatch loss and hot-spot heating. Perhaps the most well-known method is the use of bypass diodes (which are discussed further in Section 11.1.2). The other technique is to use series-parallel hybrid module layouts. Figure 11.5 shows a comparison of layouts for a module with 72 full cells connected in series and two configurations of half cells. Each layout employs three bypass diodes. The half-cell layout in Figure 11.5 b has three blocks in series with two substrings consisting of 24 half cells each [11], whereas the layout in Figure 11.5 c has six blocks in series with two parallel substrings of 12 half cells each [12]. These two series-parallel-series designs allow a similar *I-V* response to the more conventional layout shown in Figure 11.5 a; however, the hybrid layouts allow for greater tolerance to partial shading.

11.2.2 Module-level electronics

Bypass diodes

Bypass diodes have historically been used to safeguard against extreme cases of current mismatches. As shown in Figure 11.5, the diodes are connected in parallel to substrings in the module with opposite polarity. When the module is operating without mismatch, the bypass diodes are reverse-biased and will effectively be at open circuit. However, they become forward-biased when their substring is reverse-biased due to mismatch, thereby providing a low-resistance current pathway and limiting the power dissipation in the block.

The usage of bypass diodes is illustrated in the example in Figure 11.6 a and b, which show the *I–V* and *P-V* responses of a 72-cell all series connected module as depicted in Figure 11.5 a without bypass diodes for different shading ratios of one cell. Both I_{sc} and P_{max} reduce with increased shading. However, when three bypass diodes are added to the module, the module responses are significantly changed as shown in Figure 11.6 c and d. With bypass diodes, the loss is constrained in the shaded substring and P_{max} is significantly higher for both partial and full shading conditions. Figure 11.6 e and f, which depict the response for the half cell module layout shown in Figure 11.5 b, demonstrate the improved tolerance to shading provided by hybrid series and parallel layouts.

Module optimizer and module inverter

Similar to the mismatch of cells in a module, mismatches can occur in a PV system due to manufacturing variations and environmental differences, including uneven shading across an array. Module-level power electronics can be used to address these losses by utilizing the maximum power of each individual module, and include DC/DC power optimizers and DC/AC microinverters. Both employ module-level maximum power point tracking (MPPT). Conventionally, MPPT is performed with string or central inverters. When modules perform their own MPPT, each module in an array can operate at its local maximum and improve performance as long as the power gain is greater than the losses resulting from the additional conversion step [13]. Module-level power electronics offer greater flexibility in system layout in shading conditions. They also enable module-level monitoring and control, which is required in some countries where module voltage needs to be reduced to a safe level whenever instructed or failure is detected.

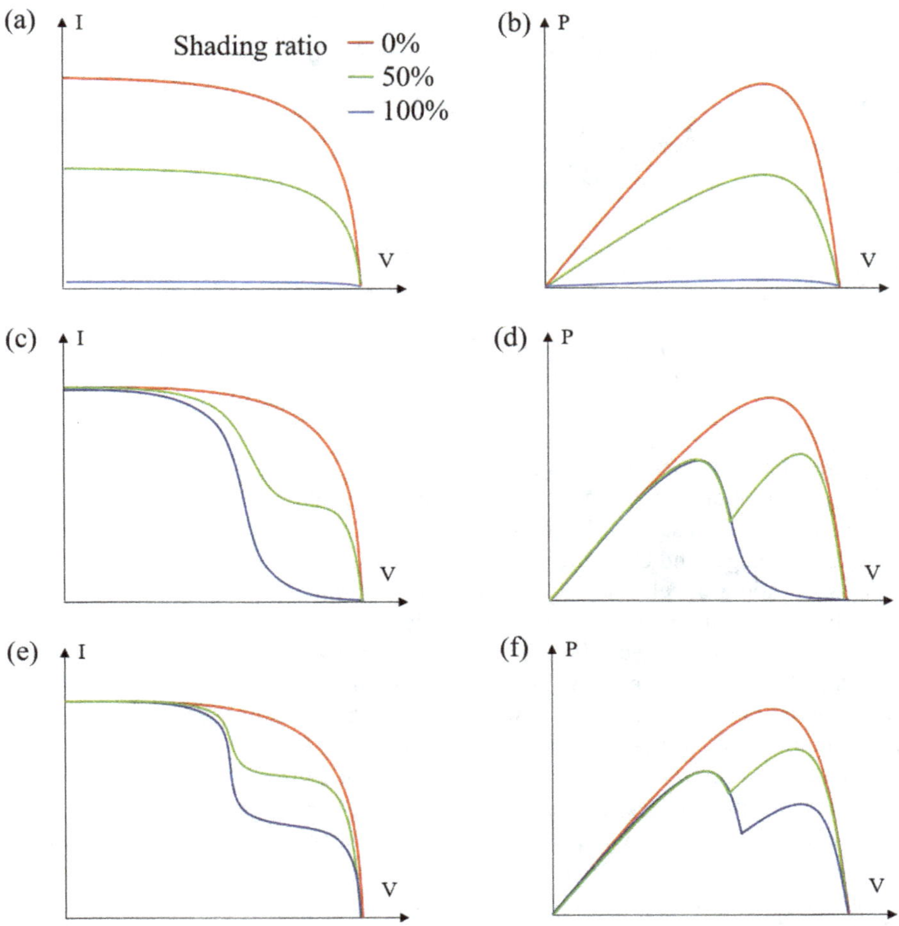

Figure 11.6 *(A) and (B)* I–V *and* P–V *responses of a module with 72 series connected cells without bypass diodes, with the unshaded curves plotted in red and the shaded curves plotted in green and blue. The shading ratios are labelled near the I–V curves. (C) and (D)* I–V *and P–V curves of a module with three bypass diodes and the layout in Figure 11.5 (A). (E) and (F) I–V and P–V curves of a 144 half-cell module with three bypass diodes and the layout in Figure 11.5 (B). In each case, one cell in each module is shaded with the labelled ratios (adapted from Qian et al. [11]).*

11.2.3 Modelling and simulation

Electrical design of modules requires a consideration of many factors and consequently can be made more efficient by modelling and simulation. Analytical solutions and finite element methods can be used to calculate resistive losses in cell

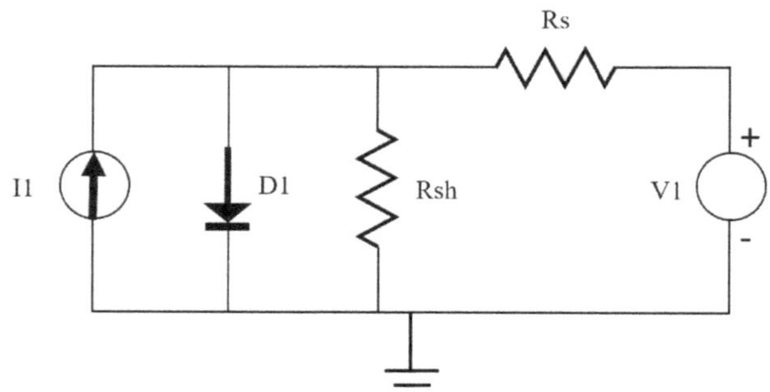

Figure 11.7 *An example of a single-diode equivalent circuit of a solar cell for SPICE simulation*

metallization and ribbons with power losses being expressed analytically for the individual components of a metallization pattern. A number of dedicated software design tools can be used. One example is Griddler [14] which can import grid geometries from images and CAD files, and enable optimization of metallization patterns using simulations of cell and module electrical performance [15, 16]. Anther module modelling tool with combined electrical, optical and thermal models is SmartCalc. CTM [17]. This software can be used for CTM loss analysis and flexible virtual prototyping [6] using 15 loss and gain factors based on material properties and module set-up with individual modelling. SmartCalc.CTM has been used for module design [18, 19], analysing test results and market prediction [19].

In the above analysis, the characteristics of solar cells are modelled in parallel. However, the module electrical properties can also be analysed with individual cells represented as composite electrical components in an equivalent circuit as shown in Figure 11.7. This enables the module to be simulated with electronic design automation (EDA), such as Simulink®, SPICE [8, 20] and SPICE-compatible software, like PSPICE [21], LTSPICE [4] and Micro-Cap [22]. However, this electrical analysis has limitations as it is sometimes insufficient to accurately represent the power delivered by the individual cells in a module. These EDA software tools can be used to analyse the dynamic, larger-scale and more complex electrical properties of a PV module with module-level electronic components.

11.3 Optical design

Silicon PV module design has evolved to now include new materials and strategies that can improve the areal power generation and electricity yield. New materials such as textured glass, glass anti-reflection layer, reflective ribbon, round wire and transparent backsheet (for bifacial modules) are directed at increasing the optical efficiency of the modules, whereas strategies such as reducing the gap between cells aim to improve the areal power generation. Incorporation of these new features and combinations of features requires careful design and, for this, tools which can accommodate the rapidly evolving list of new materials and module configurations are required. Optical design should target both maximum power output under the STCs and maximum electricity yield under realistic field conditions.

This section will introduce simulation methods and software packages that can be used for systematic optical design of Si PV modules. Working principles and design considerations of the most commonly used optical improvement features/strategies of active area improvement, light-trapping and bifacial design will be discussed in detail.

11.3.1 Modelling approaches

Optical modelling and simulation provide information needed to design a module with maximum-effective light absorption and light-to-electricity conversion. Photocurrent, or light-generated current, is a commonly used parameter to quantify the optical performance in simulation. It refers to the current or current density generated by solar cells within a PV module with an assumption of 100% collection efficiency. Photocurrent can then be converted into a working current by incorporating thermal, resistance and recombination losses.

The modelling process always starts with the collection of optical properties (i.e. refractive indices, absorption coefficients and dimensions of the individual layers of a multilayer structure) for the module. As feature sizes and layer thicknesses can range from tens of nanometres to centimetres or even metres, the layers and structures may be small or big compared to the coherence length of the incoming light source. Thus, the modelling method needs to coordinate the application of ray and wave optics to analyse the desired configuration. Ideally, the simulation output includes the reflectance and transmittance of the overall module structure as well as the absorbance for active layers. Modelling dimensions, polarization, angle of incidence and spectrum coverage should also be considered in the simulation, especially for electricity yield predictions.

If a simulation is focused on optimizing a module design prior to production, comparing different materials and quantifying losses in individual components, then a static module simulation under specific test conditions can be performed. However, if complex optical structures are used and the simulation involves correlations between system designs and module structures, a yield simulation should also be performed to determine performance in the field.

Ray-tracing

Ray-tracing is a ray optics method for calculating the path and distribution of light (rays) in a system or device. It can resolve the measurable attributes of reflection, transmission and absorption by propagating a statistically large number of light rays through a median with an arbitrary number of interfaces. Depending on the complexity of problems, ray-tracing can be performed using analytical or numerical solutions. However, in ray-tracing, light is represented as a photon flux and consequently phase information and hence the electromagnetic wave properties of light are not fully considered [23]. Optical characteristics of thin film layers can be calculated

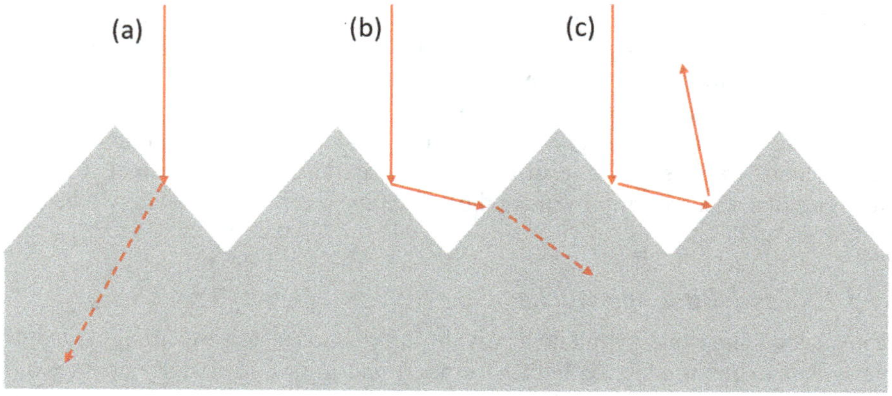

Figure 11.8 *Schematic depicting the ray-tracing method with (A), (B) and (C) showing three possible paths of light rays incident on a textured Si surface*

from wave optics separately. Surface structures much larger than the wavelength of the light are suitable to be analysed with ray-tracing, for example, an adjacent pyramid as shown in Figure 11.8 on the wafer surface or the front glass of a module.

Ray-tracing has been used to optimize a number of module properties including the separation between backsheet and cells [24], cell spacing [25], encapsulant properties [26] and effects of antireflection coating (ARC) and textured glass [27]. A key issue faced by ray-tracing processes that also attempt to take into account the multi-scale features of modules (e.g. narrow 40 μm wide fingers incorporated into a module which has an area in the order of 1.9 m^2) is the number of rays that must be simulated to accurately represent the optical contributions of all the features of a module. SunSolve, a commercially available modelling package developed by McIntosh and Abbot [28–30] and available through PV Lighthouse, utilizes cloud computing to manage the computational load of having to compute large numbers of rays to achieve a statistically valid result.

Another ray-tracing framework is called DAIDALOS [25, 31]. This framework has been used to evaluate new module designs and determine the optical losses in different module components. It achieves this by separately simulating features (e.g. textured, the cell ARC, rear aluminium contacts) and then combining the outputs of these features in a loss analysis. Module loss analysis has been extended by Ernst *et al.* who used SunSolve to develop a comprehensive CTM loss analysis for Si PV modules.

Matrix formalisms

As discussed in the section above, although increasing the number of rays in ray-tracing reduces the statistical uncertainty, it increases the computation time and memory required for purely ray-tracing solutions. Additionally, the optical properties of nano-scale structures (e.g. black silicon texturing on cells and nano-textured glass) cannot be accurately simulated using ray-tracing. To address these challenges, matrix-based methods have been proposed. The changes in intensity and direction of light at each interface in the module can be calculated using appropriate methods (i.e. electromagnetic methods for nano-structures and ray-tracing for larger features) and these values can be represented as angular response matrices by discretizing the ray-travelling direction range into angular intervals. The absorption of light in the various module materials can also be calculated and stored in absorption matrices. Once these matrices are formulated, the optical response of the whole structure can be determined through multiplication of these matrices requiring more modest computational resources.

Matrix-based methods that adopt this approach include the multi-directional net-radiation method by Santbergen *et al.* [32], scattering matrix method by Mellor *et al.* [33], OPTOS (optical properties of textured optical sheets) formalism by Eisenlohr *et al.* [34, 35] and the AMF (angular matrix framework) by Li *et al.* [36]. These matrix formalisms allow each of the module interfaces to be modelled using a method that is suitable for the structure's optical regimes. For example, Eisenlohr *et al.* used OPTOS to combine ray-tracing results for the front texturing and rigorous coupled wave analysis (RCWA) for the grating mirror on the back [35], while ray-tracing and finite-difference time-domain (FDTD) methods were combined to study solar cells with front texturing and nano arrays on the back using the AMF [36]. A key advantage of these matrix approaches is that pre-calculated matrices can be reused when some design parameters are changed. This is particularly useful for module optimisation problems, and also when a module's performance under field conditions must be evaluated.

Module performance under realistic illumination conditions

Financial decisions based on the generation capacity of a new installation require accurate estimations of a PV system's yield or electricity generation over a period of time. Although a module's photocurrent and power can be simulated under STC conditions using a number of available software packages [37, 38], evaluation of a module's performance in the field under realistic illumination conditions is more challenging. Not only does the incident angle of the light on the module surface change in real time, it has direct and diffuse components that also need to be considered.

In addition to changes in incident angle, the power spectral distribution of the irradiance on the module varies depending on site location and time. The air mass determines the scattering and absorption of sunlight by the atmosphere and therefore it is unlikely that the air mass of a particular location at a given time is exactly reproduced by the AM1.5 spectrum. Thus, the intensity and spectral power distribution of irradiance also changes over time and needs to be considered when accurately estimating the power output of the modules in the field.

There are two approaches for including realistic irradiance into the optical evaluation of modules. In a top-down approach, the annually averaged (i.e. representative) solar irradiance is first determined and then used as the irradiance source in the module power estimation. Alternatively, in a bottom-up approach the module or system yield is computed as the sum of power generated at each time step using temporal irradiance data. The top-down approach has the advantage that the representative solar irradiance can be reused in simulations involving different module configurations. This can accelerate the optimization procedure. However, it is easier to analyse the seasonal and monthly variations based on the bottom-up approach.

Both approaches require measured irradiance and weather data of the proposed module installation site. If the data are not available for the tilted plan, the horizontal irradiance data needs to be first used to compute the plane of the array (POA) irradiance, considering the direct component, diffuse component and ground reflection [39].

If yield modelling for a module or a system of modules is required, the optical simulations need to be combined with an electrical model to determine the operating current and voltage of the PV module based on cell temperature estimations. When multiple modules are simulated as an array, the DC power and yield can be then calculated with consideration of DC wiring and mismatch losses.

11.3.2 Maximising packing density

Module areal efficiency (W m^{-2}) is calculated by the power rating divided by solar irradiance on total module area. Although the increasing efficiency of solar cells is the fundamental driver of module efficiency, improvements in packing density can be an effective means by which to boost module areal efficiency. This ratio can be increased by a number of approaches, including increasing wafer size, increasing the number of cells in a module and reducing the gap between cells. These approaches also bring other major benefits of improving the power produced per module and thus reducing the cost per watt for all manufacturing processes.

Wafer sizes and the number of cells in modules have continued to increase since the 1960s where typically modules comprised 36 round solar cells with a diameter of 1.9 cm [40]. It evolved to first 60 cell modules with 125 mm pseudo square wafers, then 156 mm wafers, and now to 72 cell modules comprising the larger 166 mm (M6) and 210 mm (M12) wafers [41] with smaller rounded wafer corners.

Packing density can also be improved by reducing the gap between interconnected cells. One of the better-known methods of reducing the gap between cells, with a long history, is shingling. As illustrated in Figure 11.9 b, solar cells are cut into stripes and then subsequently assembled in strings with electrically conductive adhesive (ECA) [42]. Production of shingled modules has been demonstrated by a number of companies, including Sunpower [43] and Solaria [44]. The gap between cells in a string in modules has also been reduced by interconnection technologies known as paving and tiling that have also been used in mass production, including Jinko [45, 46] and LONGi solar [47]. With these technologies, cells are interconnected using ribbons (or wires) with a small gap between cells or a slight overlap, like roof tiles. However, unlike shingling, cells are not cut in the busbar direction and ribbons are still used. Tiling/paving requires specific processes for the ribbon and encapsulant layers to avoid cell cracks from the extra induced stress in the overlap region [48, 49]. There are still challenges, not only in cell preparation, but also

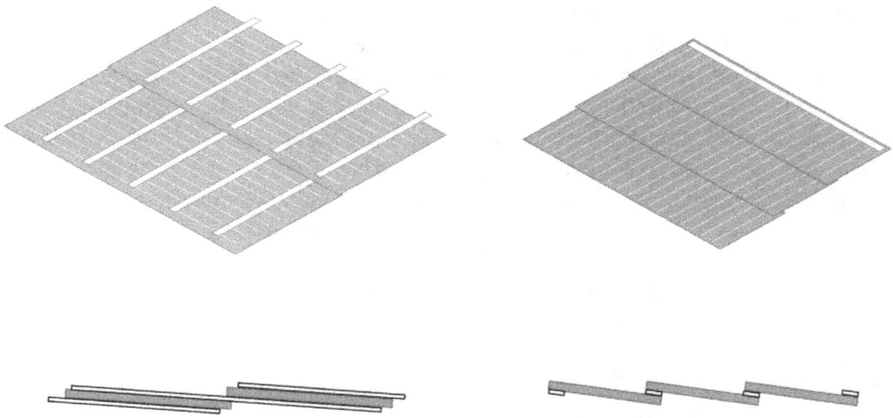

(a) (b)

Figure 11.9 Schematic and side views of (A) tiled interconnection and (B) shingled interconnection of solar cells

in module design, assembly and potentially module reliability with these reduced cell gap methods [50].

To place these advances in wafer size, number of cells and alternative interconnection methods in perspective, the landmark Philips BPX 47A module released in 1976 had a rated power of 11 W and a module efficiency of 6.4% [2]. However, 41the power and module efficiency of modules comprising 72 larger 166 mm (M6) cells, have increased to 475 W and 21.2% in 2020 [46].

11.3.3 Light-trapping in modules

Glass texturing and coatings

In order to convert more sunlight into electricity efficiently, the reflection and transmission of solar glass play a significant role. Over the years, the transmission of glass has been increased by reducing the iron content [51]. In addition, the optical properties of glass have been further improved through the use of anti-reflective (AR) coatings and surface structuring or texturing.

AR coatings are commonly utilized to reduce the reflection losses. There are two main strategies for minimising reflection: (i) depositing thin films such as a single- or multi-layer film on the glass surface and (ii) incorporating graded refractive index coatings with the aid of porous and nanostructured arrays (e.g. moth-eye structures [52]), as texturing in sub-wavelength scale acts as an AR coating. The thin-film coating suppresses the reflection occurring through destructive interference, achieved by controlling the thickness and refractive index of each layer. For example, porous SiO_2 layers deposited using a sol-gel method [53] reduce reflection ensuring that the refractive index of the structure reduces progressively from the index of the air to that of glass. In current industrial practice, the transmission of

AR coated glasses is ~94%, about 3% higher than non-coated glasses [41]. Another advantage of AR coatings is that the glare from the glass can be reduced.

Glass texturing (also referred to as patterned or structured glass), is another way to reduce front-surface reflection. Patterns can be produced in line with the glass, fabricated separately in an additional process or applied as a superstrate layer [40, 54]. A shallow pattern may increase the adhesion of encapsulant and help with anti-glare. The later makes the modules less visible and able to blend more easily with their surroundings, thereby allowing modules to be installed in urban applications, such as near airports. However, as the pattern becomes deeper, transmission can be increased due to light-trapping. The transmitted light may also be trapped between the front glass and rear backsheet.

When AR coatings and surface texturing are used on glass, the broadband anti-reflectivity, omnidirectional anti-reflectivity, polarisation sensitivity, durability and cost need to be balanced for optimised module performance. As both coatings and texturing are normally applied on the front glass surface, it requires a stable effectiveness in harsh operating environments during the entire lifetime of the module. Some AR coatings are claimed to be hydrophobic and impart a self-cleaning property [55, 56]; however, research into degradation characteristics of anti-reflectivity and impact of damage by soiling are required in different environment and climatic conditions.

Shading reduction of interconnectors

Interconnectors (e.g. ribbons and wires) connect the positive terminal of one cell to the negative terminal of the adjacent cell. As current is collected from multiple fingers, the cross-sectional area of the interconnectors needs to accommodate the current without introducing series resistance losses and increasing shading of the cell. Shading can be reduced by using narrower ribbons with higher aspect ratios, however, the benefit of this strategy is offset by the required additional thickness and hence cost of the encapsulant. Recently, module designs utilizing light re-directing surfaces to reduce the reflection loss of ribbons has attracted interest due to the availability of products for commercial production, such as light scattering ribbons (LSR) [57, 58], light scattering films (LSF) [59, 60], light diverting ribbons (LDR) [61, 62] and wires [63, 64].

One of the more widely used commercial products are LSFs (see Figure 11.10). These films are covered with highly reflective V-grooves which can be attached (via an adhesive) to the flat ribbons during the cell stringing process. The structured surfaces scatter the incident light back to the front internal surface of the glass at a high incident angle. Due to the higher refractive index of the glass compared to air, the scattered light is internally reflected back to the wafer surface increasing the probability of reabsorption by the cells in the module. The LSRs operate according to the same principle with patterns directly formed on the ribbons. LDRs typically have a triangular cross-sectional shape and act by directly diverting incident light into the wafer [65].

Round wires combine the properties of LSRs and LDRs. The top surface of the wire reflects light which escapes from the front surface of the module. While

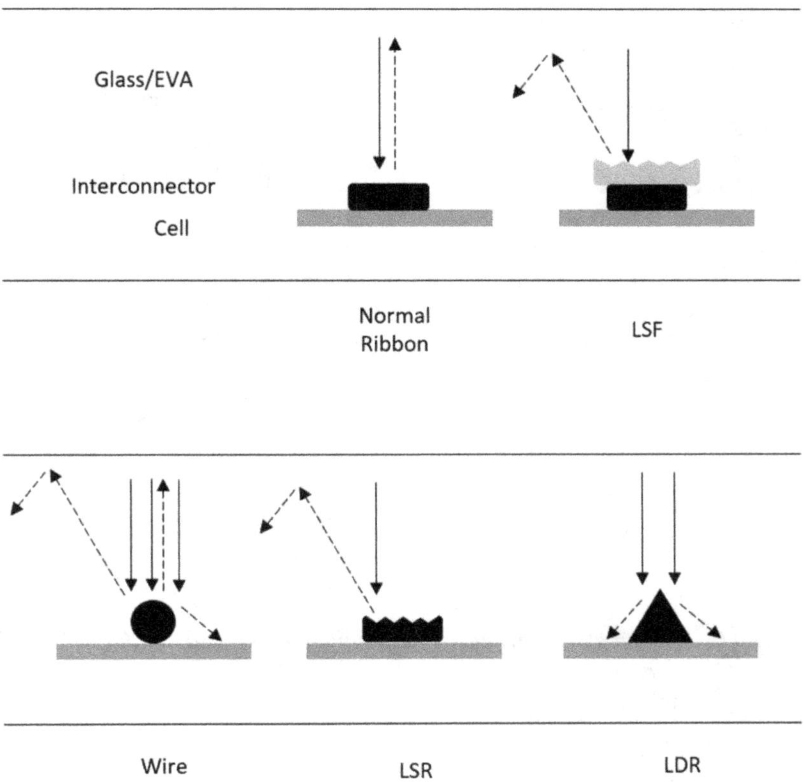

Figure 11.10 *Schematics of reflection mechanisms of normal ribbon, light scattering film (LSF), round wire, light scattering ribbon (LSR) and light diverting ribbon (LDR)*

the adjacent side surfaces can reflect the light in trajectories similar to LSRs and LDRs, respectively. The effective optical width is defined as the width of a hypothetical black interconnector which causes the same shading loss as the interconnector. When these optical structured ribbons and wires are used, their effective optical width is reduced.

Module power gains of 1.0%–2.5% have been reported depending on the ribbon width. However, when such structures are used in modules, their limitations need to be considered. First, LSFs and LSRs perform best when the incident light is within a range of angles from the normal direction of the module surface. As a result, the yield gains can be reduced from their power gains. Second, use of asymmetric LSR and LDR interconnector geometries requires ribbons to be manufactured in sections with flat ribbons (for cell rear surfaces) and special alignment during soldering to ensure maximal optimal performance. The LSFs incur the additional challenge of correct alignment when they are affixed to the pre-soldered ribbons [65].

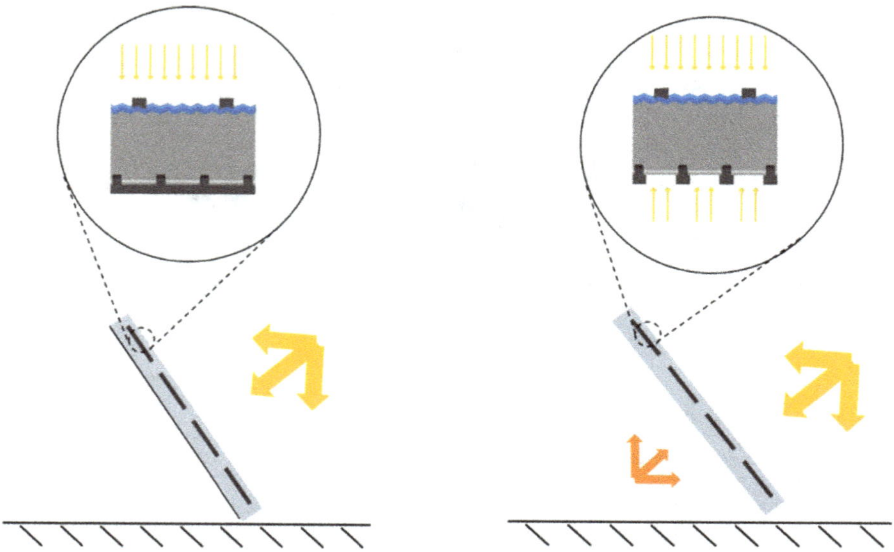

Figure 11.11 *Schematic showing: (A) a mounted monofacial module with full rear metal contact solar cells; and (B) a mounted bifacial module with bifacial cells*

Light-trapping in exposed non-cell areas

Although the CTM area performance ratio can be improved using specially shaped interconnectors as discussed above, there remain non-active areas in a typical PV module (e.g. cell gaps and string separations) and incident light falling on those module areas is not converted to electricity. LSFs can also be applied in these areas to improve the module's maximum power and annualized energy gain. Similarly, the backsheet itself can be structured with grooves to achieve the same goal [66]. The gains achieved will depend on the patterns and gap sizes. The properties of the encapsulant can also be tuned to enhance optical performance. For example, highly reflective white EVA can be used as a rear encapsulant film to promote reflectance by the non-cell regions for monofacial modules [67].

11.3.4 Bifacial module design

Bifacial modules (see Figure 11.11) can increase yield without significantly increasing process complexity and manufacturing cost. In addition to the increased light capture, bifacial modules can operate at a lower temperature due to the transmission of light below the bandgap (i.e. reduced infrared absorption that occurs in the full-area rear metallization of monofacial cells). This factor increases maximum power output of the module correspondingly.

While bifacial module technology has great potential and a long developmental history, they have not, until recently, been commercially embraced due to their added expense and difficulties arising from the availability of bifacial cells. The recent uptake of bifacial solar modules in the market has been increased by the transition to higher efficiency PERC cells. Although PERC cells are not bifacial from their original architecture, it is relatively straightforward to adapt them to be bifacial. As PERC cells have become dominant in the market, the International Technology Roadmap for Photovoltaics (ITRPV), predicts that the market share of bifacial modules will increase in the next decade and exceed 50% by 2029 [7].

Bifacial modules can be benchmarked according to their bifaciality, or bifaciality factor. This factor is defined as the ratio of the rear-side maximum power or efficiency to the front-side power or efficiency:

$$B = \frac{P_{m,rear}}{P_{m,front}} \tag{11.10}$$

Bifaciality factors for manufactured modules can vary between 60% and 95% [68]. Another metric, which is of interest to solar farm developers, is the bifacial gain or additional energy obtained from bifacial solar modules compared with a monofacial counterpart with identical configurations over a period of time. This metric is defined as:

$$BG = \frac{Y_{bifacial}}{Y_{mono}} - 1 \tag{11.11}$$

where $Y_{bifacial}$ and Y_{mono} are the electricity yields in kWh for bifacial and monofacial solar modules, respectively. It is reported that the bifacial gain ranges from 15% to 25% for small and demonstration projects and 5% to 15% for large commercial systems [69], with an averaged theoretical projection of below 10% [68]. It should be noted that the bifacial gain varies significantly due to differences in modules, racking types, surface albedo, rear-side inhomogeneity, mutual (front-side) shading and mounting structure. Optimisation needs to be performed from module designs to system implementation to unlock the full benefits from bifacial PV modules.

Although rear transmission of sub-bandgap infra-red light is advantageous from a module cooling perspective, bifacial modules have a higher rear escape optical loss of longer wavelength above bandgap light compared to monofacial modules. It should also be noted that bifacial solar cells can be encapsulated into monofacial modules with improved performance under STC, as light that passes through cell gaps can be reflected back to the open rear surface of bifacial solar cells by the backsheet [70].

Traditional monofacial solar modules have an opaque backsheet on the backside. For bifacial modules, there are two options: glass/glass (or double-glass, dual-glass) and glass/transparent backsheet designs. Since glass has a higher transmittance for the short and ultraviolet (UV) light, it offers a wider light transmittance and potentially can result in more UV damage to the encapsulant [71]. In addition to the differences in optics, glass/glass modules are generally considered to be more resistant to mechanical load and moisture, while transparent backsheets reduce module weight. This brings benefits to transportation, mounting and installation costs [45].

Bifacial module performance in the field

Surface albedo and racking configuration can have a significant effect on the bifacial gain of a bifacial module array. Albedo is a complex property which involves measuring how much of the total incoming radiation incident on a surface is reflected back into a module. It can vary greatly for different materials and conditions with green grass exhibiting an albedo of ~0.24 while white sand has an albedo of ~0.67 [72].

Typically, a single effective albedo value is used in performance evaluations of bifacial module systems. However, this property ignores dependences on the directional and spectral conditions of the incoming light which can introduce selectivity and bias that impact the optimal design of bifacial modules and systems. It has been demonstrated that spectral albedo significantly affects the efficiency and power output of bifacial modules in a way that the spectral albedo of snow, white sand or green grass leads to a higher than expected power output relative to the effective albedo while red brick and construction concrete result in a lower power output than expected [72]. If directional dependency is also of interest, bidirectional reflectance distribution functions (BRDFs) more completely define reflectance of all scattering angles to each irradiation sourcing angle. However, implementation of yield calculations using BRDFs can be computationally expensive.

View factor and ray-tracing are the two most commonly used simulation methods to model the rear side illumination for bifacial module designs to quantify power and yield potential under field conditions. View factor is often used to solve for radiative heat transfer between surfaces. It determines the amount of irradiation leaving one surface reaching another one through a distance by analysing the geometrical orientations of the interfaces. Both methods can be accurate for simple configurations [73]. However, when non-uniform illumination, module frames, torque tube and other self-shading elements are considered, ray-tracing is more comprehensive with higher computational load [74, 75]. Angular and spectral rear reflection properties can then be represented in matrices which can be used to compute expected yield for the installation.

11.4 Mechanical design

An understanding of the thermal and mechanical properties of module materials, material interactions and manufacturing processes is required to ensure that Si PV modules can operate in the field for at least 25 years. Recent studies which surveyed module failure modes in the field [76–78] have identified that corrosion, cell/interconnect breakage and delamination account for 86%–89% of all module failures. An understanding of the physics of failure can assist in the mitigation of these module failures in the field and contribute to more durable module mechanical designs in the future.

Encapsulants and their material interactions with backsheets and solar cells can contribute to a number of these module failures and therefore an understanding of the relevant material interactions can inform on appropriate selections of encapsulant materials and properties (e.g. thickness). Similarly, cell and interconnect breakages can occur as a result of stress that is induced due to temperature changes and module deformation under mechanical load. Consequently, methods of identifying and measuring this induced stress can result in more durable module designs with respect to how the cells are interconnected.

This section provides a discussion on the commonly used encapsulant materials and their properties (see Section 11.4.1), then Section 11.4.2 reviews the underlying physics of induced stress and considers the effects of induced stress on cell cracking (see Section 11.4.3) and solder joint failures (see Section 11.4.4). Section 11.4.5 expands on the mechanical design issues arising from choices in backsheet material and Section 11.4.6 discusses the effect of module frame and mounting on the mechanical load tests.

11.4.1 Encapsulants

EVA is the most widely used encapsulant for Si PV modules having a 90% world market share in 2019 [7]. It has moderate water vapour transmission rate (WVTR) of 30–70 $g.m^{-2}d^{-1}$ at 90% relative humidity and 80 °C [79]. When subjected to moisture and UV irradiation, acetic acid can be released from EVA decomposition (in the absence of sufficient UV stabilisation) and corrode cell metallization, solder joints and the Cu core of ribbons. This corrosion acts to increase series resistance and can cause a brown discoloration. In addition, acetic acid promotes module delamination, potential induced degradation (PID) and accelerates the oxidation and breakdown of the co-polymer chain of EVA.

Material contamination, arising from improper cleaning of the glass or excessive residual flux on the cell surfaces [76], can reduce interface adhesion and lead to module delamination over time. Delamination is more likely at the interface between the encapsulant and the solar cells [78, 80]. The fracture energy of the EVA/Ag interface is ~60%–70% lower than that of EVA/Si and EVA/glass, and 300 hours of damp heat ageing can decrease the fracture energy at the EVA/Ag interface by 95% [80]. The separation of EVA from Ag metal can accelerate the delamination of metal grids from the cells due to corrosion (in the presence of moisture and the products of reaction such as acetic acid).

Alternative encapsulant materials for Si PV modules include PVB, thermoplastic polyolefin (TPO), POE and silicone [77, 81]. Light transmittance, UV durability, electrical insulation, WVTR, water absorbance and cost are important properties to determine suitable encapsulants. Selection of encapsulant also depends on the backsheet material and cell type. Most encapsulants are applicable to monofacial modules, whilst bifacial glass-glass modules are typically limited to TPO and POE in order to sufficiently limit PID.

11.4.2 Implications of induced thermomechanical stress

The fabrication of PV modules involves several thermal processes, including cell metallization (metal paste firing), cell interconnection (tabbing/stringing) and module lamination. Due to the temperature changes in these processes, thermomechanical stress is induced in the cells due to mismatches in the coefficient of thermal expansion (CTE) of the different materials. This induced stress can result in the fracture or cracking of cells (see Section 11.4.3 for a more detailed discussion), solder joints or ribbons. Consequently, the materials that are used in module fabrication can directly impact module durability and it is important to consider these effects in the mechanical design of modules.

Of particular interest is the possible fracture of cells by induced thermomechanical stress. The stress in Si wafers can be measured using Raman spectroscopy [82, 83], synchrotron X-ray microdiffraction (μSXRD) [84, 85] or X-ray topography (XRT) [86, 87]. However, performing these measurements in laminated modules can be complex and an alternative approach is to use finite element modelling (FEM) to identify the locations in the Si wafer that are more susceptible to induced stress. As stresses can accumulate through different thermal processes, it is important to integrate the impact of cell interconnection and module lamination in simulations and also to consider the effects during thermal cycling (TC). This can present some challenges due to the appropriate modelling of the encapsulant as a linear viscoelastic material [88, 89].

Soldering

After soldering, the highest first principal stress (tensile stress) occurs in the Si wafer at the end of the busbars [90–93] and the stress direction is parallel to the busbar [90]. The second principal stress correlates with the stress perpendicular to the busbar. As shown in Figure 11.12 a, the induced stress is very localized. Similar stress distribution developed in cells interconnected by MBB technology was simulated and reported in Rendler *et al.* and Hsiao *et al.* [91–93, 95] The predictions of the FEM simulations are consistent with electroluminescence (EL) images of modules after TC test as shown in Figure 11.12 b, which identify power losses near the cell edges where high stress regions are prone to cause more degradation [94, 96–98].

Several approaches have been proposed to mitigate the cell stress induced by interconnection. Use of low melting temperature solders [64, 99, 100] or ECAs [101–103] can decrease the temperature difference and thus thermal strain. The mechanical properties of ribbons can be modified to be low CTE [104] or low yield [91] to reduce the stress in Si. A recently developed approach to reducing thermomechanical stress in Si wafers uses wave-shaped wires [96, 97, 105] where the main contraction/expansion of wires is perpendicular to the wire direction. Alternatively , Hsiao *et al.* [93] proposed by aligning the outer edges of front busbar (or contact pads) and rear pads, reduction of >50% of maximum tensile stress in Si was achieved in both conventional tabbing and MBB.

Lamination

Most reported tensile stress in Si induced by encapsulation/lamination without considering the stress incurred by soldering is generally lower than by soldered interconnection. This was experimentally confirmed by Walter *et al.* [94], who showed that less EL degradation was found in modules fabricated with an inert film to separate rear EVA from the solar cell than in standard glass-backsheet modules.

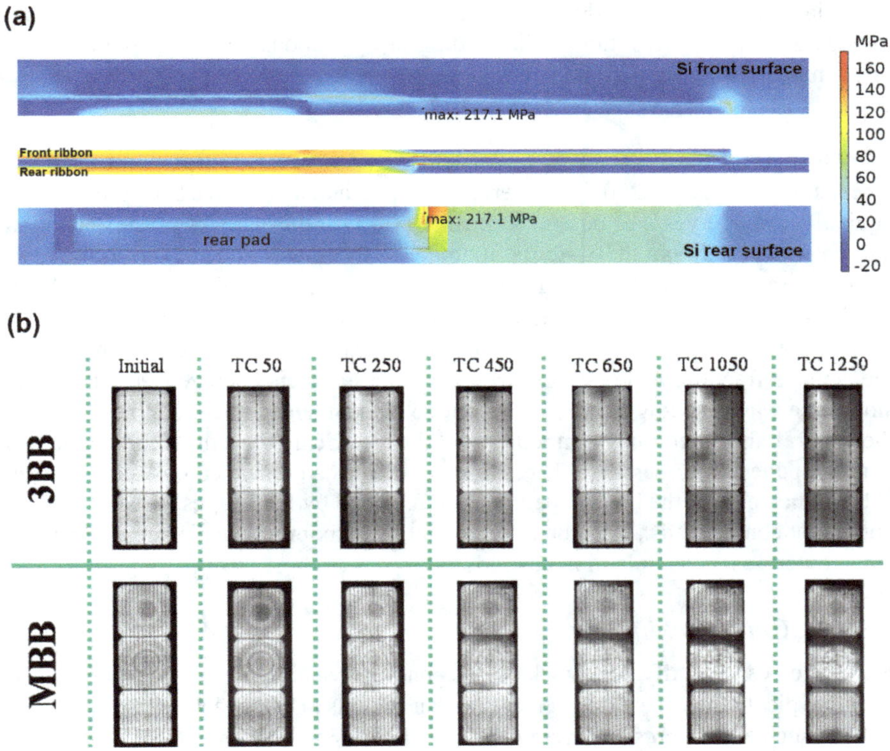

Figure 11.12 *(A) Tensile stress mapping of monofacial Si solar cells interconnected by conventional tabbing. The middle mapping is the cut plane through the centre of the ribbons along the ribbon direction. The wafer edge is towards the right (adapted from Hsiao et al.[93]); (B) electroluminescence (EL) images of modules subjected to thermal cycling (TC) tests (adapted from Rendler et al. [94]).*

Lamination of glass-backsheet modules results in expanded regions of large tensile stress in Si under the front busbars. In addition, the stress at the Si rear surface at the busbars transforms from compressive to tensile stress due to the different encapsulant contraction arising from the different Si-to-glass and ribbon-to-glass thicknesses. It indicates the occurrence of cell bending in the thickness direction [90]. If the cell gap is reduced from 5 mm to 1 mm, the probability of failure (see Section 11.4.3 for details) increases by a factor of 1.23 due to increased strain in the encapsulant.

Stress induced by lamination can be addressed by adopting accelerated cooling, which can lead to 22%–27% reduction in residual stress in the solar cells [106]. Dietrich *et al.* [107] showed that by increasing the encapsulant thickness, the stress in the Si wafer can be reduced due to less module deflection and higher shear deformation. However, the drawback is increased module cost. With large increases in encapsulant thickness, the tensile loading of Cu ribbons becomes dominant and this can increase the maximum stress in Si.

Thermal cycling

Accelerated TC tests of 200 cycles between −40 °C and 85 °C have been proposed as equivalent to ~10 years of outdoor exposure [108]. In the analysis of the deflection and change of cell gap in TC-tested modules, a hysteresis is observed in the temperature range of 20–85 °C owing to the plastic deformation of the Cu ribbons [107]. Pander *et al.* [109] found that decreasing the cell gap leads to increased stress and plastic strain energy density of the ribbons, indicating a decreased lifetime before the ribbons fracture. Although experimental measurements showed that the adhesion force was very low after TC tests due to detachment of the metallization from the wafer, it appeared that on module level the electrical properties degrade far less than the adhesion forces [94]. The 800-cycle TC tests conducted by Rendler *et al.* [110] indicated that the contact degradation revealed by EL images started from the outermost contact pads. It can be inferred that the maximum tensile stress remained at the outermost pads in laminated modules.

11.4.3 Cell cracking

EL images can identify cell cracks that cannot be visually detected. Cracks in Si wafers typically manifest in irregular dark EL regions [111]. Cracks can cause other failures such as hot spots and arcing [112], and more importantly impact the longevity of solar modules. Cell cracks induced by tensile stress can account for PV module degradation rates of up to 8% relative power loss per year [113]. As shown in Figure 11.13, cracks can be classified according to the orientations [114, 115], which can have a different impact on the module power output. Dendritic, several and parallel cracks can lead to electrically isolated cell areas and hence are considered as being critical to stable module performance. The severity of cracks is expected to increase with reduced cell thickness and increased wafer size.

The physical strength of Si is 15.8 GPa in 110 and 111 planes [116]; however, due to processing microscopic defects (such as microscopic cracks, inclusions or dislocations) it can be reduced to ~100 to 1 000 MPa [90, 112]. Microscopic defects are inhomogeneously distributed across the surface and/or the bulk of Si

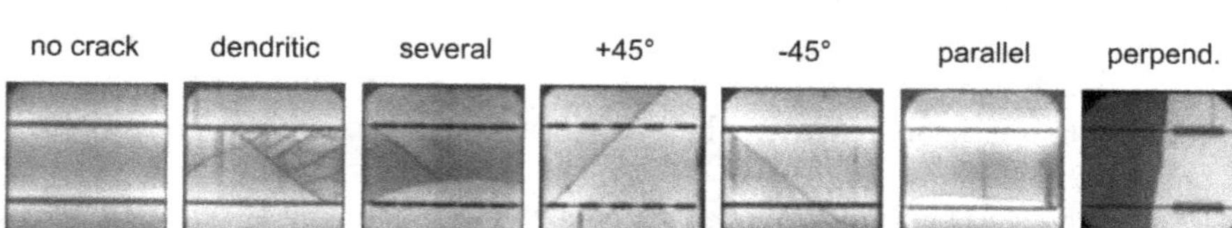

Figure 11.13 Classification of cracks based on crack orientations (adapted from Köntges et al. [114])

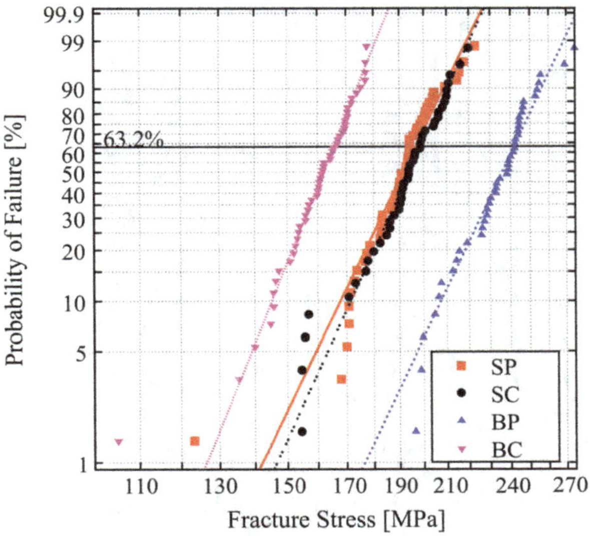

Figure 11.14 *Weibull plot of crack occurrence in c-Si solar cells under the four-point bending test (adapted from et al. Kaule [120])*

with scattered defect size. To evaluate the fracture stress in Si, a Weibull distribution is typically used. The probability of failure P_f at a specific stress σ is described by:

$$P_f = 1 - exp\left(-\left(\frac{\sigma}{\sigma_\theta}\right)^m\right) \tag{11.12}$$

where the Weibull modulus m is a measure for the scatter of the strength values; and the characteristic strength σ_θ represents the stress for a probability of failure of 63.2%.

The fracture stress in Si is influenced by the saw-damage etch process [117], cell type [118] (monocrystalline silicon, c-Si, and grain boundaries of multicrystalline silicon, mc-Si, wafers), screen-printed paste firing [119] (improving the fracture strength from induced plasticity by Al-Si eutectic layer), the surface under tension and the tensile orientation [118, 120]. An example Weibull distribution is shown in Figure 11.14 and calculated Weibull parameter values can be found [118, 120, 121].

The simulated strain distribution of uniformly loaded modules reported by Kajari-Schroder *et al.* [115] shows that the first principal strain is mostly oriented parallel to the module short edge, and in the diagonal directions at the module corners. Therefore, cracks parallel to the busbars are expected to be dominant in modules subjected to mechanical loads, confirmed to be ~50% from experimental statistical data. The study by Köntges *et al.* [114] concluded that cracks in transported modules occurred mostly in the middle of the modules, while after a mechanical load test, cracked cells additionally appeared in the four corners. The spatial distribution of cracked cells, whose number was significantly increased (mostly in

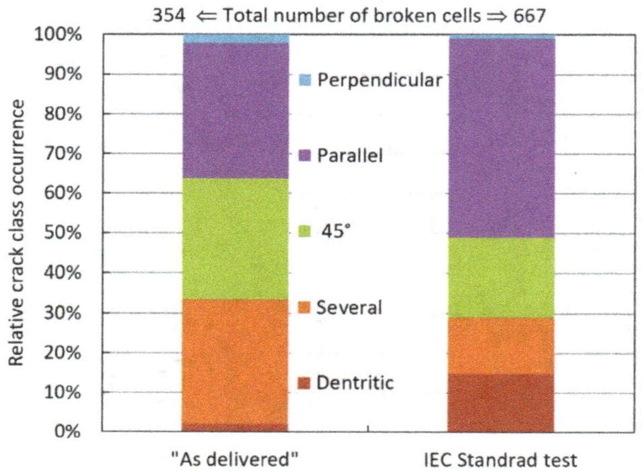

Figure 11.15 *Comparative percentage of crack class occurrences of the 'As delivered' modules and after the IEC mechanical load test (adapted from Köntges et al. [114])*

parallel cracks as shown in Figure 11.15) by the static load test, is consistent with the simulated first principal strain distribution. In addition to the consistent findings of crack propagation/orientation, Dietrich *et al.* [90, 122] found a significant impact of cell metallization layouts and load ramp rate on P_f, which was higher under the perpendicular load test and there were greater variations between metallization layouts.

11.4.4 Solder joint failures

Solder joints in PV modules can fail due to (i) Ag or Cu leaching into solder and/or (ii) long-term solder joint fatigue [123]. In both cases, a crack forms and a DC arcing discharge may occur at the crack. As solder is bonded to screen-printed Ag contacts, a brittle Ag_3Sn metallic compound is formed, which can facilitate the crack formation at the interface of Ag_3Sn and solder. When subjected to cyclic loading (thermal or mechanical), the solder undergoes progressive structural degradation [123, 124]. Furthermore, the accumulated creep strain energy acts a driving force for the formation of intermetallic compound layers and micro-voids [125, 126].

Solder joints have been reported to degrade faster at the cell rear side than on the front surface [127, 128]. Due to fewer number of contacts on the rear, the soldering quality on the rear side is of paramount importance to solder reliability [129]. Simulation studies [126, 130, 131] show that solder thickness has the most significant impact on the damage accumulation of solder, with twice as much damage estimated to occur when the solder is half as thick. Decreased thickness of Cu ribbons and Si can reduce the accumulation rate of solder damage [131]. The fatigue life of solder joints is also influenced by temperature and stored strain energy. The study of different TC profiles performed by Jabarullah *et al.* [125] demonstrated that solder damage increases with both temperature swing and the average temperature (i.e.

the difference and average of maximum and minimum temperatures, respectively), which is experimentally confirmed by the intensified Cu_3Sn phase of X-ray diffraction (XRD) measurement. Walter *et al.* [94] reported that after TC-400, cracks can be detected in the solder at the cell edges where the thermomechanical stress predicted by FEM is higher.

If solder is in contact with metal fingers, disconnected fingers from busbars may occur under thermal loading due to mismatch of CTE, resulting in rectangular dark regions in EL images [111]. Kumar *et al.* [132] showed that the damage accumulation at the finger-solder interface linearly decreases with solder thickness and finger spacing.

11.4.5 Rear module design

In 2019, glass-backsheet modules comprised ~90% of the world market share and it is predicted that they will retain ~65% of the market in 2030 [7]. However, growing adoption of bifacial modules is expected to increase the usage of glass as the backside cover material in the coming years. Glass-glass modules are a preferable configuration in terms of mechanical stability due to their symmetric design with glasses of the same thickness on both surfaces [109]. One key advantage of glass-glass modules is the reduced moisture penetration due to the improved barrier provided by the glass. Moisture penetration is one of the main failure mechanisms of glass-backsheet EVA encapsulated modules [81].

Si PV modules are typically laminated on flat-bed vacuum-membrane laminators. The lamination process of glass-backsheet modules typically consists of three phases. The temperature and pressure profiles of these phases are illustrated in Figure 11.16. In the pre-heat phase, the laminator chamber is vacuumed to extract the trapped air and the volatile organic compounds (VOCs) released from the chemical reactions of the additives in the encapsulant when the temperature of the

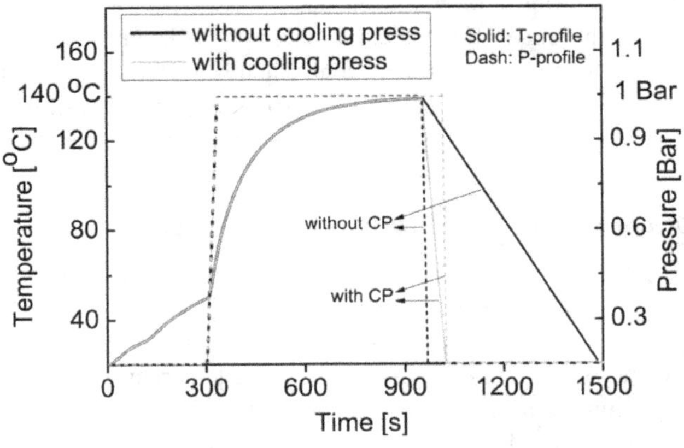

Figure 11.16 *Temperature and pressure profiles of the lamination process of silicon photovoltaic (PV) modules (adapted from the IEA PVPS report [76])*

pin-lifted module lay-up increases. During the cure phase, pressure is applied via the flexible membrane to induce adherence between different layers. In the last cooling phase, modules are cooled naturally in air or under accelerated cooling conditions with applied pressure. Accelerated cooling is advantageous because it: (i) decreases module bending; (ii) increases throughput; (iii) results in 22%–27% reduction in residual stress in the solar cells [106]; (iv) increases adhesion strength between glass and encapsulant; and (v) reduces haze of the encapsulant [133]. Despite these advantages, natural cooling remains more frequently used in the industry due to its lower cost.

Lamination of glass-glass modules requires some careful attention. Encapsulants which release substantial VOCs are not suitable for glass-glass modules. Encapsulant viscosity plays a crucial role during lamination. Encapsulants with high viscosity tend to result in unlaminated patches which can potentially cause module delamination, and therefore prolonged lamination times are required. On the contrary, low-viscous encapsulants can flow out the laminate, leading to edge compression of the glass which increases the likelihood of glass breakage during thermal cycles. These issues can be addressed by (i) taping the module edges, (ii) adding frames having a similar thickness to the laminate, or (iii) using a laminator with a symmetrical structure [81].

Glass-glass modules can result in higher electricity yields arising from the albedo effect (see Section 11.3.4); however, they also introduce some disadvantages. The heavier glass-glass modules can increase transportation cost and pose challenges for roof installation. Due to their lower thermal cooling rate, glass-glass modules operate at higher temperatures than glass-transparent backsheet modules. Under NOCT conditions, the temperature difference between the two module types can range from 0.4 °C to 5.0 °C [134–137], equivalent to a relative maximum power output loss of 1.85% assuming a power temperature coefficient of −0.37% $°C^{-1}$ [138, 139]. Although the symmetric glass-glass structure reduces the stress induced in the cells during lamination, the increased rigidity tends to increase the occurrence of delamination at the encapsulant-glass interface [137]. Furthermore, acetic acid released from EVA decomposition can vent out through the permeable backsheet but not the impermeable glass. Therefore, corrosion or EVA yellowing can be a more significant problem in glass-glass than glass-backsheet modules.

11.4.6 Module frame, mounting and mechanical load

Unlike glass-backsheet modules, glass-glass modules can be unframed (or frameless). Several issues, however, have been observed in the field for such modules. Although edge sealants are applied to frameless modules, moisture ingress gradually occurs and the water content in the encapsulant increases [140, 141], leading to power degradation from the cells along the edges. During module transport and installation, the corners of frameless glass-glass modules are prone to cracking or damage [137]. As such, glass-glass modules are mostly equipped with frames.

Several mounting methods are available for the module installation. One most common method is clamping at one quarter along the long edge with supporting beams. Such mounting provides substantial support that generally meets the criteria of both static mechanical load tests (2400 Pa and 5400 Pa) defined in IEC 61215. However, the shading by the two supporting beams is not ideal for bifacial modules to harvest the rear albedo. Alternatively, mounting at the centre of the long edge can be fixed to the main beam, when 1-axis tracking is implemented. Another option for residential roof installations is to clamp the four module corners to supporting beams along the short edge. This advantageously offers savings in material cost, but with a compromise of withstanding lower mechanical load.

Finite element analysis has been applied to evaluate different mounting scenarios. As the module length/width dimension is three to four orders of magnitude higher than the material thicknesses, a global model is simulated to determine the deflection boundary conditions which are then applied to more detailed sub-models to evaluate stress and deformation. Dietrich *et al.* [112] compared four mounting types and two different frames. The difference in maximum deflection between using stiff and soft frames is more prominent when mounting along the short edge. As expected, deflection is less if modules are mounted along their long edge rather than along their short edge. The maximum principal stress in Si is qualitatively similar between line support (i.e. support along the whole edge) and clamps support. In terms of the probability of failure, the difference between frames is insignificant with long edge mounting. However, use of stiff frame is necessary when mounting along the short edge, especially under a 5400 Pa static load test.

The stiffness and material thickness of the encapsulant used in modules can also influence the probability of fracture. An encapsulant with lower stiffness in general is preferred since the module deflection increases while the cell stress decreases [112, 142]. Depending on the shear stiffness ratio of glass to encapsulant, PV modules may behave more like a laminate structure (soft encapsulant) or a sandwich structure (stiff encapsulant) [143]. This determines the impact of encapsulant thickness on the module reliability.

When modules are under mechanical load at different temperatures, module deflection decreases with temperature because encapsulant is stiffer at low temperatures. Modules with thinner cells have larger deflection as the total module stiffness decreases with cell thickness [107]. At low temperatures, the encapsulant transfers more strain to the cell, leading to higher stress in Si. Due to reduced module stiffness, stresses in thinner cells are increased. Increasing the encapsulant thickness of modules can reduce the deflection at low temperatures, at which the increased modulus of the encapsulant can contribute to the total module stiffness, whereas there is barely any difference at high temperatures. In contrast, the effect of encapsulant thickness on the cell stresses is only significant at high temperatures. Consequently, it can be concluded that using thicker encapsulant can reduce the cell stress during lamination, but not in modules under mechanical load since the stress is more critical at low temperatures.

Schicker *et al.* [142] reported the dependence of the stress and deflection on the embedment depth (i.e. the depth of the module laminate inserted and affixed in the frame) and the frame cross-sectional shape as illustrated in Figure 11.17. Longer embedment depth and frames with two vertical beams are desirable to increase the

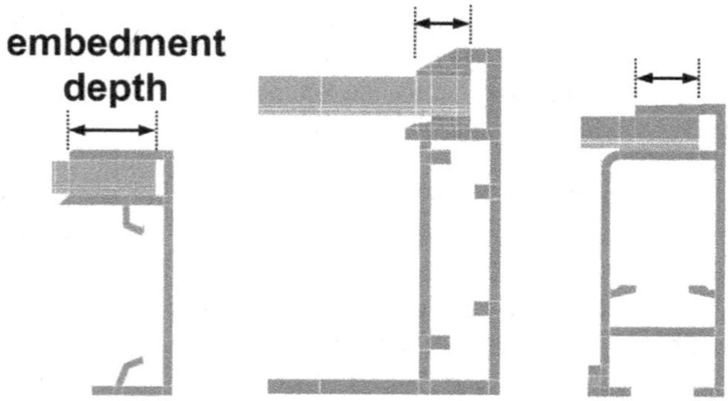

Figure 11.17 *Schematic diagram showing the frame cross-sectional shape and embedment depth (adapted from Schicker et al. [142])*

frame support to the module laminate. Under certain load conditions, the backside of the module was pressed against the supporting beams. Regardless of the boundary conditions of mounted modules, Schicker's simulations show slightly stiffer module behaviour than the tested modules because the corner joints of actual modules are to some extent movable. As the boundary restraints decrease, the module deflection and stress levels increase.

If the framed modules are fixed at the clamping point to the supporting beams, solar cells are predominantly under tensile stress on the rear surface. The direction of principal stress of the cells within the two supporting beams, located at the second row of cell strings, is parallel to the short edge, but abruptly changes 90° in the first row of solar cells. The corner cells have a principal stress direction of 45° relative to the module frame [115]. Cells near the centre of the modules experience the highest principle stress [142]. As fracture of brittle materials typically arises under tension, cell breakage would occur normal to the direction of the maximum tensile stress.

Beinert *et al.* [144] integrated the FEM simulation of (1) lamination cooling from 150 °C to 25 °C; and (2) static load of 2400 Pa or 5400 Pa. They observed that, after lamination, a maximum compressive stress of 81 MPa is located at the rear surface of the cells, which is slightly higher at the module edges than at the centre. Framed modules showed less deformation and lower first principal stress than unframed modules with clamp. In the latter the Si tensile stress increases up to >300 MPa, which may be overestimated as the supporting beams were not included in the simulation.

In conclusion, silicon PV modules may degrade due to several different failure or degradation modes and the rate of degradation can depend on the module design and/or the environment. Material interactions are the reason for most module degradation. Corrosion, delamination, thermal and mechanical stress can all accelerate cell and interconnect breakages. Corrosion can be initiated by water ingress in combination with generated chemicals from encapsulant decomposition, whilst delamination primarily occurs at the encapsulant-cell interface due to reduced interface adhesion from thermal, moisture and corrosive effects. Module deflection from temperature changes and mechanical stress leads to fatigue in solder joints and interconnect ribbons, eventually inducing cracks in solder joints and ribbons.

Temperature changes during soldering, lamination and module operation introduce stress in solar cells. Cell stress induced by mechanical loads is determined by the back cover material, material thickness, encapsulant stiffness, frame geometric shape, frame embedment depth and mounting methods. Module deflection can cause cell cracks and parallel cracks dominate in the modules subjected to mechanical loads. Predicting degradation of PV modules due to cell cracks remains a challenging task and will become more critical as wafer thicknesses reduce.

References

[1] Kazmerski L.L. 'Photovoltaics: a review of cell and module technologies'. *Renewable and Sustainable Energy Reviews*. 1997;**1**(1):71–170.

[2] Green M.A. 'Silicon photovoltaic modules: a brief history of the first 50 years'. *Progress in Photovoltaics: Research and Applications*. 2005;**13**(5):447–55.

[3] Haedrich I., Surve S., Thomson A. 'Cell to module (CTM) ratios for varying industrial cell types'. 2015 Asia Pacific Solar Research Conference; 2015. pp. 1–6.

[4] Hanifi H. 'Investigation of cell-to-module (CTM) ratios of pv modules by analysis of loss and gain mechanisms'. *Photovolt. Int*. 2016;**32**:89–99.

[5] Yang H., Wang H., Cao D., Sun D., Ju X. 'Analysis of power loss for crystalline silicon solar module during the course of encapsulation'. *International Journal of Photoenergy*. 2015;**2015**:1–5.

[6] Haedrich I., Eitner U., Wiese M., Wirth H. 'Unified methodology for determining CTM ratios: systematic prediction of module power'. *Solar Energy Materials and Solar Cells*. 2014;**131**:14–23.

[7] ITRPV. 'International Technology Roadmap for Photovoltaic (ITRPV) - 2019 Results'. *VDMA Photovoltaic Equipment, Eleventh Edition*; 2020.

[8] Roche D., Outhred H., Kaye R.J. 'Analysis and control of mismatch power loss in photovoltaic arrays'. *Progress in Photovoltaics: Research and Applications*. 1995;**3**(2):115–27.

[9] Evans R., Boreland M., Green M.A. 'A holistic review of mismatch loss: from manufacturing decision making to losses in fielded arrays'. *Solar Energy Materials and Solar Cells*. 2018;**174**(11):214–24.

[10] Evans R., Kim K.H., Wang X., *et al*. 'Simplified technique for calculating mismatch loss in mass production'. *Solar Energy Materials and Solar Cells*. 2015;**134**(4):236–43.

[11] Qian J., Clement C.E., Ernst M., Khoo Y.S., Thomson A., Blakers A. 'Analysis of hotspots in half cell modules undetected by current test standards'. *IEEE Journal of Photovoltaics*. 2019;**9**(3):842–8.

[12] Hanifi H., Schneider J., Bagdahn J. 'Reduced shading effect on half-cell modules—mMeasurement and simulation'. 31st European Photovoltaic Solar Energy Conference and Exhibition; 2015. pp. 2529–33.

[13] Deline C., Meydbray J., Donovan M. *Photovoltaic shading testbed for module-level power electronics: 2016 Performance Data Update*. Golden, CO (United States: National Renewable Energy Lab.(NREL); 2016.

[14] Wong J. 'Griddler: Intelligent computer aided design of complex solar cell metallization patterns'. 39th IEEE Photovoltaic Specialists Conference; Tampa; 2013.

[15] Singh J.P., Chai J., Saw M.H., Khoo Y.S. 'Bifacial solar cell measurements under standard test conditions and the impact on cell-to-module loss analysis'. *Japanese Journal of Applied Physics*. 2017;**56**(8S2):08MD04.

[16] Guo S., Singh J.P., Peters M., Aberle A.G., Wong J. 'Two-dimensional current flow in stringed PV cells and its influence on the cell-to-module resistive losses'. *Solar Energy*. 2016;**130**(389–395):224–31.

[17] Mette A., Richter P.L., Hörteis M., Glunz S.W. 'Metal aerosol jet printing for solar cell metallization'. *Progress in Photovoltaics: Research and Applications*. 2007;**15**(7):621–7.

[18] Mittag M., Vogt L., Herzog C., Neuhaus H. 'Thermal modelling of photovoltaic modules in operation and production'. 36th European Photovoltaic Solar Energy Conference and Exhibition (EUPVSEC); 2019.

[19] Mittag M., Zech T., Wiese M., Blasi D., Ebert M., Wirth H. 'Cell-to-module (CTM) analysis for photovoltaic modules with shingled solar cells'. 2017 IEEE 44th Photovoltaic Specialist Conference (PVSC); 2017. pp. 1531–6.

[20] Sarkar M.N.I. 'Effect of various model parameters on solar photovoltaic cell simulation: a spice analysis'. *Renewables: Wind, Water, and Solar*. 2016;**3**(1):13.

[21] Castaner L., Silvestre S. *Modelling photovoltaic systems using PSpice®*. John Wiley and Sons; 2002.

[22] Lu F., Guo S., Walsh T.M., Aberle A.G. 'Improved PV module performance under partial shading conditions'. *Energy Procedia*. 2013;**33**(7):248–55.

[23] Krc J. *Optical modeling and simulation of thin-film photovoltaic devices*. CRC Press; 2013.

[24] McIntosh K.R., Swanson R.M., Cotter J.E. 'A simple ray tracer to compute the optical concentration of photovoltaic modules'. *Progress in Photovoltaics: Research and Applications*. 2006;**14**(2):167–77.

[25] Holst H., Winter M., Vogt M.R., *et al.* 'Application of a new ray tracing framework to the analysis of extended regions in Si solar cell modules'. *Energy Procedia*. 2013;**38**:86–93.

[26] McIntosh K.R., Cotsell J.N., Cumpston J.S., Norris A.W., Powell N.E., Ketola B.M. 'An optical comparison of silicone and EVA encapsulants for conventional silicon PV modules: a ray-tracing study'. 2009 34th IEEE Photovoltaic Specialists Conference (PVSC); 2009. pp. 000544–9.

[27] Blieske U., Doege T., Gayout P., Neander M., Neumann D., Prat A. 'Light-trapping in solar modules using extra-white textured glass'. *3rd World*

Conference onPhotovoltaic Energy Conversion, 2003. Proceedings of. **1**; 2003. pp. 188–91.

[28] Haedrich I., Jordan D.C., Ernst M. 'Methodology to predict annual yield losses and gains caused by solar module design and materials under field exposure'. *Solar Energy Materials and Solar Cells*. 2019;**202**:110069.

[29] Ernst M., Haedrich I., Li Y., Lennon A. 'Impact of (multi-) Busbar design on the PERC Cell-to-Module Yield Under Realistic Conditions'. Asia-Pacific Solar Research Conference (APSRC); Canberra; 2019.

[30] Abbott M. 'Annual energy yield analysis of solar cell technology'. 2019 IEEE 46th Photovoltaic Specialists Conference (PVSC); 2019. pp. 3046–50.

[31] Holst H., Altermatt P.P., Brendel R. 'DAIDALOS – A plugin based framework for extendable ray tracing'. 25th European Photovoltaic Solar Energy Conference and Exhibition; 2010. pp. 2150–3.

[32] Santbergen R., van Zolingen R.J.C. 'The absorption factor of crystalline silicon pv cells: a numerical and experimental study'. *Solar Energy Materials and Solar Cells*. 2008;**92**(4):432–44.

[33] Mellor A., Tobías I., Martí A., Mendes M.J., Luque A. 'Upper limits to absorption enhancement in thick solar cells using diffraction gratings'. *Progress in Photovoltaics: Research and Applications*. 2011;**19**(6):676–87.

[34] Tucher N., Eisenlohr J., Kiefel P., *et al.* '3D optical simulation formalism OPTOS for textured silicon solar cells'. *Optics Express*. 2015;**23**(24):A1720–34.

[35] Eisenlohr J., Tucher N., Höhn O., *et al.* 'Matrix formalism for light propagation and absorption in thick textured optical sheets'. *Optics Express*. 2015;**23**(11):A502–18.

[36] Li Y., Chen Y., Ouyang Z., Lennon A. 'Angular matrix framework for light trapping analysis of solar cells'. *Optics Express*. 2015;**23**(24):A1707–19.

[37] Mermoud A., Wittmer B. *PVSYST user's manual*. Switzerland: PVSYST SA; 2014.

[38] Dobos A.P. *PVWatts version 5 manual*. Golden, CO (United States): National Renewable Energy Lab; 2014.

[39] Ernst M., Holst H., Winter M., Altermatt P.P. 'SunCalculator: a program to calculate the angular and spectral distribution of direct and diffuse solar radiation'. *Solar Energy Materials and Solar Cells*. 2016;**157**(4):913–22.

[40] Jaus J., Pantsar H., Eckert J., Duell M., Herfurth H., Doble D. 'Light management for reduction of bus bar and gridline shadowing in photovoltaic modules'. 2010 35th IEEE Photovoltaic Specialists Conference; 2010. pp. 000979–83.

[41] Trube J., Fischer M., Woodhouse M., Herritsch S. 'International technology roadmap for photovoltaic (ITRPV)'. *VDMA photovoltaic equipement*. 2020.

[42] Tonini D., Cellere G., Bertazzo M., Fecchio A., Cerasti L., Galiazzo M. 'Shingling technology for cell interconnection: technological aspects and process integration'. *Energy Procedia*. 2018;**150**:36–43.

[43] Morad R. *Shingled solar cell module*. Google Patents; 2015.

[44] Gibson K.R., Parilla A., Phu T. *Tiled solar cell laser process*. Google Patents; 2019.

[45] Urrejola E. 'bifiPV2020 bifacial workshop: a technology overview'. 2020.

[46] 'The Technical White Paper - Tiger Module. Jinko Solar'. 2020.

[47] 'Seamless Soldering- LONGi Announced new proprietary module encapsulation technology'. *LONGi Solar*. 2019.

[48] Schulte-Huxel H., Blankemeyer S., Morlier A., Brendel R., Köntges M. 'Interconnect-shingling: maximizing the active module area with conventional module processes'. *Solar Energy Materials and Solar Cells*. 2019;**200**(9):109991.

[49] Papet P. 'Overlap modules: a unique cell layup using smart wire connection technology'. *AIP Conference Proceedings*. 2019;**2147**(1):080001.

[50] Wilson G.M., Al-Jassim M., Metzger W.K., *et al.* 'The 2020 photovoltaic technologies roadmap'. *Journal of Physics D: Applied Physics*. 2020;**53**(49):493001.

[51] Burrows K., Fthenakis V. 'Glass needs for a growing photovoltaics industry'. *Solar Energy Materials and Solar Cells*. 2015;**132**:455–9.

[52] Shanmugam N., Pugazhendhi R., Madurai Elavarasan R., Kasiviswanathan P., Das N. 'Anti-reflective coating materials: a holistic review from PV perspective'. *Energies*. 2020;**13**(10):2631.

[53] Ballif C., Dicker J., Borchert D., Hofmann T. 'Solar glass with industrial porous SiO2 antireflection coating: measurements of photovoltaic module properties improvement and modelling of yearly energy yield gain'. *Solar Energy Materials and Solar Cells*. 2004;**82**(3):331–44.

[54] Ulbrich C., Gerber A., Hermans K., Lambertz A., Rau U. 'Analysis of short circuit current gains by an anti-reflective textured cover on silicon thin film solar cells'. *Progress in Photovoltaics: Research and Applications*. 2013;**21**(8):1672–81.

[55] Said S.A.M., Al-Aqeeli N., Walwil H.M. 'The potential of using textured and anti-reflective coated glasses in minimizing dust fouling'. *Solar Energy*. 2015;**113**(6):295–302.

[56] Verma L.K., Sakhuja M., Son J., *et al.* 'Self-cleaning and antireflective packaging glass for solar modules'. *Renewable Energy*. 2011;**36**(9):2489–93.

[57] Holst H., Schulte-Huxel H., Winter M., *et al.* 'Increased light harvesting by structured cell interconnection ribbons: an optical ray tracing study using a realistic daylight model'. *Energy Procedia*. 2016;**92**(1):505–14.

[58] Sachs E. *'Light capture with patterned solar cell bus wires'*. United States: Google Patents, US 2007/0125415 A1; 2007.

[59] Chung I., Lee W., Cho E., Moon I. 'Light capturing film for power gain of silicon PV modules'. 2014 IEEE 40th Photovoltaic Specialist Conference (PVSC); 2014. pp. 2689–92.

[60] O'neill M.B., Ma J., Votava M.J. *Light redirecting film useful with solar modules*. United States: Google Patents, US20170104121A1; 2018.

[61] Saive R. 'Effectively transparent contacts (ETCs) for solar cells'. 2016 IEEE 43rd Photovoltaic Specialists Conference (PVSC); 2016. pp. 3612–5.

[62] Mittag M., Beinert A., Rendler L., Ebert M., Eitner U. Triangular ribbons for improved module efficiency. 32th European Photovoltaic Solar Energy Conference and Exhibition; Munich; 2016.

[63] Walter J., Tranitz M., Volk M., Ebert C., Eitner U. 'Multi-wire interconnection of Busbar-free solar cells'. *Energy Procedia*. 2014;**55**:380–8.

[64] Söderström T., Papet P., Ufheil J. 'Smart wire connection technology'. 28th European Photovoltaic Solar Energy Conference; 2013. pp. 495–9.

[65] Li Y. 'Comparison of ribbon light management designs for photovoltaic modules'. IEEE 46th Photovoltaic Specialists Conference (PVSC); Chicago, IL, US; 2019.

[66] Vogt M.R., Holst H., Schulte-Huxel H., *et al.* 'PV module current gains due to structured backsheets'. *Energy Procedia*. 2017;**124**(1):495–503.

[67] López-Escalante M.C., Fernández-Rodríguez M., Caballero L.J., Martín F., Gabás M., Ramos-Barrado J.R. 'Novel encapsulant architecture on the road to photovoltaic module power output increase'. *Applied Energy*. 2018;**228**:1901–10.

[68] Sun X., Khan M.R., Deline C., Alam M.A. 'Optimization and performance of bifacial solar modules: a global perspective'. *Applied Energy*. 2018;**212**(3–4):1601–10.

[69] Deline C.A., Ayala Pelaez S., Marion W.F., Sekulic W.R., Woodhouse M.A., Stein J. *Bifacial PV system performance: Separating fact from fiction*. Golden, CO: National Renewable Energy Lab; 2019.

[70] Singh J.P., Guo S., Peters I.M., Aberle A.G., Walsh T.M. 'Comparison of glass/glass and glass/backsheet PV modules using bifacial silicon solar cells'. *IEEE Journal of Photovoltaics*. 2015;**5**(3):783–91.

[71] Wang H., Wang A., Yang H., Zhang J., Huang J. 'Performance degradation of crystalline silicon solar module in various ultraviolet radiation area'. 2016 IEEE 43rd Photovoltaic Specialists Conference (PVSC); 2016. pp. 1757–60.

[72] Russell T.C.R., Saive R., Augusto A., Bowden S.G., Atwater H.A. 'The influence of spectral albedo on bifacial solar cells: a theoretical and experimental study'. *IEEE Journal of Photovoltaics*. 2017;**7**(6):1611–18.

[73] Berrian D., Libal J. 'A comparison of ray tracing and view factor simulations of locally resolved rear irradiance with the experimental values'. *Progress in Photovoltaics: Research and Applications*. 2020;**28**(6):609–20.

[74] Pelaez S.A., Deline C., MacAlpine S.M., Marion B., Stein J.S., Kostuk R.K. 'Comparison of bifacial solar irradiance model predictions with field validation'. *IEEE Journal of Photovoltaics*. 2019;**9**(1):82–8.

[75] McIntosh K.R., Abbott M.D., Sudbury B.A., Meydbray J. 'Mismatch loss in bifacial modules due to nonuniform illumination in 1-D tracking systems'. *IEEE Journal of Photovoltaics*. 2019;**9**(6):1504–12.

[76] Köntges M. 'Assessment of photovoltaic module failures in the field'. *IEA PVPS Task 13, Report IEA-PVPS T13-09:2017*. 2017.

[77] Hasan O., Arif A.F.M. 'Performance and life prediction model for photovoltaic modules: effect of encapsulant constitutive behavior'. *Solar Energy Materials and Solar Cells*. 2014;**122**(4):75–87.

[78] Köntges M., Kurtz S., Packard C., *et al. Review of failures of photovoltaic modules*. IEA PVPS Task 13, External final report IEA-PVPS. International Energy Agency; 2014. pp. 1–140.

[79] Hülsmann P., Weiß K.-A., Köhl M. 'Temperature-Dependent water vapour and oxygen permeation through different polymeric materials used in photovoltaic-modules'. *Progress in Photovoltaics: Research and Applications*. 2014;**22**(4):415–21.

[80] Ennemri A., Logerais P., Balistrou M., Durastanti J., Belaidi I. 'Cracks in silicon photovoltaic modules: a review'. *Journal of Optoelectronics Advanced Materials*. 2019;**21**(1-2):74–92.

[81] Cattaneo G., Galliano F., Gragert M., Burger M., Grischke R. 'Lamination process and encapsulation materials for glass-glass PV module design'. *Photovoltaics International*. 2015;**27**:1–8.

[82] Büchler A., Beinert A., Kluska S., *et al.* 'Enabling stress determination on alkaline textured silicon using Raman spectroscopy'. *Energy Procedia*. 2017;**124**(10):18–23.

[83] Beinert A.J., Romer P., Büchler A., Haueisen V., Aktaa J., Eitner U. 'Thermomechanical stress analysis of PV module production processes by Raman spectroscopy and FEM simulation'. *Energy Procedia*. 2017;**124**(2):464–9.

[84] Colli A., Attenkofer K., Raghothamachar B., Dudley M. 'Synchrotron X-ray topography for encapsulation stress/strain and crack detection in crystalline silicon modules'. *IEEE Journal of Photovoltaics*. 2016;**6**(5):1387–9.

[85] Handara V.A., Radchenko I., Tippabhotla S.K., *et al.* 'Probing stress and fracture mechanism in encapsulated thin silicon solar cells by synchrotron X-ray microdiffraction'. *Solar Energy Materials and Solar Cells*. 2017;**162**(11):30–40.

[86] Meng X., Stuckelberger M., Ding L., West B., Jeffries A., Bertoni M. 'Characterization of encapsulated solar cells by X-ray topography'. Photovoltaic Specialists Conference (PVSC) 2016 IEEE 43rd; 2016. pp. 0111–14.

[87] Meng X., Stuckelberger M., Hacke P., Bertoni M. 'Process induced deflection and stress on encapsulated solar cells'. Proc. 2017 IEEE 44th Photovolt. Spec. Conf.; 2017.

[88] Eitner U., Kajari-Schröder S., Köntges M., Altenbach H. 'Thermal stress and strain of solar cells in photovoltaic modules'. *Shell-like Structures*. Springer; 2011. pp. 453–68.

[89] Paggi M., Sapora A. 'An accurate thermoviscoelastic rheological model for ethylene vinyl acetate based on fractional calculus'. *International Journal of Photoenergy*. 2015;**2015**:1–7.

[90] Dietrich S., Pander M., Sander M., Zeller U., Ebert M. 'Stress analysis of encapsulated solar cells by means of superposition of thermal and mechanical stresses'. *Reliability of Photovoltaic Cells, Modules, Components, and Systems VI*. **8825**. International Society for Optics and Photonics; 2013. p. 882505.

[91] Rendler L.C., Kraft A., Ebert C., Eitner U., Wiese S. 'Mechanical stress in solar cells with multi busbar interconnection—parameter study by FEM simulation'. 2016 17th International Conference on Thermal, Mechanical and Multi-Physics Simulation and Experiments in Microelectronics and Microsystems (EuroSimE); 2016. pp. 1–5.

[92] Rendler L.C., Kraft A., Ebert C., Wiese S., Eitner U. 'Investigation of thermomechanical stress in solar cells with multi busbar interconnection by finite element modeling'. *Proc. of the 32nd European Photovoltaic Solar Energy Conference and Exhibition*. München: WIP; 2016.

[93] Hsiao P.-C., Shen X., Wang Z., *et al.* 'Balanced contact method: reduction of thermomechanical stress in silicon solar cells induced by interconnection'. *Solar Energy Materials and Solar Cells*. 2020;**215**(1):110667.

[94] Walter J., Rendler L.C., Ebert C., Kraft A., Eitner U. 'Solder joint stability study of wire-based interconnection compared to ribbon interconnection'. *Energy Procedia*. 2017;**124**:515–25.

[95] Hsiao P.-C. 'Comparative models of induced thermomechanical stress in silicon solar cells interconnected with conventional tabbing and wire-based interconnection methods'. 2019 IEEE 46th Photovoltaic Specialists Conference (PVSC); 2019. pp. 0122–5.

[96] Rendler L.C., Walter J., Kraft A., Ebert C., Wiese S., Eitner U. 'Ultra-soft wires for direct soldering on finger grids of solar cells'. *Energy Procedia*. 2017;**124**:478–83.

[97] Rendler L.C. 'Wave-shaped wires soldered on the finger grid of solar cells: solder joint stability under thermal cycling'. *AIP Conference Proceedings*. 2018;**1999**(1):080001.

[98] Lv R., Tang J., Jaubert J.-N., Xing G. 'Highly accelerated thermal cycling test for new type of crystalline silicon photovoltaic modules'. 2019 IEEE 46th Photovoltaic Specialists Conference (PVSC); 2019. pp. 1991–4.

[99] Lalaguna B., Sanchez-Friera P., Bennett I., Sanchez D., Caballero L., Alonso J. 'Evaluation of bismuth-based solder alloys for low-stress interconnection of industrial crystalline silicon PV cells'. 22nd European Photovohaic Solar Energy Conference; Milan, Italy; 2007. pp. 2712–15.

[100] Schneider A., Rubin L., Rubin G. 'Solar Cell Efficiency Improvement by New Metallization Techniques-the day4 electrode concept'. *Photovoltaic Energy Conversion, Conference Record of the 2006 IEEE 4th World Conference on*. **1**; 2006. pp. 1095–8.

[101] Schwertheim S., Scherff M., Mueller T., Fahrner W., Neitzert H. 'Lead-free electrical conductive adhesives for solar cell interconnectors'. Photovoltaic Specialists Conference, 2008. PVSC'08. 33rd IEEE; 2008–1–6.

[102] Eitner U., Geipel T., Holtschke S.-N., Tranitz M. 'Characterization of electrically conductive adhesives'. *Energy Procedia*. 2012;**27**:676–9.

[103] Torsten G., Ulrich E. 'Electrically conductive adhesives an emerging interconnection technology for high-efficiency solar modules'. *Photovoltaics International, 21st edition, no. 0*; 2013. pp. 27–33.

[104] Zemen Y., Prewitz T., Geipel T., Pingel S., Berghold J. 'The impact of yield strength of the interconnector on the internal stress of the solar cell within a module'. Proceedings of the 25th European Photovoltaic Solar Energy Conference and Exhibition; 2010. pp. 4073–8.

[105] Walter J., Rendler L.C., Halm A., Mihailetchi V., Kraft A., Eitner U. 'Ribbon interconnection of 6" BC-BJ solar cells'. *Energy Procedia.* 2017;**124**(6):504–14.

[106] Li H.Y., Perret-Aebi L.E., Chapuis V., Ballif C., Luo Y. 'The effect of cooling press on the encapsulation properties of crystalline photovoltaic modules: residual stress and adhesion'. *Progress in Photovoltaics: Research and Applications*. 2015;**23**(2):160–9.

[107] Dietrich S., Pander M., Sander M., Schulze S.H., Ebert M. 'Mechanical and thermomechanical assessment of encapsulated solar cells by finite-element-simulation'. *Reliability of photovoltaic cells, modules, components, and systems III.* **vol. 7773**. International Society for Optics and Photonics; 2010. p. 77730F.

[108] Bosco N. 'Evaluation of dynamic mechanical loading as an accelerated test method for ribbon fatigue'. 2013 IEEE 39th Photovoltaic Specialists Conference (PVSC); 2013. pp. 3173–8.

[109] Pander M., Zeller U., Jaeckel B., Ebert M. 'Digital prototyping – application of numerical methods in module development'. Proceedings of the 36th European Photovoltaic Solar Energy Conference; Marseille, France; 2019. pp. 789–94.

[110] Rendler L.C., Romer P., Beinert A.J., *et al.* 'Thermomechanical stress in solar cells: contact pad modeling and reliability analysis'. *Solar Energy Materials and Solar Cells*. 2019;**196**(2):167–77.

[111] Chaturvedi P., Hoex B., Walsh T.M. 'Broken metal fingers in silicon wafer solar cells and PV modules'. *Solar Energy Materials and Solar Cells.* 2013;**108**:78–81.

[112] Dietrich S., Sander M., Pander M., Ebert M. 'Interdependency of mechanical failure rate of encapsulated solar cells and module design parameters'. *Reliability of Photovoltaic Cells, Modules, Components, and Systems.* **vol.8472**. International Society for Optics and Photonics; 2012. p. 84720P.

[113] Beinert A.J., Imm M., Benick J., *et al.* 'Silicon solar cell–integrated stress and temperature sensors for photovoltaic modules'. *Progress in Photovoltaics: Research and Applications*. 2020;**28**(7):717–24.

[114] Köntges M., Kajari-Schröder S., Kunze I., Jahn U. 'Crack statistic of crystalline silicon photovoltaic modules'. *26th European Photovoltaic Solar Energy Conference and Exhibition.* **26**; 2011. pp. 3290–4.

[115] Kajari-Schröder S., Kunze I., Eitner U., Köntges M. 'Spatial and directional distribution of cracks in silicon PV modules after uniform mechanical loads'. 2011 37th IEEE Photovoltaic Specialists Conference; 2011. pp. 000833–7.

[116] Umeno Y., Kushima A., Kitamura T., Gumbsch P., Li J. '*Ab initio* study of the surface properties and ideal strength of (100) silicon thin films'. *Physical Review B*. 2005;**72**(16):165431.

[117] Rupnowski P., Sopori B. 'Strength of silicon wafers: fracture mechanics approach'. *International Journal of Fracture*. 2009;**155**(1):67–74.

[118] Sander M., Dietrich S., Pander M., Ebert M., Bagdahn J. 'Systematic investigation of cracks in encapsulated solar cells after mechanical loading'. *Solar Energy Materials and Solar Cells*. 2013;**111**:82–9.

[119] Popovich V.A., Yunus A., Janssen M., Richardson I.M., Bennett I.J. 'Effect of silicon solar cell processing parameters and crystallinity on mechanical strength'. *Solar Energy Materials and Solar Cells*. 2011;**95**(1):97–100.

[120] Kaule F., Wang W., Schoenfelder S. 'Modeling and testing the mechanical strength of solar cells'. *Solar Energy Materials and Solar Cells*. 2014;**120**:441–7.

[121] Rupnowski P., Sopori B., B. J. I. jofS. 'Strength of silicon wafers: fracture mechanics approach'. *International Journal of Fracture*. 2009;**155**(1):67–74.

[122] Dietrich S., Pander M., Sander M., Ebert M. 'Mechanical investigations on metallization layouts of solar cells with respect to module reliability'. *Energy Procedia*. 2013;**38**:488–97.

[123] Itoh U., Yoshida M., Tokuhisa H., Takeuchi K., Takemura Y. 'Solder joint failure modes in the conventional crystalline Si module'. *Energy Procedia*. 2014;**55**:464–8.

[124] Zarmai M.T., Ekere N.N., Oduoza C.F., Amalu E.H. 'Evaluation of thermo-mechanical damage and fatigue life of solar cell solder interconnections'. *Robotics and Computer-Integrated Manufacturing*. 2017;**47**(3):37–43.

[125] Jabarullah N.H., Surendar A., Arun M., Siddiqi A.F., Krasnopevtseva T. 'Microstructural characterization and unified reliability assessment of aged solder joints in a PV module'. *IEEE Transactions on Components, Packaging and Manufacturing Technology*. 2020;**10**(6):1028–34.

[126] Jiang N., Ebadi A.G., Kishore K.H., Yousif Q.A., Salmani M. 'Thermomechanical reliability assessment of solder joints in a photovoltaic module operated in a hot climate'. *IEEE Transactions on Components, Packaging and Manufacturing Technology*. 2019;**10**(1):160–7.

[127] Bosco N., Silverman T.J., Kurtz S. 'Climate specific thermomechanical fatigue of flat plate photovoltaic module solder joints'. *Microelectronics Reliability*. 2016;**62**:124–9.

[128] Amalu E.H., Hughes D.J., Nabhani F., Winter J. 'Thermo-mechanical deformation degradation of crystalline silicon photovoltaic (c-Si PV) module in operation'. *Engineering Failure Analysis*. 2018;**84**(3):229–46.

[129] Heimann M., Bakowskie R., Köhler M., *et al.* 'Investigations of different soldering failure modes and their impact on module reliability'. *Energy Procedia*. 2014;**55**:456–63.

[130] Zarmai M.T., Ekere N.N., Oduoza C.F., Amalu E.H. 'Optimization of thermo-mechanical reliability of solder joints in crystalline silicon solar cell assembly'. *Microelectronics Reliability*. 2016;**59**(3):117–25.

[131] Bosco N., Silverman T.J., Kurtz S. 'The influence of PV module materials and design on solder joint thermal fatigue durability'. *IEEE Journal of Photovoltaics*. 2016;**6**(6):1407–12.

[132] Kumar S., Gupta R. 'Investigation and analysis of thermo-mechanical degradation of fingers in a photovoltaic module under thermal cyclic stress conditions'. *Solar Energy*. 2018;**174**:1044–52.

[133] H.-Y. L., Luo Y., Ballif C., Perret-Aebi L.-E. 'Effect of cooling press on the optical transmission through photovoltaic encapsulants'. *Polymer-plastics technology and engineering*. 2015;**54**(4):416–24.

[134] Alonso García M.C., Balenzategui J.L. 'Estimation of photovoltaic module yearly temperature and performance based on nominal operation cell temperature calculations'. *Renewable Energy*. 2004;**29**(12):1997–2010.

[135] Maturi L., Belluardo G., Moser D., Del Buono M. 'BiPV system performance and efficiency drops: overview on PV module temperature conditions of different module types'. *Energy Procedia*. 2014;**48**:1311–19.

[136] Rouholamini A., Pourgharibshahi H., Fadaeinedjad R., Abdolzadeh M. 'Temperature of a photovoltaic module under the influence of different environmental conditions – experimental investigation'. *International Journal of Ambient Energy*. 2016;**37**(3):266–72.

[137] Felder T.C. 'Analysis of glass-glass modules'. *New Concepts in Solar and Thermal Radiation Conversion and Reliability*. **10759**. International Society for Optics and Photonics; 2018. p. 107590E.

[138] Wang H., Cheng X., Yang H. 'Temperature coefficients and operating temperature verification for passivated emitter and rear cell bifacial silicon solar module'. *IEEE Journal of Photovoltaics*. 2020;**10**(3):729–39.

[139] Kasu M., Abdu J., Hara S., Choi S., Chiba Y., Masuda A. 'Temperature dependence measurements and performance analyses of high-efficiency interdigitated back-contact, passivated emitter and rear cell, and silicon heterojunction photovoltaic modules'. *Japanese Journal of Applied Physics*. 2018;**57**(8S3):08RG18.

[140] Kempe M.D., Dameron A.A., Reese M.O. 'Evaluation of moisture ingress from the perimeter of photovoltaic modules'. *Progress in Photovoltaics: Research and Applications*. 2014;**22**(11):1159–71.

[141] Mittag M., Eitner U., Neff T. 'TPedge: progress on costefficient and durable edge-sealed PV modules'. Proceedings of the 33rd European Photovoltaic Solar Energy Conference and Exhibition; Amsterdam, The Netherlands; 2017. pp. 48–54.

[142] Schicker J., Hirschl C., Leidl R. 'Effect of PV module frame boundaries on stresses in solar cells'. *Journal of Energy Challenges Mechanics*. 2014;**1**(3):155–62.

[143] Schulze S.-H., Pander M., Naumenko K., Altenbach H. 'Analysis of laminated glass beams for photovoltaic applications'. *International Journal of Solids and Structures*. 2012;**49**(15-16):2027–36.

[144] Beinert A.J., Ebert M., Eitner U., Aktaa J. 'Influence of photovoltaic module mounting systems on the thermo-mechanical stresses in solar cells by FEM modelling'. Proceedings of the 32nd European Photovoltaic Solar Energy Conference and Exhibition; Munich, Germany; 2016. pp. 20–4.

Index

www.ingramcontent.com/pod-product-compliance
Lightning Source LLC
LaVergne TN
LVHW011543200926
848484LV00001B/288